Population and Community Biology

THE STATISTICS OF NATURAL SELECTION ON ANIMAL POPULATIONS

Population and Community Biology

Series Editors

M. B. Usher
Reader, University of York, UK

M. L. Rosenzweig
Professor, Department of Ecology and Evolutionary Biology, University of Arizona, USA

The study of both populations and communities is central to the science of ecology. This series of books will explore many facets of population biology and the processes that determine the structure and dynamics of communities. Although individual authors and editors have freedom to develop their subjects in their own way, these books will all be scientifically rigorous and often utilize a quantitative approach to analysing population and community phenomena.

THE STATISTICS OF NATURAL SELECTION ON ANIMAL POPULATIONS

Bryan F. J. Manly

University of Otago
Dunedin, New Zealand

LONDON   NEW YORK

CHAPMAN AND HALL

First published in 1985 by
Chapman and Hall Ltd
11 New Fetter Lane, London EC4P 4EE
Published in the USA by
Chapman and Hall
29 West 35th Street, New York NY 10001

Reprinted as a paperback 1987

© *1985 Bryan F. J. Manly*

Printed in Great Britain by
J. W. Arrowsmith Ltd, Bristol

ISBN 0 412 30700 6

British Library Cataloguing in Publication Data

Manly, Bryan F. J.
  The statistics of natural selection on animal
  populations.
  1. Natural selection—Research—Statistical
  methods
  I. Title
  591.52'48   QH375
  ISBN 0-412-30700-6

Library of Congress Cataloging in Publication Data

Manly, Bryan F. J., 1944–
  The statistics of natural selection on animal
  populations.

  Bibliography: p.
  Includes indexes.
  1. Natural selection—Statistical methods.   2. Animal
  populations—Statistical methods.   I. Title.
  QH375.M36   1985   591.5'248'015195   85-5739
  ISBN 0-412-30700-6

And further, by these, my son, be admonished: of making many books there is no end; and much study is a weariness of the flesh.

Ecclesiastes 12:12

To E. L. B. for the start, and E. C. C. for the finish

Contents

viii   Contents

Examples

Preface

In the concluding chapter of his famous book on the theory of evolution by natural selection, Charles Darwin (1859) remarked that:

> When the views entertained in this volume on the origin of species, or when analogous views are generally admitted, we can dimly foresee that there will be a considerable revolution in natural history.

This proved, of course, to be completely correct. At present there is a great divergence of opinion about the general importance of natural selection in the evolutionary process. Nevertheless, biologists are, on the whole, united in their acceptance of the potential power of selection in changing populations. Given this situation, it is not surprising to find that many attempts to detect the effects of natural selection have been made since the time of Darwin. This area of study has been called ecological genetics. It involves the collection of data of various kinds and, in many cases, the development of special methods for analysing these data. This book is a summary of methods for data analysis, concentrating on those that are applicable to animal populations, particularly wild populations.

I would like to stress, however, that I intended to do more than just present a catalogue of statistical techniques. In Chapter 1 I have tried to provide an historical background to explain why interest in studying natural selection has been particularly strong at certain times. In the other chapters an attempt has been made to give readers a good idea of the scope of work that has been done. Examples have also been chosen with this idea in mind.

It was my intention that the contents of this book should be presented at a level suitable for a biologist with an interest in quantitative methods. I have assumed that readers have a background in mathematics that includes some calculus and matrix algebra. The material covered by Causton (1977) should be quite sufficient. I have also assumed a familiarity with basic statistical methods, particularly simple linear regression and analysis of variance. In this case the completion of a standard first-year university service course in statistics should be adequate. The more advanced statistical techniques that are needed are covered in the Statistical Appendix.

Although it is primarily intended for biologists, I hope that this book will also be of some interest to statisticians and biometricians. It is for this reason that I have made a point of explaining some biological matters where this is surely unnecessary as far as professional biologists are concerned. For example, I have briefly described how data are obtained by the technique of electrophoresis. This is now a commonplace process for biologists but I had to look to an elementary text in genetics to find out what is involved.

It is inevitable that some statistical analyses are only possible with the aid of a computer. In some cases it is just a question of using a standard program that is widely available. At other times a special program is needed. In the examples, I have used the program GLIM from Rothamstead Experimental Station to do many of the calculations. This very versatile program is described in Section A.10 of the Appendix. Another useful program is MAXLIK that was developed by Reed and Schull (1968) for maximum likelihood estimation of genetic parameters. The FORTRAN code of this program has not been published before and Professor Reed has kindly allowed me to provide it in Section A.16 of the Appendix. The Appendix also contains the code of some other FORTRAN programs to carry out the calculations for some special methods of analysis.

This book would never have been written if Mike Parr had not kindled my interest in applications of statistics in ecology and if Lawrence Cook had not introduced me to several problems concerned with measuring the effects of selection. Lawrence Cook has also contributed helpful comments at many stages as this book developed. George Seber commented on early versions of some chapters. John Rayner did the same for the appendix, and A. H. D. Brown for Section 10.6. The series editors, Dr M. B. Usher and Dr M. L. Rosenzweig, and their advisors, have provided many suggestions for improvements to the text. I am particularly indebted to Dr Rosenzweig for his detailed review of the complete manuscript. This was very helpful when I made the final revision. The staff of Chapman and Hall have encouraged me greatly since I first sent them some example chapters.

Much of the work on Chapter 6 was done in 1981 while I was a visitor in the Department of Experimental Statistics, Louisiana State University, Baton Rouge. I am most grateful to Dr Prentice Schilling and his staff for the hospitality and facilities that they provided me with.

The long task of typing and retyping the manuscript was carried out principally by Lenette Grant, while Mary-Jane Campbell completed the final version of the appendix; I am very thankful to both of them.

Finally, I must ask my readers to be tolerant towards a statistician writing for biologists. I am only too well aware of the shortcomings in my knowledge of even elementary aspects of biology. It is my hope that this does not show up too much in the pages that follow.

B. F. J. M

Dunedin, New Zealand
August 1984

1 The study of evidence for natural selection

This chapter provides the setting for the remainder of the book. The first few sections summarize the main historical trends in the study of natural selection. The publication of *The Origin of Species* in 1859 was followed by an optimistic period and then a decline in interest until there was a revival leading to the modern synthesis. More recent developments are related to the neutral mutation–random drift theory for molecular evolution and to the punctuated equilibria theory of long-term evolution.

The historical review is followed by a consideration of the meaning of the word 'fitness'. There is then a comparison between polymorphic and quantitative variation, and a summary of the types of evidence for natural selection that are considered in the remainder of the book.

The chapter concludes with the mention of a major practical difficulty that often arises when natural populations are studied: the difficulty of ensuring that samples are effectively random. It may well be that some of the better-known examples of 'evidence' for natural selection are in reality simply examples of effects due to sampling biases.

1.1 THE BIRTH OF EVOLUTIONARY BIOLOGY

On 1 July 1858, Charles Darwin and Alfred Wallace jointly presented their theory of evolution by natural selection to a meeting of the Linnean Society of London. In November the following year, Darwin published *The Origin of Species by Means of Natural Selection, or the Preservation of Favoured Races in the Struggle for Life*, which he had spent 20 years in writing. These events are generally taken to mark the birth of modern evolutionary biology, although the ideas of natural selection and evolution were known well before 1858. Johnson (1976) traces them back to the ancient Greeks and Romans. Indeed, Charles's grandfather, Erasmus Darwin, clearly expressed these ideas in his book *Zoonomia, or the Laws of Organic Life*, that was first published in 1794 (King-Hele, 1977). The contribution of Charles Darwin was to support the ideas with such a mass of evidence that the scientific world was convinced of their general validity.

In formulating his ideas, Darwin was influenced considerably by the book

An Essay on the Principle of Population, published by the Reverend T. R. Malthus in 1798. Malthus had pointed out that a population would increase geometrically, and soon swamp the whole world, if it were not for the fact that a large fraction of the progeny in every generation fail to survive and reproduce. Darwin realized that this extinction of some, but not of others, gave Nature a chance to select what was best. Thus in Chapter 4 of *The Origin of Species* he defined natural selection in the following way:

> . . . can we doubt (remembering that many more individuals are born than can possibly survive) that individuals having any advantage, however slight, over others, would have the best chance of surviving and of procreating their kind? On the other hand, we may feel sure that any variation in the least degree injurious would be rigidly destroyed. This preservation of favourable variations and the rejection of injurious variations I call Natural Selection.

Darwin believed that natural selection will always act extremely slowly, although it might be possible to detect evolutionary changes providing that careful records are kept over a period of perhaps 50 years for a species that reproduces annually (Ford, 1975, p. 393). However, some of his supporters were more optimistic and by early this century several attempts had been made to demonstrate natural selection. Indeed, many of these supporters clearly felt that there was a great need for examples of situations where selection could be shown to be operating. Thus Hermon Bumpus (1898) wrote:

> We are so in the habit of referring carelessly to the process of natural selection, and of invoking its aid whenever some pet theory seems a little feeble, that we forget that we are really using a hypothesis that still remains unproved, and that specific examples of the destruction of animals of known physical disability are very infrequent. Even if the theory of natural selection were as firmly established as Newton's theory of the attraction of gravity, scientific method would still require frequent examination of its claims, and scientific honesty should welcome such examination and insist on its thoroughness.

Bumpus's (1898) paper was one of the results of a severe snowstorm that hit New England on 1 February 1898. After the storm, 136 moribund house sparrows were taken to Bumpus's laboratory at Brown University and nearly half of them died as a result of the stress. Bumpus thought that it might be possible to show that '. . . the birds which perished, perished not through accident, but because they were physically disqualified, and that the birds which survived, survived because they possessed certain physical characters.' Accordingly, he took various measurements on each of the 136 birds and compared these for survivors and the non-survivors. He concluded that there were systematic differences and that natural selection had taken place. Bumpus's paper is interesting because it must have been just about the first genuine demonstration of the direct action of natural selection.

In Britain a biometric school developed towards the end of the 19th century, led by Francis Galton, a cousin of Darwin. Galton published *Hereditary Genius* in 1869 and *Natural Inheritance* in 1889. He developed the statistical technique of regression analysis in order to study the relationship between parents and offspring for characters such as stature. In 1901, Galton, Karl Pearson and W. F. R. Weldon started the journal *Biometrika*. In the editorial in the first volume they made it clear that one of the main purposes of the new journal was to publish the results of investigations of the operation of natural selection, and many of the papers published in the early volumes were on this topic. For example, Weldon (1901, 1903) and Di Cesnola (1906) realized that the earliest formed whorls in a snail's shell are preserved in the adult and they were therefore able to compare measurements on young snails with the corresponding measurements on adults. They found that mean values were almost the same for adult and young but that in some populations the adults were less variable than the young. They interpreted this as meaning that young snails with extreme measurements did not survive to become adults, and called this periodic (stabilizing) selection.

1.2   DEVELOPMENTS FROM 1910 TO 1959

Darwin believed that the inheritance of characters involved some sort of blending process and that selection, sorting over available possibilities in each generation, somehow served to maintain variability. It was not until 1900 that Mendel's (1865) famous paper that forms the basis of modern genetics became known to scientists generally, and it took a further ten years for blending theories to be completely discredited. By that time, mutation seemed more important than natural selection in the evolutionary process.

In 1924 Haldane estimated that the dark *carbonaria* form of the moth *Biston betularia* could only have spread in the Manchester area of England at the rate that it did over the years 1848 to 1901 if it had a selective advantage over the lighter *typica* form of 50% or more per generation. This calculation seems to have had little influence and when R. A. Fisher published his classic book *The Genetical Theory of Natural Selection* in 1930 this was largely based upon the idea that natural selection generally operates at a very low level of intensity. This philosophy seems to have discouraged attempts in the early part of this century to measure the magnitude of selection.

A more optimistic outlook developed in the next decade. Gordon (1935) released 36 000 *Drosophila melanogaster* at Dartington Hall in South Devon where the species was either rare or absent. Half of the released flies were 'marked' with the ebony gene. When samples of flies were taken 120 days later, after about five or six generations, the frequency of the ebony gene was down to about 11%. This was clear evidence of strong selection against the ebony gene. From studying samples of wild *Paratettix cucullatus* (formerly *P.*

texanus), Fisher (1939) concluded that certain genetic types were being eliminated by powerful selective forces. Somewhat later, Popham (1941, 1943, 1944, 1947) carried out a series of experiments in the laboratory and in the wild in which he demonstrated that predators act selectively and that the selection can be strong enough to be detected in a short period of time.

In Britain, attitudes were influenced to a large extent by the work of E. B. Ford and his colleagues, which is summarized in the book *Ecological Genetics* (Ford, 1975). An important part of this work was concerned with developing methods for estimating population sizes and survival rates by the technique of marking, releasing, and recapturing animals. This approach was used to study a population of the moth *Panaxia dominula* at Cothill in Berkshire from 1941 to 1946 (Fisher and Ford, 1947). Fluctuations in the frequencies of different genotypes seemed to be too large to be attributed to anything except selection. One of Ford's colleagues, H. B. D. Kettlewell, carried out a particularly notable series of mark–recapture experiments demonstrating selection over a very short period of time (Kettlewell, 1955, 1956).

Kettlewell's experiments concerned the moth *Biston betularia*. This moth has three genetic types: *typica*, which is greyish white sprinkled with black dots; *carbonaria*, which is almost completely black; and *insularia*, which is intermediate between *typica* and *carbonaria*. In 1848 almost all of the *B. betularia* specimens captured in Manchester were *typica*. By 1900 the moth was almost entirely *carbonaria*. As mentioned above, Haldane (1924) calculated that this change can only be explained if *carbonaria* moths produced on average 50 % more offspring than *typica* moths during this period.

To study the relative fitness of *typica* and *carbonaria*, Kettlewell began his experiments in 1953 by releasing 137 male *typica* and 447 male *carbonaria* in a bird reserve at Rubery Close in an industrial and heavily polluted area of Birmingham. The moths were then given the opportunity of being recaptured in mercury vapour and caged virgin female traps during the next few days. He recaptured 18 (13.1 %) of the *typica* and 123 (27.5 %) of the *carbonaria* moths and took this to be evidence that *typica* do not survive as well as *carbonaria* in the experimental area. He witnessed moths being eaten by birds and concluded that selective predation was the explanation for the different recapture rates. This explanation is reasonable because to the human eye *carbonaria* were far less conspicuous than *typica* in the bird reserve.

In 1955 Kettlewell repeated the release experiment in an unpolluted area in Deanend, Dorset. In this area *carbonaria* were more conspicuous to the human eye than *typica* and Kettlewell therefore expected the proportions of *typica* recaptured to be higher than the proportion of *carbonaria*. This happened: 496 *typica* were released, of which 62 (12.5 %) were recaptured, while 473 *carbonaria* were released, of which only 30 (6.3 %) were recaptured. These experiments therefore provide an explanation (selective predation by birds) for the spread of melanic forms of the moth *Biston betularia* that occurred at the same time as industrialization in England. They also explain why about 9 % of the natural moth population is *typica* at Rubery, while 95% is *typica* at Deanend.

While Ford was developing his school of ecological genetics in Britain, aided in theoretical matters by Ronald Fisher, a similar development was taking place in the United States. There Theodosius Dobzhansky was studying wild populations of the fruit fly *Drosophila*, with his theoretical aid coming from Sewall Wright. Dobzhansky began studying chromosome variation in *Drosophila pseudoobscura* in the late 1930s. At first it appeared that these are selectively neutral because different chromosome arrangements have no effect on the outward appearance of flies. Later it was discovered that there are systematic seasonal changes in the proportions of different arrangements in wild populations which, because of their rapidity, necessitate the assumption of very high selective pressures (Dobzhansky, 1943). This led to a series of investigations aimed towards an understanding of the nature of the selection. Much of Dobzhansky's work was published in a series under the general heading *The Genetics of Natural Populations* which has been reprinted, with editorial comments, by Lewontin *et al.* (1981).

1.3   THE MODERN SYNTHESIS AND ITS CRITICS

By 1959, one hundred years after the publication of *The Origin of Species*, the Darwinian theory of evolution through natural selection was firmly established for the majority of biologists. These were the days of almost complete acceptance of what Huxley (1942) called the 'modern synthesis' which is based upon two major premises: (a) that genetic mutation provides the only source of variation, with large-scale changes over long periods of time simply being the sum of many small changes occurring at a fairly constant rate; and (b) that genetic variation is raw material only, with rates and directions of change being determined by selection. In this framework, random genetic drift is considered to have relatively minor effects on evolution when compared with the effects of selection.

The study of natural populations was revolutionized in 1966 with the demonstration that electrophoresis can be used in determining the genotype of individual organisms for a wide range of genes specifying enzymes and other soluble proteins (Hubby and Lewontin, 1966; Lewontin and Hubby, 1966). This proved to be important for two reasons. Firstly, it made the study of natural variation much easier than it had been before. It was no longer necessary to restrict attention to chromosome polymorphisms in *Drosophila* and obvious visual polymorphisms such as melanism in *Biston betularia*. Secondly, it led to the discovery that a very large number of genetic loci are polymorphic in natural populations, a result that was somewhat unexpected at the time and was regarded in some quarters as evidence that the modern synthesis was wrong.

This is not the place to go into details about why the level of genetic variation observed in natural populations seemed to be too great to be maintained by selection. An account is given in most modern texts on population genetics (e.g., Spiess, 1977, p. 610; Merrell, 1981, p. 317; Wallace, 1981, p. 295). From the point of view of the present work, the important outcome was that the apparent difficulties led Kimura (1968) to propose his neutral mutation–random drift theory for the evolution of protein polymorphisms. This means exactly what it says: every now and then a new gene arises by mutation, which is functionally equivalent to existing genes at the same locus. The frequencies of genes in a population are purely due to random effects. This neutral theory has also been called non-Darwinian evolution (King and Jukes, 1969). It was modified somewhat by Ohta (1973, 1974), who suggested that some mutations may be very slightly deleterious but still random effects will be more important than selection to the evolutionary process. For recent reviews, see Kimura (1979) and McDonald (1983).

From the time of publication of Kimura's (1968) paper until the present there has been a steady stream of papers claiming to prove or disprove the neutral theory. In many cases the validity of the methods used has been disputed. The controversy seems to have generated a good deal of heat but not much light. A realistic point of view would seem to be that genetic variation associated with visible morphological features is determined largely by selection. However, some, possibly the majority, of protein polymorphisms may involve almost neutral genes.

At the same time that the modern synthesis was being attacked on the grounds that it did not explain the data on protein polymorphisms, it also began to be attacked from a completely different direction. In *The Origin of Species*, Darwin discussed the evidence from fossil remains and admitted that this did not agree with a concept of slow continuous changes guided by natural selection. He noted the large number of 'missing links' in the transition between species and the sudden way in which whole groups of species appear. The explanation that he proposed was that the fossil record is incomplete, and that a complete record would show the gradual pattern of evolution that his theory predicts. He noted at the end of his chapter 'On the imperfection of the geological record' that:

> Those who think the natural geological record in any degree perfect, and who do not attach much weight to the facts and arguments of other kinds given in this volume, will undoubtedly at once reject my theory.

The problem of reconciling the fossil record with Darwinism did not disappear with time. Many palaeontologists never accepted that it is just a matter of missing data. Recently a theory of punctuated equilibria has been put forward as a possible solution. Based upon the ideas of Simpson (1944) and Mayr (1954), this theory has been developed particularly by Eldredge and Gould (1972), Gould and Eldredge (1977), and Stanley (1979, 1981). It unties the link between natural selection and gradualism and says instead that evolutionary changes are concentrated in small populations that are rapidly forming new species. Once established, most species show very little change over millions of years. The effects of selection are therefore felt mainly during speciation. Because of small population sizes, random effects may also be important at this time (Wright, 1982).

The theory of punctuated equilibria is by no means universally accepted. This is shown, for example, by Lande's (1980a) review of the book *Macroevolution* by Stanley (1979). Possibly gradual and punctuational changes both occur. Gould (1980, 1981) has argued that gradualism is an inherent part of the modern synthesis and that therefore the synthesis breaks down if punctuational changes occur. However, others (e.g., Orzack, 1981; Stanley, 1981; Fitch, 1982; Mayr, 1982; Stebbins, 1982) have taken the view that the possibility of punctuated equilibria can be incorporated within the modern synthesis quite easily. In fact, Rhodes (1983) has recently argued that Darwin himself enunciated some of the main tenets of the theory of punctuated equilibria.

A development that introduces a new dimension into the debate is the discovery of evidence that the temporal distribution of species extinctions over the past 250 million years shows a pattern with about a 26 million year cycle (Raup and Sepkoski, 1984). It seems that most extinctions have occurred over fairly short periods of time separated by periods with much lower extinction rates. The largest extinction may have involved the disappearance of 96% of species with obvious major effects on long-term evolution.

It seems fair to sum up the situation at the present by saying that it is pretty well universally accepted amongst biologists that natural selection does occur: Darwin was right about that. However, there is no concensus of opinion about how important selection is to the evolutionary process in general.

1.4 THE MEASUREMENT OF FITNESS

Natural selection is sometimes defined as the 'survival of the fittest', an expression invented by Herbert Spencer and incorporated by Darwin into the later editions of *The Origin of Species*. Because of this expression, biologists often refer to the fitness of a certain type or organism, although it is difficult to define precisely what this means (Dawkins, 1982, Ch. 5). It is certainly not desirable to link fitness exclusively to survival, and a better approach is one which is based on the concept that the most fit individuals in one generation are those that contribute relatively the greatest proportion of progeny to the next generation. Even this is not completely satisfactory because the fittest individuals over one generation may not reproduce as well as some other individuals over two or more generations. Also, it is not straightforward to translate the concept into mathematical terms taking into account the genetics of a sexually reproducing population (Kempthorn and Pollack, 1970).

Nevertheless, it is sufficient for many practical purposes.

Measuring total fitness values is extremely difficult. The problem is that selection can occur at many different points in the life cycle of an organism. For example, Table 1.1 shows six different possible types of selection for three stages in the life cycle of a sexually reproducing species. It is obviously far from easy to get good estimates of fitness values for all of these types of selection. Most success has been achieved with *Drosophila* (Hedrick and Murray, 1983).

Table 1.1   Stages in the life cycle at which selection can act for a sexually reproducing organism.

Stage	Type of selection
Conception to maturity	1 Differential survival
Gamete (egg or sperm)	2 Differential output
	3 Differential survival
Mating	4 Non-random mating
	5 Incompatibility between mates that is not behavioural
	6 Incompatibility between progeny and parents.

Often, all that can be studied for a natural population is selection of one type, say selection relating to survival from birth or hatching to maturity. It is then convenient to refer to the survival of different forms of individual as their fitnesses, although it goes without saying that these 'fitnesses' may bear very little relationship to total fitnesses over a full generation. It might seem that information on one component of fitness is of little value, but it can be argued that this is better than nothing. Also, discovering selection at one stage of a life cycle indicates that the character being studied is not neutral, which will presumably be worth knowing.

It is an important fact that selection does not imply evolution. Selection in one stage of a life cycle can be offset by selection in the opposite direction in a later stage. Also, of course, if the character being selected is not inherited then selection cannot have any influence on the distribution of the character in future generations. It is therefore possible to have a situation where there is strong natural selection on a population with no evolutionary consequences at all.

1.5   QUANTITATIVE AND POLYMORPHIC VARIATION

In considering the many ways that natural selection has been investigated, an important distinction that has to be made is between quantitative and polymorphic variation. By *quantitative variation* is meant variation in one or more continuous or discrete variables $X_1, X_2, \ldots, X_p$. Usually the study of selection then involves comparing the distribution of these variables at different times or places. On the other hand, a morph is a distinct form of an organism that can be recognized by its colour or some other qualitative characteristic usually, but not always, under genetic control. A polymorphic population is then one in which two or more morphs are present, and by *polymorphic variation* is meant variation in the morph proportions at different times or places.

When selection operates on a single quantitative variable X it is said to be *directional* if it has the effect of either increasing or decreasing the mean value; *stabilizing* if it tends to eliminate individuals with extreme values for X; and *disruptive* if it tends to eliminate individuals with moderate values for X. Figure 1.1 illustrates these three cases for a population where X has a bell-shaped frequency distribution before selection.

The same types of selection can occur when several quantitative variables $X_1, X_2, \ldots, X_p$ are considered simultaneously. Then directional selection tends to change the vector of mean values for the X variables; stabilizing selection tends to eliminate individuals that are extreme for one or more of the variables; and disruptive selection tends to eliminate individuals with X values that are close to the population mean values. It is of course possible to have directional selection on some variables and stabilizing or disruptive selection on others.

Sometimes it will be convenient to summarize the selection related to quantitative variables by means of a *fitness function*. This is a function $w(x_1, x_2, \ldots, x_p)$ which is such that if $f(x_1, x_2, \ldots, x_p)$ is the population frequency of individuals with X-values $X_1 = x_1, X_2 = x_2, \ldots, X_p = x_p$ before selection, then the expected frequency after selection is proportional to $w(x_1, x_2, \ldots, x_p) f(x_1, x_2, \ldots, x_p)$.

With polymorphic variation it is usual to measure selection in terms of selective values, where these express the relative changes of different morphs. For example, suppose that a population comprises morphs A, B and C, with frequencies n_A, n_B and n_C before selection and frequencies n'_A, n'_B and n'_C after selection, so that absolute fitness values are n'_A/n_A, n'_B/n_B and n'_C/n_C. The selective value of morph B relative to morph A is then the relative fitness

$$w_B = (n'_B/n_B)/(n'_A/n_A),$$

and the selective value of C relative to A is the relative fitness

$$w_C = (n'_C/n_C)/(n'_A/n_A).$$

The relative fitnesses of A, B and C are then in the ratios $1:w_B:w_C$.

For some purposes it is convenient to measure relative fitnesses by the logarithms of selective values rather than the selective values themselves. The main reason for this is to obtain symmetry on an additive scale. Thus the fitness

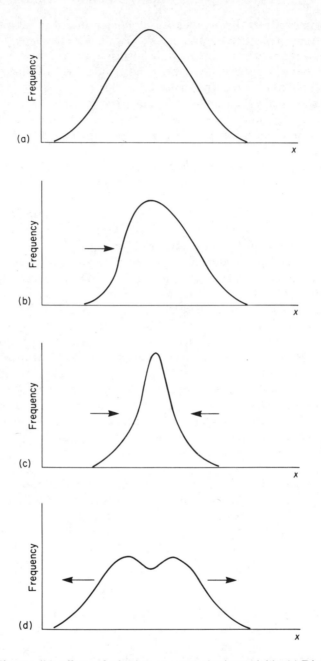

Fig. 1.1 The possible effects of selection on a quantitative variable. (a) Distribution of X before selection. (b) Distribution after directional selection against low values of X. (c) Distribution after stabilizing selection against extreme values of X. (d) Distribution after disruptive selection against moderate values of X.

of B relative to A becomes equal to minus the fitness of A relative to B, since

$$\log_e\left\{\left(\frac{n'_B}{n_B}\right)\Big/\left(\frac{n'_A}{n_A}\right)\right\} = -\log_e\left\{\left(\frac{n'_A}{n_A}\right)\Big/\left(\frac{n'_B}{n_B}\right)\right\}.$$

In practical terms this logarithmic transformation is useful since it appears generally to be the case that logarithms of estimated selective values are more normally distributed than the estimated selective values themselves.

Sections A.1 and A.2 of the Statistical Appendix give a brief outline of some basic theory of quantitative and polymorphic variation.

1.6   VARIETIES OF DATA ON NATURAL SELECTION

Table 1.2 gives a summary of the different types of data that are considered in the chapters that follow. This may be useful in placing different methods for studying selection into perspective. In this table, the nature of the evidence provided by each type of data is classified as direct or circumstantial. Evidence is considered to be direct if the observations made can be related to selection of a particular type. For example, mark–recapture experiments provide direct evidence of selection when they are specifically designed to allow the comparison of adult survival rates for several morph types. Evidence is considered to be circumstantial if the observations taken only make it possible to detect some element of non-randomness in the distribution of a population, and if the non-randomness can be explained in general terms by adaptation. For example, a significant correlation between the frequency of a gene and temperature can be explained in terms of selection related to temperature although the form of the selection might remain a mystery. On the whole, direct evidence is 'stronger' than circumstantial evidence. However, even with direct evidence there may be alternative non-selective explanations for significant results.

There are five broad categories of data shown in Table 1.2. The first of these relates to studies of the relative survival of juvenile and adult animals either in the wild or in the laboratory. These studies all provide direct evidence of selection. By contrast, the second broad category involves studies of the association between an animal's distribution and environmental variables or of the association between two related species. At best these studies provide circumstantial evidence of adaption. Both of the first two broad categories of data are straightforward in the sense that genetic considerations are not involved.

The third category is very broad indeed since it incorporates all studies of selection related to specific genetic loci. These range from the estimation of selective values to account for changes in gene frequencies from generation to generation, to tests based upon neutral mutation models for evolution. Many

Table 1.2 A summary of different approaches that have been used to study selection.

Type of selection being studied	Approach used	Nature of evidence	Type of data	Chapter reference
Adult or juvenile survival	Mark–recapture experiments	D	P	2
	Comparison of samples taken from a population before and after selection	D	P,Q	3
	Comparisons of the distribution of dead animals with the distribution of live animals	D	P,Q	4
	Comparison of all survivors with all non-survivors	D	P,Q	5
General fitness as it effects the adaption of animals to their environment	Study of the association between the distribution of an animal and various environmental variables	C	P,Q	6
	Search for an association between the distributions of two related species	C	P,Q	7
Selection over one or more generations related to one or more genetic loci	Estimation of selective values for one locus from samples taken from several generations of a population	D	P	8
	Estimation of selective values for two loci from samples taken from several generations of a population	D	P	10
	Tests for Hardy–Weinberg equilibrium	C	P	9
	Ewens–Watterson tests of the infinite allele–neutral mutation model	C	P	9
	Goodness of fit of the charge-state model	C	P	9

Population samples including mother–offspring combinations	D	P	9
Study of linkage disequilibrium	C	P	10
Association between blood groups and diseases	C	P	12
Use of demographic data with an age-structured population	D	P	12
Use of data on racial admixture	C	P	12
Tests based upon various properties of the infinite allele–neutral mutation model	C	P	12
Selection over one or more generations on quantitative variables			
Changes in the mean of a quantitative variable over two or more generations	D/C	Q	11
Comparison of changes in marginal and central populations	C	Q	11
Selection related to mating			
Laboratory experiments on mate choices	D	P	13
Samples of mating pairs and non-mating individuals in the wild	D	P	13

D = direct evidence of selection   C = circumstantial evidence of selection
Q = quantitative variation   P = polymorphic variation

of these tests have been devised for the analysis of data obtained by electrophoresis.

The fourth broad category is also concerned with studies in which genetic considerations are of major importance. However, here interest is in changes from generation to generation in quantitative variables. Some of these studies involve data on changes over many generations, possibly millions of generations with fossil data. Other studies are concerned with comparing changes from one generation to the next in marginal and central parts of a population to test the hypothesis that central parts are more adaptable than marginal ones.

The last broad category includes studies on selection related to mating. Most of these studies have been conducted on laboratory populations but in a few cases data have been collected on the frequencies of different types of mating pairs in wild populations.

1.7 SAMPLING PROBLEMS

One of the biggest problems with studying selection is the difficulty of random sampling of wild populations. This has been emphasized by Clarke et al. (1978) in a critical review of work on pulmonate molluscs (snails). They point out, for example, that the finding by Weldon (1901, 1903) and Di Cesnola (1906) that juvenile snails have more variable whorls than adults may just be an artefact due to sampling more than one population. Significant changes in morph frequencies for snails have been found over distances as short as 5 m. If juveniles were difficult to find then the collectors may well have sampled them over a larger area than that sampled for adults. This alone would tend to produce more variation for the juveniles.

Clarke et al. also point out that laboratory studies based upon animals collected from different sources can give misleading results. They give the example of Cook and O'Donald's (1971) experiment on the over-winter survival of Cepaea nemoralis in an unheated room (see Example 5.2, below). Large snails survived better than small snails. However, the snails used were collected from several different areas. It could be that the largest ones came mainly from one particular population which is characterized by the high survival of all its members. In other words, the apparent relationship between survival and size may really have been caused by population differences in survival.

Although Clarke et al. were only concerned with work on snails, it is quite clear that their remarks apply equally well with other organisms. For example, as discussed in Chapter 12, sampling problems are liable to upset studies of the association between blood groups and diseases for human populations.

2 Mark—recapture experiments

Mark—recapture methods were originally developed for estimating the size of mobile animal populations. However, it was soon recognized that the same methods could be used for estimating survival rates in the wild. It was then a short step to using the same approach for comparing the survival rates of different morphs in a polymorphic population. This application was pioneered by H. B. P. Kettlewell in the mid-1950s in a series of classical experiments on the moth *Biston betularia*.

After an introduction on the various uses of mark—recapture methods, the experimental design used by Kettlewell (the multi-sample single recapture experiment) is described, together with five different models that can be used for analysing such an experiment. These models involve different assumptions; they are compared in Section 2.10.

Following the sections on the multi-sample single recapture experiment there is a brief review of mark—recapture experiments of a more general nature. This is followed by a section on the question of how death and emigration can be separated – an important matter when it is relative survival rates that are of primary interest. The chapter concludes with a discussion of how estimates of survival rates can be converted to estimates of 'fitness'.

2.1 THE USE OF MARK—RECAPTURE EXPERIMENTS

Mark—recapture experiments have been used in many studies related to natural selection. Three particular applications can be mentioned. Firstly, these methods have been used to compare the survival of two or more morphs under natural conditions. Secondly, they have been used to estimate the size of a natural population when it has not been possible simply to count individuals. Finally, the movement of animals has been investigated.

The comparison of survival rates for different morphs is the main concern of this chapter. In this case the basic idea is very simple. For example, suppose that 100 black and 100 blue moths are released one morning with each moth having been previously marked with a small dot of paint. Suppose also that some light traps are set up in the evening and these recapture 25 blue and 10 black moths. Apart from chance, there are then three possible explanations for the different recapture rates: the two colours of moth may differ in their survival; they may differ in their tendency to migrate; or they may differ in their susceptibility to trapping. If the last two possibilities can be ruled out then it can be concluded that there is evidence of differential survival. Furthermore, since recovery rates can be expected to be roughly proportional to survival rates over the period between release and recapture, it would seem that the ratio of the black survival rate to the blue survival rate is approximately $10/25 = 0.4$. In practice, experiments have usually been more complicated than this, involving a series of releases with recapture samples taken between them.

The size of an animal population is an important parameter in any biological study. In considering natural selection it needs to be known in order to decide how much variation can be attributed to purely random fluctuations. For instance, this was the case with Fisher and Ford's (1947) study of the moth *Panaxia dominula* at Cothill which is discussed in Examples 8.1 and 8.2. The population size had to be estimated each year in order to establish that the observed fluctuations in the proportion of a mutant gene cannot be explained in terms of just random genetic changes. In cases like this the population being studied is often sampled at irregular intervals of time with rather variable sampling effort. Estimation problems are then considerable. This area of statistics is briefly reviewed in Section 2.11. No attempt has been made to cover this topic in detail since there are a number of books devoted to it.

The third use that has been mentioned for mark–recapture methods concerns the movement of animals. For example, from 1941 to 1946 very large numbers of *Drosophila pseudoobscura* were released by Dobzhansky and his colleagues to decide how far these flies can travel. In a review of these and other experiments, Powell and Dobzhansky (1976) discuss how important it is to understand the role of migration in the evolutionary process and how release and recapture experiments can help in this regard. Although this is true, these experiments are only indirectly relevant to natural selection and will not be considered further here. However, in Section 2.12 consideration is given to the problem of separating death and emigration using information on the movement of marked animals.

2.2 THE MULTI-SAMPLE SINGLE RECAPTURE EXPERIMENT

A multi-sample, single recapture experiment was used by Kettlewell (1955) to compare the survival of the three morphs *carbonaria*, *typica* and *insularia* of the melanic moth *Biston betularia* in the wild. This experiment has been described already in Section 1.2. It may be recalled that *typica* is a light-coloured morph while *carbonaria* is almost completely black. *Insularia* is much less common than the other two morphs and comes between them in terms of darkness. The polymorphism is determined by genes at a single genetic locus.

Kettlewell's experiment involved releasing batches of paint-marked male *carbonaria, typica* and *insularia* moths at daily intervals in a polluted wood over a period of ten days. Traps were set up at night to recapture them and the recapture rates obtained were: *carbonaria* 27.5%, *typica* 13.1% and *insularia* 17.4%. Kettlewell argued that these recapture rates mainly reflect survival differences and that he had therefore been able to demonstrate natural selection taking place to eliminate moths that are not well camouflaged in their natural surroundings. The selection can be explained as being the result of predation by birds.

The success of the release experiment led Kettlewell and others to carry out a number of other experiments to compare the survival of different morphs of various species of moths in different locations (Kettlewell, 1956, 1961; Kettlewell and Berry, 1969; Kettlewell *et al.*, 1969; Bishop, 1972; Bishop *et al.*, 1978). A rather ingenious variation has also been used by Brower *et al.* (1964, 1967), Cook *et al.* (1969), Sternburg *et al.* (1977), and Jeffords *et al.* (1979, 1980) to study the action of mimicry.

In analysing their results, all of these authors have made the assumption that differences in recapture rates reflect survival differences only. However, this assumption is not really necessary. It is possible to estimate a probability of survival and a probability of recapture for each type of animal released so that the survival of two morphs can be compared without assuming that they have the same probability of capture. One approach is based upon Seber's (1962) model which is for one type of animal released. Estimation is by the method of maximum likelihood which is reviewed in the appendix, Section A.3.

The model assumes that releases are made at times $t_1, t_2, \ldots, t_s$ and recapture samples are taken at times $t'_1, t'_2, \ldots, t'_s$, where $t_1 \leqslant t'_1 \leqslant t_2 \leqslant t'_2 \leqslant \ldots \leqslant t_s \leqslant t'_s$. (A zero release is possible so that there can in fact be two recapture samples without a release in the time between them.) The following assumptions are made:

(a) Every individual has the same probability p_i of being recaptured at time t'_i given that it has not been recaptured before that time and is still in the population at time t'_i. Recaptured animals are not released again.

(b) Every individual, irrespective of when it was released, has the same probability $1 - \phi_i$ of dying (or permanently emigrating) in the time interval t_i to t'_i, given that it is alive and uncaptured just after time t_i.

(c) Every animal has the same probability $1 - \chi_i$ of dying (or permanently emigrating) in the time interval t'_{i-1} to t_i, given that it is alive and uncaptured just after time t'_{i-1}.

(d) Released animals are marked in such a way that the time of release can be determined for every recaptured animal.

(e) Any emigration is permanent so that emigrants can be regarded as being 'dead'.

Seber (1962) shows that if it is assumed that the recapture samples are immediately followed by releases then it is possible to obtain explicit estimates of the survival probabilities by the method of maximum likelihood. However, in practice experimenters have often had different release and recapture times so that Seber's estimators are not appropriate. Nevertheless, his model can still be used if it is assumed that the survival probability per unit time remains constant at ψ_i between times t_i and t_{i+1} so that

$$\phi_i = \psi_i^{1-\Delta_i} \quad \text{and} \quad \chi_{i+1} = \psi_i^{\Delta_i}$$

where

$$\Delta_i = (t_{i+1} - t_i')/(t_{i+1} - t_i).$$

Release and recapture samples might or might not coincide. On this assumption, likelihood estimators of the unknown parameters are found (Manly, 1974a) by solving the equations

$$\hat{\psi}_i - \frac{m_i R_i}{(m_i + z_i)A_i} \hat{\psi}_i^{\Delta_i} = \frac{A_{i+1} z_i R_i}{A_i(m_i + z_i)R_{i+1}}, \tag{2.1}$$

and

$$\hat{p}_i = \frac{m_i R_i}{(m_i + z_i)A_i} \hat{\psi}_i^{\Delta_i} - 1, \tag{2.2}$$

for $i = 1, 2, \ldots, s-1$. Here A_i is the number of animals released at time t_i and R_i is the number of these that are recaptured. Also m_i is the number of recaptures at time t_i' while z_i is the number of animals released before time t_i' that are recaptured after time t_i'. The solution for equation (2.1) can be obtained either graphically (by plotting the left-hand side of the equation against ψ_i and seeing where it equals the right-hand side) or numerically (for example, by root bisection).

The following approximate formulae for variances and covariances can be obtained by the Taylor series method that is described in the appendix, Section A.5:

$$\text{var}(\hat{\psi}_i) \simeq \psi_i^2 \left\{ \frac{p_i^2}{E(m_i)} + \frac{(1-p_i)^2}{E(z_i)} - \frac{1}{E(m_i+z_i)} + \frac{1}{E(R_i)} - \frac{1}{A_i} \right.$$

$$\left. + (1-p_i)^2 \left(\frac{1}{E(R_{i+1})} - \frac{1}{A_{i+1}} \right) \right\} / \{1 - \Delta_i p_i\}^2,$$

while

$$\text{cov}(\hat{\psi}_i, \hat{\psi}_{i+1}) \simeq - \frac{\psi_i \psi_{i+1}(1-p_i)}{(1-\Delta_i p_i)(1-\Delta_{i+1} p_{i+1})} \left(\frac{1}{R_{i+1}} - \frac{1}{A_{i+1}} \right),$$

and

$$\text{cov}(\hat{\psi}_i, \hat{\psi}_j) \simeq 0, \quad j > i+1.$$

$$(2.3)$$

2.3 COMBINING AND COMPARING SURVIVAL ESTIMATES

If several estimated survival probabilities $\hat{\psi}_1, \hat{\psi}_2, \ldots, \hat{\psi}_n$ all relate to the same period of time (for example, they might all be one-day probabilities), then it may be useful to combine them to obtain a single more reliable estimate of the mean survival. One obvious possibility is just to calculate the usual arithmetic mean,

$$\bar{\psi}_A = \sum_{j=1}^{n} \hat{\psi}_j / n. \tag{2.4}$$

This may, however, give unreliable results because of the nature of the distribution of sampling errors. This is indicated by some estimates given in Example 2.1 below. For one morph there are 23 daily survival estimates which range from 0.05 to 1.94 (Table 2.2, the black controls). The arithmetic mean is 0.46. Hence the largest estimate is 1.48 above the mean while the smallest estimate is only 0.41 below the mean. Large overestimates are not balanced by correspondingly large underestimates. Consequently, the arithmetic mean may be unduly influenced by one or two rather large values.

Taking logarithms seems to overcome this problem. This is equivalent to using the geometric mean,

$$\bar{\psi}_G = (\hat{\psi}_1 \hat{\psi}_2 \ldots \hat{\psi}_n)^{1/n} \tag{2.5}$$

since

$$\log_e(\bar{\psi}_G) = \sum_{j=1}^{n} \log_e(\hat{\psi}_j)/n. \tag{2.6}$$

The Taylor series method gives $\text{var}\{\log_e(\hat{\psi}_i)\} \simeq \text{var}(\hat{\psi}_i)/\psi_i^2$, so that

$$\text{var}\{\log_e(\bar{\psi}_G)\} \simeq \left\{ \sum_{j=1}^{n} \text{var}(\hat{\psi}_j)/\psi_j^2 + 2 \sum_{j=1}^{n-1} \sum_{r=j+1}^{n} \text{cov}(\hat{\psi}_j, \hat{\psi}_r)/(\psi_j \psi_r) \right\} \Big/ n^2 \tag{2.7}$$

while

$$\text{var}(\bar{\psi}_G) \simeq \psi_G^2 \text{var}\{\log_e(\bar{\psi}_G)\}. \tag{2.8}$$

The value of taking logarithms is indicated by the effect on the 23 estimates mentioned above. Logarithms of estimates range from $\log_e(0.05) = -3.00$ to $\log_e(1.94) = 0.66$, with a mean of -1.21. The largest value is 1.87 above the mean while the smallest value is 1.73 below the mean. Errors of estimation now have a fairly symmetric distribution.

It may seem strange to use a value of 1.94 as an estimate of a daily survival probability since the true value cannot exceed 1. However, there is no difficulty in including estimates larger than 1 in a mean value since sampling errors should average out close to zero. Indeed, not including impossibly large estimates is liable to result in some bias.

The simple geometric mean $\bar{\psi}_G$ of equation (2.5) will be a reasonable combined estimate of survival providing that the individual values $\log_e(\hat{\psi}_1)$, $\log_e(\hat{\psi}_2), \ldots, \log_e(\hat{\psi}_n)$ have about the same variances. That is to say, the values $\text{var}(\hat{\psi}_i)/\psi_i^2$ should be approximately constant. If this is not so, then the weighted mean of $\log_e(\hat{\psi}_i)$ values with minimum variance can be used in place of $\log_e(\overline{\psi}_G)$. The calculations are described in Section A.6 of the Appendix.

For the purpose of comparing the survival of two morphs under the same conditions it can be noted that if $\hat{\psi}_a$ is an estimate with variance V_a and $\hat{\psi}_b$ is an independent estimate with variance V_b then the difference $\hat{\psi}_a - \hat{\psi}_b$ has variance $V_a + V_b$. A test statistic for the hypothesis of equal survival is therefore

$$g = (\hat{\psi}_a - \hat{\psi}_b)/(V_a + V_b)^{1/2}. \tag{2.9}$$

Assuming that g is normally distributed, a test involves seeing whether the absolute value $|g|$ is significantly large. Thus $|g| > 1.96$ corresponds to significance at the 5% level. Just how much reliance can be placed on this test is uncertain. If $\hat{\psi}_a$ and $\hat{\psi}_b$ are both averages of many individual survival estimates then it should be quite reasonable.

Example 2.1   An experiment in artificial mimicry

Some interesting mark–recapture experiments have been carried out in order to examine various aspects of the action of mimicry under natural conditions. Batesian mimicry involves two species of animal: the *model*, which has some inherent protection from predators, such as a nauseous flavour, and the *mimic* which does not have the same protection but nevertheless gains some potential advantage because of its resemblance to the model. Recapture experiments have involved painting moths to resemble conspicuous distasteful butterflies so as to produce artificial mimics. These are then released, together with painted controls. If the mimicry is effective then the artificial mimics should show a higher survival than the controls.

In one experiment (experiment A, Brower *et al.*, 1967) the mimics were males of the moth *Hyalophora promethea*, that were painted to resemble the distasteful butterfly *Parides anchises* and released together with two types of control moths. One of the control moths was inconspicuous and was painted black. The other control was conspicuous and was blue. The releases were made over a 24-day period in the Arima Valley, Trinidad, where *H. promethea* does not occur naturally. Before the experiment was carried out it was predicted that the artificial mimics would have higher survival than both types of control, and that the black controls would have higher survival than the blue controls.

The recapture data for this experiment are shown in Table 2.1. The time between release and the first opportunity for recapture has been taken to be one-half of a day because releases were made from about 9 a.m. to 10 a.m. while recaptures were made by way of assembling traps containing virgin female moths that were set at dusk each night. The estimates calculated for

Table 2.1 Releases and recaptures of *Hyalophora promethea* in Brower *et al.*'s (1967) experiment A. The moths were painted as mimics (M), black controls (C1) and blue controls (C2). These data were kindly supplied by Dr L. M. Cook. No moths were recaptured more than $3\frac{1}{2}$ days after release.

| Date | Numbers released | | | Numbers recaptured after | | | | | | | | | | | |
| | | | | $\frac{1}{2}$ day | | | $1\frac{1}{2}$ days | | | $2\frac{1}{2}$ days | | | $3\frac{1}{2}$ days | | |
	M	C1	C2	M	C1	C2	M	C1	C2	M	C1	C2	M	C1	C2
July 11	7	7	6	4	4	2	0	0	1						
12	17	17	14	8	5	4	1	0	0						
13	9	11	13	1	5	4	1	0	0						
14	13	16	15	7	6	8	0	0	3						
15	17	19	15	5	7	6	0	1	2						
16	26	24	22	3	10	6	1	0	0	0	0	1			
17	31	22	36	8	9	11	0	0	0	1	1	0	0	1	0
18	27	28	29	1	1	0	3	1	2	1	1	1			
19	23	32	26	9	10	11	0	3	0						
20	16	15	19	1	2	5	0	0	0	1	1	0			
21	22	23	22	0	2	0	1	1	2	0	0	1			
22	14	10	10	2	3	1	0	1	0						
23	16	15	14	1	0	0	0	1	1						
24	20	19	17	3	4	2	1	0	0	0	2	0			
25	17	18	18	1	1	2	0	0	2						
26	16	19	19	4	8	6	1	1	0	0	1	0			
27	13	15	18	1	4	7	1	1	0						
28	17	19	22	4	7	12	0	0	0	0	1	0			
29	15	18	18	2	3	2	0	1	1	0	0	0	0	0	1
30	12	13	16	1	3	4									
31	13	14	11	0	2	3	2	1	0						
Aug. 1	21	13	12	4	4	2	0	0	1						
2	17	16	15	7	6	6	1	1	0	1	0	0			
3	15	11	11	1	2	2									
Totals	414	414	418	78	108	106	13	13	15	4	7	3	0	1	1

these data are shown in Table 2.2. Using equations (2.5) to (2.8), the geometric mean survival probability for the mimics is estimated to be $\overline{\psi}_m = 0.144$ with standard error 0.027. For black controls the mean survival is estimated to have been rather higher at $\overline{\psi}_{C1} = 0.297$ with standard error 0.035. For blue controls the mean survival is estimated at $\overline{\psi}_{C2} = 0.254$ with standard error 0.030. The test statistic of equation (2.9) shows a difference in survival between the mimics and the black controls that is significant at about the 0.1 % level, and a

Table 2.2 Estimates of survival probabilities from the data of Table 2.1 on the recaptures of *Hyalophora promethea*.

Date	Estimates for mimics			Estimates for black controls			Estimates for blue controls		
	$\hat{\psi}_i$	SE	$\hat{r}$	$\hat{\psi}_i$	SE	$\hat{r}$	$\hat{\psi}_i$	SE	$\hat{r}$
July 11	0.33	0.21	0.00	0.33	0.21	0.00	0.90	0.68	−0.44
12	0.64	0.43	−0.44	0.09	0.07	0.00	0.08	0.07	0.00
13	0.20	0.19	−0.22	0.21	0.14	0.00	0.09	0.08	0.00
14	0.28	0.15	0.00	0.14	0.09	0.00	0.87	0.26	−0.33
15	0.09	0.07	0.00	0.34	0.19	−0.23	0.66	0.32	−0.26
16	0.18	0.16	−0.14	0.17	0.08	0.00	0.26	0.17	−0.09
17	0.30	0.22	−0.31	1.32	0.67	−0.58	0.43	0.36	−0.42
18	0.41	0.21	−0.23	0.22	0.14	−0.15	0.19	0.13	−0.21
19	0.47	0.37	−0.27	0.86	0.48	−0.46	0.34	0.20	−0.34
20	1.00	1.38	−0.56	0.29	0.29	−0.31	0.07	0.05	0.00
21	0.32	0.37	−0.56	0.19	0.14	−0.18	1.36	1.09	−0.56
22	0.02	0.03	0.00	1.39	1.16	−0.58	0.40	0.62	−0.43
23	0.004	0.008	0.00	0.12	0.14	−0.10	0.32	0.44	−0.47
24	1.00	1.26	−0.71	1.94	1.08	−0.76	0.01	0.02	0.00
25	0.004	0.007	0.00	0.08	0.08	−0.12	0.42	0.29	−0.45
26	0.60	0.56	−0.39	0.60	0.29	−0.35	0.10	0.07	0.00
27	0.27	0.28	−0.38	0.37	0.22	−0.19	0.15	0.09	0.00
28	0.06	0.05	0.00	0.44	0.30	−0.25	0.30	0.12	0.00
29	0.02	0.02	0.00	0.48	0.37	−0.58	0.52	0.38	−0.34
30	0.007	0.01	0.00	0.05	0.05	0.00	0.26	0.23	−0.26
31	0.81	0.64	−0.57	0.31	0.28	−0.40	0.40	0.37	−0.32
Aug. 1	0.04	0.03	0.00	0.09	0.08	0.00	0.25	0.21	−0.19
2	1.36	1.45	−	0.64	0.51	−	0.16	0.10	−

Note: $\hat{r}$ is the estimated correlation, $\text{cov}(\hat{\psi}_i, \hat{\psi}_{i+1})/\{\text{var}(\hat{\psi}_i)\text{var}(\hat{\psi}_{i+1})\}^{1/2}$, between $\hat{\psi}_i$ and $\hat{\psi}_{i+1}$. SE is the standard error.

difference in survival between the mimics and the blue controls that is significant at the 1 % level. The difference in survival between the black controls and the blue controls is not significant at the 5 % level. On the basis of these results there is clearly no evidence that the survival of the mimics was improved because of their resemblance to *P. anchises*.

Brower and his fellow workers continued their mark–recapture experiments using *promethea* moths (Cook *et al.*, 1969; Cook, 1969) and eventually concluded that their mimics may initially have enjoyed a selective advantage over the non-mimetic controls, but that if this was the case then the advantage rapidly converted to a disadvantage as avian predators became aware of the experimental insects (see Example 2.3). Taken as a whole their experiments were somewhat inconclusive. However, Waldbauer and Sternburg (1975) have argued that the experimental design was not satisfactory in the first place because the black-painted controls were themselves mimics of various butterflies. Therefore most of the experiments were really comparing the survival of two different mimics. (See also Waldbauer and Sternburg (1983) for further comments on the difficulty of using painted insects to study protective coloration.)

Sternburg et al. (1977) and Jeffords et al. (1979, 1980) carried out a number of mark–release experiments that they considered to be more appropriate than those of Brower and his co-workers. They were able to obtain clear evidence that painted artificial mimics are protected from bird predation relative to non-mimics.

2.4   THE CONSTANT SURVIVAL–CONSTANT RECAPTURE PROBABILITY MODEL

One problem with the method of analysing recapture data that has just been considered is that so many survival and recapture probabilities are estimated that it often happens that none of these probabilities are estimated very well. It is therefore natural to consider whether improved estimates can be obtained by making more assumptions. One possibility is to assume that the probability of surviving a unit interval of time remained constant during the experiment and that the probability of capture was the same for all recapture samples. If these assumptions are reasonable then there are only two unknowns to be estimated. Furthermore, if the times between releases and recaptures are also kept constant, and enough recapture samples are taken after the last release to ensure that just about all released animals were either recaptured or dead, then it turns out that the estimation equations are particularly simple.

In addition to assumptions (a) to (e) of Section 2.2, it will now therefore be assumed that:

(f) Releases are made at times $t_1 = 1$, $t_2 = 2, \ldots, t_s = s$, with recaptures made at times $t'_1 = 2 - \Delta, t'_2 = 3 - \Delta, \ldots, t'_{s-1} = s - \Delta, t'_s = s + 1 - \Delta, \ldots$ (continuing until all released animals are either dead or recaptured).

(g) The probability of capture (now denoted by p) is constant for all recapture samples.

(h) The probability of surviving one unit interval of time (now denoted by ψ) is constant during the experiment.

In that case, the probability of an animal being recaptured in the ith sample after its release is the probability of surviving that long, multiplied by the probability of not being captured $i - 1$ times, multiplied by the probability of being captured:

$$\alpha_i = \psi^{i-\Delta}(1 - p)^{i-1} p \qquad (2.10)$$

Maximum likelihood estimators of ψ and p are then given (Manly, 1975a) by the equations

$$\hat{\psi} - \frac{R^2}{A(R+T)}\hat{\psi}^\Delta = \frac{T}{R+T} \tag{2.11}$$

and

$$\hat{p} = \frac{R^2}{A(R+T)}\hat{\psi}^{\Delta-1} \tag{2.12}$$

where A is the total number of animals released, R is the total number of animals recaptured,

$$T = \sum_{l=1}^{\infty} (l-1)a_l,$$

and a_l is the number of animals that are captured in the lth sample after the time that they are released.

Equation (2.11) must be solved either graphically or by some numerical technique such as root-bisection.

Approximate variances for $\hat{\psi}$ and $\hat{p}$, and the covariance between $\hat{\psi}$ and $\hat{p}$, are given by

$$\left.\begin{aligned}
\mathrm{var}(\hat{\psi}) &\simeq \{\psi^2 p^2(1-\theta) + \psi(1-p)(1-\psi)^2\}/\{A\theta(1-\Delta p)^2\}, \\
\mathrm{var}(\hat{p}) &\simeq p^2(1-p)[(1-p)(1-\theta) + \{1-\Delta+\Delta\psi \\
&\quad \times (1-p)\}^2/\psi]/\{A\theta(1-\Delta p)^2\}, \\
\mathrm{cov}(\hat{\psi},\hat{p}) &\simeq p(1-p)[\psi p(1-\theta) - (1-\psi)\{1-\Delta+ \\
&\quad \Delta\psi(1-p)\}]/\{A\theta(1-\Delta p)^2\},
\end{aligned}\right\} \tag{2.13}$$

where θ stands for the total probability of recapture, that can be estimated by R/A.

2.5 INITIAL LACK OF CATCHABILITY

One problem that seems to occur quite often is that the recapture probability p is more or less constant, except in the first sample taken after an animal is released. In that sample very few recaptures are made. It is rather convenient that equations (2.11) to (2.13) can easily be modified to take this into account. Thus suppose now that the time scale of the experiment is as shown below:

Suppose also that an animal is not available for recapture until one unit of time has elapsed after its release so that, for example, the animals released at time t_3 cannot be captured until time t'_3, which is after time t_4. It turns out (Manly,

1975a) that in this situation the equations for estimation still apply, but with the Δ value negative. The few animals recaptured in the first possible sample can simply be treated as if they were never released.

2.6   THE DESIGN OF EXPERIMENTS

Consider now the problem of designing a mark—recapture experiment for which the constant survival—constant recapture probability model is going to be used for an analysis.

If the experiment is carried out simply in order to estimate the survival rate for one type of animal then the experimenter only needs to make sure that the standard error for the survival estimator is satisfactorily small. This standard error depends in a fairly complicated way on the recapture probability p, the Δ value that reflects the time interval between releases and recaptures, the total number of animals released A, and the survival probability ψ itself. Within limits the experimenter will probably have direct control of Δ and A. The recapture probability p will also usually be controllable to some extent since it will presumably depend upon the number of traps that are set. Furthermore, the experimenter will often have some idea of what the survival probability will itself be.

Table 2.3 shows the standard errors that will apply for the estimator of ψ, per animal released, for a wide range of experimental situations. The following examples illustrate its use.

Suppose it is expected that $\psi = 0.5$ per day and an initial lack of catchability of released animals means that they cannot be recaptured until $\frac{1}{2}$ day after their release. In this case Δ can be set anywhere between the limits $-\frac{1}{2}$ to $+\frac{1}{2}$ and Table 2.3 shows that the best value depends upon p. Clearly p should itself be made as large as possible. If $p \simeq 0.7$ is anticipated then Δ should be set at $-\frac{1}{2}$ in which case the standard error for $\hat{\psi}$ based on a total release of A animals will be approximately $0.48/\sqrt{A}$. A release of about 100 animals (not necessarily all at the same time) would then be needed to make the standard error as small as 0.05. On the other hand, if $p \simeq 0.3$ is the best that can be expected then Δ should be set at $+\frac{1}{2}$ and a release of 100 animals should give a standard error of about 0.07. Actually in this example the standard error depends mainly on p and the choice of Δ is not crucial.

For a second example, suppose that $\psi = 0.9$ with animals being available for recapture almost immediately after their release. Then Δ can be set between 0 and 1 and Table 2.3 suggests that 0 is appropriate. Also, p should be kept small. With $p = 0.1$ and $\Delta = 0$ the standard error for a release of A animals will be approximately $0.16/\sqrt{A}$. A badly designed experiment with $p = 0.9$ and $\Delta = 0.5$ would more than double this value.

If two types of animal are released simultaneously then each will have its own estimated daily survival rate. The difference between the two rates will have variance $\mathrm{var}(\hat{\psi}_1 - \hat{\psi}_2) = \mathrm{var}(\hat{\psi}_1) + \mathrm{var}(\hat{\psi}_2)$. Using this result, Table 2.3

Table 2.3 Standard errors (per animal released) for survival estimators from the constant survival–constant recapture probability model. If A animals are released then the standard error of $\hat{\psi}$ will be the tabulated value divided by $\sqrt{A}$. Source: Manly (1977a).

(a) $\psi = 0.1$

Δ	$p = 0.1$	0.3	0.5	0.7	0.9
$-\frac{3}{4}$	5.69	2.59	1.56	0.99	0.60
$-\frac{1}{2}$	4.37	2.07	1.29	0.84	0.51
$-\frac{1}{4}$	3.35	1.66	1.07	0.72	0.45
0	2.58	1.34	0.90	0.63	0.41
$\frac{1}{4}$	1.98	1.08	0.77	0.57	0.39
$\frac{1}{2}$	1.53	0.88	0.67	0.54	0.40
$\frac{3}{4}$	1.18	0.73	0.61	0.55	0.47

(b) $\psi = 0.3$

Δ	$p = 0.1$	0.3	0.5	0.7	0.9
$-\frac{3}{4}$	2.63	1.26	0.84	0.62	0.50
$-\frac{1}{2}$	2.32	1.16	0.79	0.60	0.49
$-\frac{1}{4}$	2.04	1.06	0.75	0.59	0.49
0	1.80	0.98	0.72	0.58	0.49
$\frac{1}{4}$	1.59	0.91	0.71	0.59	0.52
$\frac{1}{2}$	1.40	0.85	0.70	0.63	0.57
$\frac{3}{4}$	1.24	0.80	0.71	0.70	0.72

(c) $\psi = 0.5$

Δ	$p = 0.1$	0.3	0.5	0.7	0.9
$-\frac{3}{4}$	1.36	0.72	0.55	0.48	0.45
$-\frac{1}{2}$	1.27	0.70	0.55	0.48	0.46
$-\frac{1}{4}$	1.20	0.69	0.55	0.50	0.47
0	1.12	0.67	0.56	0.52	0.50
$\frac{1}{4}$	1.06	0.67	0.58	0.55	0.55
$\frac{1}{2}$	0.99	0.66	0.60	0.60	0.63
$\frac{3}{4}$	0.94	0.66	0.64	0.69	0.80

(d) $\psi = 0.7$

Δ	$p = 0.1$	0.3	0.5	0.7	0.9
$-\frac{3}{4}$	0.60	0.40	0.37	0.36	0.37
$-\frac{1}{2}$	0.59	0.40	0.38	0.38	0.39
$-\frac{1}{4}$	0.58	0.41	0.39	0.40	0.41
0	0.57	0.41	0.40	0.42	0.45
$\frac{1}{4}$	0.56	0.42	0.43	0.46	0.50
$\frac{1}{2}$	0.54	0.43	0.46	0.51	0.58
$\frac{3}{4}$	0.53	0.45	0.50	0.59	0.75

(e) $\psi = 0.9$

Δ	$p = 0.05$	0.1	0.3	0.5	0.7	0.9
$-\frac{3}{4}$	0.18	0.16	0.17	0.19	0.21	0.23
$-\frac{1}{2}$	0.18	0.16	0.18	0.20	0.22	0.24
$-\frac{1}{4}$	0.18	0.16	0.18	0.21	0.24	0.26
0	0.18	0.16	0.19	0.23	0.26	0.29
$\frac{1}{4}$	0.18	0.16	0.20	0.24	0.28	0.32
$\frac{1}{2}$	0.18	0.16	0.20	0.26	0.32	0.38
$\frac{3}{4}$	0.18	0.17	0.21	0.28	0.37	0.50

Note: ψ = probability of survival per unit interval of time
p = probability of recapture in a particular sample
Δ = time between a recapture sample and the next release.

can be of value for designing experiments to compare survival. $\text{var}(\hat{\psi}_1 - \hat{\psi}_2)$ can be calculated for various combinations of ψ_1, ψ_2, Δ and p to see how many animals need to be released in order to make this variance small enough so that a significant difference between $\hat{\psi}_1$ and $\hat{\psi}_2$ can be expected if ψ_1 and ψ_2 have a difference of, say, 0.1.

2.7 A COMMON PROBABILITY OF CAPTURE MODEL

If several morphs are released at the same time and the constant survival—constant recapture probability model is used separately for each type, then it is not being assumed that all morphs have the same capture probability. However, in some cases it will be reasonable to assume that the capture probability is the same. A single common value can then be estimated using all of the data (Manly, 1977a). Unfortunately, the estimates for a constant survival—common recapture probability model have to be calculated numerically using, for example, the Newton—Raphson iterative method for maximum likelihood that is described in the appendix, Section A.3.

Example 2.2 *Amathes glareosa* in the Shetland Islands

The moth *Amathes glareosa* is widely distributed throughout the Shetland Islands. At Unst, at the north end of the islands, about 97 % of the population has been found to be of the melanic form *edda* and 3 % of the light form *typica*. This contrasts with the situation at Dunrossness, in the south, where 98 % of *A. glareosa* have been found to be *typica* and 2 % *edda*. Indeed, Kettlewell *et al.* (1969) have shown that there is a cline in the frequency of *edda* in the Shetland Islands running from north to south. Moving south from Unst, the frequency of *edda* is about 76 % at Hillswick and 26 % at Tingwall (Fig. 2.1).

This interesting situation led Kettlewell and his colleagues to carry out mark—recapture experiments to see whether the cline is maintained by different survival rates for the two morphs. The experiments involved releasing local and 'foreign' *typica* and *edda* moths at five different locations with 'foreign' moths being captured at one location and transported to their release point. The experimental results are summarized in Table 2.4. Releases of moths were made on the north and south sides of the Tingwall valley because of the particularly fast change in morph frequencies in that area.

Moths were released at about 4 p.m. and the first opportunity for recapture occurred a few hours later (about $\frac{1}{4}$ day). It can be noted that there were rather few *edda* moths recaptured after only $\frac{1}{4}$ day of freedom. This is an example of the phenomenon of a low recapture probability immediately after release that was discussed in Section 2.5. The problem can be overcome by ignoring moths recaptured after only $\frac{1}{4}$ day, thus treating them as if they were never released, and setting $\Delta = -\frac{1}{4}$ in equations (2.11) to (2.13).

The chi-square values in Table 2.4 indicate the goodness of fit of the constant survival—constant recapture probability model for the recapture data. Table 2.5 illustrates the calculation for the moths captured and released at

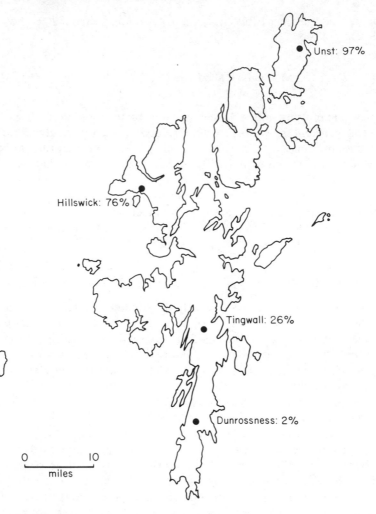

Unst: 97%

Hillswick: 76%

Tingwall: 26%

Dunrossness: 2%

0   10
    miles

Fig. 2.1 The Shetland Islands, about 100 miles off the north coast of Scotland. The map shows how the frequency of the *edda* morph of *Amathes* varies from about 97 % in the north to about 2 % in the south. Sampling points are shown approximately only.

Unst in 1960. None of the test statistics come out significantly large and therefore this model appears to be appropriate for these data. The goodness of fit test has only been carried out with the experiments with large numbers of moths released.

Table 2.6 shows estimates of ψ and p values for the 16 mark–recapture experiments, together with their estimated standard errors. It is interesting to try to relate the survival rates to the various factors that were varied in the

Table 2.4 The results of mark–recapture experiments on *Amathes glareosa* in the Shetland Islands.

Year	Morph	Captured	Released	Number released (A)	Number recaptured after (days)															R*	T*	χ²
					¼	1¼	2¼	3¼	4¼	5¼	6¼	7¼	8¼	9¼	10¼	11¼	12¼	13¼	14¼			
1960	typica	Dunrossness	Unst	2084	34	20	16	11	6	3	2	1	1	0	1					95	100	1.1 (3 d.f.)
1960	edda	Unst	Unst	2260	25	34	26	19	11	6	3	2	2	1	1	1	0	1		132	201	1.9 (5 d.f.)
1960	typica	Dunrossness	Dunrossness	1183	13	13	9	8	4	4	0	0	1							52	60	1.4 (2 d.f.)
1960	edda	Unst	Dunrossness	961	3	14	12	3	6	5	1	0	5	0	1					50	105	1.2
1961	typica	Tingwall	North Tingwall	578	2	0	4	0	3	5	6	2	1	0	1					24	91	—
1961	edda	Tingwall	North Tingwall	269	1	2	0	3	2	0	0	1	0	1	1	1				13	56	—
1961	typica	Tingwall	South Tingwall	571	7	4	2	3	3	0	0	0	1	0	1					21	31	—
1961	edda	Tingwall	South Tingwall	264	2	2	2	1	1	0	0	0	1	0	1					6	3	—
1961	typica	Hillswick	Hillswick	89	0	1	4	0	0	1										6	8	—
1961	edda	Hillswick	Hillswick	471	8	8	12	5	8	2	2	0	0	0	1	0	0	1		47	87	5.3 (3 d.f.)
1962	typica	Dunrossness	Hillswick	873	52	44	28	8	8	4	5	5	1	0	0	1				155	146	6.1 (4 d.f.)
1962	edda	Dunrossness	Hillswick	26	4	2	5	0	2											13	11	—
1962	typica	Tingwall	Hillswick	195	18	15	8	2	3	5	1	1	0	0	1					54	62	3.1 (2 d.f.)
1962	edda	Tingwall	Hillswick	92	2	6	4	0	1	1	1	0	0							14	11	—
1962	typica	Hillswick	Hillswick	164	6	5	8	7	2	0	0	0	0	0	1					34	54	—
1962	edda	Hillswick	Hillswick	565	21	22	21	14	9	6	1	1	1	1	1	0	0	0	2	100	161	2.5 (4 d.f.)

* R and T are as defined for equations (2.11) and (2.12).

Table 2.5 Calculation of a goodness of fit statistic for *edda* moths captured and released at Unst in 1960. Maximum likelihood estimates of survival and recapture probabilities are $\hat{\psi} = 0.6710$ and $\hat{p} = 0.0274$. The total number of moths released was 2260. However, because of the suspicion of a low initial probability of capture, the 25 moths captured after $\frac{1}{4}$ day have been ignored in the following calculations. The degrees of freedom (d.f.) for the chi-square value are the number of frequencies considered after pooling, minus one because the total observed and expected frequencies agree, minus two because ψ and p have been estimated. The expected number of recaptures at $(i+\frac{1}{4})$ days after release is estimated as $2235\,\hat{\psi}^{i+1/4}\,(1-\hat{p})^{i-1}\,\hat{p}$ for this model, based on equation (2.10).

Recapture time (days)	Expected recaptures (E)	Observed recaptures (O)	$(O-E)^2/E$
$1\frac{1}{4}$	37.2	34	0.3
$2\frac{1}{4}$	24.3	26	0.1
$3\frac{1}{4}$	15.8	19	0.6
$4\frac{1}{4}$	10.3	11	0.0
$5\frac{1}{4}$	6.7	6	0.1
$6\frac{1}{4}$	4.4 } 7.3	3 } 5	0.7
$7\frac{1}{4}$	2.9	2	
After $7\frac{1}{4}$	5.2	6	0.1
Not recaptured	2128.0	2128	0.0
Totals	2234.8	2235	1.9 (5 d.f.)

experiments and one way that this can be done is by carrying out a multiple regression. Thus a plausible model is one of the form

$$\hat{\psi} = \alpha_0 + \alpha_1 X_1 + \alpha_2 X_2 + \alpha_3 X_3 + \alpha_4 X_4 + \alpha_5 X_5 + \alpha_6 X_6 + \alpha_7 X_7 + e \qquad (2.14)$$

where $\hat{\psi}$ is the estimated survival in Table 2.6, the X's are indicator variables, the α's are constants to be estimated, and e represents the error of estimation. More specifically, the X's are defined as follows:

$X_1 = 0$ for *typica*, 1 for *edda*;
$X_2 = 1$ for a release at Unst, 0 otherwise;
$X_3 = 1$ for a release at North Tingwall, 0 otherwise;
$X_4 = 1$ for a release at South Tingwall, 0 otherwise;
$X_5 = 1$ for a release at Hillswick in 1961, 0 otherwise;
$X_6 = 1$ for a release at Hillswick in 1962, 0 otherwise;
$X_7 = 1$ for a release at a 'foreign' site, 0 otherwise:

With these definitions, α_0 is the daily survival probability for Dunrossness *typica* released at Dunrossness. The model then allows for a different survival

Table 2.6 Estimates and standard errors of daily survival and recapture probabilities for *A. glareosa* at various localities.

Year	Morph	Place of capture*	Place of release*	Daily survival		Recapture probability	
				Estimate	SE	Estimate	SE
1960	*typica*	D	U	0.634	0.037	0.020	0.004
1960	*edda*	U	U	0.671	0.026	0.027	0.004
1960	*typica*	D	D	0.621	0.047	0.024	0.006
1960	*edda*	U	D	0.707	0.035	0.023	0.005
1961	*typica*	T	NT	0.813	0.036	0.010	0.003
1961	*edda*	T	NT	0.832	0.044	0.010	0.004
1961	*typica*	T	ST	0.697	0.067	0.012	0.005
1961	*edda*	T	ST	0.439	0.183	0.024	0.024
1961	*typica*	H	H	0.604	0.121	0.054	0.036
1961	*edda*	H	H	0.719	0.037	0.039	0.010
1962	*typica*	D	H	0.644	0.027	0.090	0.014
1962	*edda*	D	H	0.748	0.075	0.265	0.103
1962	*typica*	T	H	0.714	0.039	0.114	0.027
1962	*edda*	T	H	0.559	0.088	0.144	0.065
1962	*typica*	H	H	0.724	0.043	0.091	0.025
1962	*edda*	H	H	0.723	0.026	0.072	0.012

* *Abbreviations*: D: Dunrossness; U: Unst; T: Tingwall Valley; NT: North Tingwall; ST: South Tingwall; H: Hillswick.

probability for *edda*, for different locations of release, and for 'foreign' moths. Year differences can only be examined at Hillswick.

Because of the different reliabilities of different survival estimates, equation (2.14) has been estimated by a weighted regression, with the weights used being the reciprocals of the variances for the survival estimates. The weighted residual mean square for the regression is 1.29, compared to an expected value of 1 (Appendix, Section A.8). The equation gives a good fit to the data. However, the coefficient of X_7 is not significantly different from zero and the coefficients of X_5 and X_6 are very similar. It seems therefore that the survival of the moths was not affected by releasing them away from their site of capture, and also that the survival at Hillswick was very similar in 1961 and 1962. Nevertheless, differences between localities and morphs do account for a significant part of the variation of the survival estimates.

The results from the regression analysis suggested that a simpler equation might fit the data. This is

$$\hat{\psi} = \alpha_0 + \alpha_1 X_1 + \alpha_2 X_2 + \alpha_3 X_3 + \alpha_4 X_4 + \alpha_5 X_5 + e$$

where X_1 to X_4 are as previously defined, while X_5 is now 1 for a release at

Hillswick but 0 otherwise. This equation does indeed fit the data (weighted residual mean square = 1.06) and is estimated as

$$\hat{\psi} = 0.65 + 0.04\,X_1 - 0.02\,X_2 + 0.15\,X_3 + 0.01\,X_4 + 0.02\,X_5 + e \quad (2.15)$$

The coefficients of X_3 and X_4, which reflect differences between the north and south sides of the Tingwall valley, are significantly different at the 5% level. It seems therefore that there was a survival difference between the two sides of the valley. The coefficient of X_1, which reflects the difference in survival between *edda* and *typica*, is almost significantly different from zero at the 5% level. Table 2.7 shows the survival rates expected at the different localities on the basis of equation (2.15). It will be observed that these survival rates do not seem to be directly related to the morph frequencies at the points of release.

Table 2.7   Expected daily survival probabilities as estimated from equation (2.15).

Location of release	Local frequency of edda (%)	Survival of edda	Survival of typica
Dunrossness	2	0.65	0.69
South Tingwall	26	0.66	0.71
North Tingwall	26	0.80	0.84
Hillswick	76	0.67	0.71
Unst	97	0.63	0.67

The regression model suggests that the *edda* morph survived better than *typica* generally, and that survival depended upon the place of release but not upon the place of origin. These conclusions conflict with those of Kettlewell *et al.* (1969), which were that: (a) populations in different parts of the Shetland Islands have undergone a degree of local adaption in that the survival of marked *typica* and *edda* was consistently higher when they were released near the site at which they were first caught rather than elsewhere; and (b) both morphs of the same origin had approximately the same survival when they were released at their site of origin or elsewhere. However, it is only fair to note that Kettlewell *et al.* based their conclusions on more than just their recapture results. For example, they directly observed the behaviour of moths. In an earlier regression analysis of the same data (Manly, 1975b) there were some differences in detail from the analysis given here, although the conclusions were the same.

A weighted multiple regression analysis on the recapture probabilities shown in Table 2.6 shows no significant differences between the probabilities for *typica* and *edda* and gives no indication that the probability of recapture depends upon whether or not a moth is released at its site of origin. Nevertheless, the recapture probabilities did seem to vary from release site to release site.

2.8 THE VARYING SELECTIVE VALUES MODEL

Another approach to the analysis of the recapture data from an experiment involving the simultaneous release of two or more morphs is available. To use this alternative approach it is necessary to assume that one of the morphs is the 'control' and all that is needed is a comparison of the survival of the other morphs relative to this. As in the models already considered, if emigration occurs then it is assumed to be permanent and equivalent to death.

Suppose that K morphs are released at time t_0 into the same environment and recapture samples are taken at later times $t_1, t_2, \ldots, t_s$. The data then take the form shown in Table 2.8. Thus A_i of the morph i are released at time t_0, a_{ij} of these are recovered in the jth recapture sample, and a total of R_i are recovered by the end of the experiment. Morph 1 is regarded as the control.

Let p_j denote the probability that an animal alive at time t_j will be recaptured at that time. Let ψ_j be the probability that one of the A_1 controls will survive from t_{j-1} to time t_j, given that it is still in the population just after the recapture sample at time t_{j-1} has been taken. For the ith morph ($i \neq 1$) the situation is the same as for the controls except that the probability of surviving from time t_{j-1} to time t_j can be taken as $w_{ij}\psi_j$: the factor w_{ij} is then the selective value of morph i for the period t_{j-1} to t_j. On this basis, maximum likelihood estimators for the w_{ij} are found to be (Manly, 1972):

$$\hat{w}_{ij} = (a_{ij}a_{1j-1})/(a_{1j}a_{ij-1}), \qquad j \neq 1, \tag{2.16}$$

and

$$\hat{w}_{i1} = (a_{i1}A_1)/(a_{11}A_i). \tag{2.17}$$

The statistical properties of estimators of the form of equation (2.16) (the cross product or odds ratio) have been considered by Woolf (1955), Haldane (1956) and Edwards (1965) in the context of the relationship between blood groups and diseases (Section 12.2, below), and by Goux and Anxolabehere

Table 2.8 The information available from an experiment where K types of animal are released at time t_0 and recapture samples are taken at later times $t_1, t_2, \ldots, t_s$.

Type of animal	Number released at time t_0	Number recaptured at times $t_1 \quad t_2 \quad \ldots t_s$			Total recaptured
1 (controls)	A_1	a_{11}	$a_{12} \ldots$	a_{1s}	R_1
2	A_2	a_{21}	$a_{22} \ldots$	a_{2s}	R_2
$\vdots$	$\vdots$	$\vdots$	$\vdots$	$\vdots$	$\vdots$
K	A_K	a_{K1}	$a_{K2} \ldots$	a_{Ks}	R_K

It is assumed that recaptured animals are removed from the population so that animals can only be recaptured once. This implies that

$$R_i = \sum_{j=1}^{s} a_{ij}.$$

(1980) in the context of non-random mating (Section 13.3, below). For small expected frequencies a_{ij}, the estimator is liable to be rather biased. Indeed, strictly speaking it has infinite bias since there is a finite probability of either a_{1j} or a_{ij-1} being zero, and consequently of $\hat{w}_{ij}$ being infinite. Another problem is that the distribution of the estimates $\hat{w}_{ij}$ is not symmetric. It is therefore difficult to derive confidence intervals and carry out tests of significance. These problems with $\hat{w}_{ij}, j \neq 1$, will also occur with w_{i1}, although possibly not to the same extent.

The lack of symmetry in the distribution of $\hat{w}$ can be overcome by working with logarithms. The infinite bias can be avoided by following a suggestion of Haldane (1956) and taking

$$\hat{w}_{ij} = \{(2a_{ij}+1)(2a_{1j-1}+1)\}/\{(2a_{1j}+1)(2a_{ij-1}+1)\}, \qquad j \neq 1 \quad (2.18)$$

and

$$\hat{w}_{i1} = \{(2a_{i1}+1)A_1\}/\{(2a_{11}+1)A_i\}. \tag{2.19}$$

Using results given by Haldane it is then possible to derive the following large-sample approximations:

$$\left. \begin{aligned} \text{bias}\,(\hat{w}_{ij}) &\simeq w_{ij}(1/\gamma_{ij-1}+1/\gamma_{1j}), \\ \text{bias}\,(\hat{w}_{i1}) &\simeq w_{i1}(1/\gamma_{11}-1/A_1) \end{aligned} \right\} \tag{2.20}$$

$$\left. \begin{aligned} \text{var}\,(\hat{w}_{ij}) &\simeq w_{ij}^2(1/\gamma_{ij}+1/\gamma_{ij-1}+1/\gamma_{1j}+1/\gamma_{1j-1}), \\ \text{var}\,(\hat{w}_{i1}) &\simeq w_{i1}^2(1/\gamma_{i1}-1/A_i+1/\gamma_{11}-1/A_1), \end{aligned} \right\} \tag{2.21}$$

$$\left. \begin{aligned} \text{cov}\,(\hat{w}_{ij}, \hat{w}_{ij+1}) &\simeq -w_{ij}w_{ij+1}(1/\gamma_{ij}+1/\gamma_{1j}), \\ \text{cov}\,(\hat{w}_{ij}, \hat{w}_{rj+1}) &\simeq -w_{ij}w_{rj+1}/\gamma_{1j}, \\ \text{cov}\,(\hat{w}_{ij}, \hat{w}_{rj}) &\simeq w_{ij}w_{rj}(1/\gamma_{1j}+1/\gamma_{1j-1}), \\ \\ \text{cov}\,(\hat{w}_{i1}, \hat{w}_{r1}) &\simeq w_{i1}w_{r1}(1/\gamma_{11}-1/A_1), \end{aligned} \right\} \tag{2.22}$$

and

where $j \neq 1, r > i, \gamma_{ij} = 1/\{E(a_{ij})+1\}$, and $E(a_{ij})$ is the expected value of a_{ij}. Any covariances not shown here are zero.

If logarithms are used to make sampling distributions more symmetric then the following standard approximations apply:

$$\text{var}\,\{\log_e(\hat{w}_{ij})\} \simeq \text{var}\,(\hat{w}_{ij})/w_{ij}^2 \tag{2.23}$$

and

$$\text{cov}\,\{\log_e(\hat{w}_{ij}), \log_e(\hat{w}_{rs})\} \simeq \text{cov}\,(\hat{w}_{ij}, \hat{w}_{rs})/(w_{ij}w_{rs}). \tag{2.24}$$

2.9   THE CONSTANT SELECTIVE VALUES MODEL

A very real problem with the selective value model just described is that there are so many parameters that it is very difficult to get sufficient data to estimate them all accurately. There are various ways of combining estimates (Manly, 1972), but it is fair to say that these are not altogether satisfactory. A better solution involves assuming that survival probabilities and recapture probabilities can vary with time but that selective values remain constant. Thus for any time period (t_a, t_b) the probability of a morph i animal surviving is assumed to be $w_i^{t_b - t_a} \psi_1$, where ψ_1 is the probability of a morph 1 animal surviving the period. Then w_i is the selective value per unit time for the morph i relative to morph 1.

Under these conditions pseudo-maximum likelihood estimates of the w_i are found by solving the equations

$$\sum_{j=1}^{s} t_j (a_{ij} - A_i \hat{w}_i^{t_j} a_{ij}/A_1) = 0, \tag{2.25}$$

for $i = 2, 3, \ldots, K$, either graphically or numerically (Manly, 1973a). The variance of $\hat{w}_i$ is then approximately

$$\mathrm{var}(\hat{w}_i) \simeq w_i^2 \left[\sum_{j=1}^{s} \{ t_j^2 E(a_{1j}) w_i^{t_j}/A_1 \} (1/A_i + w_i^{t_j}/A_1) \right.$$

$$\left. - (1/A_i + 1/A_1) \left(\sum_{j=1}^{s} t_j E(a_{1j}) w_i^{t_j}/A_1 \right)^2 \right] \bigg/ \left\{ \sum_{j=1}^{s} t_j^2 E(a_{1j}) w_i^{t_j}/A_1 \right\}^2 \tag{2.26}$$

The constant selective value model is an alternative to the constant survival–constant recapture probability model of Section 2.4 since it makes rather different assumptions and does not require recapture samples to be made at equally spaced points in time. However, the assumption that all types of animal have the same recapture probability clearly needs to be considered carefully in any application.

Example 2.3   Second analysis of the experiment on mimicry

Consider again Brower et al.'s (1967) experiment on artificial mimicry with the moth Hyalophora promethea that was the subject of Example 2.1. The experimental results are shown in Table 2.1. Here the constant selective values model seems appropriate on the grounds that the three 'morphs' (mimics, blue controls and black controls) were males of the same species, handled in similar ways. Hence it is reasonable to assume that all morphs had the same recapture probability p.

Table 2.9 shows estimated selective values obtained by solving equation (2.25), taking black-painted moths as controls and treating the data separately for each release day.

Table 2.9 The constant selective value model applied separately to each day's release for the data of Table 2.1. Black-painted *Hyalophora promethea* are used as the controls: the survival of the mimics and the blue-painted moths is compared with their survival.

Release date	Mimics $\hat{w}$	Standard error	Blue controls $\hat{w}$	Standard error
11 July	1.00	0.93	2.13	1.59
12	4.84	4.03	0.94	1.07
13	0.96	0.95	0.46	0.49
14	2.06	1.70	9.13	5.67
15	0.45	0.36	1.63	0.90
16	0.31	0.27	1.44	0.93
17	0.58	0.23	0.42	0.21
18	1.31	0.53	1.09	0.47
19	0.64	0.33	0.69	0.33
20	0.89	0.48	0.73	0.43
21	0.64	0.63	2.03	1.39
22	0.17	0.22	0.09	0.18
23	0.46	0.60	1.05	0.95
24	0.59	0.31	0.24	0.29
25	1.21	3.08	64.00	128.88
26	0.55	0.30	0.37	0.25
27	0.62	0.53	0.82	0.57
28	0.35	0.31	0.89	0.40
29	0.35	0.44	1.90	1.22
30	0.13	0.28	1.17	1.56
31	1.26	0.93	0.78	0.72
1 Aug.	0.38	0.47	1.83	1.97
2	1.66	0.89	0.65	0.46
3	0.13	0.31	1.00	1.81

Figure 2.2 shows a plot of $\log_e(w)$ values for mimics, together with the fitted regression line

$$\log_e(w) = 0.20 - 0.054(\text{day}),$$

where day denotes the day of the release (from 1 to 24). The regression is significant and indicates that at the start of the experiment the mimics and black controls had very similar survival since the regression gives $\log_e(w) = 0.144$ ($w = 1.15$) for day = 1. If anything, the mimics may have had slightly higher survival. However, at the end of the experiment (day = 24) the regression line gives $\log_e(w) = -1.09$ ($w = 0.34$), which suggests that the daily survival of the mimics was only about one-third of that for the black controls.

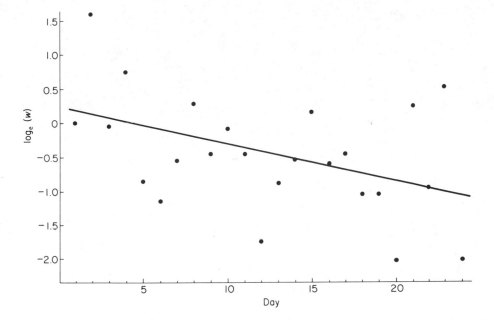

Fig. 2.2 Selective values for mimics relative to black controls for the moth *H. promethea*. The regression line shown, $\log_e(w) = 0.20 - 0.054$ (day) has a slope that is significant at the 5% level.

Figure 2.3 shows a plot of $\log_e(\hat{w})$ values for the blue controls relative to the black controls. There is no significant regression in this case and it appears that the blue controls had very similar survival to the black controls.

Logarithms of selective values have been used in Figs 2.2 and 2.3 and in the regression analyses for two reasons. Firstly, because equation (2.26) shows that estimated values of var $(\hat{w})$ will be proportional to $\hat{w}^2$, whereas var$\{\log_e(\hat{w})\} \simeq$ var $(\hat{w})/w^2$ will not be. Secondly, as has been explained in Section 1.5, the logarithm of a selective value is a more symmetrical measure of selection than the selective value itself.

This analysis of Brower *et al.*'s data has confirmed their conclusion that the artificial mimics were increasingly selected against in their experiment although they may initially have had a slight advantage compared to the controls. This is clearer from the analysis given here than it was from the analysis of Example 2.1 that was based upon the constant survival–constant recapture probability model. The main reason seems to be that the constant survival–constant probability of recapture model required recapture probabilities to be estimated separately for the three types of moth released and this had the effect of masking the trend in the survival of the mimics relative to the controls.

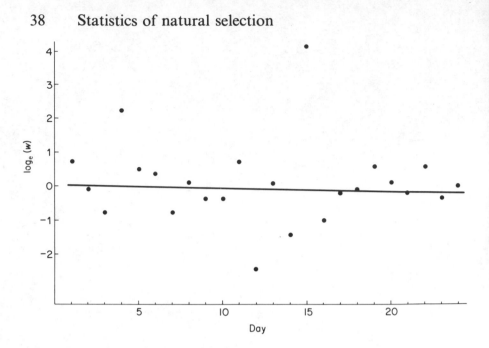

Fig. 2.3 Selective values for blue controls relative to black controls for the moth *H. promethea*. The regression line shown. $\log_e(w) = 0.02 - 0.01$ (day), has a non-significant slope.

2.10 COMPARISON OF THE MODELS

The models that have been considered for the multi-sample, single recapture experiment are summarized in Table 2.10. The results of a simulation study help in the comparison of these (Manly, 1978a).

The simulations suggest that the variances of survival estimators are not appreciably lower for the constant survival–common recapture probability model than they are for the constant survival–constant recapture probability model. In view of the computational difficulties involved with the former model, it therefore seems that this is not of great value.

The simulation results also indicate that the constant selective values model has no advantages over the constant survival–constant recapture probability model when both can be used on data. However, the constant selective values model can be useful because of its slightly different assumptions and the fact that it does not require recapture samples at equal intervals of time. In fact, the value of this model has been demonstrated in the example on the analysis of Brower *et al.*'s (1967) experiment on artificial mimicry that has just been considered.

According to the simulations, the standard error formulae that have been given should be reasonably accurate, at least for experiments involving more

Table 2.10 Summary of models for the data from a multi-sample, single recapture mark–release experiment involving K types of animal. Estimation of survival probabilities $\psi_1, \psi_2, \ldots, \psi_K$ and selective values $w_2 = \psi_2/\psi_1$, $w_3 = \psi_3/\psi_1$, $w_K = \psi_K/\psi_1$ is what is desired from the experiment. Other parameters are the recapture probabilities $p_1, p_2, \ldots, p_K$.

Model	References	Parameters	Assumptions	Comment
Seber's model	Seber (1962) Manly (1974a)	$\psi_1, \psi_2, \ldots, \psi_K$ $p_1, p_2, \ldots, p_K$	ψ's and p's change with time	Each type of animal is treated independently. A geometric mean survival rate can be estimated for each type of animal. Selective values are calculated as $\hat{w}_i = \hat{\psi}_i/\hat{\psi}_1$.
Constant survival–constant recapture probability	Manly (1975a)	$\psi_1, \psi_2, \ldots, \psi_K$ $p_1, p_2, \ldots, p_K$	ψ's and p's are constant over time	Recapture samples must be at regular intervals of time and continue until all animals are recaptured or dead. Selective values are calculated as $\hat{w}_i = \hat{\psi}_i/\hat{\psi}_1$.
Constant survival–common recapture probability	Manly (1977a)	$\psi_1, \psi_2, \ldots, \psi_K$ $p = p_1 = p_2 = \ldots = p_K$	as for the model above but with all recapture probabilities equal	Estimates are found by maximizing the likelihood function numerically.
Varying selective values	Manly (1972)	$w_2, w_3, \ldots, w_K$	w's change with time	Only one recapture sample is needed in order to estimate selective values.
Constant selective values	Manly (1973a)	$w_2, w_3, \ldots, w_K$	w's are constant over time.	Survival and recapture probabilities can change with time. The recapture probabilities are assumed to be the same for all types of animal.

than 50 animals released. The simulations were also designed to investigate various ways in which failures of assumptions could affect estimation.

2.11 MARK–RECAPTURE EXPERIMENTS IN GENERAL

The multi-sample, single recapture experiment has received special attention in this chapter because this type of experiment has frequently been used in attempts to detect differential survival under natural conditions. It is, however, only a special case of mark–recapture experiments in general.

Mark–recapture methods were originally developed mainly with the idea of estimating the size of animal populations. Parameters for survival probabilities were introduced because this was necessary in order to develop statistical models for changes in population size over time. Nevertheless, there are now a considerable number of techniques for estimating survival rates from a general mark–recapture experiment. Here only a brief review of some of the more important of these will be given. Equations for estimators and more details will be found in Seber's (1982) excellent review of the whole topic. Useful references for those mainly interested in applications are Southwood (1978), Begon (1979), and Blower et al. (1981).

The statistical model for the general mark–recapture experiment that will have most appeal to statisticians was developed by Jolly (1965) and Seber (1965). They considered the situation where a series of samples are taken from a population over a period of time and animals are marked so that their captures and recaptures can be recorded. They developed equations for estimating the size of the population at sample times, probabilities of surviving periods between samples and numbers of new animals entering the population between samples. FORTRAN computer programs for estimation have been published by White (1971a, b) and Arnason and Baniuk (1978). Bishop et al. (1978) have used the Jolly–Seber method for analysing the results of a series of experiments that were designed to investigate melanism in two species of moth in the north-west of England. They discussed the problems involved in estimating selective values and suggest that with the Jolly–Seber model the best approach is to estimate an average daily survival rate for each morph by taking the geometric mean of the series of survival estimates given by the method. Ratios of the geometric mean survival rates for different morphs then give estimated selective values.

Unfortunately, the Jolly–Seber method often fails to produce reasonable estimates with real data because there are so many parameters to be estimated. Biologists have overcome this problem by estimating parameters on the assumption that the survival rate of animals is constant over time. The method of Fisher and Ford (1947) has proved popular. This involves the numerical or graphical solution of an equation involving one unknown (the constant survival rate). See Blower et al. (1981, p. 58), for more details.

More recently, some new models have been proposed for mark–recapture experiments that allow various restrictions on parameters, such as a constant survival rate (Jolly, 1979, 1982; Crosbie and Manly, 1982, 1985). These involve rather more computation than the Jolly–Seber and the Fisher and Ford models and need special computer programs such as those described by Jolly and Dickson (1980). Given suitable programs, these new models are likely to be the best approach for the analysis of mark–recapture experiments to estimate relative survival rates.

2.12   SEPARATING DEATH FROM EMIGRATION

One problem with estimating survival rates and selective values from mark–recapture experiments is that losses from the population through death and emigration are confounded. Hence if permanent emigration occurs on a large scale then an estimated 'survival' probability may be much lower than probability of not dying. To be more precise, the probability of 'surviving' a unit interval of time as defined for the mark–recapture models is

$$\psi = \phi(1 - \varepsilon)$$

where ϕ is the probability of remaining alive and ε is the probability of permanently emigrating.

One point of view is that emigration is not important providing that ε is the same for all morphs, since in that case the ratio of two estimates of ψ will be the same as the ratio of two estimates of ϕ. That is to say, the ratio of survival probabilities estimated from a mark–recapture experiment gives an estimate of a selective value that is unaffected by emigration. This was the assumption made by Bishop et al. (1978).

Various ad hoc methods for allowing for emigration have been proposed. They will be reviewed briefly here. It is fair to say that this is an area where more work needs to be done.

One approach that has been used involves adjusting one of the standard methods for the analysis of capture–recapture data to allow for emigration. Sheppard et al. (1969) have done this for the Fisher and Ford method, and Cameron and Williamson (1977) for the Jolly–Seber method. Bishop (1972) used Sheppard et al.'s adjustment in comparing the survival of different morphs of the moth Biston betularia in the north of England.

An analysis of recapture data from a grid of traps has been proposed by Smith et al. (1972) and Jorgensen et al. (1975), particularly for sampling of small mammals. This analysis is designed to provide estimates of mortality, trap avoidance, dispersal and population density. The matching of the analysis to the field design as proposed by these authors is undoubtedly the best way of handling animal movement, although their particular design may not be practical for many applications. A more flexible approach, that is suitable for experiments involving batches of releases with any arrangement of traps, entails assuming a certain distribution for the distance travelled by individual animals (Manly, 1977b).

Jackson (1939) suggested a method for separating mortality and emigration that has the merit of being very simple. It involves dividing the area to be sampled into four small squares, as shown in Fig. 2.4. The data can then be handled in two different ways for estimating survival from the recaptures of marked animals:

(a) Only recaptures made within the small square of first release can be counted. For example, any recaptures made outside square 1 are ignored for animals released in square 1. Let the survival probability per unit time estimated in this way be denoted by $\hat{\psi}_s$.
(b) All recaptures within the large square can be counted, irrespective of the point of release. Let the survival probability per unit time estimated in this way be $\hat{\psi}_1$.

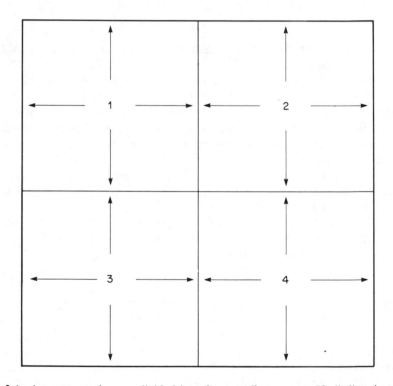

Fig. 2.4   A square study area divided into four smaller squares. If all directions of movement are equally likely then half of the movements out of one small square are into another small square.

Also, let ϕ be the probability of not dying and ε be the probability of (permanently) emigrating from a small square in a unit interval of time. Then the relationships

$$\hat{\psi}_s = \hat{\phi}\,(1 - \hat{\varepsilon})$$

and

$$\hat{\psi}_1 = \hat{\phi}\,(1 - \hat{\varepsilon}/2)$$

should approximately apply, where $\hat{\phi}$ and $\hat{\varepsilon}$ are estimates of ϕ and ε. The argument here is that the emigration rate for the large square should be half that for the small square because half of the emigrants from a small square move into another small square and are therefore still within the large square (Fig. 2.4). Solving the equations for $\hat{\phi}$ gives

$$\hat{\phi} = 2\hat{\psi}_1 - \hat{\psi}_s.$$

Jackson's procedure should be realistic providing that the emigration rate is low. If this is not the case then some of the migrants from a small square may move into another small square and then out of the study area altogether in a unit interval of time. In that case the emigration rate for the large square will be higher than half the rate for a small square. The procedure also implicitly involves the assumptions that the probability of capture is the same in all four squares and that emigration is not in any particular direction.

See Conroy and Bishop (1980) for an application of Jackson's procedure to the results from a mark–recapture experiment to compare the survival of typical and melanic morphs of the moth *Phigalia pilosaria* in Wales. Simulations indicate that Jackson's method may work quite well in practice (Manly, 1985a).

2.13 DERIVING FITNESS ESTIMATES FROM SURVIVAL ESTIMATES

Another problem concerns how a difference in the survival of two morphs can be interpreted in terms of the relative fitnesses of the morphs, where fitness is defined to be proportional to the contribution of offspring to the next generation.

In simple cases it will be reasonable to assume that the production of new individuals occurs at a constant rate over the reproductive period in the life cycle. Then the contribution per individual to the next generation for a morph will be proportional to the mean reproductive period for that morph. If morph 1 has a mean reproductive period of E_1, estimated by $\hat{E}_1$ with variance $\mathrm{var}(\hat{E}_1)$, and morph 2 has a mean of E_2, estimated by $\hat{E}_2$ with variance $\mathrm{var}(\hat{E}_2)$, then the lifetime fitness of morph 2 relative to morph 1 is $f = E_2/E_1$, estimated by

$$\hat{f} = \hat{E}_2/\hat{E}_1 \tag{2.27}$$

with variance

$$\mathrm{var}(\hat{f}) \simeq f^2\,\{\mathrm{var}(\hat{E}_1)/E_1^2 + \mathrm{var}(\hat{E}_2)/E_2^2\}. \tag{2.28}$$

Furthermore, Cormack (1964) has pointed out that if an individual has a constant survival probability of ϕ per unit time then the average life span of the individual is $E = -1/\log_e(\phi)$. Hence if an estimate $\hat{\phi}$ of ϕ is available then this gives an estimated mean lifetime of

$$\hat{E} = -1/\log_e(\hat{\phi}) \qquad\qquad (2.29)$$

with variance

$$\text{var}(\hat{E}) \simeq E^4 \, \text{var}(\hat{\phi})/\phi^2. \qquad\qquad (2.30)$$

The approach represented by equations (2.27) to (2.30) was used by Bishop *et al.* (1978) with moths since in this case the assumptions are approximately true. Earlier Clarke and Sheppard (1966) and Cook (1971, p. 47) in the same circumstances had subtracted one day from the mean lifespan of each morph in order to make some allowance for the fact that females do not begin to lay eggs until some time after they have emerged. Apparently Bishop *et al.* considered this correction to be unnecessary.

3 Samples taken from a population within one generation

With many populations the mark–recapture methods discussed in Chapter 2 are not practical as a means of estimating relative survival rates for different morphs. However, the populations are not affected appreciably by the ingress of new individuals through births or immigration so that changes with time can be attributed solely to selective survival. If this is the case then selection can be inferred from sampling the populations at two or more points in time. It is this type of situation that is the subject of the present chapter.

After an introductory section, early parts of the chapter are concerned with the analysis of polymorphic data. A cross-product estimator of selective values is defined. An alternative log-linear model approach is also suggested. Various measures of the intensity and effect of selection on a polymorphic population are reviewed.

In Section 3.4 tests for changes in means, variances and covariances of quantitative variables are covered. Under appropriate conditions these are tests for selection. Sections then follow on the estimation of fitness functions for sample data from univariate and multivariate populations. These lead on to the consideration of measures of the intensity and effect of selection on quantitative variables.

The chapter concludes with a discussion on approaches that have been used for studying characters subject to growth. As yet no satisfactory approach has been found.

3.1 SAMPLING A POPULATION WHILE SELECTIVE SURVIVAL IS OCCURRING

The models considered in this chapter all involve a single population which changes through time because individuals in the population do not all have the same probability of surviving. Thus the fittest individuals tend to increase in frequency relative to less fit individuals, where 'fitness' is simply interpreted as being proportional to the survival probability. It is assumed throughout the chapter that the population is not affected by the ingress of new individuals through either immigration or reproduction, and that it is large enough for selection to be thought of as acting deterministically.

Sampling a population before and after a period of selection can only give information on one component of fitness, but this may be of sufficient interest to make a study worth while. A common procedure has been to compare a sample of juveniles with a sample of adults. Early studies along these lines by Weldon (1901, 1903) on the snails *Clausila laminata* and *C. italia* and by Di Cesnola (1906) on the snail *Arianta arbustorum* have been mentioned already in Chapter 1. Other studies have been on wasps (Thomson *et al.*, 1911), snakes (Inger, 1943 and earlier), corixids (Popham, 1944, 1966), tsetse flies (Jackson, 1946, 1948; Glasgow, 1961; Phelps and Clarke, 1974), lizards (Hecht, 1952; Fox, 1975), a fossil horse (Van Valen, 1963, 1964), house mice (Van Valen, 1965b, Berry *et al.*, 1979), rats (Van Valen and Weiss, 1966), threespine sticklebacks (Hagen and Gilbertson, 1973), zooplankton (Zaret and Kerfoot, 1975), house sparrows (Lowther, 1977; Fleischer and Johnston, 1982), toads (Samollow, 1980), dog whelks (Berry and Crothers, 1968, 1970), and fossil orang-utan (Marcus, 1969). No doubt many other references could be provided. This list is sufficient to indicate the popularity of the approach.

3.2 SELECTION ON A POLYMORPHIC POPULATION

One approach to the analysis of data from a polymorphic population is by making use of equations that have already been provided in Section 2.8. Thus, suppose that at time zero a population consists of K morphs with unknown frequencies $A_1, A_2, \ldots, A_K$. Let random samples be taken from the population at times $t_1, t_2, \ldots, t_s$, in order, and denote the number of morph i in the jth sample as a_{ij}. It is then reasonable to assume that the expected value of a_{ij} is

$$E(a_{ij}) = A_i p_j \phi_{i1} \phi_{i2}, \ldots \phi_{ij} \qquad (3.1)$$

where p_j is the proportion of the population sampled at time t_j and ϕ_{ir} is the probability of a morph i animal surviving the time interval (t_{r-1}, t_r). It follows that

$$\frac{E(a_{ij}) E(a_{1j-1})}{E(a_{1j}) E(a_{ij-1})} = \frac{\phi_{ij}}{\phi_{1j}} = w_{ij},$$

where w_{ij} is the selective value of morph i relative to morph 1 for the time interval (t_{j-1}, t_j). An obvious estimator of w_{ij} is the cross-product ratio

$$w_{ij} = (a_{ij} a_{1j-1})/(a_{1j} a_{ij-1}), \qquad (3.2)$$

which is the same as the estimator given by equation (2.16).

For reasons discussed in Section 2.8, this estimator can be improved by following a suggestion of Haldane (1956) and taking

$$\hat{w}_{ij} = \{(2a_{ij}+1)(2a_{1j-1}+1)\}/\{(2a_{1j}+1)(2a_{ij-1}+1)\}. \qquad (3.3)$$

Biases, variances and covariances are then provided by equations (2.20) to (2.22).

The estimate $\hat{w}_{ij}$ relates to a time interval of length $t_j - t_{j-1}$. Sometimes it will be desirable to convert this to an estimate for a unit period of time. This can be done by recognizing that $w_i = \phi_i/\phi_1$ is the selective value for a unit period of time where ϕ_i and ϕ_1 are the survival probabilities per unit time for morphs i and 1, respectively. Hence the selective value for a time t is $w_i^t = (\phi_i/\phi_1)^t$. It follows that a selective value for a time t is converted to one for a unit time by taking the $(1/t)$th root. That is to say, $\hat{w}_{ij}$ can be converted to a unit time value $\hat{w}'_{ij}$ using the result

$$\hat{w}'_{ij} = (\hat{w}_{ij})^{1/(t_j - t_{j-1})}. \tag{3.4}$$

According to the Taylor series approximation method, variances are related as

$$\mathrm{var}(\hat{w}'_{ij}) \simeq [w'_{ij}/\{(t_j - t_{j-1})w_{ij}\}]^2 \, \mathrm{var}(\hat{w}_{ij}). \tag{3.5}$$

An alternative to an analysis based upon equation (3.3) is one that involves taking equation (3.1) as the basis for a Poisson log-linear model. Thus this equation can be rewritten as

$$E(a_{ij}) = \exp(\theta_i + \gamma_j - \lambda_i t_j), \tag{3.6}$$

where $\theta_i = \log_e(A_i)$, $\gamma_j = \log_e(p_j)$, and $\exp(-\lambda_i)$ is the survival probability per unit time for morph i. This equation can now be fitted to data using the computer program GLIM (Appendix, Section A.10) or some equivalent program, on the assumption that the a_{ij} values have independent Poisson distributions. Without knowing absolute survival rates it is not possible to estimate the absolute values of $\lambda_1, \lambda_2, \ldots, \lambda_K$. This problem is overcome in GLIM by setting $\lambda_K = 0$, which means in effect that the estimates $\hat{\lambda}_1, \hat{\lambda}_2, \ldots, \hat{\lambda}_{K-1}$ are really estimating $\lambda_1 - \lambda_K, \lambda_2 - \lambda_K, \ldots, \lambda_{K-1} - \lambda_K$. This being so, it follows that $\exp(-\hat{\lambda}_i)$ is an estimate of the selective value per unit time of morph i relative to morph K.

Example 3.1   Predation of corixids by minnows

A good example of selection on a polymorphic population is provided by Popham's (1944) experiment on a small pond near Blackburn in England. The pond was chosen for use because of the abundance of corixid insects usually found there, and the absence of fish predators. The experiment involved sampling corixid populations for one week, introducing minnows as predators, and sampling for another week. Prior to carrying out his experiment, Popham hypothesized that selection by the predators would bring about changes in the proportions of light, medium and dark-coloured corixids in the pond. Table 3.1 shows the experimental results for three species of corixids, *Sigara distincta*, *S. praeusta* and *S. venusta*, the last of these being by far the most common.

Table 3.1 Results of an experiment on the predation of corixids by minnows. Minnows were not present in the experimental pond until they were introduced after the sample was taken on the evening of 19 September. The table shows the numbers of light, medium and dark-coloured corixids found for each of three species. The corixids were returned to the pond after sampling so that the population was disturbed as little as possible. The dates are all for September 1942.

	Corixid Species								
	Sigara distincta Colour			*Sigara praeusta* Colour			*Sigara venusta* Colour		
Date	*Light*	*Med.*	*Dark*	*Light*	*Med.*	*Dark*	*Light*	*Med.*	*Dark*
13	5	17	2	5	10	1	22	95	37
14	0	5	1	4	1	0	12	87	21
15	1	2	1	2	2	1	17	106	33
16	2	3	0	5	6	1	18	110	33
17	2	3	1	4	6	1	17	125	34
18	3	5	0	3	4	0	17	100	35
19	2	4	1	2	4	2	17	103	32
	Minnows introduced on the evening of 19 September								
22	0	4	0	2	6	1	5	102	25
23	0	3	0	0	2	1	5	110	58
24	1	3	0	2	6	0	5	131	22
25	0	4	0	1	3	0	6	120	22
26	0	3	0	0	5	0	2	105	16
27	0	4	0	1	4	0	6	157	20
28	0	5	0	0	5	1	3	134	17

Table 3.2 shows estimated selective values calculated using equation (3.3) for *S. venusta*, together with standard errors. The 'control' colour has been taken as medium because this was the most common. For all estimations the first 'sample' has been taken as the combined samples for the period 13 to 19 September, during which time there were no predators present and the population apparently remained quite stable. Treating the data in this way has avoided small values for a_{ij-1} in equation (3.3) with their correspondingly large biases and standard errors for estimated selective values.

From Table 3.2 it can be seen that the selective value of light relative to medium corixids seems to have been about 0.32 for the three-day period from 19 to 22 September and then declined to about 0.16 for the nine-day period from 19 to 28 September. Since $0.82^9 = 0.16$, this corresponds to a selective value about 0.82 per day over the nine-day period. In other words, the daily

Table 3.2 Selective values for light and dark *Sigara venusta*, relative to medium-coloured ones. Estimates have been calculated using equations (3.3) on the data in Table 3.1 with the first seven samples lumped together and taken to be equivalent to a single large sample on 19 September.

Selection period	Light corixids relative to medium		Dark corixids relative to medium	
	$\hat{w}$	Standard error	$\hat{w}$	Standard error
19–22 Sept.	0.32	0.14	0.80	0.19
19–23	0.30	0.13	1.71	0.30
19–24	0.25	0.11	0.55	0.13
19–25	0.33	0.13	0.60	0.14
19–26	0.14	0.09	0.50	0.14
19–27	0.25	0.10	0.42	0.10
19–28	0.16	0.08	0.42	0.11

Standard errors were determined using equations (2.21), replacing expected values with observed values. According to equations (2.20) biases are negligible.

survival rate of light corixids was about 82 % of the rate for medium corixids. For dark relative to medium corixids the selective value for the nine-day period seems to have been about 0.42 which corresponds to a daily value of 0.91. Therefore the daily survival rate of dark corixids seems to have been about 91 % of the rate for medium corixids.

Because of the small sample frequencies it is not appropriate to estimate selective values for *S. distincta* and *S. praeusta* using equation (3.3). However, an analysis based upon equation (3.6) has been carried out on the data for all three species of corixids. To begin with, the nine combinations of colour and species were regarded as dividing the population into nine morphs, and equation (3.6) was fitted to the data using the computer program GLIM with all the λ values set at zero, which corresponds to a no-selection model. The fit is very poor with a GLIM deviance of 228.3 with 104 degrees of freedom. See the Appendix, Section A.10, for an explanation of 'deviance' as a measure of goodness of fit. Here two particular points should be noted. Firstly, it is generally true for a Poisson log-linear model that the deviance can be treated as a chi-square variate, with a significantly large value indicating that the model being considered does not fit the data. (In the present case the deviance is significantly large at the 0.1 % level.) Secondly, the deviance found by fitting equation (3.6) with all the λ values set at zero must be approximately the same as the chi-square value that would be obtained by doing an ordinary test for association on the 14 by 9 contingency table of morph frequencies given in Table 3.1 (Steel and Torrie, 1980, p. 496).

There is very clear evidence that the morph frequencies were not constant over time in the experimental pond, so the next stage in the analysis involved estimating the λ values. The first seven samples were taken to have $t_j = 0$, while for the other samples t_j was taken as the number of days after the introduction of the minnows (19 September). The model fits the data quite well with a deviance of 107.4 with 96 degrees of freedom. Table 3.3 shows the estimated λ values and the selective values to which these correspond. It will be seen that standard errors are large for the S. distincta and S. praeusta estimates because of the small sample numbers. Nevertheless, the estimates for light S. distincta and S. praeusta are more than two standard errors below zero, which indicates that these morphs survived rather less well than the dark S. venusta. For S. venusta the analysis suggests that the three morphs had daily survival rates in the ratios light : medium : dark of 0.86 : 1.09 : 1.00, which are the same as 0.79 : 1.00 : 0.92. These agree very well with the ratios 0.82 : 1.00 : 0.91 that were calculated earlier using the estimated selective values given in Table 3.2.

Before leaving this example there is another possible model for the data that can be mentioned. Looking again at the morph frequencies shown in Table 3.1, it will be seen that there was a very abrupt change in these upon the introduction of the minnows. Perhaps the morph frequencies were more or less constant up to 19 September, changed rapidly between 19 and 22 September because of predation by the minnows, and then remained more or less constant from 22 September on. This can be investigated by fitting equation (3.6) to the data taking $t_j = 0$ for the first seven samples and $t_j = 3$ for the last seven samples. If this is done, then the model gives a GLIM deviance of

Table 3.3   Estimated λ_i values from the fitting of equation (3.6) to the data in Table 3.1 on the predation of corixids by minnows. The computer program GLIM has automatically taken morph 9 (dark *Sigara venusta*) as the standard with $\lambda_9 = 0$. The estimates $\hat{\lambda}_1, \hat{\lambda}_2, \ldots, \hat{\lambda}_8$ account for survival differences between this and the other morphs. Also, $\exp(-\hat{\lambda}_i)$ gives the estimated daily selective value for morph i relative to morph 9.

Morph	$\hat{\lambda}$	Standard error	$\exp(-\hat{\lambda})$
1 Light *S. distincta*	−0.44	0.21	0.64
2 Medium *S. distincta*	0.01	0.04	1.01
3 Dark *S. distincta*	−2.45	8.94	0.09
4 Light *S. praeusta*	−0.20	0.09	0.82
5 Medium *S. praeusta*	0.05	0.04	1.05
6 Dark *S. praeusta*	−0.07	0.12	0.93
7 Light *S. venusta*	−0.15	0.04	0.86
8 Medium *S. venusta*	0.09	0.02	1.09
9 Dark *S. venusta*	0	0	1

120.1 with 96 degrees of freedom which is significantly large at the 5% level. Hence this abrupt change model does not seem as good as the model assuming steady changes.

3.3 THE INTENSITY AND EFFECT OF SELECTION ON A POLYMORPHIC POPULATION

A question of some interest concerns the best way to measure the intensity of selection on a population. In discussing this for a polymorphic population it will be assumed that any index must be some function of the selective values per unit time, $w_1, w_2, \ldots, w_K$, for the K morphs. None of the indices that will be considered here are changed if all the selective values are multiplied by the same positive constant. Consequently, all that is necessary is that w_i should be proportional to the probability of surviving a unit time for the ith morph.

Haldane (1954a, 1959) proposed the index

$$I_H = \log_e (w^*/\bar{w}) \tag{3.7}$$

for the intensity of selection, where w^* is the maximum selective value in the population and $\bar{w}$ is the mean selective value. This index has a minimum of zero when all morphs have the same fitness, since in that case $w^* = \bar{w}$. There is no theoretical upper limit.

When w^* and $\bar{w}$ are not greatly different, so that selection is not strong, I_H becomes approximately equal to

$$I_V = (w^* - \bar{w})/w^*, \tag{3.8}$$

which was proposed by Van Valen (1965a) as an index of the intensity of selection in its own right. To understand I_V, note that for a population of size N before selection, the expected total number of survivors is $N\alpha\bar{w}$, where α is the multiplicative constant needed to convert a selective value w to a survival probability. However, if all individuals were the optimum morph then the expected number of survivors would be $N\alpha w^*$. The difference $N\alpha w^* - N\alpha\bar{w}$ is the number of selective deaths and I_V is this number divided by $N\alpha w^*$. Haldane (1959) called I_V the fraction of deaths which are selective.

O'Donald (1970) adopted a somewhat different approach. He argued that if $\bar{w}$ is the mean selective value before selection, and $\bar{w}'$ is the mean value after selection, then

$$I_O = (\bar{w}' - \bar{w})/\bar{w} \tag{3.9}$$

is 'the best single measure of the action of selection on a population'. This does not need the determination of the optimum morph, which is an advantage because in practice sampling errors in determining selective values will tend to result in w^* of equations (3.7) and (3.8) being overestimated, and hence I_H and I_V being overestimated. (The largest estimated selective value may only be largest because of sampling errors.)

It can be argued that I_H and I_V are measures of the intensity of selection when the selective values involved relate to a unit period of time. However, I_O will only measure intensity if selection is for one unit of time. Therefore I_O is perhaps best thought of as an index of the effect of selection. Dividing by the selection time will make I_O an index of the intensity of selection.

Another index of the effect of selection is the change in the mean of $u = \log_e(w)$. That is to say, let $\bar{u}$ be the mean of $\log_e(w)$ before selection and $\bar{u}'$ be the mean after selection. Then this index is just

$$I_u = \bar{u}' - \bar{u}. \tag{3.10}$$

Dividing this by the selection time should give a measure of the intensity of selection.

No formulae are available at present for biases and variances for the indices I_H, I_V, I_O and I_u when they are estimated from sample data. However, bias may be important with all of them. This is because even when there is no selection in a population, two samples taken at different times will result in selective values that are not equal because of sampling errors. Hence all of the above indices will be positive, indicating selection.

A simple way to investigate the properties of the indices is by simulation. This can be done by assuming multinomial distributions for sample frequencies and simulating many sets of data using the observed frequencies as mean values. For example, with Popham's data shown in Table 3.1 the sample taken on 22 September contained 5 light, 102 medium and 25 dark-coloured *Sigara venusta*. Simulated samples of size 132 therefore have to be taken from a multinomial distribution with three classes having probabilities 5/132, 102/132 and 25/132. This type of simulation can be done very quickly on an electronic computer, particularly if the multinomial distributions are approximated by multivariate normal distributions. The simulated data can then be generated using Bedall and Zimmermann's (1976) algorithm, or something similar. (See the Appendix, equations (A9)).

Example 3.2   The intensity of selection on *Sigara venusta*

Estimates of selective values for different colours of *Sigara venusta* are shown in Table 3.2. These selective values relate to the selection periods indicated, which vary from three to nine days, and they need to be standardized to a unit period of time before they can be used to measure selection intensity. This can be done using equation (3.4) taking one day as the unit of time, with the results shown in Table 3.4. The daily selective values are quite consistent. Medium-coloured corixids come out as having the highest survival, except in one case where dark corixids are estimated to have a selective value of 1.14. Standard errors calculated by using equation (3.5) are shown and also standard errors determined by simulation. There is good agreement, with a slight tendency for the calculated standard error to be low.

Table 3.4   Selective values per day for the three colours of *Sigara venusta*.

Selection period	Light relative to medium			Dark relative to medium		
	$\hat{w}$	SE1	SE2	$\hat{w}$	SE1	SE2
19–22 Sept.	0.69	0.12	0.12	0.93	0.07	0.07
19–23	0.74	0.09	0.10	1.14	0.06	0.06
19–24	0.76	0.07	0.08	0.89	0.05	0.05
19–25	0.83	0.06	0.08	0.92	0.03	0.03
19–26	0.76	0.06	0.07	0.91	0.04	0.04
19–27	0.84	0.04	0.06	0.90	0.03	0.03
19–28	0.81	0.04	0.06	0.91	0.02	0.03

SE1 is the standard error for $\hat{w}$ as determined by using equation (3.5) on the standard errors given in Table 3.2.
SE2 is the standard error determined from 100 simulated samples. In all cases biases are negligible.

Table 3.5 shows estimates $\hat{I}_H$, $\hat{I}_V$, $\hat{I}_O$ and $\hat{I}_u$ of the indices defined by equations (3.7) to (3.10) for these data. Haldane's index and Van Valen's index are in good agreement. They suggest that about 4% of deaths were selective each day. O'Donald's index and the index of the change in the mean of $\log_e (w)$ give almost the same values for the effect of selection, with the results being rather surprising. Both of these indices (I_O and I_u) should increase with the selection time, but no increase is indicated. It seems therefore that the main effect of selection was achieved between 19 and 22 September, at least as far as these indices are concerned. This conclusion does not necessarily contradict the previous conclusion that about 4% of deaths were selective each day. The original data shown in Table 3.1 show that the main change in morph frequencies occurred between 19 and 22 September, but still the relative frequency of medium *Sigara venusta* increased somewhat from 22 to 28 September.

In interpreting the indices, note should be taken of the biases and standard errors. A rough correction for bias can be made by subtracting the estimated bias from the index value, which always has the effect of reducing the index. The indices cannot be expected to be normally distributed so therefore the standard errors may be used only to give a very approximate idea of accuracy. For example, $\hat{I}_H$ for 19 to 28 September is 0.039 after making a bias adjustment, with a standard error of 0.009. This is more than four standard deviations from zero so there seems little doubt that the intensity of selection is significant in the statistical sense.

Table 3.5 Indices of the intensity and effect of selection on three colours of *Sigara venusta*, together with biases and standard errors based upon 100 simulated sets of data.

Selection period	$\hat{I}_H$	Bias	Standard error	$\hat{I}_v$	Bias	Standard error	$\hat{I}_O$	Bias	Standard error	$\hat{I}_u$	Bias	Standard error
19–22 Sept.	0.051	0.010	0.022	0.053	0.011	0.024	0.026	0.008	0.020	0.029	0.011	0.026
19–23	0.133	0.002	0.039	0.141	0.003	0.044	0.039	0.005	0.018	0.042	0.006	0.021
19–24	0.052	0.000	0.016	0.053	0.000	0.017	0.029	0.002	0.015	0.031	0.003	0.018
19–25	0.037	0.003	0.012	0.038	0.003	0.013	0.018	0.004	0.012	0.019	0.005	0.014
19–26	0.048	0.001	0.013	0.049	0.001	0.014	0.032	0.002	0.013	0.034	0.003	0.015
19–27	0.040	0.004	0.012	0.041	0.004	0.013	0.024	0.005	0.013	0.025	0.006	0.014
19–28	0.041	0.002	0.009	0.042	0.003	0.010	0.027	0.003	0.010	0.029	0.004	0.011

3.4 TESTS FOR CHANGES IN DISTRIBUTIONS OF QUANTITATIVE VARIABLES

Turning now to selection on quantitative variables, it is appropriate to begin by considering how to test for a change in a population over a period in which two or more samples are taken. Clearly, if there is no evidence of any changes then there is little point in trying to estimate fitnesses. The tests that will be described are likelihood ratio tests, based on the assumption of normally distributed populations. See Kendall (1975, Ch. 9) for a proper justification.

It is convenient initially to present tests for multivariate samples. Tests for a single variable are then described as a special case. Thus suppose that samples with sizes $n_1, n_2, \ldots, n_s$ are taken at times $t_1, t_2, \ldots, t_s$ from a population, and variables $X_1, X_2, \ldots, X_p$ measured on each sample member. In the jth sample let the values for variable X_i be $x_{ij1}, x_{ij2}, \ldots, x_{ijn_j}$, with mean

$$\hat{\mu}_{ij} = \sum_{k=1}^{n_j} x_{ijk}/n_j,$$

and variance

$$\hat{V}_{iij} = \left(\sum_{k=1}^{n_j} x_{ijk}^2 - n_j \hat{\mu}_{ij}^2 \right) \Big/ n_j$$

The covariance of X_i and X_r in the jth sample is then

$$\hat{V}_{irj} = \left(\sum_{k=1}^{n_j} x_{ijk} x_{rjk} - n_j \hat{\mu}_{ij} \hat{\mu}_{rj} \right) \Big/ n_j$$

The sample mean vector and covariance matrix for the jth sample are in that case

$$\hat{\mu}_j = \begin{bmatrix} \hat{\mu}_{1j} \\ \hat{\mu}_{2j} \\ \vdots \\ \hat{\mu}_{pj} \end{bmatrix} \quad \text{and} \quad \hat{V}_j = \begin{bmatrix} \hat{V}_{11j} & \hat{V}_{12j} & \cdots & \hat{V}_{1pj} \\ \hat{V}_{21j} & \hat{V}_{22j} & \cdots & \hat{V}_{2pj} \\ \vdots & \vdots & & \vdots \\ \hat{V}_{p1j} & \hat{V}_{p2j} & \cdots & \hat{V}_{ppj} \end{bmatrix},$$

where $\hat{V}_j$, like all covariance matrices, is symmetric. The division by n_j for the variances and covariances is done on purpose here; for many applications it is, of course, more usual to divide by $n_j - 1$.

Assuming that all of the covariance matrices $\hat{V}_1, \hat{V}_2, \ldots, \hat{V}_s$ are estimating the same thing, they can be combined together to form the pooled within-sample covariance matrix

$$\hat{V}_0 = \sum_{j=1}^{s} n_j \hat{V}_j/n,$$

where $n = \sum n_j$ is the total of all the sample sizes. Also, if all the samples are lumped together then the overall mean of X_i will be

$$\hat{\mu}_{iT} = \sum_{j=1}^{s} n_j \hat{\mu}_{ij}/n$$

while overall variances and covariances can be determined as

$$\hat{V}_{irT} = \left(\sum_{j=1}^{s} \sum_{k=1}^{n_j} x_{ijk} x_{rjk} - n\hat{\mu}_{iT}\hat{\mu}_{rT} \right) \bigg/ n.$$

Then the total sample covariance matrix, which can be denoted by $\hat{V}_T$, has $\hat{V}_{irT}$ in the ith row and rth column.

Based upon the matrices just defined, a statistic for testing the hypothesis that all s samples come from the same multivariate normal distribution is

$$\lambda = \sum_{j=1}^{s} n_j \log_e \left(|\mathbf{V}_T|/|\mathbf{V}_j| \right), \tag{3.11}$$

where $|\mathbf{V}|$ indicates the determinant of $\mathbf{V}$. This can be compared with the chi-square distribution with $\frac{1}{2}(s-1)p(p+3)$ degrees of freedom; a significantly large value indicates that the samples come from different populations. It is possible to separate λ into two parts as $\lambda = \lambda_1 + \lambda_2$ where

$$\lambda_1 = \sum_{j=1}^{s} n_j \log_e \left(|\hat{V}_0|/|\hat{V}_j| \right), \tag{3.12}$$

with $\frac{1}{2}(s-1)p(p+1)$ degrees of freedom, and

$$\lambda_2 = n \log_e \left(|\hat{V}_T|/|\hat{V}_0| \right), \tag{3.13}$$

with $(s-1)p$ degrees of freedom. Then a significantly large chi-square value for λ_1 indicates that the s samples came from populations with different covariance matrices, while a significantly large chi-square value for λ_2 indicates that the samples came from populations with different means, assuming that the covariance matrices are equal.

With a single X variable, equations (3.11) to (3.13) can still be used, with $p = 1$. However, they simplify since the matrices $\hat{V}_i$, V_0 and V_T reduce down to single variances. Thus let $x_{j1}, x_{j2}, \ldots, x_{jn_j}$ denote the values of X for the jth sample, with mean and variance

$$\hat{\mu}_j = \sum_{k=1}^{n_j} x_{jk}/n_j, \quad \text{and} \quad \hat{V}_j = \left(\sum_{k=1}^{n_j} x_{jk}^2 - n_j\hat{\mu}_j^2 \right) \bigg/ n_j.$$

The pooled within-sample variance is then

$$\hat{V}_0 = \sum_{j=1}^{s} n_j \hat{V}_j/n$$

while the mean and variance for the lumped total sample of size n are

$$\hat{\mu}_T = \sum_{j=1}^{s} n_j\hat{\mu}_j/n \quad \text{and} \quad \hat{V}_T = \left(\sum_{j=1}^{s} \sum_{k=1}^{n_j} x_{jk}^2 - n\hat{\mu}_T^2 \right) \bigg/ n.$$

Then

$$\lambda = \sum_{j=1}^{s} n_j \log_e \left(\hat{V}_T/\hat{V}_j \right), \tag{3.14}$$

with $2(s-1)$ degrees of freedom;

$$\lambda_1 = \sum_{j=1}^{s} n_j \log_e (\hat{V}_0/\hat{V}_j), \qquad (3.15)$$

with $s-1$ degrees of freedom; and

$$\lambda_2 = n \log_e (\hat{V}_T/\hat{V}_0), \qquad (3.16)$$

also with $s-1$ degrees of freedom.

The chi-square approximations for λ, λ_1 and λ_2 are based upon the assumption that samples are large. Just exactly what 'large' means in this context seems a little uncertain, but results should be reasonable for samples of 25 or more, unless p is large. In the single-variable case with small samples, λ_1 can be improved by multiplying with a constant close to one and replacing n_j with $n_j - 1$ (Bartlett's test – Steel and Torrie, 1980, p. 471). Also, for a single variable the λ_2 test can be replaced by a one-factor analysis of variance.

Since the test statistics λ_1 and λ_2 are based upon the assumption of normality, the possibility of significant results through non-normality of a constant population must be kept in mind. The λ_1 test is particularly liable to be upset in this way. For this reason, Van Valen (1978) has proposed some alternative tests for changes in dispersion. (See also Schultz, 1983; Manly, 1985b.)

3.5   ESTIMATING A QUADRATIC FITNESS FUNCTION

The simplest situation for estimating a fitness function occurs when there are two samples, one before and one after selection, and a single variable X. This was the situation considered by O'Donald (1968, 1970, 1971), who assumed a quadratic function for the probability that an individual with $X = x$ survives selection, of the form

$$1 - \alpha - K(\theta - x)^2.$$

He showed that if, before selection, the population mean of X is μ, the variance is V and the third and fourth moments about the mean are μ_3 and μ_4, then the expected proportion of survivors after selection is

$$\phi = 1 - \alpha - K(\theta - \mu)^2 - KV$$

and the expected changes in the mean and variance of X resulting from selection are

$$\Delta\mu = K\{2V(\theta - \mu) - \mu_3\}/\phi,$$

and

$$\Delta V = K\{V^2 - \mu_4 + 2\mu_3(\theta - \mu)\}/\phi - (\Delta\mu)^2.$$

On this basis, the following equations relate the parameters θ, K and α to ϕ, $\Delta\mu$ and ΔV:

$$\theta = \mu + \frac{(V^2 - \mu_4)\Delta\mu + \mu_3\{(\Delta\mu)^2 + \Delta V\}}{2V\{(\Delta\mu)^2 + \Delta V\} - 2\mu_3\Delta\mu}, \tag{3.17}$$

$$K = \frac{\Delta\mu\,\phi}{2V(\theta - \mu) - \mu_3}, \tag{3.18}$$

and

$$\alpha = 1 - \phi - K\{(\theta - \mu)^2 + V\}. \tag{3.19}$$

These equations can be simplified somewhat if the population being sampled has a normal distribution before selection, since in that case $\mu_3 = 0$ and $\mu_4 = 3V^2$, which means that

$$\theta = \mu - V\Delta\mu/\{(\Delta\mu)^2 + \Delta V\}, \tag{3.20}$$

and

$$K = \Delta\mu\,\phi/\{2V(\theta - \mu)\}. \tag{3.21}$$

If a sample is taken from a population before selection, then it can be used to estimate μ, V, μ_3 and μ_4 in the usual way. If a sample is also taken after selection has taken place, then $\Delta\mu$ can be estimated as the difference between the two sample means and ΔV can be estimated as the difference between the two sample variances. All these estimates can be substituted into equation (3.17) or equation (3.20) in order to estimate θ.

The proportion of the population surviving selection, ϕ, will not be known from sample data. However, equations (3.18) and (3.19) show that both $1 - \alpha$ and K are proportional to ϕ. Hence, setting $\phi = 1$ produces a quadratic fitness function

$$w(x) = 1 - \alpha - K(\theta - x)^2 \tag{3.22}$$

which is proportional to survival probabilities. The parameter K can be found by putting $\phi = 1$ in equation (3.18) or equation (3.21), together with sample estimates of the other statistics. Likewise, α can be determined by putting $\phi = 1$ in equation (3.19).

O'Donald's method for estimating a fitness function has two particular advantages: it can be used with any distribution, normal or otherwise; and it is relatively simple to use. There are also two disadvantages: the fitness function is liable to produce negative fitness values for extreme values of X; and no formulae are available at present for finding standard errors of estimators.

Example 3.3   Selection on gill raker numbers for threespine sticklebacks

As an example of the use of O'Donald's quadratic fitness function we can consider the data on the threespine stickleback, *Gasterosteus aculeatus* that are

Table 3.6 The distribution of the number of gill rakers on the
left-hand side of threespine sticklebacks before and after selection.

Gill raker number	Frequency before selection	Frequency after selection
11	0	3
12	2	0
13	0	5
14	14	40
15	42	112
16	132	284
17	367	704
18	759	1203
19	751	1127
20	441	618
21	170	244
22	35	45
23	13	2
24	0	2
Sample size	2726	4389
Mean ($\hat{\mu}$)	18.559	18.338
Variance ($\hat{V}$)	2.072	2.255
Third moment ($\hat{\mu}_3$)	-0.401	-0.955
Fourth moment ($\hat{\mu}_4$)	15.524	18.196

shown in Table 3.6. These were first published by Hagen and Gilbertson
(1973). A population was sampled when the sticklebacks were aged about $2\frac{1}{2}$
months and then again when they were adults about nine months later. The
question to be considered is how the survival from $2\frac{1}{2}$ months to one year is
related to the number of gill rakers.

The distribution of gill raker numbers cannot be normal since it is a discrete
distribution. Furthermore, the skewness ($\hat{\mu}_3/\hat{V}^{1.5} = -0.13$) and the kurtosis
($\hat{\mu}_4/\hat{V}^2 = 3.62$) of the sample before selection are significantly different from
the normal distribution values at the 1 % level of significance (Pearson and
Hartley, 1966, Table 34). Nevertheless, it is of some interest to see the results
from applying the likelihood ratio tests of Section 3.4 to these data.

Lumping the two samples together to form a single sample of size 7115
results in a variance of $\hat{V}_T = 2.195$. Using this together with $\hat{V}_1 = 2.071$ and
$\hat{V}_2 = 2.253$ in equation (3.14) produces $\lambda = 44.05$, with 2 degrees of freedom.
This is significantly large at the 0.1 % level so there is a clear indication of a
difference between the population at age $2\frac{1}{2}$ months and the population at age

one year. The pooled within-sample variance is

$$\hat{V}_0 = (2726\,\hat{V}_1 + 4389\,\hat{V}_2)/7115 = 2.183.$$

Then from equations (3.15) and (3.16), $\lambda_1 = 5.05$ and $\lambda_2 = 39.00$, both with 1 degree of freedom. Therefore λ_1 is significantly large at the 5% level while λ_2 is significantly large at the 0.1% level. It appears that most of the population changes are related to the mean rather than the variance.

Substituting the sample estimates of means, variances and higher moments given in Table 3.6 into equations (3.17) to (3.19), taking $\phi = 1$, gives the estimates $\hat{\theta} = 21.63$, $\hat{K} = -0.017$ and $\hat{\alpha} = 0.20$. The estimated fitness function is therefore

$$\hat{w}(x) = 0.80 + 0.017(21.63 - x)^2.$$

Table 3.7 shows the fitness values that are estimated from this function. This seems to be a situation where selection was directional, with the minimum fitness close to the maximum possible gill raker number. Hagen and Gilbertson commented that this was a very surprising result since it suggests that the population was not in equilibrium.

Table 3.7   The results obtained from fitting a quadratic fitness function to data on selection related to gill raker numbers of sticklebacks.

Gill raker numbers	Relative fitness
11	2.72
12	2.38
13	2.07
14	1.79
15	1.55
16	1.34
17	1.16
18	1.02
19	0.92
20	0.85
21	0.81
22	0.80
23	0.83
24	0.90

3.6   SELECTION ON A UNIVARIATE NORMAL DISTRIBUTION

Suppose that the variable X is normally distributed at time zero in a population, with mean μ_0 and variance V_0, and that the probability of an individual with $X = x$ surviving selection is proportional to the fitness function

$$w_t(x) = \exp\{(lx + mx^2)t\}, \tag{3.23}$$

where l and m are constants. Then at time t the distribution will still be normal, but with mean and variance

$$\mu_t = (\mu_0 + lV_0 t)/(1 - 2mV_0 t)$$

and

$$V_t = V_0/(1 - 2mV_0 t).$$

This result, expressed in a somewhat different form, was first used to estimate a fitness function by O'Donald (1970) and Cavalli-Sforza and Bodmer (1972, p. 614). Note that from these equations it follows that

$$\mu_t/V_t = \mu_0/V_0 + lt, \tag{3.24}$$

and

$$1/V_t = 1/V_0 - 2mt. \tag{3.25}$$

Let random samples with sizes $n_1, n_3, \ldots, n_s$ be taken from the population at times $t_1, t_2, \ldots, t_s$, with the X values in the jth sample being x_{j1}, $x_{j2}, \ldots, x_{jn_j}$. Then the mean and variance at the time of the jth sample, μ_j and V_j, can be estimated by

$$\hat{\mu}_j = \sum_{k=1}^{n_j} x_{jk}/n_j \quad \text{and} \quad \hat{V}_j = \left(\sum_{k=1}^{n_j} x_{jk}^2 - n_j \hat{\mu}_j^2 \right) \Big/ n_j,$$

these being the same as for the λ test statistics of Section 3.4. It is then well known that

$$\hat{B}_j = (n_j - 3)/(n_j \hat{V}_j) \tag{3.26}$$

is an unbiased estimator of $B_i = 1/V_i$, with variance

$$\mathrm{var}(\hat{B}_j) = 2/\{V_j^2 (n_j - 5)\} \tag{3.27}$$

(Press, 1972, p. 112). Also,

$$\hat{a}_j = \hat{\mu}_j \hat{B}_j \tag{3.28}$$

is an approximately unbiased estimator of $a_j = \mu_j/V_j$ with a variance that is given by a Taylor series approximation (Appendix, Section A.5) as

$$\mathrm{var}(\hat{a}_j) \simeq B_j^2 \, \mathrm{var}(\hat{\mu}_j) + \mu_j^2 \, \mathrm{var}(\hat{B}_j)$$

$$\simeq 1/(n_j V_j) + 2\mu_j^2/\{V_j^2 (n_j - 5)\}. \tag{3.29}$$

From these results and equations (3.24) and (3.25) it can be seen that the value m in the fitness function can be estimated by a weighted regression of $\hat{B}_j$ values against $2t_j$, with $\hat{B}_j$ being given the regression weight $1/\mathrm{var}(\hat{B}_j)$. Similarly, the value l can be estimated by a weighted regression of $\hat{a}_j$ values against t_j, with $\hat{a}_j$ being given the regression weight $1/\mathrm{var}(\hat{a}_j)$. Estimated values of V_j and μ_j will have to be used in calculating the regression weights, but this should not have much effect unless the sample sizes are rather small. (Weighted regression is summarized in the appendix, Section A.8).

Two special cases can be mentioned. The first of these occurs when there are only two samples taken at times 0 and t. Then l and m can be estimated as

$$\hat{l} = (\hat{a}_2 - \hat{a}_1)/t, \tag{3.30}$$

with variance

$$\text{var}(\hat{l}) = \{\text{var}(\hat{a}_1) + \text{var}(\hat{a}_2)\}/t^2, \tag{3.31}$$

and

$$\hat{m} = (\hat{B}_1 - \hat{B}_2)/2t, \tag{3.32}$$

with variance

$$\text{var}(\hat{m}) = \{\text{var}(\hat{B}_1) + \text{var}(\hat{B}_2)\}/4t^2. \tag{3.33}$$

The second special case occurs when the variance is not changed by selection, and is known accurately. Equation (3.24) then becomes

$$\mu_t/V_0 = \mu_0/V_0 + lt, \tag{3.34}$$

so that a regression of sample values $\hat{\mu}_j/V_0$ against t_j will produce an equation for which the coefficient of t is an estimate of l. The variance of $\hat{\mu}_j/V_0$ is $\text{var}(\hat{\mu}_j)/V_0^2 = 1/(n_j V_0)$, so that the appropriate regression weight for $\hat{\mu}_j/V_0$ is $n_j V_0$.

If the character being studied is clearly not normally distributed, then the equations provided in this section will give nothing more than a very approximate analysis. A transformation may overcome this difficulty. For example $\log_e(X)$, $1/X$ or $\sqrt{X}$ could be analysed rather than X itself. Alternatively, it is possible to assume a gamma distribution for X and a gamma-like fitness function, or a beta distribution for X and a beta-like fitness function (Manly 1977c). On the other hand, the method described below in Section 3.8 for estimating a fitness function with a non-normal multivariate population can be used with a single X variable. As will be seen in that section, this method has the advantage of not requiring any assumptions at all about the distribution of X.

Chen (1979) has discussed a model for directional selection in which the probability of survival for an individual with the value x is the integral from minus infinity to x of a normal probability density function. He assumes a normal distribution before selection and provides tables of changes to means, variances, and other moments due to selection.

Example 3.4   Selection on size with tsetse flies

In practice there are often considerable difficulties involved in being able to compare a sample of survivors with a sample from a population before selection. Sometimes the main problem is in being able to age animals in order to decide how much selection they have undergone. It may only be possible to do this rather roughly on the basis of size or on signs of wear.

Signs of wear were used by Glasgow (1961) with females of the tsetse fly *Glossina swynnertoni* that he studied at Shinyanga, Tanganyyika (now Tanzania). He was not able to determine the exact ages of the flies. However, it is well known that wing fray increases with age, and if tsetse flies are classified into Jackson's system of wing fray classes then the class numbers should roughly correspond with age. (For male *G. morsitans* the age increment between wing fray classes is about nine days (Jackson, 1946, 1948). The corresponding relationship for female *G. swynnertoni* is not apparently known but it is presumably similar.)

Glasgow sampled his population over a long period of time and then classified the flies into wing fray classes with the results shown in Table 3.8. The measurement that he used for the size of flies was the length of the middle part of the fourth longitudinal wing vein. Taking the wing fray class number as the selection 'time', samples are then available for 'ages' 1, 2, 3, 4 and 5 where the sample at age 5 actually includes individuals with a wing fray class of 5 or 6. Glasgow also gave the distribution of size for a sample of flies with what

Table 3.8   The distribution of the length of the middle part of the fourth longitudinal wing vein for five samples of female *Glossina swynnertoni*. Samples 1 to 5 are for flies in Jackson's wing fray classes 1 to 5+. The class frequencies shown were calculated from Glasgow's (1961) Figure 4 and some might be subject to small errors as a result. The measurements are in the arbitrary units of a micrometer eyepiece (millimetres ÷ 0.0349).

	Sample				
X	1	2	3	4	5
$41\frac{1}{4}$–$42\frac{1}{4}$	1	2	2	0	0
–$43\frac{1}{4}$	15	12	4	4	1
–$44\frac{1}{4}$	47	12	23	29	3
–$45\frac{1}{4}$	137	56	52	66	21
–$46\frac{1}{4}$	179	97	52	114	28
–$47\frac{1}{4}$	164	75	73	120	38
–$48\frac{1}{4}$	62	35	28	58	15
–$49\frac{1}{4}$	16	13	18	24	9
–$50\frac{1}{4}$	2	0	2	9	1
–$51\frac{1}{4}$	0	0	0	0	1
–$52\frac{1}{4}$	0	0	1	0	0
Sample size	623	302	255	424	117
Mean	45.853	45.949	46.060	46.241	46.382
Variance	1.697	1.868	2.363	1.953	1.839
Skewness	−0.09	−0.31	0.09	0.11	0.27
Kurtosis	2.94	3.29	3.23	2.90	3.29

amounts to an age of zero. This sample has not been considered here because the sampling method was different from that for the other samples and this seems to have resulted in some bias in the results (Manly, 1977c).

To test for the normality of the data, the skewness $(\hat{\mu}_3/\hat{V}^{1.5})$ and kurtosis $(\hat{\mu}_4/\hat{V}^2)$ values have been calculated for each of the five samples. These show no trends. The skewness for the second sample is -0.31 which is significantly different from the normal distribution value of zero at the 5% level. All the other skewness and kurtosis values are close to the normal distribution values and it seems therefore that a normal distribution model fits the data quite well.

It is a good idea to start an analysis that involves the estimation of a fitness function by coding the data to have a mean of zero and a variance of one in the first sample. This then avoids the problem of having very large or very small numbers occurring during the calculations. It also makes the fitness equal to one for an average individual in the first sample. In the present case, the coded variable is

$$X = (\text{wing vein length} - 45.853)/1.3026,$$

where 45.853 and 1.3026 are the mean and standard deviation in the sample for wing fray class 1. Coded sample means are then $\hat{\mu}_1 = 0$, $\hat{\mu}_2 = 0.074$, $\hat{\mu}_3 = 0.159$, $\hat{\mu}_4 = 0.298$, and $\hat{\mu}_5 = 0.406$. The corresponding coded sample variances are $\hat{V}_1 = 1.000$, $\hat{V}_2 = 1.101$, $\hat{V}_3 = 1.392$, $\hat{V}_4 = 1.151$ and $\hat{V}_5 = 1.084$. These values are plotted on Fig. 3.1.

The variance of the lumped sample of $n = 1721$ flies is $\hat{V}_T = 1.138$, while the pooled within-sample variance is $\hat{V}_0 = \sum n_i \hat{V}_i/n = 1.119$. Using these values it is found that the test statistic λ of equation (3.14) is 40.02 with 8 degrees of freedom, which is significantly large at the 0.1% level. There is therefore very strong evidence that the distribution of the length of the wing vein changed as the flies got older. Most of the λ value can be attributed to changes in the mean ($\lambda_2 = 28.98$ with 4 degrees of freedom), but the differences in the variances are also significant at the 5% level ($\lambda_1 = 11.04$ with 4 degrees of freedom).

Although the sample variances show significant differences, there is no obvious trend (Fig. 3.1). It is therefore reasonable to make the assumption that $m = 0$ in the fitness function (3.23) and hence estimate l using equation (3.34). It seems fair enough to take $V_0 = 1.119$ as an accurate value for the within-sample variance since this value is based on over 1700 flies.

For estimating equation (3.34) the following sample values are available:

t_j	$\hat{\mu}_j/V_0$	Variance	1/Variance
1	0	0.00143	697.1
2	0.066	0.00296	337.9
3	0.142	0.00350	285.3
4	0.266	0.00211	474.5
5	0.363	0.00764	130.9

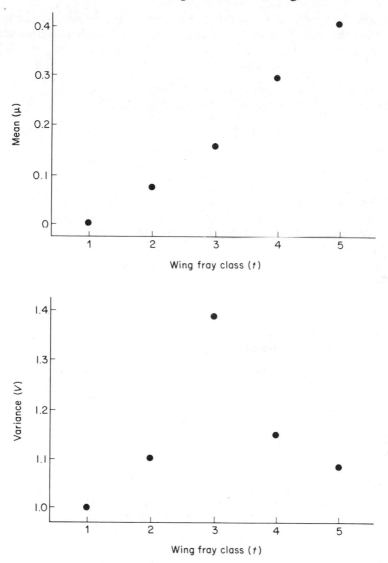

Fig. 3.1 Sample means and variances for the coded wing vein length of *Glossina swynnertoni*.

Here the variance of $\hat{\mu}_j/V_0$ is $1/(n_j V_0)$. A weighted regression analysis produces the equation

$$\mu_t/V_0 = -0.098 + 0.089\, t.$$

This gives an excellent fit to the data, with a weighted residual mean square of 0.13, which is less than the expected value of one (Appendix, Section A.8). The estimated value of l for the fitness function is the regression coefficient $\hat{l} = 0.089$, this having a standard error of $SE(\hat{l}) = 0.017$. Hence the estimated fitness function is

$$\hat{w}_t(x) = \exp(0.089\,xt).$$

The Taylor series approximation method can be used here to obtain variances for the values given by the function. Thus

$$\text{var}\{\hat{w}_t(x)\} \simeq \{w_t(x)\}^2 \, \text{var}[\log_e\{\hat{w}_t(x)\}]$$
$$\simeq \{w_t(x)\}^2 \, \text{var}(\hat{l})\,x^2 t^2$$
$$\simeq \{0.017\,xt\,w_t(x)\}^2.$$

In using this last equation it is necessary to use $\hat{w}_t(x)$ in place of the unknown true value $w_t(x)$. Table 3.9 shows estimates and their standard errors for $t = 1$. It appears that the largest flies had a survival rate about twice as great as that for the smallest flies in the time required to move from one wing fray class to the next. Thus a fairly high level of directional selection seems to have taken place.

Table 3.9   Estimated relative fitness values from data on the length of the fourth longitudinal wing vein for female *Glossina swynnertoni*.

Vein length	Coded x	$\hat{w}_1(x)$	Standard error
41	−3.726	0.72	0.05
43	−2.190	0.82	0.03
45	−0.655	0.94	0.01
47	0.881	1.08	0.02
49	2.416	1.24	0.05
51	3.951	1.42	0.10

3.7   SELECTION ON A MULTIVARIATE NORMAL DISTRIBUTION

The regression method for estimating a fitness function can be generalized for selection on several variables at the same time. Thus suppose that the variables $X_1, X_2, \ldots, X_p$ have a multivariate normal distribution at time zero with mean vector $\boldsymbol{\mu}_0$ and covariance matrix $\mathbf{V}_0$. Assume also that the fitness function for the time interval $(0, t)$ has the form

$$w_t(\mathbf{x}) = \exp\{(\mathbf{L}'\mathbf{x} + \mathbf{x}'\mathbf{M}\mathbf{x})t\} \tag{3.35}$$

(ignoring any constant term that does not involve x), where

$$
\mathbf{L} = \begin{bmatrix} l_1 \\ l_2 \\ \cdot \\ \cdot \\ \cdot \\ l_p \end{bmatrix}
\quad \text{and} \quad
\mathbf{M} = \begin{bmatrix} m_{11} \, m_{12} \ldots m_{1p} \\ m_{21} \, m_{22} \ldots m_{2p} \\ \cdot \\ \cdot \\ \cdot \\ m_{p1} \, m_{p2} \ldots m_{pp} \end{bmatrix}
$$

are unknown, with $\mathbf{M}$ being symmetric. Then at time t the distribution will be multivariate normal with a mean vector

$$
\mu_t = \mathbf{V}_t (\mathbf{L}t + \mathbf{V}_0^{-1} \mu_0)
$$

and a covariance matrix

$$
\mathbf{V}_t = \{ \mathbf{V}_0^{-1} - 2\mathbf{M}t \}^{-1}
$$

(Felsenstein, 1977; Manly, 1981b). These equations can be rewritten as

$$
\mathbf{V}_t^{-1} \mu_t = \mathbf{V}_0^{-1} \mu_0 + \mathbf{L}t \tag{3.36}
$$

and

$$
\mathbf{V}_t^{-1} = \mathbf{V}_0^{-1} - 2\mathbf{M}t . \tag{3.37}
$$

Hence, given s samples taken at times $t_1, t_2, \ldots, t_s$ from the population, with estimated mean vectors $\hat{\mu}_1, \hat{\mu}_2, \ldots, \hat{\mu}_s$, and estimated covariance matrices $\hat{\mathbf{V}}_1, \hat{\mathbf{V}}_2, \ldots, \hat{\mathbf{V}}_s$, the fitness function can be estimated by regressing estimates of the elements of $\mathbf{V}_t^{-1} \mu_t$ against t to determine $\mathbf{L}$, and estimates of the elements of $\mathbf{V}_t^{-1}$ against $2t$ to determine $\mathbf{M}$ (Manly, 1981b).

Here the discussion will be restricted to two special cases of this model. The first of these occurs when there are only two samples. These can then be assumed to be taken at times $t_1 = 0$ and $t_2 = 1$, and equations (3.36) and (3.37) give

$$
\mathbf{M} = (\mathbf{V}_1^{-1} - \mathbf{V}_2^{-1})/2, \tag{3.38}
$$

and

$$
\mathbf{L} = \mathbf{V}_2^{-1} \mu_2 - \mathbf{V}_1^{-1} \mu_1 . \tag{3.39}
$$

Sample estimates can be substituted into these equations to estimate $\mathbf{M}$ and $\mathbf{L}$.

Let $\hat{\mu}_j$ and $\hat{\mathbf{V}}_j$ be the sample mean vector and the sample covariance matrix for the jth sample, where these are calculated as described in Section 3.4. The inverse of $\hat{\mathbf{V}}_j$ is then $\hat{\mathbf{V}}_j^{-1}$, which is a biased estimator $\mathbf{B}_j = \mathbf{V}_j^{-1}$. An unbiased estimator is

$$
\hat{\mathbf{B}}_j = (n_j - p - 2) \, \hat{\mathbf{V}}_j^{-1} / n_j \tag{3.40}
$$

(Press, 1972, p. 112). If the value in the ith row and kth column of $\mathbf{B}_j$ is denoted by b_{ikj}, with an estimate $\hat{b}_{ikj}$ from equation (3.40), then

$$
\mathrm{var}(\hat{b}_{ikj}) \simeq (b_{iij} b_{kkj} + b_{ikj}^2)/(n_j - p - 4). \tag{3.41}
$$

Using these results, M can be estimated as

$$\hat{M} = (\hat{B}_1 - \hat{B}_2)/2. \tag{3.42}$$

The element in the ith row and the kth column of $\hat{M}$ is then

$$\hat{m}_{ik} = (\hat{b}_{ik1} - \hat{b}_{ik2})/2,$$

with variance

$$\text{var}(\hat{m}_{ik}) = \{\text{var}(\hat{b}_{ik1}) + \text{var}(\hat{b}_{ik2})\}/4. \tag{3.43}$$

A sample estimator of $a_j = V_j^{-1}\mu_j$ is

$$\hat{a}_j = \hat{B}_j \hat{\mu}_j. \tag{3.44}$$

A Taylor series approximation then shows (Manly, 1981b) that the covariance matrix for the elements of a_j is

$$C_j \simeq V_j^{-1}/n_j + (V_j^{-1}\mu_j' a_j + a_j a_j')/(n_j - p - 4). \tag{3.45}$$

Therefore L can be estimated as

$$\hat{L} = \hat{a}_2 - \hat{a}_1, \tag{3.46}$$

with the covariance matrix of the elements of L being

$$V_L = C_1 + C_2. \tag{3.47}$$

The second special case that will be considered occurs when the covariance matrix does not change with selection and is known accurately. In that case $M = 0$ in the fitness function and equation (3.36) becomes

$$V_0^{-1}\mu_t = V_0^{-1}\mu_0 + Lt. \tag{3.48}$$

The elements of L can now be estimated by regressing the elements of $\hat{a}_j = V_0^{-1}\hat{\mu}_j$ against t_j values. In this case, the Taylor series method shows that the covariance matrix for the elements of $\hat{a}_j$ is

$$C_j \simeq V_0^{-1}/n_j, \tag{3.49}$$

and the appropriate regression weight for an element of $\hat{a}_j$ is just the reciprocal of its variance as obtained from this matrix. When there are only two samples, L can be estimated simply as

$$L = \hat{V}_0^{-1}(\hat{\mu}_2 - \hat{\mu}_1) \tag{3.50}$$

where the covariance matrix for the elements of $\hat{L}$ is

$$V_L = V_0^{-1}(1/n_1 + 1/n_2). \tag{3.51}$$

Example 3.5   Selective Predation on *Bosmina longirostris*

Bosmina longirostris is a small zooplankton that is eaten by the fish *Melaniris changresi* in Lake Gutun, Panama. The selection involved in this predation was investigated by Zaret and Kerfoot (1975) by making use of the diurnal migration pattern of *M. changresi*. In one area of Lake Gatun, the fish only feed in the afternoon. Therefore a comparison between a morning sample (before selection) and an afternoon sample (after selection) of *B. longirostris* should indicate the amount of selection taking place. Table 3.10 shows the results for three such comparisons, although the present example will be concerned only with the one made on 5 May 1969.

Zaret and Kerfoot were particularly concerned with the question of whether selection was mainly based upon general body size (measured as the total length, in microns) or mainly upon the eye-pigmentation diameter (also measured in microns). They argued that the latter might well be more important because *B. longirostris* is highly transparent, with the most conspicuous feature being the relatively large, black compound eye. This question can be addressed by estimating a fitness function using the equations of the previous section. In doing this it has to be assumed that the two variables studied follow a bivariate normal distribution, although this assumption cannot be checked from the information given in Zaret and Kerfoot's paper.

In the calculations that follow it is convenient to work in terms of variable values that are coded to have means of zero and standard deviations of one in the sample before selection. This ensures that the fitness function gives a fitness of unity to an individual that has a body size and an eye pigmentation diameter equal to the mean values in the sample before selection. Thus the two variables to be analysed are

$$X_1 = (\text{body size} - 265.0)/22.2,$$

Table 3.10 Comparison of body size (BS) and eye pigmentation diameter (EPD) measurements, in microns, in morning and afternoon samples of *Bosmina longirostris* taken from Gatun Lake in 1969.

Sample day	Time of sample	Size of sample	Sample means		Sample std. devs.		Sample covariance
			BS	EPD	BS	EPD*	
5 February	a.m.	41	281	20.2	32.5	2.40	23.0
	p.m.	24	283	19.9	27.5	2.64	16.0
5 May	a.m.	55	265	22.2	22.2	1.88	19.0
	p.m.	44	252	18.2	36.8	2.73	32.0
4 September	a.m.	76	284	22.2	29.2	1.82	24.8
	p.m.	45	283	20.7	28.5	2.61	30.0

* The standard deviations for EPD given by Zaret and Kerfoot (1975) have been multiplied by 10 as this is necessary to make them correct (Confer *et al.*, 1980). The same correction has been made here to the covariance between BS and EPD.

and

$$X_2 = (\text{eye pigmentation diameter} - 22.2)/1.88.$$

The mean vector and the covariance matrix for the sample before selection, of size $n_1 = 55$, are

$$\hat{\mu}_1 = \begin{bmatrix} 0 \\ 0 \end{bmatrix} \quad \text{and} \quad \hat{V}_1 = \begin{bmatrix} 1.0000 & 0.4552 \\ 0.4552 & 1.0000 \end{bmatrix},$$

while the mean and covariance matrix in the sample after selection, of size $n_2 = 44$, are

$$\hat{\mu}_2 = \begin{bmatrix} -0.260 \\ -2.128 \end{bmatrix} \quad \text{and} \quad \hat{V}_2 = \begin{bmatrix} 2.7478 & 0.7667 \\ 0.7667 & 2.1087 \end{bmatrix}.$$

From these results it can be calculated that lumping the data into a single sample of size $n = 99$ gives an overall mean vector and an overall covariance matrix of

$$\hat{\mu}_T = \begin{bmatrix} -0.116 \\ -0.946 \end{bmatrix} \quad \text{and} \quad \hat{V}_T = \begin{bmatrix} 1.7935 & 0.7303 \\ 0.7303 & 2.6109 \end{bmatrix}.$$

A pooled within-sample covariance matrix is

$$\hat{V}_0 = \sum_{j=1}^{2} n_j \hat{V}_j / n = \begin{bmatrix} 1.7768 & 0.5942 \\ 0.5942 & 1.4928 \end{bmatrix}.$$

From these covariance matrices the following determinants are found: $|\hat{V}_1| = 0.7928$, $|\hat{V}_2| = 5.2065$, $|\hat{V}_T| = 4.1493$ and $|\hat{V}_0| = 2.2993$.

The statistic of equation (3.11) is found to be $\lambda = 81.05$, with 5 degrees of freedom, which is significantly large at the 0.1 % level when tested against the chi-square distribution. There seems no doubt that the population changed between the two sample times. The statistic of equation (3.12) for the change in the variance is $\lambda_1 = 22.60$, with 3 degrees of freedom, while the statistic of equation (3.13) for the change in the mean is $\lambda_2 = 58.44$, with 2 degrees of freedom. Both λ_1 and λ_2 are significantly large at the 0.1 % level. It seems clear, however, that changes in means were more important than changes in variances and covariances.

Because there are only two samples, equations (3.38) to (3.47) can be used in this example to estimate the fitness function. The inverses of the matrices $\hat{V}_1$ and $\hat{V}_2$ are found to be

$$\hat{V}_1^{-1} = \begin{bmatrix} 1.261 & -0.574 \\ -0.574 & 1.261 \end{bmatrix} \quad \text{and} \quad \hat{V}_2^{-1} = \begin{bmatrix} 0.405 & -0.147 \\ -0.147 & 0.528 \end{bmatrix}.$$

These can be converted to unbiased estimates of V_1^{-1} and V_2^{-1} using equation (3.40) to give

$$\hat{B}_1 = \begin{bmatrix} 1.170 & -0.532 \\ -0.532 & 1.170 \end{bmatrix} \quad \text{and} \quad \hat{B}_2 = \begin{bmatrix} 0.368 & -0.134 \\ -0.134 & 0.480 \end{bmatrix}.$$

The estimate of the matrix $\mathbf{M}$ in the fitness function is then given by equation (3.42) as

$$\hat{\mathbf{M}} = \tfrac{1}{2}(\hat{\mathbf{B}}_1 - \hat{\mathbf{B}}_2) = \begin{bmatrix} 0.401 & -0.199 \\ -0.199 & 0.245 \end{bmatrix}.$$

Also, from equation (3.44) it is found that

$$\hat{\mathbf{a}}_1 = \begin{bmatrix} 1.170 & -0.532 \\ -0.532 & 1.170 \end{bmatrix} \begin{bmatrix} 0 \\ 0 \end{bmatrix} = \begin{bmatrix} 0 \\ 0 \end{bmatrix},$$

and

$$\hat{\mathbf{a}}_2 = \begin{bmatrix} 0.368 & -0.134 \\ -0.134 & 0.480 \end{bmatrix} \begin{bmatrix} -0.260 \\ -2.128 \end{bmatrix} = \begin{bmatrix} 0.189 \\ -0.987 \end{bmatrix}.$$

so that the estimate of the vector $\mathbf{L}$ in the fitness function is

$$\hat{\mathbf{L}} = \hat{\mathbf{a}}_2 - \hat{\mathbf{a}}_1 = \begin{bmatrix} 0.189 \\ -0.987 \end{bmatrix}.$$

Taking the above values for $\mathbf{L}$ and $\mathbf{M}$, the fitness function of equation (3.35) can now be estimated as

$$\hat{w}(\mathbf{x}) = \exp(\hat{\mathbf{L}}'\mathbf{x} + \mathbf{x}'\hat{\mathbf{M}}\mathbf{x})$$

or

$$\hat{w}(x_1, x_2) = \exp(0.189\, x_1 - 0.987\, x_2 + 0.401\, x_1^2$$
$$- 0.398\, x_1 x_2 + 0.345\, x_2^2),$$

for a unit period of time.

Using sample values in equations (3.45) and (3.47) produces a covariance matrix for the elements $\hat{l}_1$ and $\hat{l}_2$ of $\hat{\mathbf{L}}$ of

$$\mathbf{V}_L \simeq \begin{bmatrix} 0.0457 & -0.0232 \\ -0.0232 & 0.0770 \end{bmatrix}.$$

The standard errors of $\hat{l}_1$ and $\hat{l}_2$ are therefore approximately $\sqrt{(0.0457)} = 0.21$ and $\sqrt{(0.0770)} = 0.28$, respectively. Also using sample estimates of population values, equations (3.41) and (3.43) show that the standard errors associated with the elements $\hat{m}_{11}, \hat{m}_{12}$ and $\hat{m}_{22}$ of $\hat{\mathbf{M}}$ are approximately 0.13, 0.10 and 0.13, respectively.

Fig. 3.2 shows a plot of the fitness surface that the estimated function gives. This is a surface where the height for particular values of eye pigmentation diameter and body size corresponds to the fitness of individuals with those values, relative to a fitness of unity for individuals with average values for the variables in the sample before selection. The range shown on the figure for body size is approximately the mean before selection plus and minus two standard deviations, $265 \pm 45\ \mu$m. Likewise, for the eye pigmentation diameter the range is $22.2 \pm 3.8\ \mu$m.

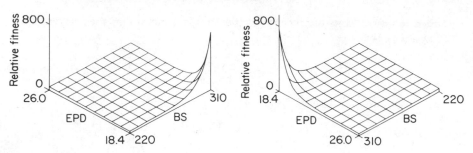

Fig. 3.2   A fitness function for *Bosmina longirostiris*. The height of the surface gives the relative fitness for individuals with particular values for the eye pigmentation diameter (EPD, in μm) and body size (BS, also in μm). Both graphs are for the same surface, viewed from different angles.

Fig. 3.2 shows that individuals with large body sizes and small eye pigmentation diameters were very much fitter than other individuals. Thus a body size of 310 μm and an eye pigmentation diameter of 18.4 μm gives an estimated fitness of about 766. The positive correlation between the two measurements has meant that selection decreases the mean body size even though large body sizes are apparently fitter than small body sizes. What has happened is that the change in the mean body size is less than the change expected on the basis of selection on eye pigmentation diameter alone and the correlation between the two measurements. Therefore the fitness function has had to include an element of selection for large body sizes in order to account for the relatively small decrease in the mean body size as a result of selection.

If the estimated fitness function is accepted as being realistic then it seems to completely confirm Zaret and Kerfoot's idea that predator selection is mainly on the basis of a large area of eye pigmentation. There is certainly no evidence of a large body size being associated with low fitness. However, it must be mentioned that Confer *et al.* (1980) have questioned Zaret and Kerfoot's experimental results on several grounds, the most important being their finding that the eye pigmentation diameter of *B. longirostris* may decrease between morning and afternoon samples as a physiological reaction to increasing light levels. Despite the objections of Confer *et al.*, Kerfoot (1980) has maintained that the experimental results are valid indicators of selective predation.

Another point that must be noted with this example is the rather inconsistent results for different sampling days. In particular, the large reduction in body size between the morning and afternoon samples on 5 May is not matched by similar reductions on 5 February and 4 September. There are insignificant changes on the latter dates, with an increase for the first of them (Table 3.10). For eye pigmentation diameter there is a drop in the mean from the morning to the afternoon sample for all three sample days. However, the drop is rather small on 5 February. It does seem that whatever factor was responsible for the apparent selection on 5 May is not always present, or does not always have the same effect on the population.

3.8 SELECTION ON A NON-NORMAL POPULATION

The method for estimating a multivariate fitness function that has been described in the last section is not suitable for a population that is clearly non-normal. Furthermore, it is not obvious what exactly the effects of non-normality will be. It is therefore useful to have an alternative method of estimation that makes no assumptions about the parametric form of the probability density function of the sampled population. One approach will now be considered.

Suppose that there are a number of different types of individuals in a population, with correspondingly different values for one or more variables $X_1, X_2, \ldots, X_p$. Suppose also that at time $t = 0$ there are f_r type r individuals while at a later time t_j there are $w_{rj} f_r$ of this type, so that w_{rj} is the fitness for the type r individuals over the time period $(0, t_j)$.

Let random samples be taken from the population at times $t_1, t_2, \ldots, t_s$ in such a way that the probability of an individual being included in the sample at time t_j is γ_j, where this is very small. Then the number of type r individuals in the sample at time t_j can be assumed to follow a Poisson distribution with mean $w_{rj} f_r \gamma_j$, in which case it can be shown that the probability of one of this type of individual being included in the sample at time t_j, given that it is in one of the samples, is

$$P_{rj} = w_{rj} f_r \gamma_j \bigg/ \sum_{u=1}^{s} w_{ru} f_r \gamma_u$$

$$= \theta_j w_{rj} \bigg/ \sum_{u=1}^{s} \theta_u w_{ru},$$

where $\theta_1 = 1$ and $\theta_u = \gamma_u/\gamma_1$.

Various assumptions can be made about the relationship between the fitness and the X values for an individual. Five models are of particular interest, as follows.

With *model 0*, w_{ij} is constant. The maximum likelihood estimator of θ_u is then $\hat{\theta}_u = n_u/n_1$ where n_u is the number of individuals in the uth sample. This is a no-selection model. With *model 1*,

$$w_{ij} = \exp\{(\alpha_1 x_{1i} + \alpha_2 x_{2i} + \ldots + \alpha_p x_{pi})t_j\}$$

so that the effective selection time is equal to the actual time t. *Model 2* is the same as model 1 except that the selection time is assumed to correspond to an unknown power of t, so that

$$w_{ij} = \exp\{(\alpha_1 x_{1i} + \alpha_2 x_{2i} + \ldots + \alpha_p x_{pi})t_j^v\}$$

where v is estimated from the data. If there are only two samples then this reduces to model 1. *Model 3* is the same as model 1 except that no connection between the selection time and the actual time is assumed. Hence

$$w_{ij} = \exp\{(\alpha_1 x_{1i} + \alpha_2 x_{2i} + \ldots + \alpha_p x_{pi})g_j\}$$

where g_j is estimated from the data. If there are only three samples then this is equivalent to model 2; for two samples it reduces to model 1. For *model 4* the fitness function is assumed to vary with time so that

$$w_{ij} = \exp(\alpha_{1j} x_{1i} + \alpha_{2j} x_{2i} + \ldots + \alpha_{pj} x_{pi})$$

This amounts to estimating a separate function for each period between two samples. With only two samples this is equivalent to model 1; with three samples it is a generalization of model 2 and model 3, providing that there are two or more X variables.

Given one of these models for fitness, the likelihood function for the data takes the form

$$L = \prod_{q=1}^{m} P_{qh_q}$$

where m is the total number of individuals seen, with the qth individual being seen in sample h_q. On this basis the models 1 to 4 can be fitted in sequence to a set of data using the method of maximum likelihood (Manly, 1981b).

Likelihood ratio tests can be used to compare the fit of the different models. Thus if l_i is the maximum of the logarithm of the likelihood function for model i, then a significantly large value for $2(l_i - l_{i-1})$ shows that model i gives a significantly better fit to the data than model $i-1$. The significance is determined with reference to the chi-squared distribution with $v_i - v_{i-1}$ degrees of freedom, where v_i is the number of estimated parameters in model i (Appendix, Section A.4).

3.9   THE VARIANCE OF FITNESS ESTIMATES

Suppose that an estimated fitness function takes the form

$$\hat{w}_t(x_1, x_2, \ldots, x_p) = \exp\{(\hat{\alpha}_1 x_1 + \hat{\alpha}_2 x_2 + \ldots + \hat{\alpha}_p x_p)t\}.$$

Then the variance of $\log_e(\hat{w}_t)$ is simply the variance of a linear function:

$$\text{var}\{\log_e(\hat{w}_t)\} = \sum_{i=1}^{p} \sum_{j=1}^{p} x_i x_j \, \text{cov}(\hat{\alpha}_i, \hat{\alpha}_j) t^2,$$

and there is a Taylor series approximation

$$\text{var}(\hat{w}_t) \simeq w_t^2 \, \text{var}\{\log_e(\hat{w}_t)\}. \tag{3.52}$$

There is nothing in the derivation of this approximate result to stop it being applied in situations where some of the X variables are powers and products of other X variables.

3.10   THE CASE OF STABILIZING SELECTION

When stabilizing selection takes place there is a tendency for individuals with extreme values for quantitative variables to be eliminated. Therefore one relatively simple approach for studying this type of selection involves defining some measure of the extent to which individuals differ from the average individual and then relating survival to this measure. See Section 5.5 for a discussion of possibilities along these lines.

3.11   KARL PEARSON'S WORK AND DEVELOPMENTS FROM IT

Recently Lande and Arnold (1983) have pointed out that Karl Pearson showed in an important but neglected paper (Pearson, 1903) how multivariate statistics can be used to disentangle the direct and indirect effects of selection to determine which traits in a correlated set of variables are the focus of selection. For example, Pearson obtained equations (3.36) and (3.37) for changes in means, variances and covariances due to selection on a multivariate normal distribution. However, his equations are a good deal more complicated since they are not expressed in the compact matrix form.

Lande and Arnold have extended and generalized Pearson's results to produce an approach for studying selection within one generation on a multivariate population that is somewhat different from the one given in previous sections of this chapter. They applied their methods to data on selection of pentatomid bugs (*Euschistus variolarius*) knocked into Lake Michigan during a storm and also to Bumpus' (1898) data on the selection of house sparrows (Example 5.3, below). See Lande and Arnold's paper for more details, also Arnold and Wade (1984a, b).

3.12   THE INTENSITY AND EFFECT OF SELECTION ON QUANTITATIVE VARIABLES

Four indices of the strength of selection have been defined in Section 3.3. These can be used equally well when a fitness function is used to determine the fitnesses of individuals. It may be recalled that Haldane's (1954a, 1959) index is $I_H = \log_e(w^*/\bar{w})$, Van Valen's (1965a) index is $I_V = (w^* - \bar{w})/\bar{w}$, O'Donald's (1970) index is $I_O = (\bar{w}' - \bar{w})/\bar{w}$, and the fourth index is $I_u = \bar{u}' - \bar{u}$. Here $\bar{w}$ is the mean fitness before selection, $\bar{w}'$ is the mean fitness after selection, w^* is the maximum fitness, $\bar{u}$ is the mean of the logarithm of fitness, $u = \log_e(w)$, before selection, and $\bar{u}'$ is the mean of u after selection. Thus I_H and I_V compare the maximum fitness with the mean fitness, while I_O and I_u measure the change in mean fitness caused by selection. If the fitness values w relate to a unit period of time then I_H and I_V are measures of the intensity of selection while I_O and I_u are measures of the effect of selection.

If there is a single quantitative variable X, which is normally distributed with mean and variance μ_1 and V_1 before selection, and mean and variance μ_2 and V_2 after selection, then Haldane (1954a) showed that his index becomes

$$I_H = \tfrac{1}{2}\log_e(V_1/V_2) + \tfrac{1}{2}(\mu_1 - \mu_2)^2/(V_1 - V_2). \qquad (3.53)$$

This is then related to Van Valen's index as

$$I_V = \{\exp(I_H) - 1\}/\exp(I_H).$$

Marcus (1964) considered the generalization of equation (3.53) for a multivariate normal distribution. Unfortunately, the resulting equation failed to produce a sensible result on test data.

Another result of some interest occurs when there are p variables with a multivariate normal distribution with a mean vector and covariance matrix of $\hat{\mu}_1$ and V before selection, and a fitness function

$$w_t(\mathbf{x}) = \exp(\mathbf{L} \times t),$$

where $\mathbf{L}' = (l_1, l_2, \ldots, l_p)$ and $\mathbf{x}' = (x_1, x_2, \ldots, x_p)$. In this case the covariance matrix does not change through selection, but the change in the mean vector after selection for a time t is given by equation (3.36) to be

$$\mu_2 - \mu_1 = \mathbf{V}\mathbf{L}t.$$

An estimate of $\mathbf{L}$ is then

$$\hat{\mathbf{L}} = \mathbf{V}^{-1}(\hat{\mu}_2 - \hat{\mu}_1)/t,$$

where $\hat{\mu}_1$ is the mean vector of a sample at time 0, $\hat{\mu}_2$ is the mean vector of a sample at time t, and the covariance matrix $\mathbf{V}$ is assumed to be known. In this case the logarithm of the fitness function for a unit interval of time is estimated as $\hat{u}(x) = \hat{\mathbf{L}}'\mathbf{x}$, with a mean of $\hat{\mathbf{L}}'\mu_1$ before selection and a mean of $\hat{\mathbf{L}}'\mu_2$ after selection. Consequently, an estimate of I_u is

$$\hat{I}_u = \hat{\mathbf{L}}'(\hat{\mu}_2 - \hat{\mu}_1) = \hat{\mathbf{L}}'\mathbf{V}\mathbf{L}t, \qquad (3.54)$$

or, alternatively,

$$\hat{I}_u = (\hat{\mu}_2 - \hat{\mu}_1)' \mathbf{V}^{-1}(\hat{\mu}_2 - \hat{\mu}_1)/t. \qquad (3.55)$$

Equation (3.54) shows that for the model being assumed $\hat{I}_u$ should increase linearly with the selection time. Equation (3.55) shows that if there is no selection, so that $\mu_1 = \mu_2$, then

$$t n_1 n_2 \hat{I}_u/(n_1 + n_2)$$

follows a chi-square distribution with p degrees of freedom (Anderson, 1958, p. 56), where n_1 and n_2 are the sample sizes before and after selection.

Therefore the mean and variance of $\hat{I}_u$ are

$$E(\hat{I}_u) \simeq np/(n_1 n_2 t), \tag{3.56}$$

and

$$\text{var}(\hat{I}_u) \simeq 2n^2 p/(n_1^2 n_2^2 t^2), \tag{3.57}$$

where $n = n_1 + n_2$.

It may be considered something of a drawback with the above indices that they depend directly on what assumptions are made in estimating a fitness function. It is in some ways better to use indices which depend only on the change in the distribution of the quantitative variables being considered. One obvious possibility is to base indices on the λ_1 and λ_2 statistics of Section 3.4. If there is no selection on a multivariate normal distribution, then these statistics have approximately chi-square distributions. Since the mean of a chi-square variate is equal to the number of degrees of freedom and its variance is twice the number of degrees of freedom,

$$\phi_1 = (\lambda_1 - d_1)/(2d_1)^{1/2} \tag{3.58}$$

is a measure of changes in dispersion and

$$\phi_2 = (\lambda_2 - d_2)/(2d_2)^{1/2} \tag{3.59}$$

is a measure of changes in mean values, where d_i is the number of degrees of freedom for λ_i. Obviously if there is no selection on a multivariate normal distribution then the indices ϕ_1 and ϕ_2 will have expected values of zero and standard errors of one. Here λ_1 and λ_2 are given by equations (3.12) and (3.13).

Other indices of the effects of selection are discussed by Manly (1981b). Marcus (1969) used Mahalanobis' distance measure for this purpose.

Example 3.6   The intensity of selection on
Bosmina longirostris

Zaret and Kerfoot's (1975) study of selective predation on the zooplankton *Bosmina longirostris* has been described in Example 3.5. It may be recalled that this study involved sampling *B. longirostris* in Gatun Lake, Panama, on mornings and afternoons. Changes in the bivariate distribution of body size and eye pigmentation diameter between two sampling times on one day were attributed to selective predation by the fish *Melaniris changresi*. Table 3.10 shows sample results for samples taken on 5 February, 5 May and 4 September 1969. Example 3.5 was concerned with the estimation of a fitness function for 5 May. The present example addresses the question of how the intensity of selection compares for the three sampling days.

Unfortunately, Zaret and Kerfoot's (1975) paper does not give body size and eye pigmentation diameter values for individual *B. longirostris* in the samples before and after selection. However, these values have to be known in order to calculate some of the indices of selection that have been discussed in

Section 3.12. It has therefore been necessary to invent some values for illustrative purposes. Thus Table 3.11 shows sample distributions before and after selection for 5 May, where these are random samples from bivariate normal distributions with mean vectors and covariance matrices equal to the sample values found by Zaret and Kerfoot. Variables are coded to have zero means and unit variances in the sample before selection.

In terms of coded values it has been shown in Example 3.5 that the estimated fitness function for 5 May is

$$\hat{w}(x_1, x_2) = \exp(0.189\,x_1 - 0.987\,x_2 + 0.401\,x_1^2$$
$$- 0.398\,x_1 x_2 + 0.345\,x_2^2).$$

When this function is evaluated for each of the 55 individuals in the sample before selection, it is found that the mean fitness is $\bar{w} = 2.94$ and the mean of $\hat{u} = \log_e(\hat{w})$ is $\bar{u} = 0.564$. The maximum fitness is $w^* = 23.43$ for an individual with $x_1 = 0.74$ and $x_2 = -1.54$. On this basis, Haldane's index of the intensity of selection is estimated at $\hat{I}_H = \log_e(w^*/\bar{w}) = 2.08$, while Van Valen's index is estimated as $\hat{I}_V = (w^* - \bar{w})/\bar{w} = 6.97$. Using the same fitness function on the 44 individuals in the sample after selection gives an estimated mean fitness of $\bar{w}'$ $= 5690.0$ and an estimated mean log fitness of $\bar{u}' = 4.94$. Therefore O'Donald's index is estimated as $\hat{I}_O = (\bar{w}' - \bar{w})/\bar{w} = 1934.4$, while the change in the mean of u is estimated at $\hat{I}_u = \bar{u}' - \bar{u} = 4.38$. From equation (3.12), the log-likelihood statistic for the difference in dispersion in the samples before and after selection is $\lambda_1 = 22.60$, with 3 degrees of freedom. From equation (3.58) the index of selection on the variance is therefore $\phi_1 = 8.00$. From equation (3.13) the log-likelihood statistic for selection on means is $\lambda_2 = 58.44$ with 2 degrees of freedom. From equation (3.59) the index of selection on means is therefore $\phi_2 = 28.22$.

In interpreting these index values it is desirable to take account of biases and standard errors and these have therefore been determined by computer simulation. In this simulation, 100 random samples of size 55 were generated from a bivariate normal distribution using Bedall and Zimmermann's (1976) algorithm. The bivariate distribution was made to have the same mean vector and covariance matrix as Zaret and Kerfoot's sample before selection. Another 100 random samples of size 44 were generated from a bivariate normal distribution with a mean vector and a covariance matrix the same as for Zaret and Kerfoot's sample after selection. The 100 sample pairs were then used to estimate 100 fitness functions, with corresponding selection indices. In this way, standard errors for the selection indices were calculated. The bias of an index can be defined as its expected value when there is no selection. Therefore to evaluate biases the 100 before-selection samples of size 55 and the 100 after-selection samples of size 44 were all generated from a bivariate normal distribution with a mean vector and a covariance matrix the same as Zaret and Kerfoot's sample before selection.

Table 3.11 Some hypothetical data for Zaret and Kerfoot's (1975) samples of *Bosmina longirostris* taken in the morning and afternoon of 5 May 1969. These data have been coded as X_1 = (BS − 265.0)/22.2, and X_2 = (EPD − 22.2)/1.88, where BS is the body size and EPD is the eye pigmentation diameter, both measured in microns. These data were computer generated from bivariate normal distributions and made to have the same means, variances and covariances as the real samples taken by Zaret and Kerfoot.

Range of X_2		−3.5 to −3.0	−2.5	−2.0	−1.5	−1.0	−0.5	0.0	0.5	1.0	1.5	2.0	2.5	3.0	3.0 to 3.5	Totals
−5.5 to −5.0	b															0
	a				1											1
−4.5	b															0
	a			1		1										2
−4.0	b															0
	a				1											1
−3.5	b															0
	a				1		1	1			1					4
−3.0	b															0
	a					2		1								3
−2.5	b															0
	a				1	2	1	2	1							7

Table 3.11 (Contd.)

Range of X_2		-3.5 to -3.0	-2.5	-2.0	-1.5	-1.0	-0.5	0.0	0.5	1.0	1.5	2.0	2.5	3.0	3.0 to 3.5	Totals
-2.0	b															0
	a		1			1	1				1			1		5
-1.5	b							1		2						3
	a	1				1	1		4			1			1	9
-1.0	b					1	2	1		1						5
	a							1			1					2
-0.5	b			1		1	1	5	2	2	1					13
	a					1							1	1		3
0.0	b				2		1	3	1		1					8
	a	1				1	1	1	1							5
0.5	b				2		1	3	1	3	1					11
	a															0

		1	2	3	4	5	6	7	8	9	10	11	12	13	14	Total
1.0	b							3	2	1	1					7
	a															0
1.5	b						1	1	1							3
	a					1										1
2.0	b					1	1	1								3
	a							1								1
2.5	b								1							1
	a															0
2.5 to 3.0	b									1						1
	a															0
Totals	b	0	0	1	4	3	5	17	11	5	3	1	0	0		55
	a	2	1	4	3	7	3	6	5	2	4	2	2	2	1	44

b = frequency in sample before selection (a.m.)
a = frequency in sample after selection (p.m.)

Table 3.12 Indices of the intensity and effect of selective predation on *Bosmina longirostris*.

Sample date	$\hat{I}_H$	Bias	SE	$\hat{I}_V$	Bias	SE	$\hat{I}_O^*$	$\hat{I}_u$	Bias	SE	$\phi_1^\dagger$	SE	$\phi_2^\dagger$	SE
5 February	1.18	0.97	0.43	2.26	1.93	1.57	0.06	0.09	0.29	0.25	−0.72	1.24	−0.81	1.28
5 May	2.08	0.89	0.51	6.97	1.74	7.61	1934.4	4.38	0.20	1.30	8.01	3.76	28.21	6.34
4 September	1.34	0.85	0.56	2.81	1.56	6.15	8.34	0.90	0.17	0.45	2.69	2.70	6.62	3.87

* Biases and standard errors are not given for $\hat{I}_O$ since simulation shows this is an extremely variable index.
† The biases of ϕ_1 and ϕ_2 are known to be approximately zero when there is no selection from the way that they are calculated.

The simulation results are shown in Table 3.12. This table also shows index values with biases and standard errors for Zaret and Kerfoot's samples taken on 5 February and 4 September where the procedure adopted for calculating the results for these two days was the same as the procedure that has been described for 5 May. Inspection of the table indicates that the indices $\hat{I}_H$, $\hat{I}_V$ and $\hat{I}_O$ are very biased in the situation being considered, and rather variable. Indeed, the simulated estimates of $\hat{I}_O$ are so variable that it makes no sense to even state biases and standard errors. For example, the largest simulated estimate of $\hat{I}_O$ for determining a standard error for the 5 May sample is 3.47×10^{20}, and this completely overwhelms the calculation. The index $\hat{I}_u$ seems the best indication of the change in fitness resulting from selection. It shows little change in February, a very large change in May, and a moderately large change in September. The indices ϕ_1 and ϕ_2 are in agreement with this: there is no indication of selection in February; selection is very strong on both means and dispersion in May; selection is quite strong on means and dispersions in September. The negative values for ϕ_1 and ϕ_2 in February indicate that the differences between the samples before and after selection are less than can be expected because of sampling errors only.

3.13   THE PROBLEM OF GROWTH

In all of the discussions so far in this chapter it has been implicitly assumed that the quantitative variables which describe individuals remain constant. However, in practice many of the variables that are of interest change because of growth. Unfortunately, growth and selection can have similar effects. For example, Bell (1974) has pointed out that variation in human intelligence test scores declines with age, presumably because of genetic effects, and this mimics stabilizing selection. It is a common phenomenon that a population becomes less variable as individuals age, and this has been called developmental canalization (Waddington, 1948).

Perhaps the first attempt to allow for growth when studying selection was due to Weldon (1895). He measured the mean frontal breadth of about 8000 female crabs (*Carcinus maenas*) in units of one thousandths of the carapace length. This form of measurement went a long way towards eliminating the effect of growth although there was a steady decrease in the mean with age. Weldon measured the frequency of crabs with abnormal values for the mean frontal breadth by the quartile deviation from the mean, separately for crabs in eight groups based upon the size of their carapace length. He found that for the smallest crabs the quartile deviation of mean frontal breadth was 9.42. This increased to 10.79 for near-adult crabs and then decreased for adults to 9.96. Thus it appears that the frequency of abnormal crabs first increased with age and then decreased. Weldon assumed that the decrease was due to stabilizing selection and estimated that this resulted in the selective elimination of 77 in every 1000 crabs. Unfortunately, his calculations do not seem very reliable.

The problem is that there is no way to know how the mean frontal breadth of the crabs changed with age. It seems very possible that an increase in variation followed by a decrease has nothing at all to do with selection.

Another approach was adopted by Berry and Crothers (1968). They found that the length divided by the cube root of the weight remains more or less constant with age for the dog whelk *Nucella lapillus*. However, the variance of this character tends to diminish as the size increases. The reduction was greatest in samples from exposed populations and Berry and Crothers concluded that this was due to stabilizing selection. The obvious criticism of this method is the same as it was for Weldon's study: there is no way of knowing what happens to the character being studied in the absence of selection. In a later paper, Berry and Crothers (1970) failed to obtain the same result.

The same criticism applies to a study by Cook (1979). He considered the body length and the leg length of the Madeiran lizard *Lacerta dugesii*. He used the second principal component of these variables as the relative leg length and considered how its variance changed with the first principal component, which was a general measure of size. There was a general decline in variance but again this may just have been due to the effects of growth.

Bell (1974) measured the head-plus-body length (HB) and the total length (TL) for samples from three populations of newts *Triturus vulgaris*. He took TL as a measure of age and calculated the coefficient of variation of HB for newts classified into 0.5 mm intervals of TL. He found that the coefficients of variation decreased with increasing TL. This decrease was not observed in a laboratory population and Bell concluded that it was due to stabilizing selection.

Bell's study is certainly superior to the others that have been mentioned because of the use of laboratory controls. However, the decrease in the coefficient of variation with age can be explained without selection. This can be seen as follows. Figure 5 of Bell's paper shows that to a fair approximation,

$$HB = A \, TL,$$

where A is a constant of about 0.5. If individuals are placed into 0.5 mm classes of TL, then within each of these classes the variance of TL, say V_T, will be approximately constant. Hence within each of these classes the variance of HB, say V_H, will be

$$V_H \simeq A^2 V_T.$$

This means that the coefficient of variation of HB (the standard deviation divided by the mean) will be approximately equal to

$$A V_T^{1/2} / (A \, TL) = V_T^{1/2} / TL$$

in the class corresponding to a total length of TL. This argument shows that the coefficient of variation of HB for one of Bell's total length classes should be roughly inversely proportional to the average total length in the class. In particular, the coefficient of variation should decrease with total length, which is just what Bell observed.

Actually, things are a little more complicated than that. A reduction in the coefficient of variation was not found with the laboratory population. Indeed, when Bell made up classes of newts in the laboratory population based on their real ages he found that the coefficient of variation increased with age. It is this increase, combined with the relationship $V_H \simeq A^2 V_T$, that seems to have resulted in a fairly constant coefficient of variation for classes based upon total length. Whatever is the true explanation, it does seem that the difference between the laboratory population and the field population may be taken as evidence of selection.

Parker (1971) also used controls to investigate selection under somewhat artificial conditions. He set up 11 aquariums each containing about 500 juvenile pink (*Oncorhynchus gorbuscha*) and chum (*O. keta*) salmon. Seven of the aquariums also contained three juvenile coho (*O. kisutch*) salmon to prey on the pinks and chums. The aquariums were left for about 16 days, by which time the predators had eaten many of the prey. In this example a comparison between what happened in the four control aquariums and what happened when predators were present allows the selectivity of the predators to be studied. The growth of the prey is not particularly important since it should have been about the same in all the aquariums. Using regression methods, Parker showed that the mean size of fish increased more in the populations subjected to predation than in the populations without predators. Also, the coefficient of variation decreased more in the populations exposed to predation than in others. The experiments therefore demonstrated directional and stabilizing selection.

With fish, 'Lee's phenomenon' is also usually taken to be evidence of selection related to size. The idea here is that the age of fish can be determined from growth marks on their scales. Their lengths at earlier ages can then be estimated from a known relationship between total lengths and scale lengths. This method for studying the growth of fish has been practised for many years (Jones, 1958; Ricker, 1969). When it is carried out on fish of various ages it is often found that the back-calculated lengths of the oldest fish are less than the equivalent lengths for younger fish. This is Lee's phenomenon, which is probably due to selection against the largest fish of a given age.

It seems fair to sum up this section on the problem of growth with three comments:

1. When measurements $X_1, X_2, \ldots, X_P$ are available which change as animals age, it may be possible to find indices that remain constant, for example $X_1/\Sigma X_i$. In doing this, it is not sufficient to demonstrate that the mean value of an index does not change with age. It must be shown that the index remains constant for individuals. This will not be possible from field samples alone. Even if a suitable index can be constructed, the study of it

may not help very much in determining whether or not there is any selection on the original variables $X_1, X_2, \ldots, X_P$.

2. Laboratory controls are essential to determine how measurements change in the absence of selection.

3. More research is needed on appropriate statistical models for selection on measurements that change with time. A start in this direction has been made by McGilchrist and Simpson (1979).

4 Comparison of live and dead animals

In the previous chapter the concern was with the comparison of samples of live animals taken after selection has operated for various amounts of time on a population. The idea was that selective survival will cause population changes that can be detected from the samples. Another similar type of approach to studying selection involves sampling dead animals and seeing how their morph distribution compares with that of live animals. In this case the idea is that if a morph is proportionately more frequent among the dead animals than among the live ones, then this morph is being selected against.

This chapter begins with the description of several examples of experiments that have involved the comparison of live and dead animals. These indicate five different situations that can arise according to whether or not the population size is decreasing appreciably during the sampling period, whether the morph distribution in the initial population is known exactly or only estimated from a sample, and whether variation is polymorphic or quantitative. Sections 4.2 to 4.7 cover these various cases in turn. The final section in the chapter considers the question of how the intensity and effect of selection can be measured from a comparison of morph distributions among live and dead animals.

4.1 COMPARING DISTRIBUTIONS FOR LIVE AND DEAD ANIMALS

The situations considered in this chapter occur when selection is to be detected and measured on the basis of random samples of dead animals from a population. The basic idea is that selective mortality can be inferred if samples of dead animals contain a distribution either of morphs or of certain quantitative variables that differs from the distribution in the population of live animals. To make this idea clearer it is useful to consider some examples.

The first concerns predation of snails by thrushes. It is known that thrushes break snail shells on stone 'anvils' so that they can eat the soft parts. The remains of shells can then be picked up and these give a sample of dead snails which can be compared with the population being eaten. This idea was first used early this century (Trueman, 1916; Haviland and Pitt, 1919). More recently, Sheppard (1951a) used it to test whether thrushes choose different colours of snail at random. He collected 1358 *C. nemoralis* snails from various locations, marked them with cellulose paint, and scattered them at random near some thrush anvils in Ten Acre Copse, Wytham Woods, England, on 26 April 1950. He then collected the broken marked shells from around the anvils on 14 occasions over the following two months.

There were 747 yellow and 611 pink and brown snails in the marked population. Table 4.1 shows the colours that were found for broken shells. Initially the proportion of broken yellow shells was quite high but it diminished as the experiment progressed. Sheppard related this change to changes in the colour of background vegetation. He suggested that yellow snails were initially at a disadvantage compared with pink and brown snails but that this turned to an advantage as the experiment progressed.

Whilst carrying out his Ten Acre Copse experiment, Sheppard also recorded information on broken shells at some thrush anvils in Marley Wood. In this case he did not have a known population of marked shells. Rather, at about the beginning of the experiment (on 14 April) he took a sample of the live snails from the area of the anvils. This can be used to indicate the colour composition of the population and the samples of broken shells can be compared with it. Sheppard took a further sample from the population at the end of the experiment (on 26 May) to see whether the population remained stable over

Table 4.1 Collections at Ten Acre Copse of broken snail shells from a population consisting of 747 yellow and 611 pink and brown *Cepaea nemoralis* on 26 April 1950.

Date	Day number	Broken shells collected		
		Pink and brown	Yellow	Total
28 April	2	0	2	2
1 May	5	0	1	1
2	6	1	1	2
5	9	1	3	4
8	12	3	9	12
11	15	0	1	1
12	16	7	4	11
16	20	0	1	1
17	21	0	1	1
20	24	1	0	1
22	26	1	1	2
30	34	0	2	2
3 June	38	2	1	3
5	40	3	0	3
	Totals	19	27	46

Table 4.2 Results from an experiment on *Cepaea nemoralis* in Marley Wood. This was different from the Ten Acre Copse experiment because the colour composition of the prey population was only approximately known from the samples taken on 14 April and 26 May. The experiment began on 6 April when the thrush 'anvils' were cleared of all broken shells.

| | | | Shells collected | | |
| | | | Pink and | | |
Date	Day Number	Sample type	brown	Yellow	Total
11 April	5	Broken shells	4	3	7
14	8	Live snails	250	80	330
23	17	Broken shells	10	7	17
30	24	Broken shells	21	11	32
7 May	31	Broken shells	25	9	34
19	43	Broken shells	16	3	19
22	46	Broken shells	6	1	7
26	50	Broken shells	12	2	14
26	50	Live snails	147	57	204

the experimental period. Table 4.2 shows the results of this Marley Wood experiment. The proportion of broken yellow shells decreased with time, just as it did with the marked population at Ten Acre Copse.

Another experiment that involved comparing a sample of dead individuals with a sample of live ones is described by Wong and Ward (1972). They sampled *Daphnia publicaria* in the stomachs of yellow perch (*Perca flavescens*) fry and also in plankton in West Blue Lake, Manitoba, on several occasions over the period 1 July to 25 August 1969. The size distributions that they found are shown in Table 4.3. It can be seen that the stomach and plankton samples had rather different distributions, with a relationship that apparently changed with time. Wong and Ward explain this in terms of changes in the relative sizes of the predators and the prey.

In the three examples that have been considered so far it seems safe to assume that only a relatively small proportion of the animals in the initial population will have died while the dead animals were being collected. Thus for any analysis, the population from which the dead animals came can be treated as having been constant and this will be a fair approximation. The situation becomes more complicated when dead animals are collected for such a long period of time that the population of live animals is reduced to only a small fraction of its initial size. This is what happens with bird banding experiments.

Thus, Table 4.4 shows the recoveries of dead Dominican gulls (*Larus dominicanus*) that were banded on Somes Island, New Zealand. These data

Table 4.3 Distribution of the lengths of *Daphnia publicaria* in plankton samples and in the stomachs of perch fry at West Blue Lake. This table was constructed from Fig. 1 of Wong and Ward (1972) and might therefore contain some small errors.

Length (mm)	1 July		15 July		29 July		12 August		25 August	
	Plankton	Stomach	Plankton	Stomach	Plankton	Stomach	Plankton	Stomach	Plankton	Stomach
0.5–0.7	20	59	28	20	2	0	1	0	6	27
0.7–	22	84	49	40	11	12	2	0	2	42
0.9–	20	154	59	101	21	61	7	34	2	124
1.1–	18	138	62	126	33	95	9	127	0	138
1.3–	26	44	46	146	59	172	17	230	3	261
1.5–	24	10	33	60	31	233	28	241	12	303
1.7–	22	5	28	2	24	168	14	218	35	604
1.9–	24	0	33	5	22	78	12	218	63	606
2.1–	26	0	13	2	16	21	4	92	36	289
2.3–	16	0	13	2	11	9	6	34	15	193
2.5–	11	0	7	0	7	1	4	11	5	55
2.7–	7	0	7	0	2	0	1	5	0	58
2.9–3.1	1	0	2	0	1	0	0	6	0	0
Total	237	494	380	504	240	850	105	1216	179	2710
Mean	1.56	1.03	1.39	1.23	1.58	1.56	1.66	1.70	1.94	1.82
Std. dev.	0.59	0.25	0.55	0.29	0.47	0.31	0.44	0.36	0.38	0.43

Table 4.4 Records of the year of death for Dominican gulls banded as chicks on Somes Island, New Zealand.

Banding season	1961–62	1962–63
Number banded	574	728
Recovered after		
1 year	16	20
2 years	10	12
3 years	10	4
4 years	6	5
5 years	7	5
6 years	5	—

were extracted from Table 6 of Fordham (1970). It will be seen that in the season 1961–62 a total of 574 chicks were banded and of these there are 54 for which the year of death is known because the bands were recovered. Likewise in the season 1962–63, 728 chicks were banded and the year of death is known for 46 of these. For the 1961–62 bandings recovery data are available for six years, while for the 1962–63 bandings data are only available for five years. The purpose of an analysis might be to determine whether the survival was the same for the 1961–62 and 1962–63 cohorts. The important thing to note with the data in Table 4.4 is that the returns of bands decreased steadily as the number of years after banding increased. This is mainly due to the reduction in the number of birds still alive at the start of each year for successive years after banding. Obviously any analysis of data will have to take this factor into account.

Another bird banding example is shown in Table 4.5. This is for the recoveries of dead starlings (*Sturnus vulgaris*) as recorded by Coulson (1960). In this case results are available separately for males and females for five years of banding and it is interesting to consider whether there was a difference in survival between the two sexes. This example differs from the previous one in that now it is not known how many birds were banded in different years.

Of course these bird banding examples have not got much to do with natural selection. They are only used here as illustrations of certain types of data that can occur. The two years of banding shown in Table 4.4 or the two sexes shown in Table 4.5 could equally well be two phenotypes in a polymorphic population. The methods of analysis presented below would still be relevant.

The above examples suggest that there are five different situations that need to be considered for data involving the returns of dead animals, as follows.

(a) The population from which the deaths come is very large in comparison with the number of deaths during the experimental period. This means that

Table 4.5 Band returns from dead starlings, with separate results for males and females. For example, this table shows that of the males banded in 1950, 37 deaths were recorded in 1951, 26 were recorded in 1952, and so on.

Year of banding	Returns in years after banding								Total
	1	2	3	4	5	6	7	8	
Males									
1950	37	26	15	8	2	0	1	0	89
1951	40	32	20	9	4	3	1		109
1952	24	13	7	3	5	1			53
1953	23	14	5	3	1				46
1954	15	9	5	10					39
Females									
1950	36	26	7	7	3	0	0	0	79
1951	29	19	6	4	3	4	2		67
1952	25	7	9	5	2	3			51
1953	16	6	1	2	0				25
1954	22	10	7	3					42

the counts of dead animals do not change appreciably due to a reduction in the population size. The population consists of $K (\geqslant 2)$ different morphs and it is desired to estimate their relative mortality rates. To do this the proportions of different types of dead animals can be compared with the proportions in the population, which are known exactly. Sheppard's Ten Acre Copse experiment yielded this type of data (Table 4.1).

(b) This is the same as (a) except that the proportions of different types of animal in the population are only known from random sampling of live individuals. Interest still centres on the estimation of relative survival rates. Sheppard's Marley Wood experiment provides this type of data (Table 4.2).

(c) This is the same as (a) except that now the population decreases through death at an appreciable rate. The initial numbers of different morphs are known and the data available for analysis consist of counts of dead animals in different time periods. This is the situation with the Dominican gull data (Table 4.4) if time is measured for each animal from the year of banding, and 'morphs' are defined according to their year of banding. In this case it is possible to estimate an absolute survival rate for each morph.

(d) This is like (c) except that the initial morph numbers are unknown. This is the situation with the starling data (Table 4.5). It is still possible to estimate an absolute survival rate for each morph.

(e) The population from which the deaths come is very large compared with the number of deaths. Each individual in the population is characterized by the values that it possesses for the variables X_1, X_2, ..., X_p. The multivariate distribution of the X's in the population is estimated from a random sample and this can be compared with the distribution in a random sample of dead individuals. The problem is to see how the survival varies for individuals with different values for the X's. Wong and Ward's data on predation of *Daphnia publicaria* (Table 4.3) give an example of this type of situation with one X variable only (length).

In all five cases it is important to note that permanent emigration is considered as being equivalent to death. None of the models discussed in this chapter are appropriate if there is an appreciable amount of temporary emigration.

Another factor which must be borne in mind is that a sample of dead animals will usually only contain individuals that have died from one or more specific causes. For example, in Sheppard's experiments at Ten Acre Copse and Marley Wood, only thrush-predated snails were recorded. It is obviously possible that these were survivors from other causes of mortality that had even more effect on the population. Studies comparing samples of dead and live animals involve the implicit assumption that causes of death for which animals cannot be recovered are of relatively minor importance compared with the causes for which recovery is possible.

4.2   SAMPLES FROM A LARGE POPULATION WITH KNOWN MORPH PROPORTIONS

The analysis of data from situation (a) will be considered to begin with. It will be assumed that the dead animals come from a population in which there are K morphs with known relative frequencies P_1, P_2, ..., P_K. It will also be assumed that the samples of dead animals are accumulated over n periods of time, and that d_{ij} dead morph i individuals are recovered in the interval (t_{j-1}, t_j), which is of length l_j. Thus the notation is as shown in Table 4.6.

It is possible to set up a Poisson log-linear model for this case. Think of a morph i animal that is alive at the start of the jth time interval. The probability of surviving the interval may be taken to be of the form $\exp(-\lambda_i l_j)$, where λ_i is a parameter specific to the ith morph, and $\lambda_i l_j$ is small. The probability of this animal dying in the interval must be small because it is being assumed that mortality has negligible effects on the population being selected from. Hence the probability of death is $1 - \exp(-\lambda_i l_j) \simeq \lambda_i l_j$. It then follows that the expected number of morph i animals found dead during the jth time interval is of the form $\gamma \lambda_i l_j P_i B$, where B is the total population size, so that $P_i B$ is the number of morph i animals at risk, and γ is the probability of an animal being recovered if it dies. On this basis the expected values of the d_{ij} values can be written as follows (allowing γ and λ_i to vary with time):

$$E(d_{ij}) = \gamma_j \lambda_{ij} l_j P_i B \qquad (4.1)$$

Table 4.6 The general form of experiment when samples of dead animals are collected from a population.

	Type of animal			
	1	2	...	K
Population relative frequency	P_1	P_2	...	P_K
Dead animal sample from period (t_0, t_1)	d_{11}	d_{21}	...	d_{K1}
Dead animal sample from period (t_1, t_2)	d_{12}	d_{22}	...	d_{K2}
$\vdots$	$\vdots$	$\vdots$		$\vdots$
Dead animal sample from period (t_{s-1}, t_s)	d_{1s}	d_{2s}	...	d_{Ks}

The jth sampling period is of length $l_j = t_j - t_{j-1}$.

This can then be rewritten as

$$E(d_{ij}) = \exp\{\alpha_j + \pi_{ij} + \log_e(l_j P_i) + \log_e B\}, \qquad (4.2)$$

where $\gamma_j = \exp(\alpha_j)$ and $\lambda_{ij} = \exp(\pi_{ij})$. It is possible that the γ_j values, and hence the α_j values, are the same for all j. Selection is indicated by varying values for $\pi_{1j}, \pi_{2j}, \ldots, \pi_{Kj}$.

The model of equation (4.2) can be fitted to data by the computer program GLIM (Appendix, Section A.10). Other programs for fitting log-linear models to frequency data could also be used. It has to be assumed that the d_{ij} values follow Poisson distributions about their expected values. This is a reasonable assumption for the present application where there is a source population of animals and a small probability that any particular animal will die in any time interval. In fitting equation (4.2) using GLIM it is necessary to use the OFFSET directive in order to allow for the known value of $\log_e(l_j P_i)$.

An approach that is simpler than fitting a log-linear model involves noting from equation (4.1) that

$$\{E(d_{ij})/P_i\} \bigg/ \sum_{r=1}^{K} \{E(d_{rj})/P_r\} = \lambda_{ij} \bigg/ \sum_{r=1}^{K} \lambda_{rj} = \beta_{ij} \qquad (4.3)$$

where β_{ij} is a relative death rate for type i animals for the time interval t_{j-1} to t_j. Thus β_{ij} can be estimated by

$$\hat{\beta}_{ij} = (d_{ij}/P_i) \bigg/ \sum_{r=1}^{K} (d_{rj}/P_r). \qquad (4.4)$$

On the assumption that the d_{ij} are independent Poisson variates, the biases, variances and covariances of the β_{ij} are then found by the Taylor series method (Appendix, Section A.5) to be as follows:

$$\text{bias}(\hat{\beta}_{ij}) \simeq \beta_{ij} \sum_{r=1}^{K} \alpha_{rj} - \alpha_{ij}, \qquad (4.5)$$

$$\text{var}\,(\hat{\beta}_{ij}) \simeq \beta_{ij}^2 \sum_{r=1}^{K} \alpha_{rj} + (1 - 2\beta_{ij})\alpha_{ij}, \tag{4.6}$$

and

$$\text{cov}\,(\hat{\beta}_{ij}, \hat{\beta}_{kj}) \simeq \beta_{ij}\beta_{kj} \sum_{r=1}^{K} \alpha_{rj} - \beta_{ij}\alpha_{kj} - \beta_{kj}\alpha_{ij}, \tag{4.7}$$

for $i \neq k$, where $\alpha_{ij} = \beta_{ij}^2/E(d_{ij})$.

Another possibility with $K = 2$ morphs is to carry out a logit or probit analysis to test for a trend in the proportions of the different types returning. This was the approach that was adopted by Sheppard (1951a).

Example 4.1   Dead snails at Ten Acre Copse

As an example consider Sheppard's data on predation of *Cepaea nemoralis* by thrushes at Ten Acre Copse (Table 4.1). In this case $K = 2$, corresponding to two colour classes of snails, with $s = 14$ samples of dead snails.

Equation (4.2) was fitted to the data using GLIM. Various models were tried, as follows:

(a) It was assumed that the α_j values (the recovery rate parameters) and the π_{ij} values (the mortality rate parameters) were constant throughout the experiment. This gives a very poor fit to the data. The deviance is 84.78 with 27 degrees of freedom which is significantly large at the 0.1 % level when compared with chi-square tables.

(b) It was assumed that the α_j values were not equal but the π_{ij} values were. This model gives a fair fit to the data. The deviance is 20.83 with 14 degrees of freedom which is nearly significantly large at the 5 % level.

(c) It was assumed that the α_j values were not equal and the π_{ij} values varied with the two colour classes of snail but were constant over time for snails in the same class. This gives about the same fit as model (b) having a deviance of 20.58 with 13 degrees of freedom.

(d) Finally, it was assumed that the α_j values were not equal and the π_{1j} values changed linearly with time relative to the π_{2j} values, so that there is a relationship of the form

$$\pi_{1j} = C_1 + C_2 t_{j-1} + \pi_{2j}$$

where C_1 and C_2 are constants and t_{j-1} is time in days. This model gives considerably better fit than models (a) to (c) with a deviance of 14.31 with 12 degrees of freedom.

(The assumption of a linear change in π_{1j} relative to π_{2j} with time that is incorporated in model (d) is somewhat arbitrary. It has been made because this is the simplest way of allowing for a changing relative survival rate with the program GLIM.)

Model (d) seems appropriate for the data. It gives the expected number of pink and brown shells in the jth sample to be of the form

$$\exp\left[\alpha_j - 9.23 + 0.079t_{j-1} + \log\{P_1(t_j - t_{j-1})\}\right]$$

while the expected number of yellow shells is

$$\exp\left[\alpha_j - 7.78 + \log\{P_2(t_j - t_{j-1})\}\right].$$

Equation (4.3) then gives relative death rates of

$$\beta_{1j} = \exp(-1.36 + 0.079t_{j-1})/\{1 + \exp(-1.36 + 0.079t_{j-1})\},$$

and $\beta_{2j} = 1 - \beta_{1j}$, for the two colour classes. This says that at the start of Sheppard's experiment ($t_{j-1} = 0$) the β_{1j} value was 0.20, while at the end of the experiment ($t_{j-1} = 40$) it was 0.86. In other words, at the start of the experiment the death rate was four times higher for yellow snails than for pink and brown snails, while by the end of the experiment pink and brown snails had a death rate about six times higher than that of yellow snails. This is a very large change, which Sheppard attributed to changes in the colour of background vegetation.

An alternative to this analysis with GLIM would have been to estimate the β_{ij} values directly using equation (4.4) and then use multiple regression to relate changes in these values to time. However, this would give a rather unsatisfactory analysis with this particular example because the small numbers of dead animals would make the individual β_{ij} estimates very unreliable. Indeed, the small number of dead animals means that the procedure of testing the GLIM deviances against the chi-square distribution needs to be regarded with some caution too. Strictly speaking, this procedure is also only valid for large expected frequencies $E(d_{ij})$.

Sheppard's experimental procedure at Ten Acre Copse has been used by several workers since 1951. There seems little doubt that thrushes practise selective predation on the shell colour of C. nemoralis. However, Sheppard's conclusion that a changing preference is related to the colour of background vegetation has been questioned. See Jones et al. (1977) and Clarke et al. (1978) for reviews of other experiments.

4.3 SAMPLES OF DEAD ANIMALS COMPARED WITH A POPULATION SAMPLE

We can now consider the same situation as in the previous section except that the relative morph frequencies $P_1, P_2, \ldots, P_K$ in the population are now not known exactly. Instead, a random sample from the population is available for estimating them.

Equation (4.2) will still hold for the samples of dead animals. However, it must be written as

$$E(d_{ij}) = \exp\{\alpha_j + \pi_{ij} + \delta_i + \log_e(l_j) + \log_e(B)\}, \tag{4.8}$$

say, where $\delta_i = \log_e(P_i)$, which is now unknown. As before, α_j takes into account the probability of a dead animal being recorded during the period of the collection of the jth sample of dead animals, π_{ij} reflects the death rate for the ith morph, and l_j is the time period covered by the jth sample of dead animals.

For a sample from the population of live animals, the frequencies of the different morphs should be approximately proportional to the population relative frequencies $P_1, P_2, \ldots, P_K$. Hence if a_i denotes the frequency of morph i animals in the population sample, then

$$E(a_i) = \theta P_i,$$
$$= \exp\{\log_e(\theta) + \log_e(P_i)\},$$
$$= \exp(\phi + \delta_i), \qquad (4.9)$$

where $\phi = \log_e(\theta)$, with θ being a parameter which takes into account the sample size. Between them, equations (4.8) and (4.9) provide a log-linear model for the data. If the population of live animals is much larger than any of the samples, and this population changes little because of mortality during the experimental period, then it will be reasonable to assume that the observed d_{ij} and a_i values will have independent Poisson distributions, so that the model can again be fitted by GLIM, or any other computer program for log-linear models.

Alternatively, equation (4.4) can be modified to

$$\hat{\beta}_{ij} = (d_{ij}/a_i) \Big/ \sum_{r=1}^{K} (d_{rj}/a_r) \qquad (4.10)$$

to allow for estimated population relative frequencies $a_1, a_2, \ldots, a_K$ for the different types of animal. Equations (4.5) to (4.7) then become

$$\text{bias}(\hat{\beta}_{ij}) \simeq \beta_{ij} \sum_{r=1}^{K} \chi_{rj} - \chi_{ij} \qquad (4.11)$$

$$\text{var}(\hat{\beta}_{ij}) \simeq \beta_{ij}^2 \sum_{r=1}^{K} \chi_{rj} + (1 - 2\beta_{ij})\chi_{ij}, \qquad (4.12)$$

and

$$\text{cov}(\hat{\beta}_{ij}, \hat{\beta}_{kj}) \simeq \beta_{ij}\beta_{kj} \sum_{r=1}^{K} \chi_{rj} - \beta_{ij}\chi_{kj} - \beta_{kj}\chi_{ij}, \qquad (4.13)$$

for $i \neq k$, where $\chi_{ij} = \beta_{ij}^2 \{1/E(d_{ij}) + 1/E(a_i)\}$. These equations have been derived on the assumption that the d_{ij} and a_i values are Poisson distributed, using the Taylor series method.

Example 4.2   Dead snails in Marley Wood

The model of equations (4.8) and (4.9) can be fitted to Sheppard's Marley Wood data (Table 4.2). Here there are seven samples of broken shells covering a period of 50 days. There are also two samples from the population. Because the live population is assumed to have been constant during the experimental period, the two samples of live snails can be lumped together. (A chi-square test for a difference between these two samples gives a test statistic of 0.92 with one degree of freedor . This non-significant result confirms that the population of live snails was more or less constant.) In analysing the data it is of course necessary to assume that the samples of live shells are random samples, although there is some doubt whether humans can take random samples of different colours of snails (Cain and Currey, 1968).

Three versions of the model were fitted to the data using GLIM. The OFFSET directive was used with the d_{ij} values to take into account the known values for $\log_e (l_j)$ in equation (4.8). The models fitted were as follows:

(a) The model with all $\pi_{ij} = 0$ implies that the different coloured snails do not have equal population frequencies but that there is no selection by the thrushes. This model gives a good fit to the data, with a deviance of 7.65 with 8 degrees of freedom.

(b) As for (a) but with $\pi_{ij} = \pi_i$ in equation (4.8) for the samples of dead snails. This is still a good fit, with a deviance of 7.43 with 7 degrees of freedom. However, the reduction in deviance in moving from model (a) to model (b) is only 0.22 which is not significantly large. Therefore there is no evidence that different π_i values for different morphs are needed in the model. It is perhaps worth noting that this model can be defined for the program GLIM by using three factors: a sample factor at one level for each of the nine samples, a type factor at two levels for the two colours of snail, and a type of sample factor at two levels for dead snails and live snails. Model (b) is then defined in terms of these three factors plus a time variable.

(c) This model is the same as (b) except that a linear time trend in π_{ij} values is allowed. The fit to the data is excellent, with a deviance of 1.44 with 6 degrees of freedom. The reduction in deviance in moving to this model from (b) is 5.99 with 1 degree of freedom, which is significantly large. There is therefore evidence of a trend in the π_{ij} values during the experimental period.

Model (c) seems best. This is of the same form as the model fitted previously to the Ten Acre Copse data. It says that the expected frequency for pink and brown snails in a population sample has the form

$$E(a_1) = \exp (\pi + 5.02),$$

while for yellow snails the expected frequency is

$$E(a_2) = \exp (\pi + 3.96).$$

On the other hand, the expected frequencies for broken shells are

$$E(d_{1j}) = \exp \{\alpha_j + 5.02 - 0.94 + 0.0387 t_{j-1} + \log (l_j)\}$$

for pink and brown, and

$$E(d_{2j}) = \exp\{\alpha_j + 3.96 + \log(l_j)\}$$

for yellow, where t_{j-1} is the time in days since the start of the experiment. This means that according to equation (4.3) the relative death rates of pink and brown snails to yellow snails were in the ratio of β_{1j} to β_{2j}, where

$$\beta_{1j} = \exp(-0.94 + 0.0387t_{j-1})/\{1 + \exp(-0.94 + 0.0387t_{j-1})\}$$

and $\beta_{2j} = 1 - \beta_{1j}$.

At the start of the experiment ($t_{j-1} = 0$), this gives $\beta_{1j} = 0.28$, so that a pink and brown snail was rather less likely to be eaten than a yellow snail. At the end of the experiment ($t_{j-1} = 50$), $\beta_{1j} = 0.73$, so that the reverse was true. Thus, the Marley Wood results confirm the trend found at Ten Acre Copse of an initial advantage for pink and brown snails that changed to a disadvantage as time went on, possibly due to changes in the colour of the background vegetation.

For these data an analysis based upon equation (4.10) is possible because of the reasonable numbers of broken shells collected. The live samples consisted of $a_1 = 397$ pink and brown and $a_2 = 137$ yellow snails. Pooling the collections of broken shells to a certain extent gives the following table:

Collection period	d_1	d_2
6–17 April	14	10
18–24 April	21	11
25 April–7 May	25	9
8–26 May	34	6

Here d_1 is the number of pink and brown and d_2 is the number of yellow shells. For 6–17 April, equation (4.10) then provides

$$\hat{\beta}_{11} = (14/397)/(14/397 + 10/137) = 0.33$$

for the estimated relative death rate of pink and brown snails, and $\hat{\beta}_{21} = 1 - \hat{\beta}_{11} = 0.67$ for the estimated relative death rate of yellow snails. Using these estimates and the observed values of a_i and d_{ij} in equation (4.11) gives bias$(\hat{\beta}_{11}) = -$ bias$(\hat{\beta}_{21}) \simeq 0.01$. Equation (4.12) gives var$(\hat{\beta}_{11}) = $ var$(\hat{\beta}_{21}) \simeq 0.00886$ so that standard errors are SE$(\hat{\beta}_{11}) = $ SE$(\hat{\beta}_{21}) \simeq 0.09$. Carrying out similar calculations for all the collection periods gives the results:

Collection period	j	$\hat{\beta}_{1j}$	Bias	SE
6–17 April	1	0.33	0.01	0.09
18–24 April	2	0.40	0.01	0.09
25 April–7 May	3	0.49	0.01	0.10
8–26 May	4	0.66	0.01	0.10

Clearly bias is not an important problem. The estimates here agree very well with those from the log-linear model analysis. An initial low relative death rate for pink and brown snails was converted over the experimental period into a high rate.

Clarke *et al.* (1978) have reviewed other experiments that have been made that are similar to the one done by Sheppard at Marley Wood. They point out that the difficulty with these experiments is in establishing the morph proportions in the population being predated. A small difference in the areas sampled by the experimenter and the thrushes can give spurious evidence of selection. Also, if thrushes change the area where they hunt then this can give the impression of changing selection.

4.4 ANALYSES DEVELOPED FOR BIRD BANDING DATA

Consider next the type of situation that is illustrated by the Dominican gull data (Table 4.4) where the decreasing numbers of dead animals as time goes on is due mainly to the declining numbers of live animals. Consider a single morph and suppose that at time $t_0 = 0$ there are A of this morph in a population. Let the probability of surviving the period from time t_{j-1} to time t_j be ϕ_j for all animals still alive at time t_{j-1}. Then the probability of an animal surviving until time t_{j-1}, dying in the interval t_{j-1} to t_j, and its death being recorded, is

$$\alpha_j = \phi_1 \phi_2 \ldots \phi_{j-1}(1 - \phi_j)\gamma_j$$

where γ_j is the probability of the death being recorded. It then follows that the expected number of dead animals recorded for the time period (t_{j-1}, t_j) is

$$E(d_j) = A\phi_1\phi_2 \ldots \phi_{j-1}(1 - \phi_j)\gamma_j. \tag{4.14}$$

This equation has been the basis for many of the models that have been used for bird banding data with the sample counts $d_1, d_2, \ldots, d_s$ being assumed to follow a multinomial distribution for the birds banded in one year. There is then a separate multinomial probability for each year of banding and the likelihood function for the full set of data is obtained by multiplying these yearly probabilities together.

Explicit formulae for maximum likelihood estimates of survival and recording probabilities are only available for a few situations. Seber (1970) gives formulae for the case when ϕ_j and γ_j are time-specific so that the probability of survival and the probability of recording a death are the same for all birds in any particular calendar period, irrespective of when they were banded. Seber (1971) has also found explicit estimators of survival probabilities on the assumption that these are age-specific and that the recovery probability remains constant over time, but these are known to have rather poor properties. North and Cormack (1981) have considered various modifications to Seber's (1971) approach that improve matters somewhat. One possibility is to follow Cormack (1970) and assume that the survival probability becomes constant for birds over a certain age. Explicit formulae are then available for survival estimators based upon the recaptures from a single batch of releases (see Section 4.6 below). Note, however, that Lakhani and Newton (1983) have demonstrated that this procedure is unreliable unless sampling is continued until nearly all birds are dead. (Indeed, Lakhani and Newton go as far as to suggest that all published estimates of age-specific survival rates based upon only the recoveries of dead birds are liable to be untrustworthy.)

Brownie *et al.* (1978) have produced a comprehensive handbook on the analysis of bird banding data, including details of computer programs that they have written for carrying out calculations. Allowance is made for animals of different ages to have different survival rates and also for survival and recovery rates to vary from year to year. Brownie *et al.* were mainly concerned with the analysis of band recoveries from birds shot by hunters. However, their computer programs can be used to analyse data coming from natural deaths with $(1 - \phi_j)\gamma_j$ of equation (4.14) replaced by a new parameter f_j, where this can be interpreted as the probability of a death being recorded in the jth year for an individual alive at the start of the year.

By using one of the estimation procedures developed for bird banding data it is possible to estimate a separate survival probability for each morph of interest. These can then be compared using methods that have been discussed in Sections 2.3 and 2.13.

4.5 A LOG-LINEAR MODEL FOR A DECREASING POPULATION

Log-linear models can be used for many sets of data on recoveries of dead animals from a decreasing population. It usually happens that only a small fraction of a population is recorded as dead so that the numbers recorded in different time intervals can reasonably be assumed to follow independent Poisson distributions.

An analysis starts with the assumption that the expected deaths recorded for morph i in the time interval (t_{j-1}, t_j) is

$$E(d_{ij}) = A_i \phi_{i1} \phi_{i2} \cdots \phi_{ij-1} f_{ij}$$

where A_i is the number of this morph at time $t_0 = 0$, ϕ_{ir} is the survival probability for the time period (t_{r-1}, t_r), and f_{ij} is the probability of an animal alive at time t_{j-1} being found dead in the interval (t_{j-1}, t_j). This can then be rewritten as

$$E(d_{ij}) = \exp\{-\lambda_{ij-1} t_{j-1} + \delta_{ij} + \log_e(A_i)\}, \tag{4.15}$$

where $\exp(-\lambda_{ij-1})$ is the survival rate per unit time over the period $(0, t_{j-1})$

and $\delta_{ij} = \log_e(f_{ij})$. This model can be fitted to data using the program GLIM. The OFFSET directive is needed to account for the known value of $\log_e(A_i)$.

There are three versions of model (4.15) that are of particular interest here. In the first, it can be assumed that all morphs have the same survival and recovery probabilities: $\lambda_{ij} = \lambda$ and $\delta_{ij} = \delta$. This is a no-selection model. In the second, it can be assumed that all morphs have the same survival probabilities but recovery probabilities vary: $\lambda_{ij} = \lambda$ and $\delta_{ij} = \delta_i$. This is another no-selection model. In the third, it can be assumed that survival and recovery probabilities vary from morph to morph: $\lambda_{ij} = \lambda_i$ and $\delta_{ij} = \delta_i$. This is a selection model.

If none of these three models seems correct, then it must be assumed that the true situation is more complicated. Perhaps the recovery probabilities vary with time. Alternatively, maybe the survival probabilities vary with time. The second of these possibilities often occurs because juveniles have lower survival rates than adults. This can be taken into account by assuming that for morph i the survival probability per unit time is $\exp(-\lambda_{1i})$ for the interval $(0, t_1)$ and $\exp(-\lambda_{2i})$ from time t_1 on. Then equation (4.15) can be rewritten as

$$
\begin{aligned}
E(d_{i1}) &= \exp\{\delta_{i1} + \log_e(A_i)\} \\
\text{and} \quad E(d_{ij}) &= \exp\{-\lambda_{i1}t_1 - \lambda_{i2}(t_{j-1} - t_1) + \delta_{i2} + \log_e(A_i)\} \\
&\quad \text{for } j \neq 1.
\end{aligned}
\quad (4.16)
$$

Note that there have to be two δ parameters since the recovery probability f_{i1} for the period $(0, t_1)$ must be different from the recovery probability f_{i2} in subsequent periods. In practice it will only be possible to estimate $\delta_{i2} - \lambda_{i1}t_1$ as a single parameter.

Variation with time in recovery rates can also be handled in some cases without too much difficulty. Thus, suppose that equation (4.15) holds with $\lambda_{ij} = \lambda_i$, so that the survival probability per unit time is constant. Then it can be assumed that $\delta_{ij} = \tau_i + \alpha_j$, say, where α_j accounts for changes in recovery probabilities from year to year, while τ_i accounts for morph differences in death rates. The model then becomes

$$
E(d_{ij}) = \exp\{-\lambda_i t_{j-1} + \tau_i + \alpha_j + \log_e(A_i)\}. \quad (4.17)
$$

There is, however, a certain difficulty here due to a relationship that exists between λ_i and τ_i. It will be recalled that $\delta_{ij} = \log_e(f_{ij})$, where f_{ij} itself can be written as

$$
f_{ij} = (1 - \phi_{ij})\gamma_j. \quad (4.18)
$$

Thus

$$
\begin{aligned}
\delta_{ij} &= \log_e(1 - \phi_{ij}) + \log_e(\gamma_j) \\
&= \log_e[1 - \exp\{-\lambda_i(t_j - t_{j-1})\}] + \log_e(\gamma_j)
\end{aligned}
$$

since $\phi_{ij} = \exp\{-\lambda_i(t_j - t_{j-1})\}$. It follows, then, putting $\delta_{ij} = \tau_i + \alpha_j$, that

$$
\tau_i = \log_e[1 - \exp\{-\lambda_i(t_j - t_{j-1})\}] \quad (4.19)
$$

while
$$\alpha_j = \log_e(\gamma_j).$$

Now, this relationship for α_j is fine; this parameter is simply the logarithm of the probability that a dead animal will be recovered in the jth period, which is assumed to be the same for all morphs. However, τ_i is the parameter λ_i transformed in a particular way. In other words, equation (4.17) is not a log-linear model. It is not linear in all the independent parameters.

There are two ways around this problem. Equation (4.17) can be fitted to data using a general maximum likelihood computer program with the relationship of equation (4.19) incorporated. Alternatively, it can be argued that in reality the relationship (4.19) will only hold approximately for real populations, so that it is fair enough to regard λ_i and τ_i as independent parameters. The point here is that in practice there will be many different causes of death, each with a different recovery probability. Therefore f_{ij} is really given by a much more complicated equation than (4.18). On this basis there is no problem in fitting (4.17) using a program like GLIM for log-linear models.

It is possible to modify the log-linear model of equation (4.15) for the case when the A_i values are not known. It is just a question of incorporating their estimation into the model. Thus the equation can be rewritten as

$$E(d_{ij}) = \exp(-\lambda_{ij-1}t_{j-1} + \delta_{ij}) \tag{4.20}$$

where $\delta_{ij} = \log_e(f_{ij}) + \log_e(A_i)$. If all the morphs have the same survival and recovery probabilities then δ_{ij} will just vary with i, reflecting the different A_i values. This will still be true if different morphs have different recovery probabilities providing that these recovery probabilities are constant over time. Different survival probabilities for different intervals can be handled in the way indicated by model (4.16). Different recovery probabilities can be handled in the way indicated by model (4.17). In all cases, all that needs to be done is to regard the $\log_e(A_i)$ values as being unknown and to incorporate them into the other parameters that vary from morph to morph.

Example 4.3   Relative survival of two cohorts of Dominican gulls

Consider Fordham's (1970) data on Dominican gulls from Somes Island, New Zealand (Table 4.4). Here it can be asked whether there is any evidence that the survival varied for chicks born in the 1961–62 and the 1962–63 seasons. A log-linear model analysis is reasonable because only a small number of banded birds were recovered after death.

Based upon equation (4.15), the following two models were fitted:

(a) It was assumed that λ_{ij} and δ_{ij} were the same for all years and both cohorts, so that $\lambda_{ij} = \lambda$ and $\delta_{ij} = \delta$. This model fits the data very well, with a deviance of 8.93 with 9 degrees of freedom.

(b) It was assumed that λ_{ij} and δ_{ij} varied with the cohort but were constant over years, so that $\lambda_{ij} = \lambda_i$ and $\delta_{ij} = \delta_i$. This gives only a slight improvement on the fit of model (a) with a deviance of 4.05 with 7 degrees of freedom.

From these results it seems that model (a) is most reasonable. This gives

$$E(d_{1j}) = \exp(-3.71 + 6.35 - 0.29t_{j-1})$$

for the 1961–62 cohort, and

$$E(d_{2j}) = \exp(-3.71 + 6.59 - 0.29t_{j-1})$$

for the 1962–63 cohort. The implication is that both cohorts were subject to a $\exp(-0.29) = 0.75$ yearly survival rate.

Model (b) does not give a fit that is much better than the fit of model (a). The difference in deviance is 4.89 with 2 degrees of freedom which is significant at the 10% level but not at the 5% level. However, it is interesting to see what model (b) says about the data. This is that

$$E(d_{1j}) = \exp(-3.67 + 6.35 - 0.22t_{j-1}),$$

while

$$E(d_{2j}) = \exp(-3.68 + 6.59 - 0.43t_{j-1}),$$

which gives an estimated yearly survival of $\exp(-0.22) = 0.80$ for the 1961–62 cohort and an estimated yearly survival of $\exp(-0.43) = 0.65$ for the 1962–63 cohort. Thus this model suggests a fairly large survival difference, although this difference is not significant.

Note that the values 6.35 and 6.59 that are shown in the above expressions for $E(d_{1j})$ and $E(d_{2j})$ are $\log_e(574)$ and $\log_e(728)$. They have been included in the fitted models using the OFFSET directive in GLIM in order to account for the initial numbers of chicks marked.

4.6 EXPLICIT SURVIVAL ESTIMATION WITH A CONSTANT RECOVERY PROBABILITY

Consider again the situation with a single morph. A special case of some interest occurs when the recovery probabilities $\gamma_1, \gamma_2, \ldots, \gamma_s$ of equation (4.14) are all equal and the survival probability becomes constant for animals over a certain age. Haldane (1953, 1955) consider this model with all survival probabilities constant. Seber (1971) used it with only the two oldest age groups having the same survival rate. North and Morgan (1979) used it when analysing some heron data. They assumed that first-year survival rates varied from year to year, whilst the survival rate was constant from year to year for older birds. An interesting aspect of North and Morgan's work is the way that they have been able to relate first-year survival to the severity of the winter.

An advantage of this model is that it provides explicit equations for estimation of survival probabilities from the recoveries of a single batch of released animals. To see this, suppose that the number of animals in a population at time $t_0 = 0$ is unknown but large relative to the total number of dead animals recovered. Also, suppose that samples of dead animals all relate to a unit period of time (such as a year) so that d_j is the number of animals recorded as dying between time $t_{j-1} = j - 1$ and time $t_j = j$. Then d_1, $d_2, \ldots, d_s$ will approximately be independent Poisson variates with the mean value of d_j given by equation (4.14) to be

$$E(d_j) = \phi_1 \phi_2 \ldots \phi_{j-1}(1 - \phi_j)B,$$

where $B = A\gamma$, with γ being the constant recovery probability. The particular situation that is of interest occurs when $\phi_{r+1} = \phi_{r+2} = \ldots = \phi_s = \phi$, for some value r. With $r = 2$ this corresponds to North and Morgan's (1979) assumptions.

If the usual procedure for obtaining maximum likelihood estimates of $\phi_1, \phi_2, \ldots, \phi_r, \phi$ and B is followed, then the estimates $\hat{\phi}_1, \hat{\phi}_2, \ldots, \hat{\phi}_r, \hat{\phi}$ and $\hat{B}$ are found (Manly, 1981a) to satisfy the equations

$$T_r/R_r - 1/(1 - \hat{\phi}) + (s - r)\hat{\phi}^{s-r}/(1 - \hat{\phi}^{s-r}) = 0, \tag{4.21}$$

$$\hat{B} = R_0 + R_r \hat{\phi}^{s-r}/(1 - \hat{\phi}^{s-r}), \tag{4.22}$$

and

$$\hat{\phi}_i = (\hat{B} - R_0 + R_i)/(\hat{B} - R_0 + R_{i-1}), \tag{4.23}$$

for $i = 1, 2, \ldots, r$. Here

$$R_i = \sum_{j=i+1}^{s} d_j \quad \text{and} \quad T_i = \sum_{j=i+1}^{s} (j - i)d_j.$$

If $s - r$ is large so that $\phi^{s-r} \simeq 0$, then these equations reduce to

$$\hat{\phi} = 1 - R_r/T_r, \quad \hat{B} = R_0, \quad \text{and} \quad \hat{\phi}_i = R_i/R_{i-1}. \tag{4.24}$$

If ϕ^{s-r} is not near zero, then the solution for equation (4.21) can be read easily from Table 4.7 and estimates of B and ϕ_i follow directly from equations (4.22) and (4.23). With $r = 0$, equations (4.24) gives Lack's estimate of survival (Seber, 1982, p. 247). Equation (4.21) can also be solved using a table given by Robson and Chapman (1961) and reproduced by Seber (1982, Appendix A6).

The variances and covariances of the estimators of survival are given approximately by the following equations:

$$\operatorname{var}(\hat{\phi}) \simeq \frac{\phi(1 - \phi)^2}{\phi_1 \phi_2 \ldots \phi_r B}$$

$$+ \frac{\phi^{s-r+1}(1 - \phi)^3}{\phi_1 \phi_2 \ldots \phi_r B} \left\{ \frac{1}{1 - \phi} + \frac{(s - r)^2(1 - \phi)}{\phi} \right\}, \tag{4.25}$$

Table 4.7 A table for the solution of equation (4.21) for the survival probability estimate $\hat{\phi}$. The entries in the table are values of T_r/R_r. As an example of the use of the table, suppose that an experiment with $s - r = 5$ samples yields $T_r/R_r = 2.15$. Then the table shows that $0.6 < \hat{\phi} < 0.7$ since $T_r/R_r = 2.08$ for $\hat{\phi} = 0.6$ and $T_r/R_r = 2.32$ for $\hat{\phi} = 0.7$. Linear interpolation determines that the value of $\hat{\phi}$ for $T_r/R_r = 2.15$ is

$$\hat{\phi} = 0.6 + \frac{2.15 - 2.08}{2.32 - 2.08} \, 0.1 = 0.63.$$

This is the maximum likelihood estimator of the survival rate, correct to two decimal places.

$\hat{\phi}$	2	3	4	5	6	7	8	9	10	11	12	13	14	15	∞
								$s - r$							
0.0	1.00	1.00	1.00	1.00	1.00	1.00	1.00	1.00	1.00	1.00	1.00	1.00	1.00	1.00	1.00
0.1	1.09	1.11	1.11	1.11	1.11	1.11	1.11	1.11	1.11	1.11	1.11	1.11	1.11	1.11	1.11
0.2	1.17	1.23	1.24	1.25	1.25	1.25	1.25	1.25	1.25	1.25	1.25	1.25	1.25	1.25	1.25
0.3	1.23	1.35	1.40	1.42	1.42	1.43	1.43	1.43	1.43	1.43	1.43	1.43	1.43	1.43	1.43
0.4	1.29	1.46	1.56	1.61	1.64	1.66	1.66	1.66	1.67	1.67	1.67	1.67	1.67	1.67	1.67
0.5	1.33	1.57	1.73	1.84	1.90	1.94	1.97	1.98	1.99	1.99	2.00	2.00	2.00	2.00	2.00
0.6	1.37	1.67	1.90	2.08	2.21	2.30	2.36	2.41	2.44	2.46	2.46	2.48	2.49	2.49	2.50
0.7	1.41	1.77	2.07	2.32	2.53	2.71	2.84	2.95	3.04	3.11	3.16	3.21	3.24	3.26	3.33
0.8	1.44	1.85	2.22	2.56	2.87	3.14	3.39	3.60	3.80	3.97	4.11	4.24	4.36	4.45	5.00
0.9	1.47	1.93	2.37	2.79	3.19	3.58	3.95	4.31	4.65	4.97	5.28	5.57	5.85	6.11	10.00
1.0	1.50	2.00	2.50	3.00	3.50	4.00	4.50	5.00	5.50	6.00	6.50	7.00	7.50	8.00	∞

$$\text{var}(\hat{\phi}_i) \simeq \frac{\phi_i(1 - \phi_i)}{\phi_1\phi_2 \dots \phi_{i-1} B} + \frac{\phi_1\phi_2 \dots \phi_r\phi^{s-r}(1 - \phi_i)^2}{(\phi_1\phi_2 \dots \phi_{i-1})^2 B}, \tag{4.26}$$

$$\text{cov}(\hat{\phi}, \hat{\phi}_i) \simeq \phi^{s-r}(1 - \phi)^2(s - r)(1 - \phi_i)/(\phi_1\phi_2 \dots \phi_{i-1} B), \tag{4.27}$$

and

$$\text{cov}(\hat{\phi}_i, \hat{\phi}_j) \simeq \phi_1\phi_2 \dots \phi_r\phi^{s-r}(1 - \phi_i)(1 - \phi_j) \left/ \left(\prod_{k=1}^{i-1} \phi_k \prod_{k=1}^{j-1} \phi_k B \right) \right. \tag{4.28}$$

The terms multiplied by the factor ϕ^{s-r} represent corrections to take into account the fact that sampling has not continued until all animals are dead (i.e., until $\phi^{s-r} \simeq 0$). These terms should be small compared with the other terms. If they are not, then these variance and covariance formulae may be rather inaccurate (Manly, 1981a). Indeed, if ϕ^{s-r} is not small then the results of Lakhani and Newton (1983) suggest that the model of the present section may give very poor estimates.

Example 4.4   Survival of male and female starlings

Table 4.8 shows estimates of survival probabilities for the starling recovery data of Table 4.5. These have been calculated by treating each year of release

Table 4.8   Estimates of yearly survival probabilities for male and female starlings banded in different years, based upon the data of Table 4.5.

	Banding year	Yearly survival prob. est. $\hat{\phi}$	Standard error*	Chi-square for the model†	d.f.
Males	1950	0.52	0.038	4.65	3
	1951	0.59	0.036	3.44	4
	1952	0.58	0.053	1.51	3
	1953	0.48	0.058	0.58	2
	1954	0.82	0.064	3.32	2
Females	1950	0.49	0.041	4.67	3
	1951	0.60	0.046	3.23	3
	1952	0.61	0.054	3.58	3
	1953	0.37	0.079	0.03	1
	1954	0.53	0.062	0.24	1

* All standard errors were calculated using the mean survival estimate of 0.56, as explained in the text.
† None of the chi-square values are significant, indicating that the assumption of a constant survival probability for each banding year/sex combination is realistic. Expected values for these tests are calculated as $E(d_j) = \hat{\phi}^{j-1}(1 - \hat{\phi})\hat{B}$, with the degrees of freedom being the number of d_j values (after pooling to ensure that all the $E(d_j)$ values are larger than about five) minus two.

and each sex separately, using equations (4.21) and (4.22), with $r = 0$. The assumption that the yearly survival probability was constant for each release year seems realistic according to the chi-square tests shown in the table.

The variance for an estimated survival probability given by equation (4.25) depends rather critically on the true survival probability. For this reason it was considered best to evaluate variances using an average estimate of the survival probability rather than the yearly estimates themselves. The unweighted mean of the yearly estimates is $\bar{\phi} = 0.56$. If it is assumed that this is the correct survival probability, then the B value for any release can be estimated from equation (4.22) as

$$\hat{B} = R_0/(1 - 0.56^s).$$

This value of B, together with $\phi = 0.56$, can then be substituted into equation (4.25). This results in the standard errors shown in Table 4.8.

A weighted two-factor analysis of variance has been used to test whether the survival probability estimates vary significantly from year to year or between males and females (Table 4.9). The weight used for an estimate $\hat{\phi}$ was $1/\text{var}(\hat{\phi})$. It will be seen that there is no significant difference in survival between the sexes, although there is between years. Furthermore, the residual mean square is significantly large which indicates that a simple additive model of year and sex effects is too simple for these data. It is possible, of course, that the recovery rate varied with time. Table 4.10 shows the estimated yearly survival probabilities according to the additive model.

When the log-linear model of equation (4.20) was fitted to the data, a more or less similar conclusion was reached. A number of possibilities were tried, as follows:

(a) All releases for males and females were assumed to result in the same yearly survival probability. This model gives a good fit to the data, with a deviance of 56.05 with 49 degrees of freedom.

Table 4.9 Analysis of variance on the survival probability estimates of Table 4.8. A weighted analysis was carried out so that the expected value of the residual mean square is 1.00 (Appendix, Section A.8).

Source	Sum of squares	Degrees of freedom	Mean square
Sex, after allowing for years	2.4	1	2.4
Years, after allowing for sex	19.16	4	4.79*
Residual	9.93	4	2.48†
Error		∞	1.00

* Significant at the 0.1 % level
† Significant at the 5 % level

Table 4.10 Estimates of survival obtained by
fitting a two-factor analysis of variance model
to the survival estimates shown in Table 4.8.

| Year of | Yearly survival | |
banding	Males	Females
1950	0.53	0.48
1951	0.61	0.56
1952	0.62	0.57
1953	0.46	0.41
1954	0.70	0.65

(b) Survival was assumed to vary according to the sex of birds. This gives very little improvement over the fit of model (a), with a deviance of 55.34 with 48 degrees of freedom.

(c) Survival was assumed to vary according to the year of banding. This gives a substantial improvement over the fit of model (a), with a deviance of 44.51 with 45 degrees of freedom. The reduction in deviance is 11.54, with 4 degrees of freedom, which is significantly large at the 5% level.

(d) Survival was assumed to vary with the year of banding and sex, with these factors having additive effects. This model fits only slightly better than model (c), with a deviance of 43.66 with 44 degrees of freedom.

It appears again that survival varied according to the year of banding but not according to sex.

The survival estimates from model (d) are almost identical to the estimates shown in Table 4.10 based on analysis of variance.

It is interesting that a GLIM analysis has indicated that a model with year-to-year differences in survival fits these data quite well, whereas this is not the case according to an analysis of variance on survival estimates. The reason for the discrepancy is quite clearly the data from the 1954 bandings, which contribute 7.37 of the weighted residual sum of squares in Table 4.9. In the GLIM analysis these data are also somewhat anomalous but not enough to result in a large deviance.

On the whole, the GLIM analysis seems more reliable than the analysis of variance because it is based directly on the initial data. The analysis of variance has to rely on estimated standard errors that may not be all that accurate. The GLIM analysis also involves far more 'observations' than the analysis of variance. On the other hand, some of the expected frequencies are rather low in the GLIM model which means that testing deviances against the chi-square distribution may be rather an approximate procedure.

4.7 ESTIMATION OF A RELATIVE DEATH RATE FUNCTION

The methods that have been described in Chapter 3 for estimating a fitness function by comparing samples of survivors taken at different times from a population can also be used when the available data come from a population sample and a separate sample of dead animals. All that is necessary is that the population is large enough for the deaths to have only a negligible effect on it.

Thus suppose that the individuals in a population are characterized by their values for a single variable X, which has a probability density function $f_0(x)$ before selection. Let $\exp\{-\lambda(x)t\}$ be the probability of an individual with $X = x$ surviving selection for a time t, where $\lambda(x)$ is some positive function of x. Since for small t the probability of death is

$$\exp\{-\lambda(x)t\} \simeq 1 - \lambda(x)t \qquad (4.29)$$

it follows that $\lambda(x)$ is an instantaneous death rate.

Now, the probability density function of survivors to time t will be proportional to $f_0(x)$ and $\exp\{-\lambda(x)t\}$, so that it will be of the form

$$f_t(x) = Af_0(x)\exp\{-\lambda(x)t\}, \qquad (4.30)$$

where A is a constant. Likewise, the probability density function for non-survivors will be of the form

$$g_t(x) = Bf_0(x)[1 - \exp\{-\lambda(x)t\}],$$

where B is a constant. Therefore, for small selection times (so that result (4.29) holds),

$$g_t(x) \simeq Btf_0(x)\lambda(x). \qquad (4.31)$$

A comparison of equation (4.30) with equation (4.31) shows that these are of the same form. With (4.30) there is the initial probability density function multiplied by a probability of survival. With (4.31) there is the initial probability density function multiplied by an instantaneous death rate. The significance of this is that the methods for estimating fitness functions that have been discussed in Chapter 3 can also be used to estimate a relative death rate function. By comparing a sample from a population before selection with a sample from the same population after selection for a time t, it is possible to estimate $\alpha\exp\{-\lambda(x)t\}$, where α is an unknown constant. By comparing the sample before selection with a sample of dead animals instead, using exactly the same method of estimation will result in an estimate of $r(x) = \alpha\lambda(x)$, where again α is an unknown constant.

If there are several variables $X_1, X_2, \ldots, X_p$ rather than just one, then the same principle holds. Equations (4.30) and (4.31) just become

$$f_t(x_1, x_2, \ldots, x_p) = Af_0(x_1, x_2, \ldots, x_p)\exp\{-\lambda(x_1, x_2, \ldots, x_p)t\}$$

for survivors, and

$$g_t(x_1, x_2, \ldots, x_p) = Af_0(x_1, x_2, \ldots, x_p)\lambda(x_1, x_2, \ldots, x_p)$$

for non-survivors. Hence the methods of Chapter 3 estimate the fitness function $\alpha \exp\{-\lambda(x_1, x_2, \ldots, x_p)t\}$ if the after-selection sample is of survivors, or $\alpha\lambda(x_1, x_2, \ldots, x_p)$ if this sample is of dead animals.

Example 4.5   Comparison of living and eaten
Daphnia publicaria

As an example of the estimation of a relative death rate function using a sample of dead animals, consider Wong and Ward's (1972) data on plankton (population) samples of *Daphnia publicaria* compared with samples found in the stomachs of perch fry (Table 4.3). Here there is only a single X value involved, which is the length of the *D. publicaria*.

Consider the 1 July samples. From an inspection of the distributions for the plankton and stomach samples it seems clear that neither sample could be from a normal distribution. Therefore the test statistics λ_1 and λ_2 that have been given in Section 3.4 for testing for changes in means and standard deviations are not valid. They have been calculated nevertheless, and give very highly significant values ($\lambda_1 = 283.9$ with 1 degree of freedom; $\lambda_2 = 259.6$ with 1 degree of freedom, to be compared with the chi-square distribution).

The general maximum likelihood method described in Section 3.8 for estimating a fitness function was applied to the data. Since there are two samples only, model 1 is appropriate. A function of the form

$$r(x) = \exp(\alpha x)$$

was first fitted to the data. This gave a much better fit than model 0, the no-selection model. A likelihood ratio test gives a value of 233.80 with 1 degree of freedom for the improvement of fit, which is very significantly large compared to the chi-square distribution. A model with

$$r(x) = \exp(\alpha_1 x + \alpha_2 x^2)$$

gave a considerable further improvement of the fit, corresponding to a likelihood ratio test statistic of 83.59 with 1 degree of freedom. However, the improvement of a cubic model

$$r(x) = \exp(\alpha_1 x + \alpha_2 x^2 + \alpha_3 x^3)$$

was minimal, with a likelihood ratio test statistic of only 0.13 with 1 degree of freedom. Therefore the quadratic model seems appropriate. It is estimated as

$$\hat{r}(x) = \exp\left\{-5.01\left(\frac{x - 1.605}{0.625}\right) - 2.54\left(\frac{x - 1.605}{0.625}\right)^2\right\}$$

where 1.605 is the mean and 0.625 is the standard deviation in the plankton

Table 4.11   Relative death rates estimated from the July 1 plankton and stomach samples of *Daphnia publicaria* given in Table 4.3.

Length of D. publicaria (mm)	Estimated death rate $\hat{r}(x)$	Standard error*
0.5–0.7	4.42	1.39
–0.9	9.39	1.93
–1.1	11.82	3.12
–1.3	8.85	3.07
–1.5	3.94	1.73
–1.7	0.96	0.52
–1.9	0.16	0.10
–2.1	0.02	0.01
–2.3	0.00	0.00
–2.5	0.00	0.00
–2.7	0.00	0.00
–2.9	0.00	0.00
–3.1	0.00	0.00

* Calculated using equation (3.52).

sample. The standard errors associated with the coefficients -5.01 and -2.54 are 0.54 and 0.33, respectively. Estimated relative death rates for individuals with different values of X are shown in Table 4.11.

In Table 4.12 will be found a summary of the results of estimating fitness functions for all of Wong and Ward's sample times. Quadratic functions have provided a satisfactory fit except for the 29 July and 25 August samples, for which cases cubic terms in x were needed. For the first four sample times there appeared to be minimum survival for individuals with lengths of about 1.0, 1.2, 1.5 and 2.0 mm, respectively. It seems that the predators had an 'optimum' size of prey which gradually increased with time. The result on 25 August is not so easy to understand. Very small prey (about 0.5 mm long) had high survival, slightly larger prey (about 1.0 mm long) had low survival, then survival was higher for somewhat larger individuals (about 2.0 mm long), and finally was very low for large individuals. The fitness function is simply reflecting the data, and the relationship between survival and length seems to have been rather complex at the time of this sample.

4.8   THE INTENSITY AND EFFECT OF SELECTION

The methods of analysis that have been discussed in this chapter fall into two categories. On the one hand, the methods of Sections 4.4 to 4.6 provide

Table 4.12 Estimated relative death rates for the *Daphnia publicaria* data of Table 4.3.

Length of Daphnia (mm)		1 July samples $\hat{r}(x)$		15 July samples $\hat{r}(x)$		29 July samples $\hat{r}(x)$		12 August samples $\hat{r}(x)$		25 August samples $\hat{r}(x)$	
0.5		2.50		0.13		0.16		0.08		0.11	
1.0		11.82		0.84		0.35		0.39		5.74	
1.5		2.16		0.84		0.92		0.89		3.98	
2.0		0.02		0.09		0.75		1.05		0.87	
2.5		0.00		0.00		0.22		0.63		1.31	
3.0		0.00		0.00		0.00		0.20		2.99	
Test statistics	λ_1	283.9*	(1 d.f.)	183.1*	(1 d.f.)	70.0*	(1 d.f.)	7.3‡	(1 d.f.)	4.2‡	(1 d.f.)
	λ_2	259.6*	(1 d.f.)	29.5*	(1 d.f.)	0.4	(1 d.f.)	1.2	(1 d.f.)	15.8*	(1 d.f.)
Chi-square values for improvement of fit over no-selection model	Linear	233.8*	(1 d.f.)	29.3*	(1 d.f.)	0.4	(1 d.f.)	1.2	(1 d.f.)	16.3*	(1 d.f.)
	Quadratic	317.4*	(1 d.f.)	132.0*	(2 d.f.)	66.4*	(2 d.f.)	10.3†	(2 d.f.)	17.3*	(2 d.f.)
	Cubic	317.5*	(3 d.f.)	132.1*	(3 d.f.)	71.4*	(3 d.f.)	13.3†	(3 d.f.)	71.4*	(3 d.f.)

* significant at the 0.1% level   † significant at the 1% level   ‡ significant at the 5% level

estimates of survival probabilities per unit time for different types of individuals. On the other hand, other methods only give estimates of relative death rates.

When estimates of survival probabilities are available, these can be used as fitness values for indices I_H, I_V, I_O and I_u of the effect and intensity of selection, as discussed in Sections 3.3 and 3.12. Unfortunately, these indices cannot be used when only relative death rates are known, for the simple reason that there is no way of calculating fitness values per unit time. The problem is that individuals have survival probabilities $\exp(-\lambda)$ per unit time but all that can be estimated are relative death rates $\alpha\lambda$, where α is an unknown constant. For example, Table 4.11 gives relative death rates for *Daphnia publicaria* of different lengths. There is no way of knowing what these have to be multiplied by to give λ values.

Given this situation, it seems clear that any index of the intensity of selection has to be of such a form that it is unaffected when all the λ values are multiplied by a constant. This fact, together with the idea that high selection is implied by variable λ values, suggests the index

$$I_R = \operatorname{var}(\lambda)/\bar{\lambda}^2,$$

where $\operatorname{var}(\lambda)$ is the variance and $\bar{\lambda}$ is the mean of λ in the population before selection. Clearly, this index can be estimated by using relative death rates $r = \alpha\lambda$ in place of λ values, to give

$$I_R = \operatorname{var}(r)/\bar{r}^2. \qquad (4.32)$$

The unknown constant α cancels out.

In assessing a somewhat arbitrary index like I_R there is perhaps some value in seeing how it works in model situations. Thus assume that λ is distributed normally in a population, with mean μ and variance V. Then taking equations (3.23) to (3.25) with $x = \lambda$, $l = -1$ and $m = 0$ shows that at time t, λ will still be normally distributed, with mean $\mu - Vt$ and variance V. Consequently, at time t the index becomes

$$I_R(t) = V/(\mu - Vt)^2$$
$$= I_R(0)/\{1 - I_R(0)\mu t\}^2.$$

This shows that with this model the intensity of selection increases as μt increases, tending to infinity as $I_R(0)\mu t$ tends to one.

Since λ cannot be negative it may be argued that the normal distribution is not altogether appropriate and that it is better to assume something like a gamma distribution. Suppose therefore that this is the case, with λ having a mean μ and variance V before selection. Then it can be shown (Manly, 1977c) that after selection for a time t the distribution is still gamma but with mean $\mu^2/(\mu + Vt)$ and variance $\mu^2 V/(\mu + Vt)^2$. Hence at time t the index of the intensity of selection is

$$I_R(t) = \{\mu^2 V/(\mu + Vt)^2\}/\{\mu^2/(\mu + Vt)\}^2$$
$$= V/\mu^2,$$

which is constant. Therefore for this model, which is perhaps more realistic than the normal distribution model, the intensity of selection remains constant. This seems a desirable property since it means that it does not matter when a population is studied; I_R is always the same.

The bias and variance of the index I_R can be determined in particular cases by simulation, following the approach of Example 3.6. This has not been done in the example that follows because of the extensive computations that would have been involved.

Example 4.6   The intensity of selection on
Daphnia publicaria

When I_R is calculated for Wong and Ward's (1972) *Daphnia publicaria* data, using relative death rates that are estimated as discussed in Example 4.5, the following estimated values are obtained: 1 July, $\hat{I}_R = 1.45$; 15 July, $\hat{I}_R = 0.39$; 29 July, $\hat{I}_R = 0.21$; 12 August, $\hat{I}_R = 0.08$; and 25 August, $\hat{I}_R = 0.55$. It seems that the intensity of selection declined from 1 July through to 12 August, and then became somewhat higher towards the end of August. (It has been noted already, in Example 4.5, that the pattern of apparent selection on 29 August is not easy to understand.)

5 Complete counts of survivors

The previous three chapters have been concerned with various aspects of the analysis of samples from populations, these consisting of either living or dead animals. In the present chapter attention is turned to situations where the action of selection can be observed directly on a group of individuals. The particular interest is in cases where the survivors are recorded after certain specific amounts of selection time. By its nature, this type of data is unlikely to be available for a wild population. However, as will be seen, this can occur. More commonly, such data will arise from a laboratory experiment, or from a controlled field experiment.

The first sections of this chapter are devoted particularly to selection on quantitative variables. After a review of alternative approaches, the usefulness of the proportional hazards model is emphasized. Consideration is then given to the particular case of stabilizing selection, which occurs when selection is against those individuals that are far from the average.

Sections 5.6 to 5.8 are concerned with a model for a selection process in which individuals are selected one by one in a competitive manner. This may occur in predation experiments, for example. Section 5.9 contains a discussion on the difference between competitive and non-competitive selection. It is suggested that this distinction may not be of crucial importance when it comes to the question of how to analyse data.

The chapter concludes with another look at the problem of measuring the intensity and effect of selection, a topic that has already been discussed in the previous two chapters.

5.1 SELECTION ON A COUNTABLE POPULATION

The situations that are considered in this chapter all involve a finite population of identifiable individuals and a selection process that removes some of them. There are no sampling errors. The only source of random variation is the selection process itself, so that these are in a way the simplest situations where selection can be studied. They are like those that have been discussed in Chapter 3. The only difference is that in Chapter 3 only samples of survivors from selection were considered to be available; here it is assumed that all survivors are counted at certain particular points in time. If selection is occurring then the survivors will be a non-random sample of the original members of the population. In a polymorphic population they will have morph proportions that are different from those of the original members. In a population in which each individual has values for certain quantitative variables, it is the distribution of these variables that will be different.

A number of models have been proposed for selection on quantitative variables. These will be reviewed in the following section.

5.2   MODELS FOR SELECTION ON QUANTITATIVE VARIABLES

The identification of variables that are related to the probability of survival of human babies is of obvious importance to the medical profession. One early study was made by Karn and Penrose (1951) using records on babies born at University College Hospital, London, during the period 1935–46. Table 5.1 shows their data on neonatal survival (survival to 28 days after birth) for male babies with different birth weights and gestation times.

Karn and Penrose (1951) analysed their data on the assumption that the survivors were a random sample from a population for which the distribution of birth weight and gestation time is bivariate normal, and the non-survivors were a random sample from another population for which the distribution of birth weight and gestation time is also bivariate normal but with different parameters. The parameters (means, variances and correlations) were estimated from the data in the usual way by Karn and Penrose and they were then able to estimate fitness functions that give the survival probability for a baby with any given birth weight and gestation time.

The main problem with Karn and Penrose's model is that it gives a very poor fit to their data. For example, according to their equations the probability of survival for a male baby with a birth weight of 7 lbs and a gestation time of 230 days was only about 0.1. However, 18 out of the 19 infants with approximately these values for birth weight and gestation time did in fact survive (Table 5.1). This is by no means an isolated example of the poor agreement between their model for survival and the actual survival rates.

It is the assumption of a bivariate normal distribution for survivors and non-survivors that seems to have upset Karn and Penrose's calculations. For non-survivors in particular this assumption is far from reasonable. However, even if their model did fit the data there is a theoretical objection that can be made to it. The survival period of 28 days that was used is fairly arbitrary. Some of the babies died after this. Thus Karn and Penrose's model involves the concept of a bivariate normal distribution for the survivors gradually being depleted by death, with the deaths becoming part of a growing bivariate normal distribution of non-survivors. This seems a very peculiar model for

Table 5.1 Survival to 28 days after birth for males born at University College Hospital, London, 1935–46, classified according to their birth weight and gestation time. In the table, *A* denotes the initial number and *r* the number still alive after 28 days. There were a small number of babies with birth weights and gestation times outside the ranges shown in the table.

Birth weight (lbs)		Gestation time (days)							
		200–219	220–239	240–259	260–279	280–299	300–319	320–339	*Totals*
$2\frac{1}{2}-$	A	15	19	12	5	–	–	–	51
	r	2	6	2	2	–	–	–	12
$3\frac{1}{2}-$	A	13	34	32	19	7	–	–	105
	r	5	21	24	16	6	–	–	72
$4\frac{1}{2}-$	A	–	22	67	114	52	9	–	264
	r	–	18	55	101	51	6	–	231
$5\frac{1}{2}-$	A	–	24	136	490	374	47	8	1079
	r	–	23	129	473	358	45	8	1036
$6\frac{1}{2}-$	A	–	19	123	928	1152	160	20	2402
	r	–	18	119	905	1119	157	19	2337
$7\frac{1}{2}-$	A	–	10	73	577	1216	177	26	2079
	r	–	10	69	566	1194	172	24	2035
$8\frac{1}{2}-$	A	–	–	27	159	484	73	10	753
	r	–	–	26	156	474	68	10	734
$9\frac{1}{2}-$	A	–	–	–	36	122	26	7	191
	r	–	–	–	36	117	23	5	181
$10\frac{1}{2}-$	A	–	–	–	6	21	11	–	38
	r	–	–	–	6	19	9	–	34
Total	A	28	128	470	2334	3428	503	71	6962
	r	7	96	424	2261	3338	480	66	6672

selection and it seems doubtful whether it could ever be even approximately correct.

The same objection applies to the generalization of Karn and Penrose's model that has been proposed by Sansing and Chinnici (1976). They consider birth weight only and allow this to have various distributions for survivors and non-survivors. However, survival is still treated as if it is a phenomenon that has no relationship to time.

Van Valen and Mellin (1967) reviewed the results from a number of earlier studies of the relationship between infant survival and birth weight. They examined the simple model

$$m = aX^2 + bX + c \qquad (5.1)$$

where m is the mortality rate, X is the birth weight, and a, b and c are constants. They also considered other models of the same type, fitting them to data by normal regression methods. They showed that for Karn and Penrose's data the equations

$$m = 0.0155 \, e^{-1.64D}$$

for males, and

$$m = 0.0095 \, e^{-1.82D},$$

for females, give a good fit. Here D represents the absolute deviation from the optimum birth weight, in kilograms.

The fitness functions that have been discussed in Chapter 3 for sample data can be used just as well in the present situation. Thus O'Donald's (1968, 1970, 1971) papers are all relevant. In particular, he suggested using the quadratic fitness function

$$\phi(x) = 1 - \alpha - K(\theta - x)^2 \qquad (5.2)$$

to relate survival to a single quantitative variable. This is essentially the same as Van Valen and Mellin's (1967) equation (5.1) but with a different form of parameterization. Here $\phi(x)$ represents the probability of survival for an individual with $X = x$. Individuals with $X = \theta$ have the highest possible survival, which is $1 - \alpha$. The parameter K reflects the extent to which survival is reduced for individuals with $X \neq \theta$. Equations given in Section 3.5 can be used to estimate θ, K and α.

The 'nor-optimal' fitness function

$$\phi(x) = (1 - \alpha) \exp\{-K(\theta - x)^2\}$$

is essentially the same as the function (3.23) of Section 3.6. It seems to have been first used by Cavalli-Sforza and Bodmer (1972) to represent the relationship between birth weight and survival for Karn and Penrose's data. It is important to note that the methods that have been proposed for fitting this function to data all assume that samples are available from a large normal population before and after selection. Hence they are not really appropriate for data of the form being discussed in the present chapter. The assumption of a normal distribution is particularly questionable.

Of course, the survival of human infants is not the only situation of interest for the present chapter. Indeed, it should perhaps be mentioned that in recent years there has been an upsurge of interest in the analysis of survival data in general. This is reflected in a number of books that have been published on this topic; see, for example, Elandt-Johnson and Johnson (1980) and Kalbfleish and Prentice (1980).

5.3   THE PROPORTIONAL HAZARDS MODEL

It is always possible to represent the probability of an individual surviving a period of time $(0, t)$ by the expression

$$\psi(t) = \exp\{-\lambda g(t)\},$$

where $g(t)$ is a positive, non-decreasing function of time, with $g(0) = 0$. If the probability of surviving a unit period of time remains constant then $g(t) = t$. More generally, $g(t)$ reflects the 'rate' at which time is passing in terms of survival. The function $\lambda g(t)$ is called the cumulative hazard function.

The basis of the proportional hazard model is the assumption that $g(t)$ is the same for all individuals in a population, although λ may vary. That is to say, if there is a population of N individuals then the probability of survival to time t for individual i is

$$\psi_i(t) = \exp\{-\lambda_i g(t)\}.$$

The cumulative hazard functions for all individuals will then remain in the same proportions $\lambda_1 g(t):\lambda_2 g(t):\ldots:\lambda_N g(t)$ or $\lambda_1:\lambda_2:\ldots:\lambda_N$. An alternative form of the last equation that is more useful for estimation is

$$\psi_i(t) = \exp[-\exp\{\pi(t)+\gamma_i\}], \tag{5.3}$$

where

$$\pi(t) = \log_e\{g(t)\} \quad \text{and} \quad \gamma_i = \log_e(\lambda_i). \tag{5.4}$$

The main concern in the present chapter is with situations where the individuals in a population are counted before selection, at time t_0, and at subsequent times $t_1, t_2, \ldots, t_s; s \geqslant 1$. Equation (5.3) shows that in this case the probability of the ith individual surviving to time t_j, given that it has survived until time t_{j-1}, is

$$\phi_{ij} = \psi_i(t_j)/\psi_i(t_{j-1})$$

$$= \exp\{-\exp(\beta_j+\gamma_i)\}, \tag{5.5}$$

where $\beta_j = \log_e\{g(t_j)-g(t_{j-1})\}$. If the ith individual is described by the values $x_{i1}, x_{i2}, \ldots, x_{ip}$ that it possesses for the quantitative variables $X_1, X_2, \ldots, X_p$, then a simple assumption is that

$$\gamma_i = \alpha_0 + \alpha_1 x_{i1} + \alpha_2 x_{i2} + \ldots + \alpha_p x_{ip}$$

so that equation (5.5) becomes

$$\phi_{ij} = \exp\{-\exp(\beta_j+\alpha_0+\alpha_1 x_{i1} + \ldots + \alpha_p x_{ip})\}. \tag{5.6}$$

This is the proportional hazards fitness function, which is a special case of a model proposed by Cox (1972). There are no restrictions on relationships between the variables $X_1, X_2, \ldots, X_p$ so that equation (5.6) could just involve a single variable X and, for example, be of the form

$$\phi_{ij} = \exp\{-\exp(\beta_j + \alpha_0 + \alpha_1 x_i + \alpha_2 x_i^2)\}.$$

In that case there can be an optimum value of X for which survival is at a maximum.

Fitting the model to data is discussed in the Appendix, Section A.11. Maximum likelihood estimation can be carried out using a computer program such as GLIM. Alternatively, a multiple regression approach is possible. GLIM has been used in all the examples that follow. With this program there is no difficulty in allowing survival to vary for different morphs in a polymorphic population as well as allowing for the effects of quantitative variables.

5.4   GRAPHICAL ANALYSIS OF THE PROPORTIONAL HAZARDS MODEL

In some cases the assumptions behind the proportional hazards model can be tested by a graphical analysis. Thus suppose that there are K morphs and that the survival of these morphs is known after selection for increasing amounts of time $t_1, t_2, \ldots, t_s$. Then from equation (5.3) the probability of an individual of morph i surviving until at least time t_j can be written as

$$\psi_{ij} = \exp\{-\exp(\pi_j + \gamma_i)\} \tag{5.7}$$

where $\gamma_1, \gamma_2, \ldots, \gamma_K$ account for survival differences between morphs, and $\pi_1 < \pi_2 < \ldots < \pi_s$ account for the increasing survival time. From this last equation it follows that

$$y_{ij} = \log_e\{-\log_e(\psi_{ij})\} = \pi_j + \gamma_i. \tag{5.8}$$

Hence if y_{ij} is plotted against time then the plots given for different values of i should be parallel.

In practice, y_{ij} can be estimated from the observed proportion surviving. If there are initially A_i individuals of morph i, and r_{ij} of these survive until at least the end of the jth survival period, then an almost unbiased estimator of y_{ij} is

$$\hat{y}_{ij} = \begin{cases} \log_e[-\log_e\{1/2A_i\}], & \hat{\psi}_{ij} = 0 \\ \log_e[-\log_e[\hat{\psi}_{ij} + (1 - \hat{\psi}_{ij})\{\log_e(\hat{\psi}_{ij}) \\ \quad + 1\}/\{2A_i \log_e(\hat{\psi}_{ij})\}]] & 0 < \hat{\psi}_{ij} < 1 \\ \log_e[-\log_e\{1 - 1/2A_i\}], & \hat{\psi}_{ij} = 1 \end{cases} \tag{5.9}$$

where $\hat{\psi}_{ij} = r_{ij}/A_i$ (Manly, 1978b). These estimates can be calculated from the data and plotted against time. If the plots are not parallel for different values of i then this suggests that the proportional hazards model does not hold.

Actually, it is better to plot $\hat{y}_{ij}$ against $\log_e(t_j)$ rather than t_j. This is because equations (5.4) shows that $\pi_j = \log_e\{g(t_j)\}$, where $g(t)$ is the function which reflects the 'rate' at which time is passing as far as survival is concerned. Consequently, if the survival probability per unit time is constant, so that $g(t) = t$, then equation (5.8) becomes

$$y_{ij} = \log_e(t_j) + \gamma_i.$$

Therefore a plot of $\hat{y}_{ij}$ against $\log_e(t_j)$ with a slope of one indicates that the survival probability per unit time is constant.

Example 5.1   Selection related to size of *Cepaea nemoralis*

Consider the data shown in Table 5.2 for the over-winter survival of *Cepaea nemoralis* (L. M. Cook, private communication). This is an example of a situation where the model of equation (5.6) is appropriate, with a single quantitative variable X, the maximum shell diameter.

The first thing to note is that the data can be regarded as consisting of 48 independent survival proportions: there are eight different size classes for each of six different survival periods. (The justification for this way of looking at the data is given in the Appendix, Section A.11). Thus the initial survival period was 29 September to 31 December. For this period the survival proportions for the size classes 21, 22, . . . , 28 of snail were 15/21, 72/93, . . . , 1/2, respectively. The second survival period was 1 January to 7 February. For this period the survival proportions for the different size classes were 12/15, 57/72, . . . , 1/1, respectively. Continuing in this way, we get to the last survival period, 22 May to 14 June, for which the survival proportions for the different size classes were 1/3, 6/14, . . . , 1/1.

Table 5.2   The survivors at various times from 1160 *Cepaea nemoralis* left over the winter of 1968/69 in an unheated room without food. The snails were collected from Cledes in the Garonne valley, a few kilometres south of the border of France and Spain.

Maximum shell diameter (mm)	Initial nos. on 29/9/68	Survivors on day/month					
		31/12	7/2	14/3	16/4	21/5	14/6
21	21	15	12	7	4	3	1
22	93	72	57	42	33	14	6
23	255	228	194	152	129	85	44
24	343	326	286	239	192	133	70
25	289	270	239	203	175	132	62
26	128	120	107	99	94	63	36
27	29	29	27	22	21	18	7
28	2	1	1	1	1	1	1

If ϕ_{ij} denotes the probability of surviving the jth time period for a snail with shell diameter $X = x_i$, then a no-selection model for the data is

$$\phi_{ij} = \exp\{-\exp(\beta_j + \gamma)\},$$

where $\beta_1, \beta_2, \ldots, \beta_6$ allow for different survival probabilities in the six time periods. This model fits the data very badly, with a GLIM deviance of 112.2 with 42 degrees of freedom. (This deviance can be treated as a chi-square variate for a goodness of fit test. See the Appendix, Section A.10).

A model for directional selection is

$$\phi_{ij} = \exp\{-\exp(\beta_j + \alpha_0 + \alpha_1 x_i)\}.$$

The β parameters still account for the different selection times and the assumption is being made that the effect of size, allowed for by the term $\alpha_1 x_i$, is the same for all selection periods. The deviance is now 68.2 with 41 degrees of freedom, which represents a considerable improvement over the fit of the no-selection model. However, the deviance is still significantly large at the 1% level. Adding a squared term in X gives

$$\phi_{ij} = \exp\{-\exp(\beta_j + \alpha_0 + \alpha_1 x_i + \alpha_2 x_i^2)\}$$

which has a deviance of 61.5 with 40 degrees of freedom. This is still significantly large at the 5% level.

It would be possible to continue in this way, adding higher and higher powers of X. However, this cannot result in a good fit to the data because it turns out that the model

$$\phi_{ij} = \exp\{-\exp(\beta_j + \gamma_i)\}$$

gives a deviance of 57.4 with 35 degrees of freedom, which is significantly large at about the 1% level. This model is one that allows the different sizes of snail to have different survival probabilities without assuming any particular relationship between survival and X, since each size class has its own parameter γ_i. All that the model assumes is that the effect of size is the same for all time periods.

Since the last model does not fit the data, it must be concluded that the fitness function was not constant over the full experimental period. In other words, a separate fitness function is needed for each of the six survival periods. This is achieved with the model

$$\phi_{ij} = \exp\{-\exp(\beta_j + \alpha_0 + \alpha_j x_i)\}$$

which gives a deviance of 49.3 with 36 degrees of freedom. This is still significantly large at about the 5% level. However, adding a squared term gives

$$\phi_{ij} = \exp\{-\exp(\beta_j + \alpha_0 + \alpha_{1j} x_i + \alpha_{2j} x_i^2)\} \tag{5.10}$$

which gives a deviance of 38.7 with 30 degrees of freedom. This is not significantly large at the 5 % level, so that this last model may be considered to give an acceptable representation of the data.

The fitted equation (5.10) gives the estimated survival probabilities that are shown in Table 5.3. For selection periods 1, 2, 3, 5 and 6 it appears that maximum survival occurred for snails with maximum shell diameters in the range 25–28 mm. However, for survival period 4 there is no estimated maximum but snails with a maximum shell diameter of about 22 mm are estimated to have the minimum possible survival.

Cook and O'Donald (1971) based an analysis of these data on quadratic and nor-optimal functions and reached very similar conclusions. In particular, they noted that the selection during the fourth period did not appear to follow the same pattern as the selection in the other periods. One advantage of an analysis based upon the proportional hazards fitness function is that it has been very simple to test the hypothesis that the relationship between fitness and size was constant for the six selection periods, making allowance for the fact that these intervals were of different lengths.

A graphical approach for examining the fit of the proportional hazards model confirms the results of the computer analysis. Table 5.4 shows the estimates of $\hat{y}_{ij}$ calculated for the different size groups of snails using equation (5.9). A plot of these against the logarithm of the survival time is shown in Fig. 5.1.

An examination of this figure highlights two particular aspects of the data. The plots for different size classes are not parallel. They have moved together as the survival time has increased, indicating that the effect of selection has diminished somewhat with increasing time. The other aspect of the data which shows up clearly is the somewhat erratic survival of the sixth size class of snails (shell diameter 26 mm). Initially these had similar survival to that for size class

Table 5.3   Estimated survival probabilities for Cepaea nemoralis of different sizes.

Maximum shell diameter (mm)	Selection period					
	1	2	3	4	5	6
21	0.62	0.76	0.61	0.81	0.43	0.44
22	0.82	0.82	0.72	0.79	0.55	0.47
23	0.90	0.85	0.79	0.80	0.64	0.50
24	0.93	0.87	0.83	0.83	0.69	0.51
25	0.95	0.88	0.86	0.87	0.73	0.51
26	0.95	0.89	0.88	0.92	0.74	0.51
27	0.93	0.89	0.89	0.96	0.75	0.49
28	0.90	0.88	0.89	0.98	0.74	0.46

Table 5.4 The values $\hat{y}_{ij}$ for a graphical analysis of the proportional hazards model with the data on the over-winter survival of *Cepaea nemoralis*. The values have not been estimated for the 28 mm size group containing only two snails.

Maximum shell diameter (mm)	Size class (i)	$\hat{y}$ values for different survival times in months					
		3.00	4.25	5.50	6.50	7.75	8.50
21	1	−1.03	−0.56	0.09	0.48	0.63	1.02
22	2	−1.35	−0.71	−0.23	0.04	0.63	0.99
23	3	−2.17	−1.29	−0.66	−0.38	0.09	0.56
24	4	−2.95	−1.70	−1.02	−0.54	−0.05	0.46
25	5	−2.66	−1.65	−1.04	−0.69	−0.24	0.43
26	6	−2.68	−1.70	−1.35	−1.17	−0.34	0.24
27	7	−4.05	−2.43	−1.24	−1.09	−0.72	0.34

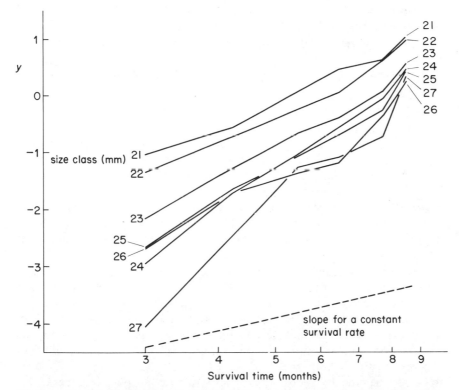

Fig. 5.1 Graphical analysis of the data on the over-winter survival of *Cepaea nemoralis*. For the proportional hazards model with constant relative fitnesses the plots for the size classes 21 to 27 mm should be approximately parallel. In addition, a plot with a slope parallel to the broken line would indicate a constant survival rate per unit time.

five. However, their survival was rather high for a survival time between about four to six months and then rather low for a survival time of about six to eight months. The plot for size class seven also fluctuates somewhat. However, there were initially only 29 snails in class seven, compared with 128 in class six. Therefore the fluctuations for group seven are of less significance than those for group six.

The final point to note about Fig. 5.1 is that the plots have somewhat greater slopes than a plot of a constant plus $\log_e$ (time) (the broken line on the figure). This indicates that the survival probability per month was decreasing over the course of the experiment.

There is a problem with this example that was mentioned in Section 1.7. Clarke *et al.* (1978) have pointed out that the snails used in the experiment may have come from several populations. If the larger snails tended to come from a population with high survival then the observed 'selection' related to size may really just be reflecting population differences. There might be no relationship at all between survival and size for snails from a single population.

See Bantock and Bayley (1973) and Bantock *et al.* (1976) for examples of field experiments on the selection of *Cepaea nemoralis* and *C. hortensis* by thrushes in relationship to the size, shell colour and shell banding of the snails.

Example 5.2   Climatic selection on *Cepaea nemoralis*

The next example also concerns *Cepaea nemoralis*. This time it is a question of climatic selection related to different colour and banding morphs. The shells of the snails are yellow, pink or brown and can have up to five dark bands. There is some evidence from laboratory experiments to suggest that different morphs have different tolerances to temperature extremes, although this evidence is open to criticism (Jones *et al.*, 1977; Clarke *et al.*, 1978).

In one field experiment, Bantock (1974) set up six population cages for the period 30 May to 7 September 1973. On the latter date he counted the number of live and dead snails in each cage. The results are shown in Table 5.5. Six different morphs were used: B00000, brown unbanded; P00300, pink mid-banded; P12345, pink fully banded; Y00300, yellow mid-banded; Y12345, yellow fully banded; and Y00000, yellow unbanded. In the area where the cages were set up, only B00000 and P00300 occur naturally. Bantock hypothesized that these morphs should therefore survive better than the other morphs due to climatic selection. In the analysis that follows, the results for yellow unbanded snails have been ignored because there were only 11 of them and they all died. All that can be said about this morph is that the probability of survival must have been rather low.

Let ϕ_{ij} denote the probability of survival for morph i in cage j. One possibility is then that this probability was the same for all cages, but varied from morph to morph, so that for the proportional hazards model

$$\phi_{ij} = \exp\{-\exp(\gamma + \alpha_i)\}$$

Table 5.5   The results from an experiment on climatic selection on the snail *Cepaea nemoralis.*

Morph		Population cage					
		1	2	3	4	5	6
Y00000*	Initial number	5	4	2	0	0	0
	Survivors	0	0	0	–	–	–
Y12345	Initial number	19	16	11	7	8	3
	Survivors	5	0	0	2	2	3
Y00300	Initial number	20	15	12	7	10	3
	Survivors	2	2	2	2	4	1
P12345	Initial number	88	69	50	29	19	10
	Survivors	6	4	11	2	2	6
P00300	Initial number	84	70	70	29	31	10
	Survivors	9	13	12	5	11	7
B00000	Initial number	89	72	50	32	68	10
	Survivors	11	19	11	9	24	7

* Ignored in the analysis

for $i = 1, 2, 3, 4, 5$. This model gives a GLIM deviance of 93.8 with 25 degrees of freedom, which is very highly significantly large when compared with the chi-square distribution. Alternatively, it may be assumed that survival was the same for all morphs, but varied from cage to cage, so that

$$\phi_{ij} = \exp\{-\exp(\gamma + \beta_j)\}.$$

This fits better than the first model, with a deviance of 44.4 with 24 degrees of freedom. However, this is still significantly large at the 1% level.

A more plausible model is perhaps one which allows survival to vary with both morph and cage, so that

$$\phi_{ij} = \exp\{-\exp(\gamma + \alpha_i + \beta_j)\}.$$

This fits quite well, with a deviance of 29.8 with 20 degrees of freedom, and is therefore acceptable.

The estimates of the morph parameters α_i of this last model are shown in Table 5.6, together with standard errors. The fifth morph B00000 (brown unbanded) has been taken by GLIM as a 'standard' morph with $\alpha_5 = 0$. The other α values are defined relative to this. Since these other values are all positive, it appears the standard morph was the one that had the highest probability of survival. (A large value for $\gamma + \alpha_i + \beta_j$ gives a low probability of survival.) However the morph P12345 (pink, five-banded) is the only one that has an α value that is significant in the sense of being more than two standard

Table 5.6　Parameter estimates obtained using GLIM for the experiment on climatic selection on *Cepaea nemoralis*.

Parameter	Estimate	Standard error
Constant γ	-0.10	0.30
α_1 (for Y12345)	0.19	0.18
α_2 (for Y00300)	0.17	0.17
α_3 (for P12345)	0.42	0.11
α_4 (for P00300)	0.15	0.10
α_5 (for B00000)	0.00	0.00
β_1 (for cage 1)	1.71	0.30
β_2 (for cage 2)	1.55	0.30
β_3 (for cage 3)	1.42	0.31
β_4 (for cage 4)	1.44	0.32
β_5 (for cage 5)	1.13	0.31
β_6 (for cage 6)	0.00	0.00

Notes:
1. The method of parameterization used by the computer program GLIM is explained in the Appendix, Section A.10. For these data the brown unbanded morph (B00000) was taken as a standard by making $\alpha_5 = 0$. Similarly, cage 6 was taken as the standard cage with $\beta_6 = 0$.
2. An approximately 95% confidence interval for the true value of one of the parameters is estimate ± 2 (standard error). If this range does not include zero then the parameter estimate is significantly different from zero at the 5% level. In order to see whether two parameter estimates (e.g., α_3 and α_4) are significantly different from each other, it is necessary to know the correlation between the parameter estimators. GLIM can provide this.

errors away from zero. Of course, the sample sizes for the yellow morphs were so small that significance cannot really be expected. Table 5.7 shows the survival probabilities that held for Bantock's experiment, according to the fitted model.

Table 5.7　Probabilities of survival for *Cepaea nemoralis* according to the proportional hazards model.

Morph	Population cage					
	1	2	3	4	5	6
Y12345	0.11	0.15	0.19	0.18	0.29	0.67
Y00300	0.11	0.15	0.19	0.19	0.29	0.67
P12345	0.06	0.09	0.12	0.12	0.21	0.60
P00300	0.12	0.16	0.20	0.20	0.30	0.68
B00000	0.16	0.21	0.25	0.25	0.36	0.72

Generally, these results are in line with what Bantock expected to occur, and he concluded that climatic selection was responsible for the survival differences between morphs. However, Jones *et al.* (1977) and Clarke *et al.* (1978) have suggested that the true explanation might be that the snails used in the experiment were collected from several different colonies. Jones and Parkin (1977) carried out a rather similar experiment on a *Cepaea vindobonensis* population in North Yugoslavia and they were unable to detect any differential survival. Further experiments have been carried out by Knights (1979) and Bantock and Ratsey (1980). See also Tilling (1983) and Jones (1982).

5.5   THE CASE OF STABILIZING SELECTION

With stabilizing selection on quantitative variables there is a tendency for unusual individuals to survive less well than average individuals. Highest survival occurs for individuals who are either average or close to average in all measurements. With a single quantitative variable X the proportional hazards model

$$\phi = \exp\{-\exp(\beta_0 + \beta_1 x + \beta_2 x^2)\},$$

can allow for this type of situation, where ϕ is the probability of survival. With two variables X_1 and X_2, this generalizes to

$$\phi = \exp\{-\exp(\beta_0 + \beta_1 x_1 + \beta_2 x_2 + \beta_{11} x_1^2 + \beta_{12} x_1 x_2 + \beta_{13} x_2^2)\}$$

where, again, this allows for the possibility that survival is highest for average individuals. With p variables the function becomes

$$\phi = \exp\{-\exp(\beta_0 + \sum_{i=1}^{p} \beta_i x_i + \sum_{i=1}^{p} \sum_{j=i}^{p} \beta_{ij} x_i x_j)\}.$$

There is a problem here caused by the number of parameters that have to be estimated to fit the relationship between survival and the quantitative variables. With a single variable there are two parameters, β_1 and β_2. With two variables there are five parameters β_1, β_2, β_{11}, β_{12} and β_{22}. With three variables there are nine parameters. Under the circumstances it may well be better to avoid estimating large numbers of parameters by adopting an alternative approach for handling stabilizing selection where there are more than one or two variables. One possibility is to decide upon a suitable measure of the extent to which individuals differ from the 'average' and see whether survival is related to this. For example, suppose that there are p variables with means $\mu_1, \mu_2, \ldots, \mu_p$ and covariance matrix $\mathbf{V}$ for the population before selection begins. Then one reasonable measure of the extent to which an individual with $X_1 = x_1$, $X_2 = x_2, \ldots, X_p = x_p$ is abnormal is the Mahalanobis distance from the individual to the centroid:

$$D = (\mathbf{X} - \mu)' \mathbf{V}^{-1} (\mathbf{X} - \mu)/p, \tag{5.11}$$

where

$$X = \begin{bmatrix} x_1 \\ x_2 \\ \cdot \\ \cdot \\ \cdot \\ x_p \end{bmatrix} \quad \text{and} \quad \mu = \begin{bmatrix} \mu_1 \\ \mu_2 \\ \cdot \\ \cdot \\ \cdot \\ \mu_p \end{bmatrix}.$$

For a multivariate normal distribution, pD is approximately a chi-square variate with p degrees of freedom. The mean value of D itself is 1.0.

A simpler measure of abnormality is

$$D^* = \frac{1}{p} \sum_{i=1}^{p} (x_i - \mu_i)^2 / \text{var}(X_i), \tag{5.12}$$

where, again, the mean value is 1.0. See Manly (1985b) for a further discussion of relatively simple ways to model stabilizing selection.

Example 5.3   Storm survival of sparrows

A well known set of data on selection with respect to quantitative variables was published by Bumpus (1898). After a severe storm on 1 February 1898, a total of 136 sparrows (*Passer domesticus*) were taken to Bumpus's laboratory. The birds were in a moribund condition and subsequently 72 revived and 64 died. Bumpus took eight morphological measurements on each bird and also weighed them. His data are reproduced here in Table 5.8, classified according to sex and the age of males. In the analyses that follow all males have been treated together in order to obtain reasonably large sample sizes.

From a comparison between the survivors and non-survivors, Bumpus reached the conclusion that:

> . . . there are fundamental differences between the birds which survived and those which perished. While the former are shorter and weigh less (i.e., are of smaller body), they have longer wing bones, longer legs, longer sternums, and greater brain capacity. These characters are in accord with our ideas of physical fitness; their defective development is evidently a mark of inferiority, and we are justified in concluding that birds so handicapped failed to pass one of Nature's rigorous tests and perished.

He also found that:

> The process of selective elimination is most severe with extremely variable individuals, no matter in what direction the variations may occur. It is quite as dangerous to be conspicuously above a certain standard of organic excellence as it is to be below the standard. It is the type that nature favours.

In this last statement, Bumpus is saying that the sparrows suffered from stabilizing selection.

Table 5.8 Measurements on 136 English sparrows classified according to sex, the age of males, and whether or not the birds survived a severe storm on 1 February 1898. The measurements are as follows: TL = total length in mm; AE = length of alar in mm; W = weight in grams; LBH = length of the beak and head in mm; LH = length of the humerus in inches; LF = length of the femur in inches; LTT = length of the tibio-tarsus in mm; WS = width of the skull in inches; and LKS = length of the keel of the sternum in inches.

	TL	AE	W	LBH	LH	LF	LTT	WS	LKS
YMS	156	246	24.6	32.0	0.741	0.735	1.167	0.592	0.849
YMS	156	245	25.5	32.1	0.761	0.717	1.147	0.620	0.816
YMS	163	248	24.8	32.2	0.742	0.733	1.165	0.606	0.854
YMS	163	248	26.3	33.0	0.736	0.704	1.148	0.609	0.839
YMS	160	250	24.4	31.5	0.746	0.715	1.173	0.604	0.893
YMS	156	237	23.3	30.6	0.692	0.664	1.011	0.588	0.774
YMS	162	253	26.7	32.0	0.759	0.734	1.197	0.630	0.878
YMS	163	254	26.4	32.0	0.766	0.750	1.165	0.605	0.886
YMS	164	251	26.9	32.0	0.755	0.742	1.171	0.620	0.886
YMS	163	244	24.3	31.3	0.718	0.680	1.082	0.610	0.892
YMS	160	247	27.0	31.5	0.764	0.732	1.177	0.617	0.846
YMS	160	250	26.8	32.5	0.764	0.729	1.123	0.635	0.842
YMS	158	247	24.9	32.4	0.745	0.724	1.139	0.588	0.865
YMS	158	249	26.1	32.2	0.742	0.736	1.148	0.602	0.817
YMS	158	243	26.6	32.4	0.747	0.711	1.163	0.612	0.891
YMS	155	237	23.3	30.2	0.685	0.653	1.011	0.587	0.794
YMD	160	249	24.2	30.4	0.740	0.717	1.130	0.620	0.840
YMD	156	236	26.8	30.0	0.690	0.671	1.067	0.563	0.832
YMD	158	240	23.5	31.0	0.715	0.702	1.113	0.595	0.805
YMD	166	215	26.9	31.7	0.715	0.695	1.107	0.601	0.847
YMD	165	255	28.6	31.5	0.766	0.744	1.175	0.613	0.854
YMD	157	238	24.7	31.2	0.680	0.677	1.156	0.599	0.769
YMD	164	250	27.3	31.8	0.764	0.726	1.171	0.588	0.860
YMD	166	256	25.7	31.7	0.752	0.751	1.187	0.595	0.858
YMD	167	255	29.0	32.2	0.765	0.745	1.197	0.638	0.855
YMD	161	246	25.0	31.5	0.739	0.707	1.123	0.587	0.850
YMD	166	254	27.5	31.4	0.760	0.742	1.124	0.604	0.914
YMD	161	251	26.0	31.5	0.731	0.707	1.122	0.589	0.828
AMS	154	241	24.5	31.2	0.687	0.668	1.022	0.587	0.830
AMS	160	252	26.9	30.8	0.736	0.709	1.180	0.602	0.841
AMS	155	243	26.9	30.6	0.733	0.704	1.151	0.602	0.846
AMS	154	245	24.3	31.7	0.741	0.688	1.146	0.584	0.839
AMS	156	247	24.1	31.5	0.715	0.706	1.129	0.575	0.821
AMS	161	253	26.5	31.8	0.780	0.743	1.144	0.607	0.893
AMS	157	251	24.6	31.1	0.741	0.736	1.153	0.610	0.862
AMS	159	247	24.3	31.4	0.728	0.718	1.126	0.609	0.793
AMS	158	247	23.6	29.8	0.703	0.673	1.079	0.602	0.820

Table 5.8   (*Contd.*)

	TL	AE	W	LBH	LH	LF	LTT	WS	LKS
AMS	158	252	26.2	32.0	0.749	0.739	1.153	0.614	0.857
AMS	160	252	26.2	32.0	0.741	0.723	1.129	0.624	0.892
AMS	162	253	24.8	32.3	0.766	0.752	1.134	0.633	0.923
AMS	161	243	25.4	31.8	0.721	0.722	1.126	0.597	0.891
AMS	160	250	23.7	29.8	0.730	0.703	1.103	0.590	0.820
AMS	159	247	25.7	31.4	0.729	0.717	1.141	0.592	0.927
AMS	158	253	25.7	31.9	0.743	0.699	1.150	0.600	0.860
AMS	159	247	26.5	31.6	0.733	0.714	1.155	0.611	0.923
AMS	166	253	26.7	32.5	0.767	0.765	1.230	0.600	0.878
AMS	159	247	23.9	31.4	0.752	0.723	1.113	0.602	0.825
AMS	160	248	24.7	31.3	0.752	0.737	1.176	0.603	0.803
AMS	161	252	28.0	31.8	0.770	0.731	1.190	0.590	0.885
AMS	163	251	27.9	31.9	0.769	0.745	1.168	0.622	0.860
AMS	156	242	25.9	32.0	0.723	0.711	1.116	0.609	0.886
AMS	165	251	25.7	32.2	0.751	0.742	1.161	0.613	0.865
AMS	160	247	26.6	32.4	0.728	0.707	1.108	0.590	0.836
AMS	158	244	23.2	31.6	0.730	0.713	1.142	0.585	0.888
AMS	160	242	25.7	31.6	0.709	0.705	1.124	0.620	0.788
AMS	157	245	26.3	32.2	0.741	0.726	1.143	0.595	0.850
AMS	159	244	24.3	31.5	0.723	0.698	1.107	0.615	0.847
AMS	160	253	26.7	32.1	0.739	0.714	1.117	0.592	0.864
AMS	158	245	24.9	31.4	0.726	0.703	1.119	0.580	0.854
AMS	161	247	23.8	31.4	0.735	0.694	1.101	0.602	0.789
AMS	160	247	25.6	32.3	0.756	0.745	1.135	0.607	0.902
AMS	160	247	27.0	32.0	0.755	0.736	1.174	0.631	0.873
AMS	153	241	24.7	32.2	0.728	0.680	1.092	0.592	0.884
AMD	165	249	26.5	31.0	0.738	0.704	1.095	0.606	0.847
AMD	160	245	26.1	32.0	0.736	0.709	1.109	0.611	0.842
AMD	161	249	25.6	32.3	0.743	0.718	1.128	0.602	0.828
AMD	162	246	25.9	32.3	0.738	0.709	1.135	0.607	0.869
AMD	163	250	25.5	32.5	0.752	0.731	1.197	0.623	0.888
AMD	162	247	27.6	31.8	0.731	0.719	1.113	0.597	0.869
AMD	163	246	25.8	31.4	0.689	0.662	1.073	0.604	0.836
AMD	161	246	24.9	30.5	0.739	0.726	1.138	0.580	0.803
AMD	160	242	26.0	31.0	0.745	0.713	1.105	0.600	0.803
AMD	162	246	26.5	31.5	0.720	0.696	1.092	0.606	0.809
AMD	160	249	26.0	31.4	0.726	0.689	1.097	0.602	0.850
AMD	161	250	27.1	31.6	0.737	0.711	1.120	0.631	0.852
AMD	162	248	25.1	31.9	0.744	0.722	1.154	0.591	0.839
AMD	165	252	26.0	32.3	0.726	0.710	1.145	0.609	0.887
AMD	161	243	25.6	32.5	0.709	0.707	1.122	0.607	0.832
AMD	161	244	25.0	31.3	0.702	0.685	1.082	0.595	0.874
AMD	162	248	24.6	31.0	0.713	0.700	1.086	0.590	0.837

Table 5.8 (*Contd.*)

	TL	AE	W	LBH	LH	LF	LTT	WS	LKS
AMD	164	244	25.0	31.2	0.703	0.690	1.074	0.608	0.795
AMD	158	247	26.0	32.0	0.729	0.710	1.145	0.607	0.803
AMD	162	253	28.3	31.8	0.752	0.718	1.152	0.600	0.857
AMD	156	239	24.6	30.5	0.659	0.658	1.042	0.570	0.810
AMD	166	251	27.5	31.5	0.720	0.691	1.118	0.612	0.847
AMD	165	253	31.0	32.4	0.765	0.750	1.183	0.613	0.905
AMD	166	250	28.3	32.4	0.754	0.718	1.179	0.607	0.916
FS	156	245	25.3	31.6	0.729	0.710	1.152	0.620	0.809
FS	154	240	22.6	30.4	0.705	0.686	1.103	0.584	0.770
FS	153	240	25.1	31.0	0.724	0.713	1.123	0.585	0.812
FS	153	236	23.2	30.9	0.698	0.678	1.132	0.596	0.795
FS	155	243	24.4	31.5	0.734	0.736	1.170	0.596	0.801
FS	163	247	25.1	32.0	0.748	0.734	1.166	0.602	0.821
FS	157	238	24.6	30.9	0.726	0.727	1.175	0.588	0.797
FS	155	239	24.0	32.8	0.732	0.742	1.175	0.601	0.835
FS	164	248	24.2	32.7	0.752	0.752	1.201	0.604	0.830
FS	158	238	24.9	31.0	0.741	0.689	1.091	0.592	0.866
FS	158	240	24.1	31.3	0.733	0.706	1.107	0.591	0.867
FS	160	244	24.0	31.1	0.731	0.730	1.152	0.589	0.808
FS	161	246	26.0	32.3	0.758	0.732	1.154	0.623	0.859
FS	157	245	24.9	32.0	0.752	0.740	1.186	0.593	0.787
FS	157	235	25.5	31.5	0.712	0.704	1.132	0.611	0.781
FS	156	237	23.4	30.9	0.708	0.691	1.123	0.613	0.798
FS	158	244	25.9	31.4	0.729	0.705	1.146	0.597	0.851
FS	153	238	24.2	30.5	0.715	0.707	1.116	0.595	0.821
FS	155	236	24.2	30.3	0.727	0.705	1.120	0.585	0.790
FS	163	246	27.4	32.5	0.732	0.711	1.163	0.630	0.862
FS	159	236	24.0	31.5	0.709	0.713	1.129	0.607	0.845
FD	155	240	26.3	31.4	0.709	0.710	1.125	0.614	0.815
FD	156	240	25.8	31.5	0.715	0.678	1.127	0.597	0.812
FD	160	242	26.0	32.6	0.740	0.732	1.157	0.597	0.854
FD	152	232	23.2	30.3	0.676	0.683	1.048	0.590	0.780
FD	160	250	26.5	31.7	0.741	0.731	1.187	0.615	0.886
FD	155	237	24.2	31.0	0.727	0.723	1.118	0.610	0.787
FD	157	245	26.9	32.2	0.766	0.751	1.227	0.620	0.841
FD	165	245	27.7	33.1	0.780	0.757	1.195	0.633	0.895
FD	153	231	23.9	30.1	0.680	0.662	1.042	0.592	0.781
FD	162	239	26.1	30.3	0.709	0.685	1.092	0.587	0.911
FD	162	243	24.6	31.6	0.741	0.729	1.162	0.605	0.840
FD	159	245	23.6	31.8	0.727	0.700	1.129	0.610	0.855
FD	159	247	26.0	30.9	0.711	0.666	1.098	0.580	0.749
FD	155	243	25.0	30.9	0.730	0.711	1.127	0.598	0.839
FD	162	252	24.8	31.9	0.752	0.738	1.180	0.615	0.875

Table 5.8 (Contd.)

	TL	AE	W	LBH	LH	LF	LTT	WS	LKS
FD	152	230	22.8	30.4	0.682	0.664	1.042	0.551	0.734
FD	159	242	24.8	30.8	0.717	0.667	1.090	0.575	0.809
FD	155	238	24.6	31.2	0.706	0.702	1.102	0.588	0.758
FD	163	249	30.5	33.4	0.767	0.767	1.207	0.640	0.896
FD	163	242	24.8	31.0	0.713	0.713	1.128	0.607	0.813
FD	156	237	23.9	31.7	0.718	0.716	1.090	0.611	0.800
FD	159	238	24.7	31.5	0.726	0.701	1.145	0.600	0.800
FD	161	245	26.9	32.1	0.751	0.704	1.142	0.607	0.819
FD	155	235	22.6	30.7	0.695	0.692	1.119	0.584	0.771
FD	162	247	26.1	31.9	0.751	0.735	1.157	0.618	0.802
FD	153	237	24.8	30.6	0.732	0.718	1.172	0.594	0.802
FD	162	245	26.2	32.5	0.728	0.731	1.102	0.614	0.832
FD	164	248	26.1	32.3	0.739	0.707	1.159	0.592	0.823

YMS = young males that survived; YMD = young males that died; AMS = adult males that survived; AMD = adult males that died; FS = adult and young females that survived; FD = adult and young females that died.

There certainly is evidence of some differences between survivors and non-survivors, as is shown in Table 5.9. For males there are significant differences between mean values for total length and weight, according to *t*-tests. For females there are significant differences between standard deviations for the length of the humerus and the length of the tibio-tarsus, according to variance ratio tests. The test statistic λ of Section 3.4 can be used to test for an overall difference between survivors and non-survivors for all nine measurements simultaneously: a significantly large value is found for both males and females. Furthermore, this statistic can be partitioned into $\lambda = \lambda_1 + \lambda_2$, where λ_1 relates to changes in dispersion and λ_2 to changes in means. It is found that λ_2 is significant for males while λ_1 is significant for females.

It appears, then, that selection has resulted in males having different mean values for survivors and non-survivors while the variation in measurements has changed little. On the other hand, for females, mean values are rather similar for survivors and non-survivors but the amount of variation has changed. Indeed, the surviving females have a smaller standard deviation than the non-survivors for all nine measurements (Table 5.9). Thus the stabilizing selection noted by Bumpus seems to have acted primarily on the females.

Bumpus's data have been re-analysed many times since 1898 (Grant, 1972; Johnston *et al.*, 1972; O'Donald, 1973; Lande and Arnold, 1983; Lowther, 1977; Manly, 1976). O'Donald concluded that survival was mainly related to the length of the humerus and the total length of birds. He therefore carried out a discriminant function analysis to find the linear combination of these two

Table 5.9 Comparison of measurements on survivors and non-survivors for Bumpus's data.

		Males		Females	
		Survivors	Non-survivors	Survivors	Non-survivors
Sample size		51	36	21	28
Total length	Mean	159.26	162.08*	157.38	158.42
	Std. dev.	2.87	2.94	3.32	3.81
Alar extent	Mean	247.41	247.55	241.00	241.57
	Std. dev.	4.17	4.83	4.18	5.70
Weight	Mean	254.78	269.69†	246.19	252.36
	Std. dev.	12.52	15.26	10.66	16.41
Length of beak	Mean	316.94	315.56	314.33	314.79
and head	Std. dev.	6.61	6.24	7.29	8.54
Length of	Mean	739.14	730.20	728.33	726.04
humerus	Std. dev.	21.12	25.44	16.37	25.93‡
Length of	Mean	716.63	709.17	714.81	709.75
femur	Std. dev.	24.10	22.79	20.20	28.38
Length of	Mean	113.61	112.72	114.38	113.21
tibio-tarsus	Std. dev.	4.31	3.92	3.04	4.84‡
Width of skull	Mean	604.16	601.95	600.10	601.57
	Std. dev.	14.19	14.76	13.08	18.17
Length of keel	Mean	855.63	844.72	819.29	820.67
of sternum	Std. dev.	36.77	33.23	29.87	45.29

		Males	Females
Test for constant dispersion	(45 d.f.)	$\lambda_1 = 43.21$	71.63†
Test for constant means	(9 d.f.)	$\lambda_2 = 53.17$*	9.99
Total	(54 d.f.)	$\lambda = 96.38$*	81.62†

Mean values have been compared using *t*-tests, variances have been compared using *F*-tests, and λ values are tested against the chi-square distribution. Significant differences between survivors and non-survivors are indicated as: * 0.1% level, † 1% level, ‡ 5% level.

variables that differs as much as possible between the surviving and dead birds. This linear combination is

$$Y = -0.002285(\text{TL}) + 0.3075(\text{LH})$$

where TL denotes the total length and LH denotes the length of the humerus. The probability of survival for the birds in terms of the composite variable Y was then estimated using the equations of Section 3.5, to be

$$\phi = 0.8400 - 708.2(0.120 + Y)^2.$$

This function suggests very strong selection indeed. For example, a bird with TL = 153 and LH = 0.728 has $Y = -0.13$ and hence $\hat{\phi} = 0.82$. However, a bird with TL = 163 and LH = 0.689 has $Y = -0.16$ and $\hat{\phi} = -0.33$. This negative value is impossible since ϕ is, of course, supposed to be a probability of survival.

O'Donald also fitted the nor-optimal function to Bumpus's data. This is better than the quadratic function because it does not give negative fitnesses. However, the calculations were based on the assumption that the distribution of Y was normal before selection. This is invalid in the sense that the 'population' before selection is a finite population of 136 birds that cannot be normal.

When proportional hazard fitness functions are fitted to Bumpus's male data, O'Donald's general conclusions are confirmed: selection appears to have been related only to the total length and the length of the humerus (Manly, 1976). Thus the following equation fits the data on the 87 male birds:

$$\phi = \exp\left[-\exp\left\{-35.8 + 0.402\,(\text{TL}) - 40.2\,(\text{LH})\right\}\right]. \tag{5.13}$$

For a bird with TL = 153 and LH = 0.728 this gives the probability of surviving the storm as 0.972. However, for a bird with TL = 163 and LH = 0.689 the probability is only 0.0005. This suggests a very high level of selection and confirms the high level suggested by O'Donald's quadratic fitness function. The function is illustrated in Fig. 5.2.

The amount of selection suggested by equation (5.13) is rather difficult to believe. However, it is shown in the data itself. Table 5.10 gives the results that are obtained when Bumpus's 87 male birds are classified according to their total length and humerus length in a frequency table. The grouping allows some observed survival rates to be calculated, and these agree well with the rates predicted by the equation.

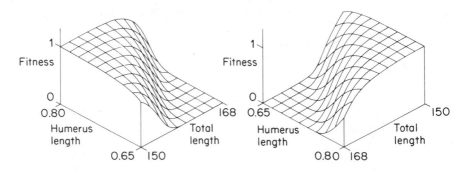

Fig. 5.2 An estimated fitness function for the storm survival of male sparrows. The height of the surface gives the probability of survival for particular values of total length (in mm) and humerus length (in inches). Both graphs show the same surface, viewed from different angles.

Table 5.10 The results obtained when Bumpus's 87 male birds are grouped into a frequency table on the basis of total length and humerus length. The expected numbers surviving were obtained using equation (5.13) as expected number surviving = number of birds × probability of survival.

Total length (mm)		*Length of humerus* (inches)			
		0.64–0.67	0.68–0.71	0.72–0.75	0.76–0.79
164–167	*Number of birds*	—	1	7	7
	Observed number surviving	—	0	1	2
	Expected number surviving	—	0.0	0.3	3.6
160–163	*Number of birds*	—	5	27	10
	Observed number surviving	—	1	12	10
	Expected number surviving	—	0.2	14.0	8.8
156–159	*Number of birds*	1	4	19	1
	Observed number surviving	0	2	17	1
	Expected number surviving	0.0	2.1	16.7	0.97
152–155	*Number of birds*	—	2	3	—
	Observed number surviving	—	2	3	—
	Expected number surviving	—	1.8	2.9	—

As mentioned above, stabilizing selection seems to have taken place on female sparrows. This is verified when the Mahalanobis distances are calculated for these birds (equation (5.11)). It can be argued that the weight variable should not be included in the calculation because a bird's weight is liable to depend upon its temporary physical condition. Distances have therefore been calculated from the eight other variables only. The relationship between survival and distance is shown in Table 5.11. It appears quite clear that birds far from the average had relatively low survival probabilities.

A relationship

$$\phi = \exp\{-\exp(-0.85 + 0.71D)\}$$

fits the data well according to a GLIM analysis. Thus average birds ($D = 0$) appear to have had a survival probability of about $\exp\{-\exp(-0.85)\}$ = 0.65. On the other hand, the most unusual bird, ($D = 2.54$) only had a survival probability of about $\exp\{-\exp(-0.85 + 2.54 \times 0.71)\} = 0.07$.

If the distance measure D^* of equation (5.12) is used instead of D, then the observed survival rates shown in Table 5.12 are obtained. Overall, the D^* values are about twice as variable as the D values. This comes about because the positive correlation between body measurements is taken into account with D but not with D^*. However, since survival seems strongly related to D^* as well as D, it does not matter much which measure is used as a deviation from average.

Actually, Johnston *et al.* (1972) have pointed out that since Bumpus used his data to formulate the hypothesis of stabilizing selection, this hypothesis cannot be tested on the data. For this reason, Baker and Fox (1978) attempted to replicate Bumpus's experiment by spraying a two-acre roost of grackles and other birds with the wetting agent Turgitol and water. This simulated a severe storm and Baker and Fox collected 16 surviving and 42 dead male grackles.

Table 5.11 Survival related to Mahalanobis distances for Bumpus's 49 female birds. A bird's distance is a measure of how far it is from the average of all birds.

Distance (D)	Initial number of birds	Survivors from selection	Percentage survival
0–0.50	4	3	75
–1.00	28	13	46
–1.50	10	3	30
–2.00	6	2	33
–2.50	0	–	–
–3.00	1	0	0
	49	21	

Table 5.12   Survival related to D^* distances for Bumpus's 49 female birds.

Distance (D^*)	Initial number of birds	Survivors from selection	Percentage survival
0–0.5	19	9	47
−1.0	14	8	57
−1.5	6	3	50
−2.0	5	1	20
−2.5	0	–	–
−3.0	1	0	0
−3.5	2	0	0
−4.0	1	0	0
−4.5	0	–	–
−5.0	1	0	0
	49	21	

They examined differences between the survivors and the non-survivors but were unable to confirm the stabilizing selection that was found by Bumpus.

See Lowther (1977) and Fleischer and Johnston (1982) for the analysis of other sets of data on *Passer domesticus* similar to those of Bumpus. Boag and Grant (1981) have provided data on the drought survival of Darwin's finches, *Geospiza fortis*, on one of the Galapagos Islands.

5.6   THE CHESSON–MANLY MODEL FOR COMPETITIVE SURVIVAL

The Chesson–Manly model is for a type of selection process that is rather different from those considered so far in this chapter. The main difference comes from the fact that individuals can no longer be considered to survive independently. Rather, they compete to survive. The following two examples should clarify what is meant by this. Both examples involve selective predation where the death of an individual meant that the probability of survival increased for all other individuals.

Table 5.13 shows the results of six selective predation experiments conducted by Popham (1943). The predators were three rudd (*Scardinius erythrophthalmus*: Osteichthyes) and their prey were corixids (*Sigara distincta*: Coleoptera). There were three colours of prey (light, medium and dark grey) and Popham was interested in the reaction of the rudd to changes in the proportions of these three colours.

All the experiments were performed in an aquarium. About ten minutes after the fish had been placed in the aquarium the insects were introduced by pouring them into a part of the aquarium not containing any fish. When an

Table 5.13   The results of six selective predation experiments. The predators were three rudd and the prey were three colours of the insect *Sigara distincta*.

	Type 1 prey (light grey)			Type 2 prey (medium grey)			Type 3 prey (dark grey)			
Experiment	Number presented	Observed number eaten	Expected number eaten	Number presented	Observed number eaten	Expected number eaten	Number presented	Observed number eaten	Expected number eaten	χ^2 (2 d.f.)
1	4	105	66.7	4	67	66.7	4	28	66.7	44.47 ($p < 0.001$)
2	3	56	50.0	6	132	100.0	3	12	50.0	39.84 ($p < 0.001$)
3	6	144	100.0	3	40	50.0	3	16	50.0	44.48 ($p < 0.001$)
4	3	55	47.0	3	25	47.0	6	108	94.0	13.74 ($0.01 > p > 0.001$)
5	2	46	40.0	2	23	40.0	6	131	120.0	9.13 ($0.05 > p < 0.01$)
6	2	29	32.7	4	62	65.3	6	105	98.0	1.08 ($p > 0.05$)

Notes: 1. Twelve prey were available to the predators throughout experiments 1 to 4 and 6; in experiment 5 there were ten prey. These numbers available were kept constant by the replacement of eaten prey.
2. The chi-square values shown are based upon a comparison between the observed and expected prey numbers taken with the expected numbers being calculated on the assumption that prey are taken in proportion to the numbers available. The significantly high chi-square values show that this assumption is not realistic.

insect was eaten by one of the rudd it was immediately replaced by another specimen of the same colour so that the composition of the prey population was kept constant. All the experiments were stopped after about 200 insects had been eaten. The aquarium was placed in a tank lined with dark grey sand so that the light grey beetles should have been the most visible to the rudd.

From his experimental results Popham concluded that: 'When the types of prey are not in equal numbers, those insects present in the largest numbers are destroyed relatively faster than the others.' Today this phenomenon is called *apostatic selection* and it has been proposed as one of the main mechanisms for maintaining variation in nature (Clarke, 1962b, 1969, 1979).

Popham's experiments were slightly unusual in that he managed to keep the prey numbers constant while the experiment was being carried out. The experimental procedure used by L. M. Cook and P. Miller to study apostatic selection in Japanese quail, *Coturnix coturnix japonica*, is more usual with selective predation experiments. They used red and blue pastry pellets as 'prey' and did not attempt to replace them as they were eaten. The experiments involved putting a pair of quail into an aviary that was strongly illuminated with artificial light. A number of red and blue pellets were scattered on the floor and the birds were allowed to eat half of these. Each pair of quail were used to repeat the experiment on several successive days with the same initial frequencies of red and blue pellets.

One study (Manly et al., 1972) involved a large number of experiments on 70 pairs of birds. In these experiments the initial total number of pastry pellets presented to the birds was always 20, but the percentage of red pellets ranged from 10% to 80%. The motivation for the experiment was to see whether the birds would exhibit apostatic selection of the type observed by Popham in his experiments with rudd. From all the experimental results it is quite clear that the birds were tending to over-eat the most common form of prey. However, this apostatic selection was superimposed on a general tendency to prefer the red to the blue pellets. In later experiments that are described by Cook and Miller (1977), it was found that the nature of the apostatic selection of the quail changed according to the density of the prey on the aviary floor.

One model for these selective predation experiments has been discussed in several papers (Manly et al., 1972; Manly, 1973b, 1974b, 1980; Chesson, 1976, 1978, 1983). It is based upon the assumption that if a selection process such as predation is acting on a population that consists of K morphs, with n_i of these being of morph i, then the probability that a type i individual will be taken first is given by

$$P_i = \beta_i n_i \bigg/ \sum_{j=1}^{K} \beta_j n_j. \qquad (5.14)$$

Here the β_j are positive constants that reflect the amount of selection taking place. If all the β's are equal then there is no selection at all; if the β's are very variable then there is a great deal of selection.

If equation (5.14) is valid then the predators or other agents of selection are behaving as if there are $\beta_i n_i$ individuals of morph i, rather than n_i of them; P_i is the probability of selecting a type i individual at random from this distorted population. It is often convenient to assume that the β values are scaled so that

$$\sum_{j=1}^{K} \beta_j = 1.$$

Then β_i can be interpreted as the probability of a type i individual being selected first if there are an equal number of each of the K morphs available.

Chesson (1978, 1983) gives a justification for equation (5.14) based upon a simple stochastic model for predator behaviour involving probabilities of encounter and capture. She also shows how the β's are related to other measures of selective predation. Cock (1978), Pearre (1982, 1983) and Vanderploeg and Scavia (1983) also review these other measures.

5.7   ESTIMATION WITH A CONSTANT POPULATION

The estimation of β values is particularly simple for experiments in which the population is kept constant by the replacement of selected individuals. Thus, consider an experiment where the relative frequencies of K morphs are P_1, $P_2, \ldots, P_K$. Suppose that individuals are selected one by one, with replacement, until n have been removed. Let d_i denote the number of morph i individuals selected, so that $\sum d_i = n$. From equation (5.14) the probability of a morph i individual being selected at any stage is always

$$\text{Prob}\,(i) = \beta_i P_i \bigg/ \sum_{j=1}^{K} \beta_j P_j.$$

The numbers removed by the predator then have a multinomial distribution, and maximum likelihood estimators of $\beta_1, \beta_2, \ldots, \beta_K$, subject to the constraint $\sum \beta_i = 1$, are given by

$$\hat{\beta}_i = (d_i/P_i) \bigg/ \left(\sum_{j=1}^{K} d_j/P_j \right), \tag{5.15}$$

which is essentially the same as the estimator of equation (4.4).

The following equations give biases, variances and covariances, based upon the usual Taylor series approximation:

$$\left. \begin{aligned} \text{bias}\,(\hat{\beta}_i) &\simeq \beta_i \sum_{r=1}^{K} \alpha_r - \alpha_i \\[2mm] \text{var}\,(\hat{\beta}_i) &\simeq \beta_i^2 \sum_{r=1}^{K} \alpha_r + (1 - 2\beta_i)\alpha_i \\[2mm] \text{cov}\,(\hat{\beta}_i,\,\hat{\beta}_j) &\simeq \beta_i \beta_j \sum_{r=1}^{K} \alpha_r - \beta_i \alpha_j - \beta_j \alpha_i \end{aligned} \right\} \tag{5.16}$$

and

where $i \neq j$, $\alpha_i = \beta_i^2 / E(d_i)$, and $E(d_i)$ is the expected value of d_i. These are just equations (4.5) to (4.7) with the subscript j omitted.

Example 5.4   Predation of corixids by rudd

Table 5.14 shows the estimated β values obtained using equation (5.15) on Popham's data on predation of corixids by rudd (Table 5.13). It will be seen that for each morph the β value seems to have been related to the P value. For example, for morph 1 (light grey) the highest estimate of β_1 is 0.56, which occurs when the proportion of this colour present was $P_1 = 0.5$. On the other hand, the lowest estimate of β_1 is 0.31 which occurs when $P_1 = 0.17$. Generally it appears that β_i increased with P_i, so that apostatic selection was taking place with the predators 'over-eating' the most common forms of prey.

Various models have been proposed for the relationship between β and P values in a situation such as this. A linear model fitted to Popham's data provides the relationships

$$\beta_1 = 0.944 P_1 + 0.233 P_2 + 0.241 P_3$$
$$\beta_2 = 0.160 P_1 + 0.936 P_2 - 0.067 P_3$$
$$\beta_3 = -0.106 P_1 - 0.169 P_2 + 0.825 P_3$$

when it is fitted by multiple regression. This type of model is relatively simple and fits experimental data quite well (Manly, 1973b). However, it has been criticized as being too simple. Greenwood and Elton (1979) and Fullick and Greenwood (1979) discuss one alternative which fits an extensive set of data better. See also Horsley $et\ al.$ (1979) and Greenwood $et\ al.$ (1981).

5.8   ESTIMATION WITH A DECREASING POPULATION

If individuals are not replaced when they are selected, then the population that is being selected will gradually change, and equation (5.15) no longer provides proper estimates of β values. Chesson (1976) has provided an expression for the probability associated with the numbers $d_1, d_2, \ldots, d_K$ of the different morphs selected in this case. Maximum likelihood estimators are found by maximizing this probability for the observed values d_i, which has to be done numerically. However, it is simpler to use the moment estimator

$$\hat{\beta}_i = \log_e (r_i / A_i) \bigg/ \sum_{j=1}^{K} \log_e (r_j / A_j), \tag{5.17}$$

where $r_j = A_j - d_j$ is the number of j morph individuals remaining after selection. Equations for biases, variances and covariances are available (Manly, 1974b) but these are rather complicated so they will not be given here. (However, see the comments in the next section of this chapter.)

Table 5.14 Estimates of β values for experiments on predation of corixids by rudd.

Experiment	Proportions of prey			$\hat{\beta}_1$	Standard error	$\hat{\beta}_2$	Standard error	$\hat{\beta}_3$	Standard error
	P_1	P_2	P_3						
1	0.33	0.33	0.33	0.53	0.035	0.34	0.033	0.14	0.025
2	0.25	0.50	0.25	0.42	0.039	0.49	0.037	0.09	0.025
3	0.50	0.25	0.25	0.56	0.039	0.31	0.038	0.13	0.029
4	0.25	0.25	0.50	0.41	0.049	0.19	0.033	0.40	0.036
5	0.20	0.20	0.60	0.41	0.042	0.20	0.037	0.39	0.035
6	0.17	0.33	0.50	0.31	0.043	0.33	0.035	0.37	0.034

Note: morphs 1 to 3 are respectively light, medium and dark-coloured Sigara distincta.

If it happens that a record is available of the actual order in which n individuals are selected, then obtaining maximum likelihood estimates of the β_i values is fairly easy and is liable to produce somewhat more precise results than using equation (5.17) (Manly, 1980). Thus let g_{ij} be the number of morph i individuals still unselected when a total of j individuals have been selected. Then it can be shown that maximum likelihood estimates of the β_i are provided by the solution of the equations

$$A_i - r_i = \sum_{j=1}^{n} \hat{P}_{ij}, \qquad i = 1, 2, \ldots, K-1 \tag{5.18}$$

where

$$\hat{P}_{ij} = \hat{\beta}_i g_{ij-1} \Big/ \sum_{t=1}^{K} \hat{\beta}_t g_{tj-1}$$

is the estimated probability that the jth selection is of morph i, and $\hat{\beta}_K = c$, any convenient constant. (Setting $\hat{\beta}_K = c$ is needed to fix the scale of the β values.)

Equations (5.18) are conveniently solved by an iterative method. They can be rewritten as

$$\hat{\beta}_i = (A_i - r_i) \Big/ \sum_{j=1}^{n} (\hat{P}_{ij}/\hat{\beta}_i), \qquad i = 1, 2, \ldots, K. \tag{5.19}$$

The idea is to start with initial approximations for the β's, substitute these into the right-hand side of equations (5.19), and hence obtain better approximations. It is convenient to have the $\hat{\beta}$'s adding to one, and this can be achieved by dividing the new approximations for the $\hat{\beta}$'s by their sum. The iteration equations are then

$$\hat{\beta}_i^{(1)} = \left\{ (A_i - r_i) \Big/ \sum_{j=1}^{n} \hat{P}_{ij}^{(0)}/\hat{\beta}_i^{(0)} \right\} \Big/ \left\{ \sum_{t=1}^{K} (A_t - r_t) \Big/ \sum_{j=1}^{n} \hat{P}_{tj}^{(0)} \hat{\beta}_t^{(0)} \right\}$$

for $i = 1, 2, \ldots, K$, where $\hat{\beta}_i^{(0)}$ is an initial approximation and $\hat{\beta}_i^{(1)}$ is an improved approximation for $\hat{\beta}_i$. Also, $\hat{P}_{ij}^{(0)}$ is $\hat{P}_{ij}$ evaluated using the $\hat{\beta}_i^{(0)}$.

Provided that equations (5.19) have a solution, the iterative method for solving them always appears to converge after three or four iterations when the initial approximations for the β's are taken as $\beta_i = 1/K$ for all i. However, the equations do not always possess a solution. For example, suppose that an experiment starts with two individuals of each of morphs A, B and C, and that morphs A, A, B and C are selected in that order. Then equations (5.18) have the form

$$2 = \hat{P}_{11} + \hat{P}_{12} + 0 + 0$$

and

$$1 = \hat{P}_{21} + \hat{P}_{22} + \hat{P}_{23} + \hat{P}_{24}.$$

These equations cannot have a solution since the first one implies that β_2 and β_3 are zero while the second one implies that β_2 is non-zero. What the experiment is really suggesting is that β_1 is large, while β_2 is small, but bigger than β_3. Under these conditions it is perhaps simplest to take $\beta_1 = 1$ and $\beta_2 = \beta_3 = 0$. This amounts to ignoring the selection results after the morph A individuals were all selected. This problem arises only when all of one morph are removed before any of another morph have been taken.

Variances and covariances for the estimators of equations (5.18), subject to the constraint $\Sigma \beta_i = 1$, are approximately as follows:

$$\left. \begin{aligned} \text{var}(\hat{\beta}_i) &\simeq \beta_i^2 \left(\frac{1 - 2\beta_i}{A_i - E(r_i)} + \sum_{t=1}^{K} \frac{\beta_t^2}{A_t - E(r_t)} \right) \\ \text{and} \\ \text{cov}(\hat{\beta}_i, \hat{\beta}_j) &\simeq -\beta_i \beta_j \left(\frac{\beta_i}{A_i - E(r_i)} + \frac{\beta_j}{A_j - E(r_j)} - \sum_{t=1}^{K} \frac{\beta_t^2}{A_t - E(r_t)} \right), \end{aligned} \right\} \quad (5.20)$$

for i and $j = 1, 2, \ldots, K$, with $i \neq j$.

5.9   HARD OR SOFT SELECTION?

The difference between the Chesson–Manly model and the proportional hazards model for survival is that in the former case the individuals are assumed to be in competition, whereas in the latter case the survival or death of an individual is independent of the survival or death of any other individual. The first type of selection has been called 'soft', because it takes into account the relative fitnesses of different individuals; the second type of selection has been called 'hard', because the probability of selection does not depend upon the relative fitnesses (Wallace, 1981, Ch. 16). The point here is that with soft selection a certain fixed proportion of a population survive, whereas with hard selection it can happen that none of a population survive.

In many cases it is easy to decide whether selection is hard or soft. For instance, with Bumpus's sparrows it is difficult to see how the death of one bird as a result of a storm could influence the probability of death of another bird. This is hard selection. On the other hand, there are cases where it is not so clear. For example, Table 5.15 shows the results of an experiment that is described by O'Donald (1971). This involved setting up a population cage containing 564 *Drosophila melanogaster* with a limited amount of food. The survivors were counted when about two-thirds of the population had died, and classified according to the number of sternopleural bristles. In this case there was presumably a good deal of competition for food. However, the total number dying by the end of the experiment was not predetermined and no doubt had something to do with the inherent ability of the flies to survive a food shortage. Thus this seems to be an example where there was a mixture of hard and soft selection.

In some respects the difference between hard and soft selection is just a question of experimental design. Thus suppose that O'Donald had decided in advance to stop his experiment on the effect of food shortage when exactly 179 *Drosophila* were still alive. Then there seems no difficulty at all in regarding the

Table 5.15 The survival of male *Drosophila melanogaster* with different numbers of sternopleural bristles.

Sternopleural bristle number	Freq. before selection (A_i)	Freq. after selection (r_i)	Proportion surviving ($\hat{p}_i$)
12	1	0	0.0000
13	1	0	0.0000
14	25	7	0.2800
15	51	8	0.1569
16	98	45	0.4592
17	104	36	0.3462
18	106	32	0.3019
19	67	22	0.3284
20	54	17	0.3148
21	24	6	0.2500
22	21	4	0.1905
23	7	1	0.1429
24	4	1	0.2500
25	0	0	–
26	1	0	0.0000
	564	179	

experiment as an example of a Chesson–Manly selection process. On the other hand, suppose that, instead, he decided to stop the experiment after five days. In that case the total number of deaths is a random variable and it does not seem unreasonable to use the proportional hazards model to analyse the data. Actually, O'Donald did neither of these things. He stopped the experiment when about two-thirds of the flies were dead. Under these circumstances either the Chesson–Manly model or the proportional hazards model might be used to analyse the data.

Luckily, the difference between assuming hard or soft selection seems fairly unimportant when it comes to estimation. In particular, it can be noted that the estimator of equation (5.17) for the β_i parameter of the Chesson–Manly selection process on a decreasing population can be derived equally well assuming independent selection. Thus, assume that a polymorphic population has initial numbers $A_1, A_2, \ldots, A_K$ for K morphs, and that the survival probability per unit time is $\exp(-\lambda_i)$ for morph i. Then the expected number of survivors for this morph at time t is

$$E(r_i) = A_i \exp(-\lambda_i t),$$

so that

$$\log_e \{ E(r_i)/A_i \} = -\lambda_i t$$

or

$$\log_e\{E(r_i)/A_i\}\bigg/ \sum_{j=1}^{K} \log_e\{E(r_j)/A_j\} = \lambda_i\bigg/ \sum_{j=1}^{K} \lambda_j.$$

This shows that

$$\hat{\beta}_i = \log_e(r_i/A_i)\bigg/ \sum_{j=1}^{K} \log_e(r_j/A_j) \tag{5.21}$$

is an estimator of $\beta_i = \lambda_i/\Sigma\lambda_j$, the relative death rate for morph i animals. This estimator is the same as the estimator (5.17) of β_i for the Chesson–Manly selection process. Therefore $\hat{\beta}_i$ values are equally meaningful for hard or soft selection.

Of course, the sampling properties of $\hat{\beta}_i$ values depend upon what model for selection is correct. It was remarked in the previous section of this chapter that the equations for biases, variances and covariances are rather complicated for the Chesson–Manly model. However, there are fairly straightforward approximations available with the model of independent selection that has just been considered. For this model (Manly, 1975c),

$$\text{bias}(\hat{\beta}_i) \simeq \left(\beta_i \sum_{s=1}^{K} \alpha_s - \alpha_i\right)(1 + H/2)/H^2, \tag{5.22}$$

$$\text{var}(\hat{\beta}_i) \simeq \left\{\beta_i^2 \sum_{s=1}^{K} \alpha_s + (1 - 2\beta_i)\alpha_i\right\}\bigg/ H^2, \tag{5.23}$$

and

$$\text{cov}(\hat{\beta}_i, \hat{\beta}_j) \simeq \left(\beta_i\beta_j \sum_{s=1}^{K} \alpha_s - \beta_i\alpha_j - \beta_j\alpha_i\right)\bigg/ H^2 \tag{5.24}$$

where $\alpha_i = (1 - \phi_i)/(A_i\phi_i)$,

$$H = \sum_{s=1}^{K} \log_e(\phi_s)$$

and $\phi_i = E(r_i)/A_i$.

In the example that follows it will be seen that the variances given by equation (5.23) are virtually the same as the variances for the Chesson–Manly model. Intuitively it seems reasonable to hope that this agreement is general. It is suggested, therefore, that the equations (5.22) to (5.24) may be used with the Chesson–Manly model in place of the rather complicated more exact formulae.

Example 5.5   Food shortage with *Drosophila melanogaster*

The purpose of this example is to indicate that from the point of view of estimation it seems to make very little difference whether selection is hard or soft. The data are from O'Donald's (1971) experiment on the effect of food shortage with *Drosophila melanogaster* (Table 5.15).

The important results are shown in Table 5.16. The sternopleural bristle numbers can be regarded as dividing the population into different morphs. Then equation (5.21) provides estimates of the parameters β for the Chesson–Manly selection model of Section 5.8, or relative death rates for the independent survival model of Section 5.9. Only flies with bristle numbers from 14 to 22 have been considered. It will be seen that the standard errors for β estimates are virtually the same for either model.

Table 5.16   Estimates of β_i values from the experiment on the effect of food shortage on *Drosophila melanogaster*.

Bristle number	$\hat{\beta}_i$	Standard error (1)	Standard error (2)
14	0.111	0.026	0.026
15	0.161	0.025	0.025
16	0.068	0.010	0.010
17	0.093	0.012	0.013
18	0.104	0.013	0.014
19	0.097	0.015	0.015
20	0.101	0.017	0.017
21	0.121	0.028	0.028
22	0.144	0.034	0.035
	1.000		

Standard error (1) is the standard error for the Chesson–Manly model (soft selection) as calculated using equations given by Manly (1974b).
Standard error (2) is calculated using equation (5.23) which is based on a model for independent survival (hard selection).
Both models give quite small biases to the $\hat{\beta}$ estimates.

5.10   THE INTENSITY AND EFFECT OF SELECTION

In Chapter 3, four indices of the intensity and effect of selection were introduced. Haldane's index I_H and Van Valen's index I_V are measures of the intensity of selection per unit time. O'Donald's index I_O and the index I_u are measures of the effect of selection for a particular amount of time on a population. With weak selection $I_H \simeq I_V$ and $I_O \simeq I_u$. Example 3.6 suggests that I_u may have better sampling properties than I_H, I_V and I_O. These indices were also considered in Chapter 4. However, in some of the situations considered there it is only possible to estimate relative mortality rates for different types of individual. This means that none of the indices I_H, I_V, I_O and

I_u can be calculated. Consequently, a new index I_R was suggested where this is a measure of the intensity of a selection process that is not related to any particular selection time.

For the situations considered in the present chapter, all five of these indices can be calculated. They are all valid measures of different aspects of the selection process, but for simplicity attention will be restricted to I_V (intensity per unit time), I_u (effect), and I_R (inherent intensity). These are defined as

$$I_V = (\phi^* - \overline{\phi})/\phi^*, \quad I_u = \overline{u}' - \overline{u} \quad \text{and} \quad I_R = \mathrm{var}(u)/\overline{u}^2, \tag{5.25}$$

where ϕ denotes a survival probability, $u = \log_e(\phi)$, ϕ^* is the maximum and $\overline{\phi}$ is the mean of ϕ before selection, $\overline{u}$ is the mean of u before selection, $\overline{u}'$ is the mean of u after selection, and $\mathrm{var}(u)$ is the variance of u before selection.

In using these indices the survival probabilities ϕ can either be determined from an estimated fitness function or simply be taken as proportions surviving. Using proportions surviving is simpler and seems preferable if it is possible. That is to say, suppose that before selection there are K morphs with frequencies $A_1, A_2, \ldots, A_K$, and of these $r_1, r_2, \ldots, r_K$ survive selection for a time t. Then

$$\hat{\phi}_i = (r_i/A_i)^{1/t} \tag{5.26}$$

is an estimate of the probability of survival per unit time that can be used in equations (5.25). There is a problem here if $r_i = 0$, which can be overcome by taking

$$\hat{u}_i = \frac{1}{t}\log_e\left\{\frac{r_i}{A_i} + \frac{A_i - r_i}{2A_i^2}\right\} \tag{5.27}$$

instead of $u_i = \log_e(\phi_i)$ for I_u and I_R. A Taylor series approximation shows that equation (5.27) provides an almost unbiased estimator of $\log_e(\phi_i)$, with variance

$$\mathrm{var}(\hat{u}_i) \simeq (1 - \phi_i)/(t^2 \phi_i A_i). \tag{5.28}$$

The Taylor series method is unable to provide valid expressions for the variances of the indices I_V, I_u and I_R. With I_V it cannot be used because this index is a function of the maximum survival probability ϕ^*. With I_u and I_R the technique can be applied but it produces zero variances for the case of no selection. However, in practice it is fairly straightforward to simulate selection in order to determine biases and standard errors. To simulate one set of data it is merely necessary to give each individual the appropriate probability of survival using random numbers on an electronic computer. For example, consider simulating O'Donald's food shortage experiment, for which the data are given in Table 5.15. In this case there were 25 flies with 14 bristles at the start of the experiment, of which 7 survived. Simulated data are therefore obtained for this number of bristles by giving each of the 25 flies a probability $\phi = 7/25 = 0.28$ of surviving. Alternatively, it can be noted that the number of survivors will be approximately distributed as a normal variate with mean

$25\phi = 7$ and variance $25\phi(1 - \phi) = 5.04$. Consequently, a simulated survival number can be determined by generating a random value from this distribution and rounding it to the nearest integer.

Finally, it can be noted that another index for the intensity of selection is $I = \text{var}\{\log_e(u)\}$, where u is as defined for equations (5.25) (Manly, 1977e). Since a Taylor series approximation shows that

$$\text{var}\{\log_e(u)\} \simeq \text{var}(u)/\bar{u}^2$$

this is virtually the same as the index I_R.

Example 5.6   Selection on human birth weight

Karn and Penrose's (1951) study of the relationship between infant survival and other variables such as birth weight and gestation time was the first of many. There has been particular interest in survival related to birth weight because this is one of the best demonstrated cases of stabilizing selection in the human species. Three recent examples are worth mentioning in particular: Chinnici and Sansing (1977) considered data from all births in the Commonwealth of Virginia, USA, for 1955 to 1973; Terrenato et al., (1981a,b) and Ulizzi et al. (1981) considered all births in Italy for 1954 to 1974; and Terrenato (1983) considered all births in the USA from 1950 to 1976.

For the present purpose a small part of Chinnici and Sansing's Virginia data will be considered, as shown in Table 5.17. These data can be used to consider the question of how the intensity and effect of selection changed over the time period studied, and how it compared for whites and non-whites. Index values are shown in Table 5.18.

Consider first of all Van Valen's index I_V of the intensity of selection per unit time. The time unit is here taken to be one week, so that $t = 4$ in equation (5.26). The pattern of results shown in Table 5.18 is then that selection intensity was higher for non-whites than for whites but in both cases there was an increase from 1955 to 1964 followed by a decrease from 1964 to 1973. Simulations indicate very little bias in $\hat{I}_V$. The standard errors are small enough for the changes in the index to be accepted as being meaningful.

Next consider the index I_u of the effect of selection. Here the difference between whites and non-whites is not so clear. The index is obviously related to the overall percentage survival so that it is highest ($\hat{I}_u = 0.0060$) when the survival is lowest (for non-whites in 1964), while it is lowest ($\hat{I}_u = 0.0017$) when the survival is highest (for whites in 1973). There is also a relationship with the intensity of selection per unit time, so that $\hat{I}_V$ and $\hat{I}_u$ tend to be high or low together.

Finally, consider I_R, the index of the intensity of selection not related to any particular selection time. Here the pattern is for higher intensity for whites than for non-whites, gradually increasing with time. This increasing trend may seem a curious result at first. The explanation is found by looking at the way

Table 5.17 Survival to 28 days for all white and non-white single live births in Virginia, USA

| Birth weight (kg) | White | | | | | | Non-white | | | | | |
| | 1955 | | 1964 | | 1973 | | 1955 | | 1964 | | 1973 | |
	A	r	A	r	A	r	A	r	A	r	A	r
less than 1.0	268	18	254	12	180	24	132	25	209	16	132	23
1.0–1.5	342	151	348	148	203	113	181	99	225	123	150	106
1.5–2.0	719	585	780	630	535	457	411	346	555	482	323	293
2.0–2.5	2 849	2 714	3 022	2 891	2 049	2 003	1 543	1 496	1 871	1 816	1 160	1 144
2.5–3.0	11 955	11 821	12 312	12 192	8 478	8 425	5 130	5 068	5 864	5 811	4 115	4 097
3.0–3.5	25 465	25 349	26 325	26 193	20 436	20 381	8 458	8 400	8 717	8 664	6 569	6 549
3.5–4.0	17 468	17 392	17 904	17 855	16 299	16 273	5 301	5 263	4 326	4 290	3 193	3 184
4.0–4.5	4 578	4 558	4 692	4 684	4 734	4 726	1 422	1 406	971	958	642	638
over 4.5	919	910	861	855	976	972	534	526	247	244	112	108
Total	64 563	63 498	66 498	65 460	52 890	53 374	23 112	22 629	22 985	22 404	16 396	16 142

A = number of live births
r = survivors to 28 days

Table 5.18 Selection indices for survival related to birth weight in Virginia, USA. The biases and standard errors shown here were all estimated from 100 simulated sets of data.

		Survival (%)	$\hat{I}_V$	Bias	SE	$\hat{I}_u$	Bias	SE	$\hat{I}_R$	Bias	SE
White	1955	98.35	0.0122	0.0005	0.0006	0.0033	0.0000	0.0003	51.5	0.1	4.3
	1964	98.44	0.0139	0.0000	0.0007	0.0035	0.0000	0.0003	60.4	0.6	5.6
	1973	99.04	0.0080	0.0002	0.0004	0.0017	0.0000	0.0002	74.8	0.8	6.6
Non-white	1955	97.91	0.0141	0.0006	0.0015	0.0025	0.0001	0.0003	24.8	0.4	3.3
	1964	97.47	0.0193	0.0012	0.0020	0.0060	0.0001	0.0006	35.1	0.1	2.7
	1973	98.45	0.0127	0.0009	0.0014	0.0031	0.0000	0.0004	45.6	−0.3	4.1

the index is calculated. For example, take non-whites in 1955. The mean of u from equation (5.27) is -0.00677 while the variance is 0.00114. This gives the index value $\hat{I}_R = 0.00114/(-0.00677)^2 = 24.8$. On the other hand, for non-whites in 1973 the mean of $\hat{u}$ is -0.00586 and the variance is 0.00157, so that $I_R = 0.00157/(-0.00586)^2 = 45.6$. The index is higher in 1973 because the variance of u has increased even though the mean has gone down. The lower mean indicates fewer deaths but the increased variance indicates more selection with the deaths that did occur.

The lower intensity of selection for non-whites compared with whites is due mainly to the fact that non-whites with low birth weights survived distinctly better than whites of the same weight.

The simulated bias estimates for $\hat{I}_R$ are not large enough to be of concern in this example. However, the standard errors are such that a fair part of the variation in the index could be due to sampling errors. Nevertheless, the difference between whites and non-whites seems quite clear, as does the general trend towards larger values.

Perhaps the main point of this example is to emphasize that different indices of selection give different results on the same data. Thus we have the situation where according to the index I_V the selection intensity was lower in 1973 than in 1955, while according to the index I_R it was higher. This is not really a contradiction. It just seems to be the case that the selection intensity per week (I_V) decreased while the inherent intensity of the selection process (I_R) increased.

6 Evidence from the spatial distribution of a population

Selection can effect the distribution of an animal population in time and space. Evidence for selection can therefore be found either from time changes in populations or from spatial variation. In earlier chapters it has been time changes that have been the main interest. In the present chapter attention is turned to spatial variation.

To begin with, the factors affecting spatial variation (adaptation, isolation by distance, genetic drift, migration and founder effects) are discussed. The main part of the chapter is then concerned with the analysis of data for cases where a number of colonies are sampled, with these being so far apart that they can reasonably be considered to be independent. Two main approaches to the analysis of such data are considered. One approach involves something like a regression analysis with the aim being to 'explain' some characteristic of the distribution of a population using environmental variables. The second approach is based upon comparing colonies in terms of environmental 'distances' and morph 'distances' using a randomization test proposed by Nathan Mantel.

Sections 6.11 and 6.12 review the question of how selection can be inferred from spatial patterns in an animal distribution at colonies that are close in space. The chapter then concludes with a mention of the Kluge–Kerfoot phenomenon which occurs when the amount of within-colony variation in a morphometric variable is positively correlated with the amount of between-colony variation.

6.1 FACTORS AFFECTING ANIMAL DISTRIBUTIONS

This chapter is concerned with how data on the distribution of a subdivided animal population can provide evidence of selection. It is assumed that a population is spread out in space, and samples are taken at a single point in time at a number of localities. At each locality the distribution of animals can be characterized by sample morph frequencies, by the sample distribution of certain quantitative variables $X_1, X_2, \ldots, X_p$, or by gene frequencies determined from morph frequencies. Whichever characterization is used, the question of interest is whether or not differences between localities have been determined by natural selection to any appreciable extent.

If a distribution is largely determined by selection then strong correlations between distributional and environmental variables should exist. Searching for such correlations is a primary test for selection that has been made with many species. A typical recent study is that of Verspoor (1983) on the water louse *Asellus aquaticus*. Samples of this organism were collected from 37 sites in western Europe. The frequencies of allozyme morphs were found to be associated with latitude and the level of water pollution, and Verspoor concluded that selection occurs related to temperature and oxygen concentration. Another recent study has shown that for man the frequency of many genes is related to the distance from the equator, again presumably because of temperature effects (Piazza, 1980).

There are, of course, some well known ecogeographical rules concerning relationships between animal distributions and climate. These are no doubt at least partly a result of adaptation through natural selection, although there may also be elements of non-genetic environmental response (Dobzhansky, 1970, p. 302). Bergmann's rule states that races of warm-blooded vertebrates living in cold areas are larger than their counterparts in warm areas. Many selective explanations are possible (Rosenzweig, 1968). Allen's rule states that the races in colder areas also have smaller protruding body parts (ears, legs, tails, etc.). A selective explanation for this is possible in terms of heat loss. Gloger's rule states that races in warmer, more humid parts of a species' range are more heavily pigmented than races in cooler, drier parts. It holds for insects as well as birds and mammals. There is no generally accepted explanation in this case.

For the present discussion it is helpful to begin by considering Sokal's (1978) four main factors that may be responsible for spatial variation:

(a) There can be local adaptations to environmental gradients of some type.
(b) There can be localized environmental patches which are homogenous within themselves but heterogenous between patch types. Local adaptations then result in a patchy distribution of animals.
(c) Sample points may be more or less isolated by distance from each other and differ because of local genetic drifts.
(d) Purely historical factors such as migration and founder effects may be important.

Here (a) and (b) imply selection, whereas (c) and (d) do not.

It is, in general, not easy to determine which of these factors have been important in particular cases. For example, the cline in the frequencies of the melanic morph of the moth *Amathes glareosa* in the Shetland Isles could be due to either (a) or (d). It may be recalled that this was the subject of Example 2.2. Kettlewell *et al.* (1969) have shown that the frequency of the melanic *edda* is 2 % in the south of the Shetland Isles and increases more or less steadily up to 97 % in the north. It seems very possible that this represents a response to an environmental gradient of some kind. On the other hand, the distribution could just be the result of historical factors. If at some time in the past a colony of mainly *edda* was founded in the north while a colony of mainly the alternative morph *typica* existed in the south, with both these morphs having the same fitness, then over the course of time limited migration could result in the cline that we see today.

A rather similar case occurs with the ABO blood grouping system in man. The frequency of the A blood group is about 44 % in the south of England and this declines more or less steadily going north until it is about 36 % in Scotland (Kopec, 1973). This could be explained as a response to an environmental gradient. However, past migration patterns can easily account for the variation (Roberts, 1973).

It is obviously important to have some idea of the magnitude of non-selective factors such as migration on the distribution of an animal population. This is a complex subject, about which there are differing viewpoints, but there are some general points that are worth noting that concern the way that genetic divergence occurs between partially isolated colonies when there is no selection.

One simple model for partial isolation assumes that there are s colonies of a population, each with the same moderate or large size N. Each generation the immigrants into a colony are a random sample of mN individuals from the total population. The mutation rate is v per generation at a particular genetic locus. Let J_0 denote the probability that two genes chosen at random from a single colony are the same. Let J_1 denote the probability that two genes chosen at random from the total population are the same. Then Nei's measure of the genetic identity between colonies is

$$I = J_1/J_0.$$

This is unity if there is no isolation between colonies, and approaches zero when colonies have been isolated for a long time.

Assuming no selection, Nei (1975, p. 122) shows that after sufficient time has elapsed for an equilibrium state to be reached,

$$I \simeq m(2-m)/\{2vs(1-m)^2 + m(2-m)\}. \tag{6.1}$$

This means that I will be close to unity providing that vs is small compared with m. This implies that the genetic difference between populations will be small unless the migration rate is exceedingly small. For example, for *Drosophila* species $v \simeq 10^{-8}$ per generation so that large population differences require m to be smaller than $10^{-8} s$.

Based upon a model which ignores mutation, Lewontin (1974, p. 213) suggests that if ten or more migrants per generation are exchanged between two colonies, then the colonies will stay similar. Christiansen and Frydenberg (1974) suggest that a fair rule of thumb is that divergence will not occur if two colonies exchange more than one individual per generation. There is thus a certain amount of disagreement about how much migration is needed to retain a similarity in the genetic composition of two populations. However, it seems quite clear that the amount needed is exceptionally low if there is no selection.

Suppose next that at a certain point in time a population splits at random into s completely isolated colonies, each with the same moderate or large size N. An interesting question is how many generations it takes for the colonies to lose their resemblance to each other just through mutation and random drift. Nei (1975, p. 125) shows that after t generations his measure of the genetic identity between colonies is

$$I = \exp(-2vt). \tag{6.2}$$

This implies that divergence requires an inordinately long amount of time. For example, for *Drosophila*, $I \simeq \exp(-2 \times 10^{-8} t)$ so that it would need 10^8 generations, which is about 10 million years, to reduce I to $\exp(-2) = 0.14$. For sibling species of *Drosophila willistoni* the present-day value of I is about 0.60 (Nei and Tateno, 1975). Substituting this value into equation (6.2) suggests that the divergence time of these species is about

$$t = -\log_e(I)/(2v) = -\log_e(0.60)/(2 \times 10^{-8})$$
$$= 26 \times 10^6 \text{ generations,}$$

or about 2.6 million years.

The validity of calculations of this type has been questioned by Ehrlich and White (1980). They point out that the results obtained depend rather critically on the value that is assumed for the mutation rate v, where this is difficult to estimate. Furthermore, they argue that using a generally accepted value of v on populations of the butterfly *Euphydryas editha* suggests that colonies that have been almost completely isolated for 7000 years should be genetically almost identical, which seems intuitively to be very unlikely. On this basis it seems that Nei's model may be somewhat unrealistic.

Leaving questions of their validity aside, the conclusion from equations (6.1) and (6.2) must be that a similarity between morph distributions in different parts of a population should cause no surprise, even when these populations have been virtually isolated for extremely long periods of time. It is variation that suggests selection since local adaptations are capable of causing far more variation between moderate or large size colonies than limited migration and genetic drift.

For small colonies, or colonies that have been small in the past, the situation is rather different. If several small and isolated colonies are started by a few individuals and then subsequently grow larger, then each colony will tend to display the characteristics of its founders. These founder effects can result in just about any patterns of distribution, including clines.

6.2 ANALYSES IGNORING SPATIAL PATTERNS

In the analysis of many sets of data, the spatial relationship between the different colonies of a population is not considered important. The colonies are so far apart that they can be regarded as providing independent data points for something like a regression analysis. The reason for an analysis may be to determine whether there is any association between morph frequencies and environmental variables such as temperature. Alternatively, the colonies may be grouped into habitat types and interest then centres on the question of whether colonies are relatively homogenous within habitats.

Two rather different approaches are possible for analysing this type of data. Firstly, there are analyses like multiple regression where each colony provides a single data point. For example, the proportion of a certain morph in colonies might be treated as a dependent variable to be 'explained' by environmental variables. A chi-square type of test might be used to compare several morph proportions simultaneously. The second approach involves distance matrices. If there are L colonies, then there are $L(L-1)/2$ pairwise comparisons. For each of these comparisons a 'distance' in terms of morph proportions can be calculated. Another 'distance' in terms of environment can also be calculated. A test can then be made to see whether the colonies that are close environmentally are also close in terms of morph proportions.

Sections 6.3 to 6.7 are concerned with the first of these two approaches. Sections 6.8 to 6.10 are concerned with the second approach and particularly with the use of a nonparametric test that is due to Nathan Mantel. The final sections of the chapter give a fairly brief discussion of methods of analysis that emphasize the spatial relationship between sample locations.

6.3 CHI-SQUARE RANDOMIZATION TEST ON GROUPED COLONIES

The first tests for environmental associations that will be considered are for polymorphic variation where a number of colonies have been grouped into several categories according to their general nature. If selection is operating then it can be expected that within categories the localities should show relatively homogenous morph proportions. For example, Table 6.1 shows Cain and Sheppard's (1950) data on the frequencies of ten morphs of the snail *Cepaea nemoralis* in 17 widely separated colonies in Southern England. The colonies have been grouped into six habitat types. We can ask whether there is any evidence that the distribution of the snail is related to the habitat.

If only one morph is considered, this question is not too difficult to answer. With a similar type of data, Johnson (1980) used a Mann–Whitney U-test to compare the proportion of banded and unbanded snails in exposed and bushy areas. However, if an answer is required without grouping morphs it is not so obvious what to do. A straightforward chi-square test is not appropriate because of the very high variation that exists within habitats. For example, a usual contingency table chi-square value (Steel and Torrie, 1980, p. 500) calculated for the five locations of mixed deciduous woods comes out at 154.1 with 36 degrees of freedom. This is very significantly large and it seems that the

Table 6.1 The colour and banding of *Cepaea nemoralis* at six different types of location. The main body of the table gives sample numbers for ten colour and band classes (UB = unbanded, MB = mid-banded, FB = fully banded, OB = other types of banding).

Habitat	Location	Yellow				Pink				Brown	
		UB	MB	FB	OB	UB	MB	FB	OB	UB	Banded
Downland beech	Hackpen	15	24	0	0	76	39	0	0	1	1
	Rockley 1	17	25	0	0	41	7	0	0	57	9
Oakwood	Broomsgrove	1	4	1	1	5	22	4	1	30	17
	Cobham Frith	0	24	1	0	51	80	7	16	0	6
	Puthall Gate	9	6	4	3	135	165	62	138	59	0
Mixed deciduous wood	Streatley S. Wood	2	1	3	0	33	4	11	4	8	0
	Wytham	1	7	33	28	15	15	54	40	24	6
	Bagley	0	0	4	2	6	5	9	5	4	0
	Burdrop Wood	1	4	18	14	17	3	8	10	23	1
	Wantage 2	7	0	3	4	7	0	12	11	8	0
Hedgerows	Wytham Lane	2	1	10	3	0	0	2	2	0	1
	Canal Bank	17	25	46	26	7	27	59	61	9	2
	Hinksey Hedge	5	2	5	6	4	1	5	3	0	0
	Parks Hedge	1	1	10	2	0	1	42	19	0	0
Downside long coarse grass	Pentridge	0	66	185	23	0	32	82	11	8	20
	C. C. Downside	0	11	21	34	19	6	33	31	0	0
Downside short turf	Rough Down	47	21	0	6	5	13	0	7	49	16

morph proportions are certainly not constant within this type of habitat.

One possibility with data like this is to use a randomization test (Manly, 1983). This involves calculating a chi-square value for each habitat separately, as has been done for mixed deciduous woods, and adding these up. This is then a measure of the amount of within-habitat variation. The distribution that is obtained for this chi-square statistic by randomly allocating locations to habitats is then found by doing the random allocations on a computer. If the observed chi-square value is significantly low when compared with the randomized distribution, then this is evidence of relatively homogenous results within habitats compared to between habitats. Furthermore, the significance of the variation for an individual morph can be considered by seeing how its contribution to the total observed chi-square value compares with the distribution of contributions obtained by randomization. The FORTRAN code for a computer program HABITATS for carrying out the calculations is provided in the Appendix, Section A.12.

Example 6.1   *Cepaea nemoralis* in Southern England

With Cain and Sheppard's (1950) data that are shown in Table 6.1 there are 17 locations in six habitats. The total within-habitat chi-square value is 824.8 with 94 degrees of freedom. This total is the sum for the different habitats, which are as follows: downland beech, 93.3 with 5 degrees of freedom; oakwood, 298.6 with 18 degrees of freedom; mixed deciduous woods, 154.1 with 36 degrees of freedom; hedgerows, 99.7 with 27 degrees of freedom; downside coarse grass, 179.2 with 8 degrees of freedom; and downside short turf, zero with no degrees of freedom.

We wish to know whether the total chi-square of 824.8 is significantly small, taking into account the considerable variation that is present within habitat types. To test this, the 17 locations can be reallocated at random to the habitats, keeping the number of locations for each habitat fixed. Two locations can be chosen at random to be 'downland beech'. Three of the remaining locations can be chosen at random to be 'oakwood', and so on. A total within-habitat chi-square value can then be calculated for this random allocation. Repeating the randomization procedure a large number of times on a computer provides the randomization distribution of chi-square.

With Cain and Sheppard's data, 1000 random allocations gave a mean chi-square of 1582.4 with a smallest value of 857.8. Since the chi-square for the proper allocation of locations is only 824.8 it can be said that this is significantly low at about the 0.1 % (one in a thousand) level. There is very clear evidence that morph proportions are relatively homogenous within habitats.

We can now ask which morphs are involved most in producing the significant result. Since there are ten morphs it is appropriate to test individual morphs using a lower level of significance than 5 % or 1 %. In general if there are m tests to be made, and a level of significance $(\alpha/m)\%$ is used for each one, then the Bonferro⁻ . inequality tells us that the probability of declaring one or more test significant by chance alone is less than $\alpha\%$ (Harris, 1975, p. 98). Therefore in the present example using a level of significance of 0.5% means that if there were in fact no habitat effects then the probability of obtaining a significant result for one or more morphs would be 5% or less.

What this means in terms of the randomization test is that the result for a particular morph can be considered significant if its contribution to the total chi-square value is lower than 99.5% of the contributions obtained by randomization. With this definition there is significance for yellow fully banded and pink fully banded snails.

Fig. 6.1 shows a scatter diagram for the percentages of yellow fully banded and pink fully banded snails at the 17 locations. There is some slight indication of grouping according to the nature of the habitat. Cain and Sheppard (1950)

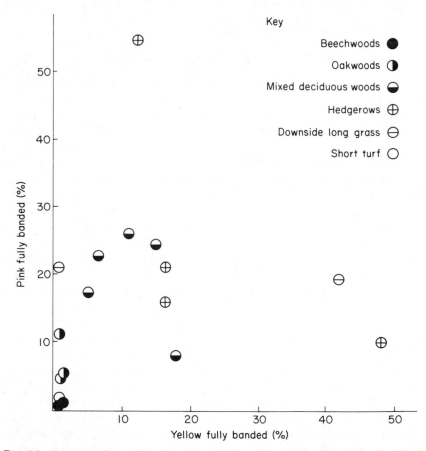

Fig. 6.1 A scatter diagram showing the percentages of yellow fully banded and pink fully banded *Cepaea nemoralis* in 17 colonies. A certain amount of clustering can be seen for the samples from beechwoods, oakwoods and mixed deciduous woods.

plotted the percentage of yellow snails against the percentage effectively unbanded which gives about the same amount of grouping. The grouping is far more apparent when there are many more locations plotted (Cain and Sheppard, 1954, Figure 1).

See Clarke *et al.* (1978) for a review of later work following the approach of Cain and Sheppard.

6.4   AN ALTERNATIVE CHI-SQUARE APPROACH FOR GROUPED COLONIES

There is an alternative chi-square type of analysis that requires far less computing than the test of the previous section. The basic idea is to adjust the data in such a way that an approximate test of significance can be carried out using simple chi-square methods.

Suppose that there are H habitat types and K morphs. Assume that L_h locations are chosen at random from habitat type h in such a way that at the lth of these locations the total number of animals in the sample is a random variable m_{lh}, with expected value θ_{lh}. Assume further that for this lth random location in habitat h the mean and variance of the observed number of morph k, n_{klh}, are

$$E(n_{klh}) = \pi_{kh}\theta_{lh},\tag{6.3}$$

and

$$\text{var}(n_{klh}) = \pi_{kh}\theta_{lh}(1 + \alpha\theta_{lh}).\tag{6.4}$$

Here π_{kh} is the mean proportion of morph k animals in habitat h, so that

$$\sum_{k=1}^{K} \pi_{kh} = 1,$$

while $\alpha > 0$ is a measure of the amount of variation that exists from location to location within a habitat. If $\alpha = 0$ then the mean and variance of n_{kh} are the same. This corresponds to the situation where n_{klh} has a Poisson distribution, which occurs when π_{kh} is the same for all locations within habitat h. On the other hand, if α is large then the variance of n_{klh} exceeds the mean. This corresponds to the situation where π_{kh} is rather variable from location to location within one type of habitat.

Equations (6.3) and (6.4) were originally derived on the assumption that sample counts follow Poisson distributions whose mean values have gamma distributions within habitats (Manly, 1983). This leads to a negative binomial distribution for n_{klh} for which the mean and variance are as given by these equations. Maximum likelihood estimation can be used to fit this negative binomial model to data. However, here a somewhat simpler approach will be adopted.

Note first of all that the expected sample size θ_{lh} can obviously be estimated by the actual sample size obtained,

$$\hat{\theta}_{lh} = m_{lh}. \tag{6.5}$$

Next, note that the observed sample proportion of morph k,

$$\hat{\pi}_{klh} = n_{klh}/m_{lh}, \tag{6.6}$$

is an obvious estimator of the mean habitat proportion π_{kh}. A Taylor series approximation shows that

$$\text{var}(\hat{\pi}_{klh}) \simeq \pi_{kh}(1 - \pi_{kh})(1 + \alpha\theta_{lh})/\theta_{lh},$$

and

$$\text{cov}(\hat{\pi}_{klh}, \hat{\pi}_{slh}) \simeq -\pi_{kh}\pi_{sh}(1 + \alpha\theta_{lh})/\theta_{lh}, \tag{6.7}$$

$s \neq k$. Hence, replacing θ_{lh} by m_{lh} gives

$$\text{var}(\hat{\pi}_{klh}) \simeq \pi_{kh}(1 - \pi_{kh})(1 - \alpha m_{lh})/m_{lh},$$

and

$$\text{cov}(\hat{\pi}_{klh}, \hat{\pi}_{slh}) \simeq -\pi_{kh}\pi_{sh}(1 - \alpha m_{lh})/m_{lh}. \tag{6.8}$$

The interesting thing here is that equations (6.8) are precisely what is obtained when sample proportions are estimated using a sample of size $m_{lh}/(1 + \alpha m_{lh})$ from a multinomial distribution (Appendix, Section A.2). In other words, in the situation being considered, the morph proportions for the lth location in habitat type h can reasonably be treated as simple proportions from a sample of size $m_{lh}/(1 + \alpha m_{lh})$. This is achieved most simply by replacing the sample counts n_{klh} with the adjusted counts

$$a_{klh} = n_{klh}/(1 + \alpha m_{lh}) \tag{6.9}$$

Of course, this procedure can only be adopted when α is known, which will not usually be the case. A reasonable approach seems to be to choose α so that the total within-habitat chi-square value for the adjusted frequencies a_{klh} is equal to its expected value, which is the chi-square degrees of freedom.

Having found the adjusted frequencies, a chi-square test on habitat total frequencies can be used to test for differences between habitats in the morph proportions.

Example 6.2   Further analysis of *Cepaea nemoralis* in Southern England

Consider again Cain and Sheppard's (1950) data that are shown in Table 6.1. From Example 6.1 it is known that if the within-habitat chi-square values are calculated for each habitat and added up, then the result is a total chi-square value of 824.8 with 94 degrees of freedom. This is very significantly large, indicating, of course, that the morph proportions vary from location to location within habitats. The frequencies therefore need to be adjusted as shown in equation (6.9).

Let $Q(\alpha)$ denote the total within-habitat chi-square value that is obtained using the adjusted frequencies a_{klh} in place of the original frequencies n_{klh}. Since $\alpha = 0$ involves no adjustment at all, it is known that $Q(0) = 824.8$. A positive value of α will make $Q(\alpha)$ smaller than this. Since a chi-square variate with 94 degrees of freedom has expected value 94, a moment estimator $\hat{\alpha}$ of α is one which satisfies the equation

$$Q(\hat{\alpha}) = 94.$$

This value for α should make the adjusted samples within a habitat behave just like repeated random samples from a population with constant morph proportions.

The value $\hat{\alpha}$ can be obtained quite simply by trial and error and interpolation. In the present case, trying $\alpha = 0.05$ gives $Q(0.05) = 100.2$, while $\alpha = 0.10$ gives $Q(0.10) = 55.5$. Further trial and error then shows that $Q(0.054) = 94.0$, so that $\hat{\alpha} = 0.054$.

Table 6.2 shows the adjusted frequencies a_{klh} obtained using equation (6.9) with $\alpha = 0.054$. It can be seen that the original samples are equivalent to random samples of a much smaller size from constant within-habitat populations. The fact that the frequencies are not integers is of no particular importance.

A simple test for differences between habitats can be based upon the habitat totals for adjusted frequencies, as shown in Table 6.3. This provides a chi-squared value of 153.29 with 45 degrees of freedom, which is significantly large at the 0.1 % level. There is clear evidence of differences between habitats.

From their contributions to the total chi-square value, it seems that the morphs that varied most from habitat to habitat were pink mid-banded, yellow fully banded, and yellow unbanded. Small expected frequencies mean that these contributions are only an indication of the relative variation of different morphs. Indeed, the contribution from pink fully banded snails is not particularly large although the randomization test of Example 6.1 indicated significant habitat effects for this morph.

6.5 THE USE OF MULTIPLE REGRESSION

Multiple regression has often been used to study the relationship between the distribution of an animal and environmental variables. However, it is well known that there are liable to be difficulties in interpretation due to correlations between the so-called 'independent' variables. Regression equations accounting for a large part of the variation between colonies may well be found. However, such equations may be of little value in determining causal factors.

There are two technical problems that are particularly worth mentioning. The first of these is the bias that is introduced by stepwise regression. A common procedure is to start with a number of environmental variables and pick a subset of these to account for some aspect of the distribution of an

Table 6.2 Adjusted frequencies for the data on the colour and banding of *Cepaea nemoralis* at six different types of location. These are the frequencies of Table 6.1 adjusted using equation (6.9) with $\alpha = 0.054$. Within habitats the locations can be treated as providing random samples from a population with constant morph proportions. (UB = unbanded, MB = mid-banded, FB = fully banded, OB = other types of banding.)

Habitat	Location	Yellow				Pink				Brown	
		UB	MB	FB	OB	UB	MB	FB	OB	UB	Banded
Downland beech	1	1.6	2.5	0	0	8.1	4.1	0	0	0.1	0.1
	2	1.8	2.7	0	0	4.4	0.7	0	0	6.0	1.0
Oakwood	1	0.2	0.7	0.2	0.2	0.9	3.9	0.7	0.2	5.3	3.0
	2	0	2.2	0.1	0	4.6	7.3	0.6	1.5	0	0.5
	3	0.3	0.2	0.1	0.1	4.2	5.1	1.9	4.3	1.8	0
Mixed deciduous wood	1	0.4	0.2	0.7	0	7.2	0.9	2.4	0.9	1.8	0
	2	0.1	0.5	2.5	2.1	1.2	1.2	4.1	3.1	1.8	0.5
	3	0	0	1.4	0.7	2.1	1.7	3.1	1.7	1.4	0
	4	0.2	0.6	2.8	2.2	2.7	0.5	1.3	1.6	3.6	0.2
	5	1.8	0	0.8	1.1	1.8	0	3.2	2.9	2.1	0
Hedgerows	1	0.9	0.5	4.7	1.4	0	0	0.9	0.9	0	0.5
	2	1.1	1.6	2.9	1.6	0.4	1.7	3.7	3.8	0.6	0.1
	3	1.9	0.7	1.9	2.2	1.5	0.4	1.9	1.1	0	0
	4	0.2	0.2	2.0	0.4	0	0.2	8.2	3.7	0	0
Downside long coarse grass	1	0	2.7	7.7	1.0	0	1.3	3.4	0.5	0.3	0.8
	2	0	1.2	2.2	3.6	2.0	0.6	3.5	3.3	0	0
Downside short turf	1	4.8	2.1	0	0.6	0.5	1.3	0	0.7	5.0	1.6

Table 6.3 A chi-square test for habitat differences in morph proportions. The habitat total adjusted morph frequencies are simply treated as sample values and an ordinary contingency table chi-square test applied. This gives

$$\sum \frac{(O-E)^2}{E} = 153.29$$

with 45 degrees of freedom. (O = adjusted morph frequencies; E = expected value assuming constant morph proportions for all habitats).

Habitat		Yellow				Pink				Brown		Total freq.
		UB	MB	FB	OB	UB	MB	FB	OB	UB	Banded	
1	O	3.4	5.2	0	0	12.5	4.8	0	0	6.1	1.1	33.1
	E	1.9	2.4	3.8	2.2	5.3	3.9	5.0	3.8	3.8	1.1	
2	O	0.5	3.1	0.4	0.3	9.7	16.3	3.2	6.0	7.1	3.5	50.1
	E	2.9	3.6	5.7	3.3	8.0	5.9	7.5	5.8	5.7	1.6	
3	O	2.5	1.3	8.2	6.1	15.0	4.3	14.1	10.2	10.7	0.7	73.1
	E	4.2	5.2	8.4	4.8	11.6	8.7	10.9	8.4	8.4	2.3	
4	O	4.1	3.0	11.5	5.6	1.9	2.3	14.7	9.5	0.6	0.6	53.8
	E	3.1	3.8	6.2	3.5	8.6	6.4	8.0	6.2	6.1	1.7	
5	O	0	3.9	9.9	4.6	2.0	1.9	6.9	3.8	0.3	0.8	34.1
	E	2.0	2.5	3.9	2.3	5.5	4.1	5.1	4.0	3.9	1.1	
6	O	4.8	2.1	0	0.6	0.5	1.3	0	0.7	5.0	1.6	16.6
	E	1.0	1.2	1.9	1.1	2.7	2.0	2.5	1.9	1.9	0.5	
Total frequency		15.3	18.6	30.0	17.2	41.6	30.9	38.9	30.2	29.8	8.3	260.8
Total of $(O-E)^2/E$		21.06	8.07	24.26	9.17	20.02	24.47	16.93	6.79	15.78	6.74	153.29

animal using a stepwise procedure. Unfortunately, chance alone is liable to produce at least one 'significant' relationship between the dependent variable and the environmental variables. Stepwise regression should discover this. In other words, there will be a bias towards obtaining significant results. In this connection a small simulation that is part of the example that follows is interesting. Using purely random dependent variables a 'significant' environmental association was discovered in seven out of ten cases.

The second problem concerns the use of several tests at the same time. Suppose that an animal's distribution is described by *m* variables (for example,

m morph proportions), and each of these is used as the dependent variable in a regression analysis. If a level of α % is used to determine significance then it can be expected that $\alpha m/100$ regressions will display significance through chance alone. This is exactly the problem that was discussed in Example 6.1 in terms of testing ten morph proportions for significance. The solution involves testing each regression using the significance level (α/m) %. In that case the probability of getting one or more significant results by chance is less than α %.

A further difficulty that may occur with some sets of data is a lack of independence in the data from different colonies. As was mentioned in Section 6.2, regression and chi-square types of analysis rely on the assumption that the colonies of an animal population that are being studied are so far apart that they can be considered to be independent. However, equations (6.1) and (6.2) suggest that complete independence between two colonies will only occur if they are so far apart that migration rates are infinitesimal, and this has been true for an extremely long period of time. This means that in many cases there will be some doubt about whether or not the independence assumption is correct.

With regression studies the assumption of independence can be checked by looking at the residuals from the fitted regression equation. This can be done either informally or by a formal test. With the informal approach it is just a case of writing the residual values on a map showing the location of different colonies and seeing whether there is any pattern. The particular thing to look for is similar residual values in colonies that are close together.

A formal test for spatial relationships in regression residuals is described by Ripley (1981, p. 98). If residuals are correlated then it is possible to estimate the correlation structure and the regression model (Ripley, 1981, p. 101; Cook and Pocock, 1983). However, this is not a simple matter and it will not be considered here. A nonparametric test for correlated residuals for colonies that are close in space is described in Section 6.10.

Example 6.3   *Euphydryas editha* in California and Oregon

The problems involved in using regression methods to study an animal distribution are well illustrated by McKechnie *et al.*'s (1975) study of 21 colonies of the butterfly *Euphydryas editha* in California and Oregon. Fig. 6.2 shows the sampling locations and Table 6.4 shows the values of 11 environmental variables at these locations. The distribution of *E. editha* is characterized by gene frequencies for various genetic loci as determined by electrophoresis. The results for three loci, Pgm, Pgi and Hk, are presented in Table 6.5. McKechnie *et al.* considered eight loci but three are sufficient for the present purpose.

The technique of electrophoresis is described in Chapter 9. Here all that needs to be said is that a sample of n butterflies provides $2n$ genes, the product of each of which shows a certain mobility in an electric field. For example, a

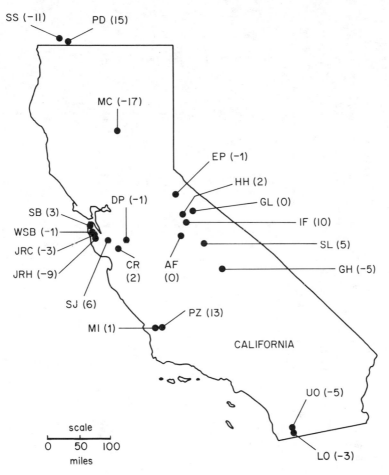

Fig. 6.2 Colonies of *Euphydryas editha* sampled in California (19 colonies) and Oregon (two colonies, SS and PD). The numerical values shown in parentheses are residuals from a regression equation that relates the percentage of mobility 1.00 Hk genes to 11 environmental variables.

sample of 33 individuals from the colony PD had 66 genes, of which 4 (6 %) had a mobility of 0.94 for the Pgm genetic locus. The remaining 62 genes (94 %) had a mobility of 1.00 for this locus.

McKechnie *et al.* used stepwise regression to relate all the mobility frequencies to the 11 environmental variables. Five of the 17 mobilities shown in Table 6.5 then had a significant regression at the 0.1 % level. It seems therefore that there is strong evidence of a relationship between the distribution of *E. editha* and the environmental variables.

Table 6.4 Environmental variables for 21 colonies of the butterfly *Euphydryas editha* in California and Oregon. Values have been rounded to the nearest integer for convenience. Temperatures are in Fahrenheit.

Colony	Altitude (feet)	Latitude (degrees)	Annual precipitation (inches)	Annual max. temperature	Daily max. temperature post-diapause	Daily max. temperature adults	Daily maximum temperature pre-diapause	Annual minimum temperature	Daily minimum temperature post-diapause	Daily minimum temperature adults	Daily minimum temperature pre-diapause
PD	500	45	72	97	59	66	74	15	36	41	47
SS	500	45	43	98	59	66	74	17	38	43	49
SB	800	38	20	92	60	63	64	32	43	44	45
WSB	570	37	28	98	61	64	66	26	42	43	44
JRC	550	37	28	98	61	64	66	26	42	43	44
JRH	550	37	28	98	61	64	66	26	42	43	44
SJ	380	37	15	99	63	66	69	28	43	44	45
CR	930	37	21	99	63	66	69	28	43	44	45
MI	480	35	24	101	65	66	66	27	43	43	44
UO	650	33	10	101	69	69	72	27	42	45	47
LO	600	33	10	101	69	69	72	27	42	45	47
DP	1500	37	19	99	60	65	72	23	41	47	53
PZ	1750	35	22	101	66	69	72	27	44	47	50
MC	2000	40	58	100	62	73	80	18	37	44	50
HH	4200	38	36	95	59	70	76	13	35	43	48
IF	2500	38	34	102	63	75	82	16	37	45	51
AF	2000	37	21	105	67	78	86	20	40	47	54
SL	6500	37	40	83	56	65	74	0	33	41	47
GH	7850	36	42	84	57	65	75	5	37	44	52
EP	8950	39	57	79	60	70	70	-7	37	43	43
GL	10500	38	50	81	58	68	67	-12	34	42	42

Table 6.5 Sample gene frequencies for *Euphydryas editha* estimated for 21 colonies. Results are given for the phosphoglucomutase (Pgm), phosphoglucose-isomerase (Pgi) and hexokinase (Hk) genes. These data were obtained by electrophoresis, which is described more fully in Chapter 9.

Colony	Pgm mobility class frequencies (%)						Pgi mobility class frequencies (%)								Hk mobility class frequencies (%)			Sample size (genes)
	0.87	0.94	1.00	1.06	1.14	1.20	0.20	0.40	0.60	0.80	1.00	1.16	1.30	1.40	1.00	1.12	1.24	
PD	0	6	94	0	0	0	0	0	2	15	63	20	0	0	100	0	0	66
SS	0	7	93	0	0	0	0	0	3	22	57	17	1	0	97	3	0	116
SB	2	4	88	6	0	0	0	0	16	20	38	13	12	1	36	64	0	100*
WSB	0	3	90	7	0	0	0	0	6	28	46	17	2	1	72	28	0	106*
JRC	0	1	88	11	0	0	0	0	4	19	47	27	3	0	70	30	0	160*
JRH	2	2	79	17	0	0	0	0	1	8	50	35	6	0	64	36	0	102*
SJ	1	6	86	8	0	0	0	0	2	19	44	32	3	0	82	18	0	168
CR	0	5	80	15	0	0	0	0	0	15	50	27	3	5	72	28	0	106*
MI	3	2	82	12	1	0	0	0	6	26	32	35	1	0	65	35	0	70*
UO	2	7	73	18	0	0	1	9	21	40	25	4	0	0	0	99	1	124*
LO	5	10	74	11	0	0	4	10	26	32	28	0	0	0	2	27	1	200*
DP	0	0	73	27	0	0	0	0	1	6	80	12	1	0	40	60	0	118*
PZ	5	3	84	6	1	1	0	1	4	34	33	22	6	0	39	61	0	120*
MC	1	13	74	12	0	0	0	0	7	14	66	13	0	0	9	91	0	102*
HH	6	15	71	8	0	0	0	2	0	19	50	27	2	0	19	81	0	48
IF	3	3	80	13	1	0	0	0	9	15	47	21	5	3	42	55	3	108*
AF	2	4	85	9	0	0	3	0	7	17	32	27	8	6	37	59	4	114*
SL	2	2	80	16	0	0	0	0	4	18	53	23	2	0	16	84	0	56
GH	25	0	69	6	0	0	0	0	5	7	84	4	0	0	4	96	0	102*
EP	0	2	82	16	0	0	0	0	2	1	86	11	0	0	1	99	0	82*
GL	0	0	76	24	0	0	0	0	3	1	92	4	0	0	4	96	0	120*

* In these cases the sample size varied somewhat with the gene. The only case of substantial variation was at the locality LO where the sample sizes were 274, 110 and 214 for Pgm, Pgi and Hk, respectively.

Some doubt is cast on this conclusion by a small simulation experiment. This involved generating 210 pseudo-random numbers from the standard normal distribution, and making these into ten blocks of 21 observations each. Each block then provided dependent variable values for the 21 colonies sampled by McKechnie *et al.* and an attempt was made to 'explain' these using the 11 environmental variables. Thus ten stepwise regression analyses were carried out using different random values for a morph distribution. This resulted in a significant regression for seven out of the ten analyses, as shown in Table 6.6. Faced with results like this, it seems clear that stepwise regression programs are liable to give a rather biased impression of how well environmental variables are able to explain an animal's distribution. In the case of McKechnie *et al.*'s data a 'significant' relationship for 5 out of 17 mobilities could well be just due to chance.

An unbiased idea of the significance of the relationship between mobilities and environmental variables can be obtained by fitting regression equations using all of the environmental variables to the mobility percentages. According to McKechnie *et al.*, this results in a significant regression for 6 of the 17 mobilities, using the 5% level of significance on each one.

Table 6.6 The results of a simulation experiment which illustrates the way that stepwise regression can produce fallacious significance. Ten stepwise regression analyses were carried out using random dependent variable values and the 11 environmental variables $(X_1, X_2, \ldots X_{11}$, in order) to 'explain' these. Although no relationship can be present, a 'significant' regression has been found for seven out of the ten cases. The BMDP2R program (Dixon, 1981) was used for analyses, using an 'F to enter' of 1.5. This is not quite the same stepwise procedure as was used by McKechnie *et al.* but the same fallacious significance is liable to occur with the procedure that they used.

Regression	Variables entered	Multiple correlation R^2	F	Significance level (%)
1	X_9	0.20	4.74	5
2	X_3, X_9, X_{10}, X_{11}	0.45	3.23	5
3	X_2	0.37	10.96	1
4	none	—	—	N.S.
5	X_2	0.13	2.95	N.S.
6	X_2	0.21	4.92	5
7	X_4, X_6, X_8	0.49	5.47	1
8	X_3, X_4, X_{11}	0.33	4.33	5
9	X_3, X_6, X_9	0.42	4.04	5
10	X_3, X_5	0.20	2.26	N.S.

N.S. = not significant at the 5% level.

For reasons explained in Example 6.1, it is better to require a higher level of significance than 5% when regression equations are being fitted to several variables related to the distribution of an animal. In the present case a more appropriate procedure is to test each equation at the $(5/17)\% = 0.3\%$ level of significance since there is then a probability of less than 5% of getting one or more significant results purely due to chance. On this basis three of the gene frequencies are significant (0.87 Pgm, 1.00 Hk and 1.12 Hk). It really does seem therefore that there is evidence of a relationship between the distribution of *E. editha* and the environmental variables.

The estimated regression equation for 1.00 Hk is as follows, where Y denotes the percentage for this mobility in a colony and the environmental variables are in the same order as in Table 6.4:

$$\begin{aligned}
Y = & -286 - 0.00116\,X_1 + 5.92\,X_2 - 0.806\,X_3 + 5.45\,X_4 \\
& - 6.10\,X_5 - 2.55\,X_6 - 3.40\,X_7 - 3.86\,X_8 + 15.9\,X_9 \\
& - 13.4\,X_{10} - 1.17\,X_{11}.
\end{aligned} \tag{6.10}$$

The coefficients of X_2 (latitude), X_4 (annual maximum temperature), X_5 (daily maximum temperature post-diapause), X_8 (annual minimum temperature) and X_9 (daily minimum temperature post-diapause) are significantly different from zero on a t-test. Note, however, that a stepwise regression chooses X_1 (altitude), X_2, X_4, X_6 (daily maximum temperature for adults) and X_7 (daily maximum temperature pre-diapause) as the best five predictors of Y. There are some quite high correlations between the environmental variables which makes it virtually impossible to be sure which of them are most related to Y.

It has been mentioned before that regression analysis is not capable of proving the existence of causal relationships. For example, the significant positive coefficient for X_2 (latitude) in equation (6.10) certainly does not mean that increasing latitude causes the percentage of 1.00 Hk genes to increase. On the contrary, the variable 'latitude' is probably best thought of simply as an index that reflects many more or less systematic changes in the environment that change from the south to the north of California.

The residuals from a regression are the differences between the observed values of the dependent variable and the values predicted by the fitted equation. With data from different localities there is some value in plotting residuals to see whether they show any patterns. For example, similar residual values for two colonies that are close in space will be an indication that they are not providing independent data. Perhaps the data from the two locations should be pooled to provide a single data point.

The residuals from equation (6.10) are displayed on Fig. 6.1. There is no really strong suggestion of similar values for colonies that are close together, although this has occurred, for example, with colonies UO and LO. It is shown in Section 6.10, using a Mantel test, that the residuals at neighbouring colonies have a positive correlation but this is not quite significant. However, the residuals for all the Hk mobilities taken together indicate that close colonies are not producing completely independent data.

6.6 THE PROBLEM OF UNEQUAL SAMPLE SIZES

One problem with the last example concerns the fact that the sample sizes were rather variable at different locations (Table 6.5). For example, at the first colony (PD) all the gene proportions were based upon a sample of 66 genes but nearly twice this number were used at the last colony (GL). It seems that different colonies should be given different weights in regression analyses.

Suppose that the true proportion of a morph or gene in colony i is π_i. Then for a sample of size n_i, the sampling variance of the sample proportion, $\hat{\pi}_i$, is

$$\text{var}(\hat{\pi}_i) = \pi_i(1 - \pi_i)/n_i.$$

This will not be the total variance of $\hat{\pi}_i$ about a regression equation. There will also be a non-sampling component of variance which reflects the fact that two colonies with identical values for environmental variables cannot be expected to have exactly the same values for π_i. Thus for regression analyses an appropriate expression for the variance of π_i is perhaps

$$\text{var}(\hat{\pi}_i) = \pi_i(1 - \pi_i)/n_i + V, \qquad (6.11)$$

where V is the non-sampling variance. If V is very much larger than $\pi_i(1 - \pi_i)/n_i$ then there will be no need to give different weights to data from different colonies.

A rough calculation can be done to ascertain the relative importance of the two components in equation (6.11). For example, consider the 1.00 Hk gene proportions shown in Table 6.5. (It is convenient to use proportions rather than percentages for the calculation to be made.) The regression of these proportions on the 11 environmental variables results in a residual mean square of 0.0128. This is made up of an average value of $\pi_i(1 - \pi_i)/n_i$ plus V, as shown in equation (6.11). The average value of $\pi_i(1 - \pi_i)/n_i$ can be determined approximately using estimated values $\hat{\pi}_i(1 - \hat{\pi}_i)/n_i$. The average of these estimates is 0.0014. It therefore seems that equation (6.11) can be written approximately as

$$0.0128 \simeq 0.0014 + V$$

or $V \simeq 0.0114$, so that V seems to be about ten times as big as $\pi_i(1 - \pi_i)/n_i$ for an 'average' colony. Weighting the results for different colonies is therefore not important.

On the other hand, if n_i is small for a particular colony then the sampling variance $\pi_i(1 - \pi_i)/n_i$ could be quite large relative to V. Some lower limit therefore needs to be put on n_i. This can be done by noting that the largest possible sampling variance is $0.25/n_i$, which occurs if $\pi_i = 0.5$. Hence for the regression of 1.00 Hk proportions on environmental variables it is required that $0.25/n_i$ is small relative to $V \simeq 0.0114$. Rather arbitrarily, it could be required that the sampling variance be less than half of V so that $0.25/n_i < 0.0057$ or $n_i > 43.9$. Since the smallest sample size with the *Euphydryas editha* data is 48 genes, this condition is met with all 21 colonies.

In cases where weighting of morph or gene proportions is necessary, the estimate $\hat{\pi}_i$ should be given the weight $1/\mathrm{var}(\hat{\pi}_i)$ as discussed in Section A.8 of the Appendix. A rough value for V can be determined for equation (6.11) as shown above. A better value is the one which makes the weighted residual mean square equal to exactly 1.0. This latter value can be determined by trial and error.

6.7   THE USE OF STANDARD METHODS OF MULTIVARIATE ANALYSIS

The use of standard methods of multivariate analysis in studying animal distributions has been reviewed by Gould and Johnston (1972). Here only a few brief comments will be made on this topic.

Canonical correlation analysis (Harris, 1975, p. 20) is a well known statistical technique for relating one set of variables $Y_1, Y_2, \ldots, Y_p$ to another set of variables $X_1, X_2, \ldots, X_q$. If the Y variables are measures related to an animal distribution and the X variables are measures of the environment, then this technique provides one way to analyse several aspects of an animal's distribution at the same time.

For example, if the BMDP6M computer program (Dixon, 1981) is used to relate the three Hk mobility percentages shown in Table 6.5 to all the environmental variables shown in Table 6.4 then this allows an overall test to be made for selection at this locus, taking into account the fact that the three mobility percentages add up to 100%. It turns out that there is a relationship which is significant at the 0.1% level between the Hk mobilities and the environmental variables. Similar analyses are possible for the other loci taken individually. Alternatively, the percentages for several loci can be analysed together.

Principal component analysis (Harris, 1975, p. 23) is another standard statistical technique that may be of value in analysing many variables that are related to an animal distribution. The idea is to replace the many variables with two or three principal components. For example, Baker (1980) studied the distribution of the sparrow *Passer domesticus* in New Zealand by using three principal components in place of 16 skeletal measurements. These three principal components accounted for about 87% of the variation in the original measurements.

Discriminant function analysis (Harris, 1975, p. 108) is also worth mentioning. This involves seeing how groups of samples can be separated using two or more variables. For example, Baker (1980) used this technique to see how sparrows from different locations in New Zealand can best be separated. Four discriminant functions could be used in place of the 16 original variables. Similarly, morph proportions of *Cepaea nemoralis* and *C. hortensis* can be used to discriminate between different types of habitat (Manly, 1983).

All of these multivariate methods rely on the assumption that data points are equally reliable. This causes no problems for studies where values of quantitative variables for individual animals are analysed. It may, however, cause difficulties when morph or gene proportions based on different sample sizes are being analysed. Luckily, the discussion in Section 6.6 suggests that the major source of variation in sample proportions will not generally be sampling errors unless sample sizes are quite small. In practice it may therefore be reasonable to use these standard methods even on proportions.

6.8   THE MANTEL NONPARAMETRIC TEST

A rather useful nonparametric test for comparing two distance matrices was introduced by Mantel (1967) as a solution to the problem of detecting space and time clustering of diseases. Sokal (1979) brought this test to the attention of biologists and pointed out that it was a more general version of the one proposed by Royaltey *et al.* (1975) for studying geographical variation in populations. The value of the test for studying selection has been emphasized by Douglas and Endler (1982). See also Dietz (1983) and Roberson and Fisher (1983).

To understand the nature of the procedure, the following simple example is helpful. Suppose that a population is sampled at four colonies and morph proportions are determined. These morph proportions are then used to construct a 4×4 distance matrix where the entry in the ith row and jth column is the 'morph distance' between colony i and colony j. The way that such a distance can be calculated will be considered later. The important thing is that colonies with similar morph proportions should have a small distance between them while colonies with very different morph proportions should have a large distance between them. The morph distance matrix might well look like

$$\mathbf{M} = \begin{bmatrix} m_{11} & m_{12} & m_{13} & m_{14} \\ m_{21} & m_{22} & m_{23} & m_{24} \\ m_{31} & m_{32} & m_{33} & m_{34} \\ m_{41} & m_{42} & m_{43} & m_{44} \end{bmatrix} = \begin{bmatrix} 0.0 & 1.0 & 1.4 & 0.9 \\ 1.0 & 0.0 & 1.1 & 1.6 \\ 1.4 & 1.1 & 0.0 & 0.7 \\ 0.9 & 1.6 & 0.7 & 0.0 \end{bmatrix}$$

It is symmetric because, for example, the distance from colony 2 to colony 3 must be the same as the distance from colony 3 to colony 2: both distances are 1.1 units. Diagonal elements are zero since these represent distances from colonies to themselves.

Suppose that certain environmental variables are also recorded at each colony. These variables can then be used to construct a matrix of 'environmental distances' between the colonies. These should be such that colonies with similar environments have a small distance between them while colonies with very different environments have a large distance between them. The environmental distance matrix for our example will be taken as

$$\mathbf{E} = \begin{bmatrix} e_{11} & e_{12} & e_{13} & e_{14} \\ e_{21} & e_{22} & e_{23} & e_{24} \\ e_{31} & e_{32} & e_{33} & e_{34} \\ e_{41} & e_{42} & e_{43} & e_{44} \end{bmatrix} = \begin{bmatrix} 0.0 & 0.5 & 0.8 & 0.6 \\ 0.5 & 0.0 & 0.5 & 0.9 \\ 0.8 & 0.5 & 0.0 & 0.4 \\ 0.6 & 0.9 & 0.4 & 0.0 \end{bmatrix}$$

Like $\mathbf{M}$, this must be symmetric with zeros down the diagonal.

Mantel's test is concerned with assessing whether the elements in $\mathbf{M}$ and $\mathbf{E}$ show correlation. The test statistic

$$Z = \sum_i \sum_j m_{ij} e_{ij} \tag{6.12}$$

is calculated and compared with the distribution of Z that is obtained by taking the colonies in a random order for one of the matrices. That is to say, matrix $\mathbf{M}$ can be left as it is; a random order can then be chosen for the colonies for matrix $\mathbf{E}$. For example, suppose that a random ordering of colony numbers turns out to be 3, 2, 4, 1. Then this gives a randomized $\mathbf{E}$ matrix of

$$\mathbf{E}_R = \begin{bmatrix} 0.0 & 0.5 & 0.4 & 0.8 \\ 0.5 & 0.0 & 0.9 & 0.5 \\ 0.4 & 0.9 & 0.0 & 0.6 \\ 0.8 & 0.5 & 0.6 & 0.0 \end{bmatrix}$$

The entry in row 1, column 2 is 0.5, the distance between colonies 3 and 2; the entry in row 1, column 3 is 0.4, the distance between colonies 3 and 4; and so on. A Z value can be calculated using $\mathbf{M}$ and $\mathbf{E}_R$. Repeating this procedure using different random orders of the colonies for $\mathbf{E}_R$ produces the randomized distribution of Z. A check can then be made to see whether the observed Z value is a typical value from this distribution.

The basic idea here is that if environmental distances and morph distances are quite unrelated then the matrix $\mathbf{E}$ will be just like one of the randomly ordered matrices $\mathbf{E}_R$. Hence the observed Z will be a typical randomized Z value. On the other hand, if morph and environmental distances have a positive correlation then the observed Z will tend to be larger than values given by randomization. A negative correlation between morph and environmental distances should not occur, but if it does then the result would be that the observed Z value would tend to be low when compared with the randomized distribution.

With L colonies there are $L!$ different possible orderings of the colony numbers. There are therefore $L!$ possible randomizations of the elements of $\mathbf{E}$, some of which might give the same Z values. Hence in the example with four colonies the randomized Z distribution has $4! = 24$ equally likely values. It is not too difficult to calculate all of these. More realistic cases might involve, say, ten colonies in which case the number of possible Z values is $10! = 3\,628\,800$. Enumerating all of these then becomes impractical and there are two possible approaches for carrying out the Mantel test. A large number of randomized $\mathbf{E_R}$ matrices can be generated on the computer and the resulting distribution of Z values used in place of the true randomized distribution. Alternatively, the mean, $E(Z)$, and variance, $\mathrm{var}(Z)$, of the randomized distribution of Z can be calculated, and

$$g = \{Z - E(Z)\}/\{\mathrm{var}(Z)\}^{1/2} \tag{6.13}$$

can be treated as a standard normal variate.

Consider a general situation where $\mathbf{M}$ and $\mathbf{E}$ are of order $L \times L$. These matrices will both be symmetric, with zero diagonal elements. The typical element of $\mathbf{M}$ in row i and column j will be m_{ij}, the morph distance from colony i to colony j, with $m_{ji} = m_{ij}$ and $m_{ii} = 0$. Similarly, the typical element of $\mathbf{E}$ will be e_{ij}, the environmental distance from colony i to colony j, with $e_{ji} = e_{ij}$ and $e_{ii} = 0$. To calculate $E(Z)$ and var (Z), the following quantities are needed:

$$A_M = \sum_{i=1}^{L} \sum_{j=1}^{L} m_{ij} \quad = \text{sum of the morph distances;}$$

$$A_E = \sum_{i=1}^{L} \sum_{j=1}^{L} e_{ij} \quad = \text{sum of the environmental distances;}$$

$$B_M = \sum_{i=1}^{L} \sum_{j=1}^{L} m_{ij}^2 \quad = \text{sum of the squared morph distances;}$$

$$B_E = \sum_{i=1}^{L} \sum_{j=1}^{L} e_{ij}^2 \quad = \text{sum of the squared environmental distances;}$$

$$D_M = \sum_{i=1}^{L} \left(\sum_{j=1}^{L} m_{ij} \right)^2 = \text{sum of squared row totals for morph distances;}$$

$$D_E = \sum_{i=1}^{L} \left(\sum_{j=1}^{L} e_{ij} \right)^2 = \text{sum of squared row totals for environmental distances;}$$

$$G_M = A_M^2; \qquad\qquad G_E = A_E^2;$$

$$H_M = D_M - B_M; \qquad H_E = D_E - B_E;$$

$$K_M = G_M + 2B_M - 4D_M;$$

and

$$K_E = G_E + 2B_E - 4D_E.$$

Mantel (1967) has shown that for the randomized distribution of Z,

$$E(Z) = A_M A_E / \{L(L-1)\}, \tag{6.14}$$

and

$$\text{var}(Z) = \left\{ 2B_M B_E + \frac{4H_M H_E}{L-2} + \frac{K_M K_E}{(L-2)(L-3)} - \frac{G_M G_E}{L(L-1)} \right\} \Big/ \{L(L-1)\}. \tag{6.15}$$

The ordinary Pearson coefficient of correlation between the elements of **M** and **E** is easily shown to be given by

$$r = \{Z - E(Z)\} \Big/ \left[[B_M - G_M^2/\{L(L-1)\}][B_E - G_E^2/\{L(L-1)\}] \right]^{1/2}. \tag{6.16}$$

Table 6.7 illustrates these calculations on the four-colony example data given above.

A certain amount of caution is needed in using equation (6.13) to test for significance against the standard normal distribution. Mantel (1967) remarks that, for large matrices, Z may be normally distributed because it is similar to

Table 6.7 The calculations for Mantel's test with four localities.

Morph distances				Row sum	Environmental distances				Row sum
0.0	1.0	1.4	0.9	3.3	0.0	0.5	0.8	0.6	1.9
1.0	0.0	1.1	1.6	3.7	0.5	0.0	0.5	0.9	1.9
1.4	1.1	0.0	0.7	3.2	0.8	0.5	0.0	0.4	1.7
0.9	1.6	0.7	0.0	3.2	0.6	0.9	0.4	0.0	1.9
				13.4					7.4

$$
\begin{aligned}
A_M &= 13.4 & A_E &= 7.4 \\
B_M &= 1.0^2 + 1.4^2 + \ldots + 0.7^2 & B_E &= 0.5^2 + 0.8^2 + \ldots + 0.4^2 \\
&= 16.06 & &= 4.94 \\
D_M &= 3.3^2 + 3.7^2 + 3.2^2 + 3.2^2 & D_E &= 1.9^2 + 1.9^2 + 1.7^2 + 1.9^2 \\
&= 45.06 & &= 13.72 \\
G_M &= 13.4^2 & G_E &= 7.4^2 \\
&= 179.56 & &= 54.76 \\
H_M &= 45.06 - 16.06 & H_E &= 13.72 - 4.94 \\
&= 29.00 & &= 8.78 \\
K_M &= 179.56 + 2(16.06) - 4(45.06) & K_E &= 54.76 + 2(4.94) - 4(13.72) \\
&= 31.44 & &= 9.76 \\
\end{aligned}
$$
$$Z = 1.0 \times 0.5 + 1.4 \times 0.8 + \ldots + 0.7 \times 0.4 = 8.86$$

From equations (6.14) and (6.15), $E(Z) = 8.26$ and $\text{var}(Z) = 0.162$. Hence the test statistic of equation (6.13) is

$$g = (8.86 - 8.26)/\sqrt{(0.162)} = 1.48,$$

which is not significantly large when compared with the standard normal distribution.

Hoeffding's (1948) U-statistic. However, Mielke (1978) notes that one of the conditions for the asymptotic normality of U-statistics does not hold in the present application. In particular, the randomized distribution of Z may be highly skewed. Mielke *et al.* (1976) and Mielke (1978) suggest overcoming this problem by calculating the first three moments of the randomization distributions of Z and approximating the distribution using a beta distribution of the first kind. See their paper for details.

The four location test data that are analysed in Table 6.7 provide a rather extreme example of a situation where the test statistic g of equation (6.13) has a distribution that is far from standard normal under randomization. There are 24 possible values, ranging from -1.35 to $+1.48$. The sample value of g is in fact the largest possible value. Since the probability of getting such a large value is $1/24 = 0.04$ it follows that it is significantly large at the 5% level although the standard normal approximation for g does not even suggest near significance.

Mantel's test procedure is really more general than the test that has been considered here. The two matrices being compared do not need to be symmetric, with zero diagonals. Mantel's (1967) paper discusses the comparison of two matrices with the only restriction being that they are the same size. Douglas and Endler (1982) give a simple explanation of the calculations for the general case.

A computer program to carry out the calculations for Mantel's test on a symmetric matrix is provided in Section A.13 of the Appendix. This includes a facility for generating the randomized distribution of Z.

Mantel's test procedure is not particularly simple, and the computer-based randomization test requires a good deal of processor time. It may well therefore be asked why the observed value of Pearson's correlation coefficient, calculated from equation (6.16), cannot be tested for significance in the usual way. The problem here is that the values within one distance matrix are not all independent. For example, in the morph distance matrix, small values for m_{12} and m_{13} imply that m_{23} is also small. The usual test for a significant correlation requires independence and it is therefore not a valid alternative to Mantel's test.

6.9 CONSTRUCTING DISTANCE MATRICES

Suppose that an animal population contains K morphs. In one colony let these be in proportions $p_1, p_2, \ldots, p_K$, where $\sum p_i = 1$, and in a second colony let them be in proportions $q_1, q_2, \ldots, q_K$, where $\sum q_i = 1$. There are then many possible functions of the p_i and q_i that can be used to measure the distance between the colonies. For example,

$$d_1 = 0.5 \sum |p_i - q_i|,$$

is a sensible measure since it takes the value 1 when there are no morphs in common in the two colonies while it takes the value 0 if both colonies have the same morph distribution. Another sensible measure is

$$d_2 = -\log_e\left(\sum p_i q_i / \{ \sum p_i^2 \sum q_i^2 \}^{1/2} \right),$$ (6.17)

which is infinite when the two colonies have no morphs in common and zero when both colonies have the same distribution.

In many cases the information available on animal distributions in different colonies will be in the form of gene counts. Assuming that this is so, consider a single genetic locus. Let p_i denote the proportion of gene i in one colony and q_i the proportion of the same gene at a second colony, with $\sum p_i = \sum q_i = 1$. Then d_1 and d_2 are sensible measures of the difference between the colonies with respect to the locus considered. With several loci, values of $j_1 = \sum p_i^2$, $j_2 = \sum q_i^2$ and $j_{12} = \sum p_i q_i$ can be calculated for each one. Let j_1, j_2 and j_{12} be totalled over the loci to give the sums J_1, J_2, and J_{12}, respectively. Then

$$I = J_{12}/(J_1 J_2)^{1/2}$$

is Nei's measure of the normalized identity of genes between the two colonies being considered, and

$$d_3 = -\log_e(I)$$ (6.18)

is the standardized genetic distance (Nei, 1975, p. 177). A recent review has shown that d_3 is the most commonly used index of the genetic distance between populations (Avise and Aquadro, 1982). It reduces to d_2 for data from a single locus. Nei (1973) gives a review of other genetic based indices of distance. See Jelnes (1983) for more recent references.

Environmental distances can be constructed in many ways. McKechnie *et al.* (1975) used a points system. The points difference between two colonies was calculated as one point for each 1000 feet of difference in altitude, plus one point for each two degrees of latitude, and so on. This approach seems somewhat arbitrary, and one of the distance measures commonly used in cluster analysis would perhaps have been better. A review of cluster analysis measures is given by Gordon (1981, Ch. 2). The simplest approach involves scaling the environmental variables $X_1, X_2, \ldots, X_m$ to have standard deviations of unity, and calculating

$$d_4 = \left\{ \sum_i (x_{i1} - x_{i2})^2 \right\}^{1/2},$$ (6.19)

where x_{i1} is the value of X_i in one colony and x_{i2} is the value in a second colony.

Mantel and Valand (1970) put forward an interesting method for calculating distances that is based upon ranking. Suppose that there are L colonies. The values for the first environmental variable X_1 can then be ranked from 1 (smallest) to L (largest), with any equal values being given an average rank. Let

r_{i1} be the rank for the ith colony. The other variables can be ranked in a similar way so that r_{ik} is the rank of variable k at colony i. A sensible measure of the distance between colony i and colony j is then the sum of the absolute rank differences:

$$d_5 = \sum_k |r_{ik} - r_{jk}|. \qquad (6.20)$$

The simplest case for making up a matrix of environmental distances occurs when locations are grouped into habitat classes. An example of this is provided by Cain and Sheppard's (1950) data on *Cepaea nemoralis* that is shown in Table 6.1 where 17 colonies are grouped into six classes. Here the matrix of environmental distances can simply have a value 1 for two colonies in different habitats and a value 0 for two colonies in the same habitat.

Example 6.4   Mantel's test on *Euphydryas editha* in California and Oregon

Table 6.8 shows McKechnie *et al.*'s (1975) environmental distance matrix between 21 colonies of *Euphydryas editha* in California and Oregon. The table also shows values for Nei's genetic distance between the colonies as calculated from equation (6.18) applied to the three genetic locus data that are given in Table 6.5.

It is perhaps worth just clarifying the calculation of the genetic distance. For example, if we consider the first two colonies (PD and SS) then we find that the first has proportions 0, 0.06, 0.94, . . . , 1.00, 0, 0 for the 17 genes shown in Table 6.5. The second colony has corresponding proportions 0, 0.07, 0.93, . . . , 0.97, 0.03, 0. Then

$$J_1 = 0^2 + 0.06^2 + 0.94^2 + \ldots + 1.00^2 + 0^2 + 0^2 = 2.3470,$$
$$J_2 = 0^2 + 0.07^2 + 0.93^2 + \ldots + 0.97^2 + 0.03^2 + 0^2 = 2.2148,$$

and

$$J_{12} = 0 \times 0 + 0.06 \times 0.07 + 0.94 \times 0.93 + \ldots + 1.00 \times 0.97$$
$$+ 0 \times 0.03 + 0 \times 0 = 2.2751.$$

Hence Nei's normalized genetic identity is

$$I = 2.2751/(2.3470 \times 2.2148)^{1/2} = 0.9979$$

so that the genetic distance is

$$d_3 = -\log_e (0.9979) = 0.002.$$

In Table 6.8 this has been multiplied by 100 and rounded to the nearest integer to give a distance of 0.

Mantel's test statistic for the two matrices is given by equation (6.12) as $Z = 185\,266$. Equations (6.14) and (6.15) give the randomization mean and

Table 6.8 Distance matrices for Mantel's test on *Euphydryas editha* data for 21 colonies. The triangular matrix above the diagonal is an environmental distance matrix based upon a points system for differences between colonies. The distances have been rounded to the nearest integer for convenience. The triangular matrix below the diagonal gives genetic distances d_3 of equation (6.18) multiplied by 100 and rounded to the nearest integer.

	PD	SS	SB	WSB	JRC	JRH	SJ	CR	MI	UO	LO	DP	PZ	MC	HH	IF	AF	SL	GH	EP	GL
PD	–	5	22	20	20	20	22	22	24	28	28	22	26	20	21	24	32	29	29	31	35
SS	0	–	21	18	18	18	19	19	21	25	25	18	22	19	19	22	28	28	27	31	34
SB	25	23	–	5	5	5	6	7	8	18	18	15	19	23	24	25	24	28	31	36	39
WSB	4	3	5	–	0	0	4	5	5	15	15	13	17	20	22	21	22	27	30	35	37
JRC	5	4	5	1	–	0	4	5	5	15	15	13	17	20	22	21	22	27	30	35	37
JRH	8	8	5	0	1	–	4	5	5	15	15	13	17	20	22	21	22	27	30	35	37
SJ	2	2	16	2	3	3	–	2	5	12	12	12	15	20	23	21	20	29	31	37	39
CR	4	4	12	1	1	1	1	–	6	13	13	13	13	20	21	19	20	28	31	35	38
MI	9	8	9	2	2	2	2	2	–	11	11	15	13	22	24	21	20	29	32	37	39
UO	81	74	12	40	41	40	56	46	35	–	0	19	15	25	25	24	22	32	33	41	44
LO	43	39	12	23	25	30	34	30	25	16	–	19	16	25	25	24	22	32	33	41	44
DP	23	23	9	13	11	8	18	11	15	25	25	–	8	16	19	16	14	27	22	31	34
PZ	24	21	2	8	8	9	13	10	6	13	14	12	–	18	20	15	15	31	27	34	37
MC	52	50	8	28	26	22	39	28	27	8	20	7	11	–	14	8	12	26	25	30	35
HH	42	40	6	21	19	16	28	20	18	9	20	9	6	2	–	10	19	15	16	22	23
IF	18	17	2	6	5	4	10	6	5	18	16	5	2	9	5	–	8	25	24	30	34
AF	23	22	1	9	7	7	13	9	6	15	16	10	1	10	5	1	–	31	29	38	42
SL	43	41	5	21	19	16	29	21	18	7	18	7	6	2	1	5	5	–	10	18	18
GH	60	60	15	37	35	30	50	37	38	15	28	9	19	3	7	15	18	6	–	16	18
EP	57	57	13	35	32	26	46	33	35	14	26	7	17	1	5	12	15	4	2	–	6
GL	57	57	15	36	33	27	47	33	37	16	28	6	20	2	7	13	18	5	2	2	–

standard error to be $E(Z) = 159\,012$ and $\mathrm{SE}(Z) = \sqrt{\mathrm{var}(Z)} = 6331$. Hence the significance of Z may be assessed by comparing

$$g = \frac{185\,266 - 159\,012}{6331} = 4.15$$

with the standard normal distribution. On this basis, g is significantly large at the 0.01 % level and there is very clear evidence of an association between genetic and environmental distances.

In this example the treatment of g as a standard normal is a fair approximation. When 1000 random orderings of the environmental distance matrix were made using the computer program in Section A.13 of the Appendix, the distribution obtained for g was fairly close to normal. The range was from -2.52 to $+3.59$, indicating some positive skewness. The median was -0.02, very close to zero. The upper 5 % point was 1.70, which compares well with the normal distribution value of 1.64. The observed Z is much higher than any of the 1000 randomized values. Consequently it can be declared significantly large at at least the 0.1 % level without making any assumptions of normality. It seems that the gene frequencies of E. editha are related to the environmental variables. It would be possible to carry out a separate test for each of the three genetic loci for which data are provided in Table 6.5 to see which ones are related, but this will not be done here.

One interpretation of an association between gene frequencies and environmental variables is that there have been local adaptations through natural selection. There is, however, the alternative possibility that isolation by distance is indirectly responsible. This is discussed further in Section 6.10.

The matrix correlation r of equation (6.16) is 0.41 for this example. As has been explained above, this cannot be tested for significance in the normal way for a correlation coefficient because the distances between different pairs of colonies are not independent. However, it does have use as a purely descriptive statistic of the amount of association between the two matrices and suggests a mild association only.

Fig. 6.3 shows the nature of the association that is present. Genetic distances are plotted against environmental distances. All the figure shows is that when an environmental distance is small, the genetic distance also tends to be small. However, this is not a normal regression type of relationship since if an environmental distance is large then the corresponding genetic distance may be large or small.

The relationship between genetic and environmental distances shown by Fig. 6.3 is precisely the type of relationship that Mantel (1967) devised his test for. He suggested that the clustering at zero can be emphasized if the two distance matrices are transformed suitably before carrying out the test. For example, a reciprocal transformation

$$\text{closeness} = 1/\text{distance}$$

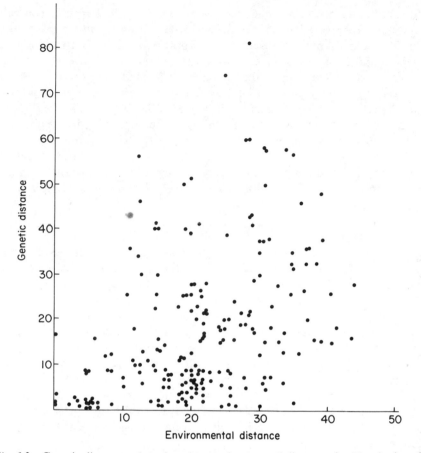

Fig. 6.3   Genetic distances plotted against environmental distances for 21 colonies of *Euphydryas editha*. Genetic distances tend to be small for small environmental distances only.

has the effect of bringing the large distances close together near the origin whilst at the same time the important small distances are spread out away from the origin. Unfortunately, zero distances (such as occur in the data for the present example) transform to infinity. Such infinities can be set equal to some arbitrary large number, but this is not completely satisfactory. A number of other transformations have been discussed by Sokal (1979). Of course, for the present example Mantel's test works well enough without any transformation.

Example 6.5   Mantel's test on *Cepaea nemoralis* in Southern England

Table 6.9 shows morph distances and environmental distances for Cain and

Sheppard's (1950) data on *Cepaea nemoralis* at 17 colonies in southern England (Table 6.1). The morph distances have been found using equation (6.17). The environmental distances are simply 0 for two colonies in the same habitat and 1 for two colonies in different habitats.

The situation where one of the two matrices being considered has 0–1 values derived from simple groupings in this manner has been discussed by Mantel and Valand (1970), Mielke *et al.* (1976) and Mielke (1978). Mielke (1978) notes in particular the fact that it may be completely inappropriate to use a normal approximation for the Mantel test statistic for this special case.

The Mantel test statistic from equation (6.12) is $Z = 2176$. Equations (6.14) and (6.15) give the mean and standard deviation of the randomization distribution to be 1970.2 and 61.6, respectively. Hence the standardized test statistic is

$$g = (2176 - 1970.2)/61.6 = 3.34.$$

Comparing this with the standard normal distribution shows it to be significant at the 0.1 % level. It is also larger than 1000 randomized Z values as determined by the computer program given in the Appendix, Section A.13. Hence it is also significantly large at the 0.1 % level on that basis. There is clear evidence that morph distributions in colonies within the same habitat type are more similar than they are for colonies in general. Thus the Mantel test agrees with the results from other tests given in Examples 6.1 and 6.2.

It is worth noting that Mielke's (1978) reservations about the use of the normal distribution test are not borne out by this example. The 1000 values of g found by randomization have a distribution very close to the standard normal. The range is from -2.99 to $+2.92$ with a median of -0.04. The upper 5 % point for the standard normal distribution of 1.64 is exceeded by 4.2 % of the randomized values.

6.10   SPURIOUS TEST RESULTS DUE TO ISOLATION BY DISTANCE

It was mentioned in Example 6.4 that the significant association between genetic and environmental distances for colonies of the butterfly *Euphydryas editha* could be due to isolation by distance between the 21 colonies. The point is that if colonies have been isolated by their distances from each other, then close colonies will tend to have more similar gene distributions than colonies that are far apart. Unfortunately, it is the nature of things that colonies that are close together will also tend to have similar environments. Consequently, isolation by distance may cause a spurious correlation between genetic and environmental distances. It seems, then, that this type of correlation is not evidence for natural selection unless the possibility of a spurious correlation can be ruled out.

Table 6.9 Morph and environmental distances between 17 colonies of *Cepaea nemoralis* in Southern England. Morph distances (bottom triangular matrix) are calculated from equation (6.17). Environmental distances (top triangular matrix) are 1 for different habitats and 0 for similar habitats.

Colony

	1	2	3	4	5	6	7	8	9	10	11	12	13	14	15	16	17
1	–	0	1	1	1	1	1	1	1	1	1	1	1	1	1	1	1
2	5	–	1	1	1	1	1	1	1	1	1	1	1	1	1	1	1
3	10	3	–	0	0	1	1	1	1	1	1	1	1	1	1	1	1
4	2	9	6	–	0	1	1	1	1	1	1	1	1	1	1	1	1
5	3	7	5	2	–	1	1	1	1	1	1	1	1	1	1	1	1
6	2	4	10	5	4	–	0	0	0	0	1	1	1	1	1	1	1
7	14	11	9	10	4	7	–	0	0	0	1	1	1	1	1	1	1
8	7	8	7	5	2	3	1	–	0	0	1	1	1	1	1	1	1
9	8	3	5	10	5	4	2	3	–	0	1	1	1	1	1	1	1
10	11	6	9	12	4	4	1	2	3	–	1	1	1	1	1	1	1
11	29	27	27	26	18	18	4	8	5	8	–	0	0	0	1	1	1
12	14	15	12	9	4	10	1	2	4	2	4	–	0	0	1	1	1
13	8	11	17	10	7	6	2	3	3	3	3	2	–	0	1	1	1
14	40	43	21	19	8	11	2	3	3	3	8	2	5	–	1	0	1
15	19	20	16	14	14	15	4	6	5	9	1	3	4	6	–	0	1
16	11	15	17	10	5	7	1	2	3	2	5	1	1	3	5	–	1
17	11	2	4	13	10	13	11	11	6	6	17	11	9	27	19	18	–

McKechnie *et al.* (1975) argued that the migration between the colonies of *E. editha* that they studied has been too low to maintain genetic similarities, since each colony has received no more than about one migrant per ten generations from neighbouring colonies. Taking a typical population size to be 500, the proportion of migrants per generation is then $0.1/500 = 2 \times 10^{-4}$, which is certainly very low. However, equation (6.1) suggests that migration rates have to be as low as mutation rates for colonies to diverge genetically. Since mutation rates are more likely to be of the order of 10^{-8}, this condition does not appear to be satisfied. Apart from this consideration, equation (6.2) shows that it may take millions of years for isolated colonies to diverge. Of course, the models upon which equations (6.1) and (6.2) are based are not very realistic for the *E. editha* colonies. Nevertheless, the equations do give an idea of the order of magnitude of effects and it does seem a possibility that the colonies are showing some effects of isolation by distance.

The obvious way to examine the possibility of isolation by distance is to construct a geographical distance matrix, **G**, for the colonies and compare this with the matrix **E** of environmental distances and the matrix **M** of genetic distances. Table 6.10 shows the test statistics g of equation (6.13) and the matrix correlations r of equation (6.16) that are obtained on this basis. It appears that genetic distances are more closely associated with geographical distances ($r = 0.52$) than they are with environmental distances ($r = 0.41$). However, because of the association between geographical and environmental distances ($r = 0.28$) it is not possible to decide what is causing the associations. It could be that isolation by distance is responsible for the association between genetic and geographical distances and this indirectly causes the association between environmental and genetic distances. On the other hand, it may be that local adaptations result in an association between genetic and environmental distances and incidentally produce an association between genetic and geographical distances. Then again, perhaps isolation by distance and local adaption have both been involved in determining gene distributions.

Table 6.10 Comparison of the geographical (**G**), environmental (**E**) and genetic (**M**) distance matrices for 21 colonies of *Euphydryas editha*. The upper triangular matrix gives values for the Mantel test statistic g. The lower triangular matrix gives the matrix correlations.

	G	E	M
G	–	2.07*	4.57†
E	0.28	–	4.15†
M	0.52	0.41	–

* significantly large at the 5 % level
† significantly large at the 0.1 % level

One way around this problem is to see whether the residuals from regressions of gene frequencies on environmental variables are similar for colonies that are close together in space. The idea here is that if environmental variables account for all the non-random variation in gene frequencies, then the regression residuals should have no tendency to be similar in neighbouring colonies. However, if isolation by distance has been the main determinant of gene frequencies then it is unlikely that regressions on environmental variables will account completely for the similarity between close colonies.

For example, equation (6.10) gives the regression of the percentage of 1.00 Hk genes on McKechnie *et al.*'s 11 environmental variables. The residuals from this regression are given in Table 6.11. These residuals are also shown on Fig. 6.2 and it seems that close colonies do sometimes have similar residuals, but this is not always the case. If colony i has a residual R_i and colony j a residual R_j, then the residual distance between the two colonies can be taken as

Table 6.11   Residuals from multiple regressions of three hexokinase (Hk) gene frequencies on the 11 environmental variables shown in Table 6.4.

Colony	Hk		
	1.00	1.12	1.24
PD	15	−13	1
SS	−11	10	−1
SB	3	−3	0
WSB	−1	−2	0
JRC	−3	0	0
JRH	−9	6	0
SJ	6	−3	−1
CR	2	0	0
MI	1	2	−1
UO	−5	37	0
LO	−3	−34	0
DP	−1	1	0
PZ	13	−6	0
MC	−17	16	−1
HH	2	−1	−1
IF	10	−12	1
AF	0	0	0
SL	5	1	0
GH	−5	0	0
EP	−1	0	0
GL	0	0	0

the absolute value $|R_i - R_j|$. A residual distance matrix **R** can be constructed in this way and compared to the geographical distance matrix **G** using a Mantel test. For the 1.00 Hk gene this results in a test statistic $g = 1.72$ which is significantly large at the 5% level when regarded as a standard normal variate. It is higher than 923 out of 1000 values of g obtained by direct randomization, so that from this point of view it is not quite significantly large at the 5% level. It seems that there is a tendency for close colonies to have a similar residual but this effect is not significant.

This idea of comparing residuals for different colonies can be extended to include several regression residuals from each colony. Thus Table 6.11 shows the residuals for the three Hk mobility genes detected by McKechnie *et al.* Denoting these by R_{i1}, R_{i2} and R_{i3} at the ith colony and R_{j1}, R_{j2} and R_{j3} at the jth colony, a residual difference between the two colonies is

$$d_{ij} = \left\{ \sum_{k=1}^{3} (R_{ik} - R_{jk})^2 \right\}^{1/2}.$$

A residual difference matrix **R** can be constructed using this measure and tested against the geographical distance matrix **G**. This results in a Mantel test statistic of $g = 2.07$, which is significantly large at the 5% level either by comparison with 1000 randomized values or by the standard normal approximation.

It does seem then that even allowing for environmental effects, there are significant similarities between close colonies with respect to gene frequencies at the Hk genetic locus. There definitely seems to be an element of isolation by distance needed to explain the gene frequencies. It follows that the regression equations seem to be based upon data that are not quite independent. Therefore the apparent significance of the environmental variables in determining gene frequencies may be overestimated to some extent.

The next stage in considering this set of data would involve removing the data on some colonies that are very close to others and repeating the analyses. Alternatively, Ripley (1981, p. 101) and Cook and Pocock (1983) give details of how regression analyses can be carried out taking into account the spatial relationship of colonies. Neither of these possibilities will be followed up here.

Example 6.6   Colour polymorphism in guppies

Douglas and Endler (1982) considered data on the colour and sizes of spots on the male fish *Poecilia reticulata* (guppies) at 41 sampling sites in Trinidad in the light of Sokal's (1978) four possibilities for explaining the observed morph distribution that were mentioned in Section 6.1. Specifically they entertained four models:

(a) Spotting patterns are a response to some type of environmental gradient, with populations being distributed along this gradient in a clinal fashion.

(b) Spotting patterns are the product of localized environmental patches, which are homogeneous within themselves yet heterogeneous between patch types.

(c) The observed phenotypic differences between guppies from different populations are due to isolation by distance and genetic drift.

(d) Spotting patterns are the result of historical factors, such as founder effects or migration.

Models (a) and (b) imply selection (local adaptation) as the evolutionary force generating differences between populations; models (c) and (d) imply stochastic or historical processes at work.

At each sample site, the biological data consisted of density estimates for six guppy predators and seven phenotypic measures of colour polymorphism for male guppies. Pairwise site distances between guppy spot distributions were calculated to form one matrix of distance. A similar matrix of predator distances was also constructed. Four further distance matrices were based upon the geographical relationship between sites. This produced six matrices, as follows:

ALT: a matrix of altitudinal differences between localities.

GEO: a matrix of kilometric distances between sites.

ASM: an asymptotic distance matrix composed of actual kilometric distances between geographically contiguous sites and a constant, very large, distance between non-contiguous sites.

STR: a binary connectivity matrix of stream-connected sites.

PRD: a matrix of predator distances between sites for six visually hunting vertebrate and invertebrate predators.

SPT: a matrix of spot distances between sample sites based upon mean spot sizes for six colours, plus the total number of spots.

To test models (a) to (d), Douglas and Endler matched up these six matrices in a pairwise fashion. They chose the pairing so that a positive Mantel test result is expected if the model is correct. Thus for model (a) they predicted that spot phenotypes should become more dissimilar and spot distances should increase moving from lowland sites to upland ones. This suggests comparing ALT to SPT. For model (b) they expected that distances between sites for spot polymorphisms would be associated with differences in predator distributions. This suggests comparing PRD to SPT. For model (c) sites separated by short kilometric distances should display small spot distances. This suggests comparing GEO to SPT. Finally, for model (d) they predicted that sites which are in the same watershed, or which are geographically contiguous, should be similar. with regard to spot polymorphisms while phenotypic differences should increase significantly for sites in different watersheds, or sites which are geographically disjunct. This suggests comparing both ASM and STR to SPT.

Douglas and Endler performed 15 pairwise statistical comparisons of the

Table 6.12 Pairwise comparisons between six distance matrices constructed for guppies. The upper triangle represents g values generated by the Mantel test, while the lower triangle displays matrix correlations from equation (6.16).

		ALT	GEO	ASM	STR	PRD	SPT
Altitudinal difference	(ALT)	–	1.333	–2.366	–1.377	2.400	4.782*
Geographic distances	(GEO)	0.077	–	–1.622	–3.666*	6.418*	0.854
Asymptotic distance	(ASM)	–0.006	–0.031	–	–0.268	–3.904*	–6.612*
Stream connectivity	(STR)	–0.083	–0.244	0.057	–	–2.990*	–3.107*
Predator densities	(PRD)	0.148	0.444	–0.021	–0.183	–	3.607*
Spot polymorphisms	(SPT)	0.302	0.060	–0.125	–0.194	0.213	–

* Significant at the 0.33% level using the normal approximation for the Mantel test statistic.

data matrices using the Mantel test. Five of the comparisons directly tested the four evolutionary models as mentioned above. The ten remaining comparisons either corroborated results from different models, or provided additional information about the geographic distribution of polymorphisms or predators. They used the Bonferroni technique (discussed in Example 6.1, above) to decide on an appropriate level of significance for tests. That is to say, since there were 15 tests, the level of significance used for each test was $(5/15)\% = 0.33\%$ in order that there was only a probability of 5% or less of getting any significant results by chance.

The results from comparing all of the data matrices are shown in Table 6.12. It can be seen that model (a) is confirmed since there is a significant positive association between ALT and SPT. Model (b) is also confirmed because of the significant positive association between PRD and SPT. Model (c) is not confirmed since there is no significant relationship between SPT and GEO. Model (d) is not confirmed since the association between ASM and SPT is negative, as is the association between STR and SPT.

Douglas and Endler argue that their results indicate that for patterns of geographic variation in colour polymorphisms of male *P. reticulata*, selection plays a more important role than stochastic processes such as the founder effect, drift or migration. Spot polymorphisms are not only distributed clinally in response to an altitudinal gradient, but they are also associated in a significant positive manner with predator communities. There is no evidence that spot polymorphisms are related to geographic distances between sites. In contrast to this result, there is evidence that predators in stream communities grow more dissimilar as geographic distances increase between study sites, thus agreeing with an isolation-by-distance model. There is also no evidence for the historical effects model since the largest phenotypic differences and the greatest disparity in predator communities occur between contiguous sites. It seems that gene flow between sites is low or nonexistent. Douglas and Endler suggest that the geographic pattern of their data is due to strong selective pressures acting on a morphologically plastic species divided into small populations that inhabit spatially heterogeneous patches within the environment.

6.11   THE EVIDENCE FROM SPATIAL
PATTERNS ALONE

We can turn now to cases where a non-random spatial pattern on its own has been claimed to provide evidence of selection. An obvious example concerns the distribution of the moth *Amathes glareosa* in the Shetland Isles. The problem here has already been discussed in Section 6.1. It is not possible to determine whether the north-to-south trend in the frequency of the *edda* morph is due to adaptation to some environmental gradient or to historical factors. This case therefore highlights the need for a non-random distribution in space to be supplemented by some further evidence of selection.

The situation with *Maniola jurtina* in England is somewhat different. There have been numerous studies concerned with the spot numbers on the underside of the hind wings of this butterfly. The distribution of spot numbers can remain constant over large areas and then change over a rather short distance to a new distribution (Ford, 1975). The boundaries of spot number distributions can change from year to year. The explanation for this is still not clear, but perhaps this is an example of adaptation to localized environmental patches. The mobility of *M. jurtina* seems to rule out isolation and historical factors for explaining the distribution. Thus a 'patchy' distribution by itself can give some evidence of selection.

Christiansen and Frydenberg (1974) note the difficulty in deciding whether it is natural selection that is responsible for a pattern in an animal distribution. However, they point out that other factors affect all variation in the same way. On this basis they claimed evidence for selection on the eelpout *Zoarces viviparus* in Danish waters.

Samples of eelpouts were taken at 46 sampling stations ranging from north of Jutland to the east of the Baltic Sea. Four genetic polymorphisms were studied. Two of these polymorphisms showed near-constant morph frequencies over the entire sampling area. The other two polymorphisms showed distinct trends in morph frequencies when the samples were arranged in geographical order. In terms of gene frequencies the important data are shown in Table 6.13.

The argument used by Christiansen and Frydenberg is as follows, simplified somewhat. The trends in the frequencies of the genes HbI^1 and $EstIII^1$ can be explained in terms of small local population sizes, very restricted migration, or long separation times. On the other hand, the constancy of the frequencies of the $PgmI^1$ and $PgmII^2$ genes can be explained by large population sizes, extensive migration or short separation times. Hence models involving no selection make either the constant or the variable polymorphisms very unlikely. It appears that selection has to be evoked to account for the geographical pattern of at least one of the two groups of polymorphisms.

Whilst this argument appears reasonable, it does not seem to be possible to rule out the explanation that the geographical patterns are simply due to

Table 6.13 Data on four polymorphisms of the eelpout *Zoarces viviparus*. The table shows the frequencies of four genes HbI^1, $EstIII^1$, $PgmI^1$ and $PgmII^2$ in two different parts of Danish waters. The frequencies for HbI^1 and $EstIII^1$ show steady, more or less parallel trends from Kattegat to the Baltic.

	HbI^1	$EstIII^1$	$PgmI^1$	$PgmII^2$
Northern region (Kattegat)	1.00	0.40	0.60	0.85
South-eastern region (Baltic)	0.10	0.05	0.60	0.85

historical factors. For example, Christiansen and Frydenberg mention that from 6000 to 4000 BC the Baltic Sea was cut off from the Kattegat, forming the Ancylus Lake. The eelpout is fairly tolerant of fresh water and an isolated Ancylus population could have existed. If this resulted from a small number of individuals then the founder effect could well have produced a population with rather low frequencies for the HbI^1 and $EstIII^1$ genes but about the same frequencies as in the Kattegat for $PgmI^1$ and $PgmII^2$. Subsequent opening of the Baltic Sea and restricted migration could then have resulted in the present-day distribution of eelpouts. Nei and Tateno's (1975) simulation results are of some relevance here. These results show that if a population splits into two then the effects of mutation and random genetic drift are such that after a long period of time the two populations will still have similar gene distributions at many genetic loci, although at other loci the distributions may be quite different.

Christiansen and Frydenberg point out that independent evidence is available which suggests that selection is operating at the Est III genetic locus on eelpout populations in the present day. This independent evidence comes from the analysis of mother–offspring data (see Example 9.7).

On the whole it seems fair to summarize matters by saying that a non-random distribution in space may provide some circumstantial evidence of selection but it will often be possible to make alternative explanations which seem equally plausible.

6.12   ANALYSIS OF SPATIAL PATTERNS BY SPATIAL AUTOCORRELATION

Sokal and Oden (1978a, b) have discussed the uses of spatial autocorrelation analysis in biology at some length. This is a technique that can be applied to variables mapped on to a geographic area. It tests whether the observed value of a variable at one locality is independent of values of the variable at neighbouring localities.

In their first paper, Sokal and Oden showed how it is possible to construct spatial 'correlograms' which show autocorrelation as a function of distance between pairs of localities. These correlograms summarize the patterns of geographic variation exhibited by a variable. While identical variation patterns will lead to identical correlograms, different patterns may or may not yield different correlograms. Similarity in the correlograms of different variation patterns suggests similarity in the generating mechanism of the pattern.

In their second paper, Sokal and Oden concentrated on the biological implications of various combinations of significant or non-significant heterogeneity of character means with significant or non-significant variation patterns for these means. They suggest that by examining and analysing variation patterns of several characters or gene frequencies for one population, or of several populations in different places or at different times, some conclusions can be reached about the nature of the populational processes generating the observed patterns.

Sokal and Oden applied their methods of autocorrelation in a variety of situations, including the gene distribution of mice in a barn and the blood distribution of humans in Ireland. They concluded that the mice are non-randomly distributed because of the genetic consequences of social structure in the mouse colony; while the human blood groups show a clinical structure, probably as a result of successive invasions of Ireland from the east.

Clearly, spatial autocorrelation analyses may be of value in assessing evidence of natural selection. However, these methods will not be considered any further here. See the papers by Sokal and Oden (1978a, b), Sokal (1978), Jones *et al.* (1980) and Sokal and Wartenberg (1983) for more details and examples.

6.13   THE KLUGE–KERFOOT PHENOMENON

Kluge and Kerfoot (1973) reported that for seven vertebrate data sets, the relative amount of between-colony variation for a given morphometric variable was positively correlated with the amount of within-colony variation. This 'Kluge–Kerfoot' phenomenon has since been demonstrated for a wide variety of organisms. Kluge and Kerfoot attributed it, at least partly, to the variability of a character within populations being inversely proportional to its effect upon fitness. It can therefore be thought of as some sort of evidence for natural selection. However, the reality of the phenomenon has recently been questioned by Rohlf *et al.* (1983). Through an analysis of earlier data, these authors show that all reported cases can be explained as statistical artefacts, caused by such things as grouping data. They conclude that the Kluge–Kerfoot relationship may exist, but previous studies do not provide adequate data to investigate it.

7 Association between related species

Chapter 6 was concerned with the question of how the distribution in space of a single species may provide evidence for selection. A natural progression from this is to consider how selection can be inferred from an association between the spatial distribution of two related species.

The present chapter begins with a general discussion of possible relationships between species distributions in a number of colonies, and how these relationships may arise. A situation of particular interest occurs when there is parallel evolution, so that two species have similar gene distributions in sympatric (mixed) colonies. Section 7.2 describes Borowsky's randomization test for parallel evolution, and alternatives to this are discussed in Section 7.3. Then follows a section on sample size requirements in order that most of the variation in data reflects differences between colonies rather than sampling variation.

Evidence for parallel evolution can only be looked for using data from sympatric colonies, and field studies suggest that it is difficult to find this. However, there are many cases where there is a difference between the morph distribution of a species in allopatric colonies and the same species in sympatric colonies. The significance of a finding like this is discussed in Section 7.5.

The chapter concludes with a brief mention of how overall comparisons of gene frequency distributions for *Drosophila* species have been put forward as evidence for selection.

7.1 PATTERNS OF SPECIES ASSOCIATION

If two closely related species have geographical ranges which overlap, then a relationship between their morph or gene frequencies may be used as evidence of natural selection. For example, suppose that a certain morph is common in some colonies of species A but rare in others. If the same morph in species B is common and rare in exactly the same colonies, then this indicates that a selective agent is affecting the two species in a similar way. On the other hand, if for species B the morph is common in colonies where it is rare for species A, and vice versa, then this may indicate competition between the individuals of the two species. There again, suppose that the morph is only common for species A in colonies where species B is absent. This also may indicate some interaction between the species.

In the simplest possible situations, the distributions of the two species being considered will be reflected in values for a single variable X for species A and a single variable Y for species B, as shown in Table 7.1. The variables X and Y might be colony proportions for a certain gene or a certain morph, or colony mean values for a quantitative variable such as length. In most cases X and Y will be the same measure but taken on the different species. It is important for the sampled colonies to be sufficiently isolated from each other to provide more or less independent data.

Even in this single variable situation there are various possible patterns that can appear in the relationship between X and Y. Fig. 7.1 shows three possibilities, as follows:

(a) There can be positive correlation between X and Y in the sympatric colonies. This can occur if species A and B are independently affected in the same way by some selective agency so that conditions which promote high values of X also promote high values of Y. Barring hybridization, parallel patterns of variation like this should arise only through selection (Borowsky, 1977).

Table 7.1 The simplest type of data on the association between the distribution of two closely related species. The variables X and Y measure the same aspect of distributions for species A and B respectively. For example, X might be the proportion of a certain morph for species A, and Y is then the proportion of the same morph for species B. In this illustration there are nine allopatric colonies (four with species A only and five with species B only) and six sympatric colonies.

Colony	X	Y	
1	x_1	–	
2	x_2	–	Allopatric colonies with
3	x_3	–	species A only
4	x_4	–	
5	x_5	y_5	
6	x_6	y_6	
7	x_7	y_7	Sympatric colonies
8	x_8	y_8	
9	x_9	y_9	
10	x_{10}	y_{10}	
11	–	y_{11}	
12	–	y_{12}	Allopatric colonies with
13	–	y_{13}	species B only
14	–	y_{14}	
15	–	y_{15}	

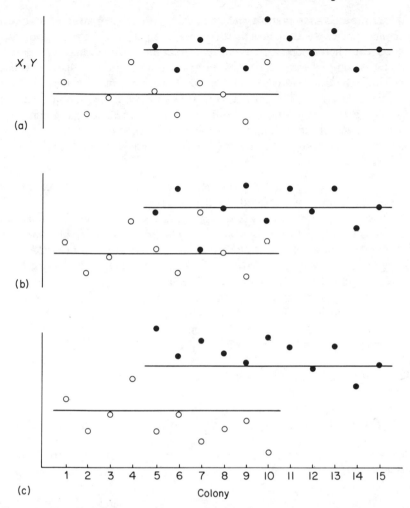

X, Y

(a)

(b)

(c)

| 1 | 2 | 3 | 4 | 5 | 6 | 7 | 8 | 9 | 10 | 11 | 12 | 13 | 14 | 15 |

Colony

Fig. 7.1   Some possible patterns in the relationship between a distribution variable X for species A (open circles) and a distribution variable Y for species B (solid circles). The horizontal lines show average values of X and Y for allopatric colonies. Colonies 1 to 4 are allopatric, with species A only; colonies 5 to 10 are sympatric, with species A and B; and colonies 11 to 15 are allopatric, with species B only. (a) Positive correlation. (b) Negative correlation. (c) Character displacement.

(b)   There can be negative correlation between X and Y in sympatric colonies. This can occur if a selective agent such as a predator treats species A and B as if they are a single species and acts in such a way as to maintain a fairly constant morph distribution for the combined species. Then in colonies where X happens to be high, Y will tend to be low, so that $X + Y$ remains close to a stable value. Similarly, in colonies where Y is high, X will tend to be low (Clarke, 1962a).

(c) There may be *character displacement*. It has often been argued that two similar species will each face their severest competition in areas where they occur together. If similar morphs tend to compete for the same resource, then the competition will be most intense for individuals of species A that are most like individuals from species B. The result may then be that species A diverges from species B in sympatric colonies. A similar effect may or may not occur with species B. This character displacement will show up mainly in differences in the mean values of X and Y between allopatric and sympatric colonies (Brown and Wilson, 1956; Arthur, 1982a).

Clarke (1962a, b, 1969) has discussed the evolutionary consequences of the interaction between two related species, particularly in situations where *apostatic selection* by predators occurs. These are situations where predators develop a 'searching image' for common morphs, with the result that rare morphs are at a selective advantage. Then if the predators cannot tell species A and B apart, it follows that selection will act in one species against any morph that is common in the second species. Clarke has suggested apostatic selection as a general mechanism for maintaining variation in nature. His most important conclusion as far as the present chapter is concerned is that when apostatic selection occurs there will be negative correlations between similar morph proportions for two species in a series of colonies with the same type of habitat.

There is a certain difficulty involved in comparing data from sympatric and allopatric colonies. The problem is that differences between species A in these two types of colony may have nothing to do with species B. They may be due to environmental differences which alter the distribution of species A slightly but are sufficient to prevent species B from occurring at all. Because of this difficulty it is convenient to begin this chapter by considering tests on data from sympatric colonies only, particularly Borowsky's (1977) randomization test.

7.2 BOROWSKY'S TEST FOR PARALLEL VARIATION

Borowsky (1977) has devised a randomization test for parallel variation in gene proportions for two closely related sympatric species. He was particularly concerned with situations like that shown in Figure 7.1(a), but with several gene frequencies being considered at the same time. The principle behind the test is not restricted to use with gene proportions. However, it is convenient to describe the test in terms of these.

The basic question that the test considers is whether the distribution of species A, as measured by gene frequencies $X_1, X_2, \ldots, X_r$, is associated in sympatric colonies with gene frequencies $Y_1, Y_2, \ldots, Y_r$ at the same genetic locus for species B. More precisely, it is a question of whether the gene frequencies for species A and B in a sympatric colony are 'closer' together than can be expected on the basis of the variation in gene frequencies at other colonies. The test involves seeing how the observed 'closeness' of species A and B in sympatric colonies compares with the 'closeness' that is found by randomly assigning colony numbers to the gene frequencies of species B.

Table 7.2 shows the form of the data required for the test when there are r alleles at n sympatric colonies. (The alleles at a genetic locus are the different possible versions of the gene.) The proportion of allele j in colony i is x_{ij} for species A and y_{ij} for species B. The statistic

$$D_{ik} = \frac{1}{2} \sum_{j=1}^{r} |x_{ij} - y_{kj}| \tag{7.1}$$

can then be used to measure the 'distance' between species A in colony i and species B in colony k. This gives $D_{ik} = 1$ if A and B have no alleles in common and $D_{ik} = 0$ if the two species have the same gene distribution. In practice, other measures of distance could be used equally well, for example Nei's standard genetic distance as defined in Section 6.9.

Borowsky used the statistic

$$S_{ik} = \frac{1}{2}(D_{ik} + D_{ki} - D_{ii} - D_{kk}) \tag{7.2}$$

as a measure of the parallel variation between species A and B. This varies from -1 to $+1$. A value of $+1$ is obtained when $D_{ik} = D_{ki} = 1$ and $D_{ii} = D_{kk} = 0$. In other words, in colony i, species A and B have the same gene distributions ($D_{ii} = 0$); the same is true in colony k ($D_{kk} = 0$). However, colonies i and k have completely different alleles ($D_{ik} = D_{ki} = 1$). This is a clear indication of parallel variation. A value of -1 for S_{ik} occurs when $D_{ik} = D_{ki} = 0$ and $D_{ii} = D_{kk} = 1$.

Table 7.2   Data for Borowsky's test of association between gene frequencies of two sympatric species. Here x_{ij} and y_{ij} are proportions of the jth allele in colony i, for species A and B, respectively.

$$\left(\sum_{j=1}^{r} x_{ij} = \sum_{j=1}^{r} y_{ij} = 1 \right)$$

Colony	Species A				Species B			
	1	2	...	r	1	2	...	r
1	x_{11}	x_{12}	...	x_{1r}	y_{11}	y_{12}	...	y_{1r}
2	x_{21}	x_{22}	...	x_{2r}	y_{21}	y_{22}	...	y_{2r}
⋮	⋮	⋮	⋮	⋮	⋮	⋮	⋮	⋮
n	x_{n1}	x_{n2}	...	x_{nr}	y_{n1}	y_{n2}	...	y_{nr}

In this case species A at location i has the same gene distribution as species B at location k ($D_{ik} = 0$) and, at the same time, species B at location i has the same gene distribution as species A at location k ($D_{ki} = 0$). However, species A and B have no common alleles in colony i ($D_{ii} = 1$) or in colony k ($D_{kk} = 1$). This is a clear indication of a negative association between the species.

If the gene distributions for the two species are completely unrelated then the S_{ik} values that can be calculated using different pairs of colonies should all be about zero. Given n colonies there are $\frac{1}{2}n(n-1)$ different S_{ik} values and their mean

$$
\begin{aligned}
\bar{S} &= \sum_{i=1}^{n-1} \sum_{k=i+1}^{n} S_{ik} / \{\tfrac{1}{2}n(n-1)\} \\
&= \sum_{i=1}^{n-1} \sum_{k=i+1}^{n} (D_{ik} + D_{ki} - D_{ii} - D_{kk}) / \{n(n-1)\} \\
&= \sum_{i=1}^{n-1} \sum_{k=i+1}^{n} (D_{ik} + D_{ki}) / \{n(n-1)\} - \sum_{i=1}^{n} D_{ii} / n
\end{aligned}
\tag{7.3}
$$

is a suitable statistic for testing for selection. If $\bar{S}$ is significantly different from zero then this indicates some association between the species.

Borowsky suggested a randomization test for significance, based upon the random pairing of the data from the two species, as shown in Table 7.3. The

Table 7.3 The allocation of data for a randomization test. As it occurs the pairing of species A and B data is by colony. Thus if a_i denotes the data for the ith colony for A and b_i the data for the ith colony for B, then the pairing is as shown in the first two columns. For the randomization test the ordering of B is made random (using for example a random number generator on a computer) which could give the allocations shown in columns three to seven. There are $n!$ possible pairings, including the natural one given in the first two columns. For a test of significance a large number of random allocations are made and for each one, $\bar{S}$ of equation (7.3) is calculated to provide the randomization distribution of $\bar{S}$. It can then be seen whether the $\bar{S}$ value given by the true ordering shown in columns 1 and 2 is significantly different from the mean of the randomized distribution.

Natural order of the data		(Random pairings)				
Species A	Species B	B data				
a_1	b_1	b_4	b_{11}	b_1	b_n	b_2
a_2	b_2	b_3	b_9	b_2	b_6	b_8
a_3	b_3	b_n	b_1	b_n	b_2	b_{10}
a_4	b_4	b_9	b_n	b_5	b_1	b_9
$\vdots$	$\vdots$	$\vdots$	$\vdots$	$\vdots$	$\vdots$	$\vdots$
a_n	b_n	b_1	b_2	b_6	b_4	b_5

idea here is that if the two species have no association then $\bar{S}$ as given by equation (7.3) will look like a typical value obtained by pairing up gene frequencies of the species at random. Thus a suitable procedure for testing involves making, say, 1000 random pairings of species B with species A and generating the randomization distribution of $\bar{S}$. It can then be seen whether the observed value of $\bar{S}$ is an exceptional value when compared with this distribution.

For this test to work it is necessary that the n should be quite large. For example, with $n = 3$ there are only $3! = 6$ possible pairings of the species B data to the species A data, including the one actually observed. Thus there are only six possible values of $\bar{S}$, each with probability $1/6$. The observed $\bar{S}$ value therefore cannot be significantly large at the 5% level. With $n = 4$ there are 24 possible pairings, each with probability $1/24$, so that significance at the 5% level is possible. In this connection it is perhaps worth noting that with small values of n it may be easier to enumerate all possible pairings of species B to species A locations rather than to generate a large number of random allocations on the computer. However, for large values of n the only practical approach is to generate random allocations. For example, if $n = 14$, which was one case considered by Borowsky, there are $14! \simeq 8.7 \times 10^{10}$ possible allocations.

If equations were available for the mean and variance of the randomized distribution of $\bar{S}$, then some idea of the significance of an observed $\bar{S}$ could be obtained without resorting to computer randomization. However, unfortunately this is not the case at the present time. A FORTRAN computer program BOROWSKY is therefore provided in Section A.14 of the Appendix to carry out randomizations.

So far, Borowsky's test has been described in terms of gene frequencies only. In practice, however, the gene proportions used in equation (7.1) can equally well be morph proportions. Indeed, the D_{ik} measure of equation (7.1) can be replaced by any measure of the distance between species A in colony i and species B in colony j. In particular, if x_{ij} and y_{ij} are mean values for a quantitative variable X_j in the colony i for species A and B, respectively, then equation (7.1) might more appropriately be replaced by the Euclidean distance

$$D_{ik} = \{\Sigma (x_{ij} - y_{ij})^2\}^{1/2} \tag{7.4}$$

(Gordon, 1981, p. 21). Whatever distance measure is used, it will probably be desirable to code quantitative variables to have standard deviations of unity before calculating distances.

Example 7.1   Association between enzyme genes

Borowsky (1977) applied his test to 26 different species–enzyme systems. Table 7.4 shows the data for one of these, the Pgm locus for the butterflies *Euphydryas editha* and *E. chalcedona*. Results are available for seven colonies

Table 7.4 Gene distributions for the butterflies *Euphydryas editha* and *E. chalcedona* at seven localities in California and Oregon. These distributions are for the phosphoglucomutase (Pgm) locus and were derived using electrophoretic techniques so that the 'alleles' are actually charge state classes. The sample sizes are the numbers of genes and not the number of individuals.

Location	Allele frequencies for *E. editha*						Allele frequencies for *E. chalcedona*								
	0.87	0.94	1.00	1.06	1.14	Sample size	0.70	0.80	0.87	0.94	1.00	1.06	1.14	1.22	Sample size
PD		0.06	0.94			66				0.06	0.33	0.48	0.13		104
SB	0.02	0.04	0.88	0.06		98	0.03	0.01	0.09	0.39	0.26	0.07	0.15		106
JR(JRC+JRH)	0.01	0.02	0.84	0.14		264		0.04	0.08	0.46	0.26	0.11	0.02	0.03	90
DP			0.73	0.27		116			0.15	0.38	0.25	0.22			40
MC	0.01	0.13	0.74	0.12		104		0.04	0.19	0.39	0.26	0.09	0.03		116
HH	0.06	0.15	0.71	0.08		48		0.02	0.12	0.43	0.31	0.12			100
IF	0.03	0.03	0.80	0.13	0.01	110		0.03	0.22	0.26	0.40	0.05	0.03		106

in California and Oregon. These data were originally published by McKechnie *et al.* (1975). The *E. editha* gene frequencies for the colonies shown in Table 7.4, plus data from another 14 colonies, have already been discussed in Examples 6.3 and 6.4 in terms of how they can be related to environmental variables. Fig. 6.2 shows colony locations.

Using equation (7.1) the matrix of 'distances' between species shown in Table 7.5(a) were calculated. Using these distances, the S values shown in part (b) of the table were then calculated using equation (7.2). There are 8 negative S values and 13 positive ones. This in itself shows that there is little indication of associated variation for the two species at the Pgm locus. This conclusion is made more clear by Borowsky's randomization test. The average of the S values shown in Table 7.5(b) is $\bar{S} = 0.0148$. When Borowsky generated 1000 random pairings of species data this resulted in 193 values larger than 0.0148. Thus the probability of getting an $\bar{S}$ value as large as 0.0148 is estimated as

Table 7.5 Borowsky's test for parallel variation between allele proportions of *Euphydryas editha* and *E. chalcedona.*
(a) Genetic distances D_{ik} calculated using equation (7.1) on the data of Table 7.4.

i	1	2	3	4	5	6	7
1	0.61	0.68	0.68	0.69	0.68	0.63	0.54
2	0.57	0.62	0.62	0.58	0.62	0.57	0.49
3	0.52	0.65	0.61	0.59	0.63	0.55	0.52
4	0.40	0.67	0.63	0.53	0.65	0.57	0.55
5	0.49	0.53	0.49	0.49	0.51	0.43	0.40
6	0.53	0.46	0.49	0.46	0.45	0.40	0.33
7	0.50	0.60	0.56	0.56	0.58	0.51	0.47

(b) The measures S_{ik} of parallel variation between the species at locations i and k.

i	1	2	3	4	5	6	7
1	–						
2	0.010	–					
3	−0.010	0.020	–				
4	−0.025	0.070	0.040	–			
5	0.025	0.010	0.000	0.050	–		
6	0.075	0.005	−0.010	0.050	−0.015	–	
7	−0.025	−0.005	−0.005	0.050	0.000	−0.015	–

0.193. Hence the observed value is not significantly different from zero at the 5% level.

Table 7.6 shows the results that Borowsky obtained with all his 26 tests. If

Table 7.6 Results of the tests carried out for parallel variation for 26 species–enzyme combinations. Significance levels were calculated by comparing the observed $\bar{S}$ with 1000 randomized values of $\bar{S}$.

Species/ Enzyme	Number of colonies	$\bar{S}$	Significance level One-sided*	Two-sided†
Euphydryas editha/E. chalcedona				
Ak	7	0.00119	0.252	0.504
Bdh	7	−0.0226	0.975	0.050
Got	7	0.0102	0.054	0.108
αGpd	7	−0.000238	0.503	0.994
Hk	7	0.0450	0.225	0.450
Pgm	7	0.0148	0.193	0.386
Drosophila willistoni/D. tropicalis				
Est-7	12	−0.0100	0.711	0.578
αGpd	12	0.00470	0.026	0.052
Lap-5	14	0.0241	0.029	0.058
Drosophila willistoni/D. paulistorum				
Ao-1	4	0.0608	0.143	0.286
Ao-2	4	0.00167	0.385	0.770
Est-7	4	0.0100	0.305	0.610
Hk-2	4	0.0108	0.055	0.110
Lap-5	4	0.0500	0.414	0.828
Me-2	4	0.0142	0.363	0.726
Odh-1	4	−0.00667	0.684	0.632
Pgm-1	4	0.00917	0.208	0.416
Xdh	4	0.0508	0.120	0.420
Drosophila willistoni/D. equinoxialis				
Ao-2	4	−0.0175	0.756	0.488
αGpd	12	0.00417	0.046	0.092
Hk-2	4	−0.00583	0.825	0.250
Me-2	4	−0.00000	0.517	0.966
Odh-1	4	0.00583	0.185	0.370
Pgm-1	4	0.00250	0.376	0.752
Xdh	4	0.0125	0.312	0.624
Drosophila tropicalis/D. equinoxialis				
αGpd	10	0.00356	0.131	0.262

* For a one-sided test, to see whether $\bar{S}$ is significantly large, the significance level is the proportion of randomized $\bar{S}$ values that are larger than the observed $\bar{S}$.

† For a two-sided test, to see whether $\bar{S}$ is significantly different from zero, the significance level is the proportion of randomized $\bar{S}$ values that are further from zero than the observed $\bar{S}$.

these are regarded as one-sided tests to see whether the observed $\overline{S}$ is significantly large, then there are three significant values at the 5% level of significance. On the other hand, if two-sided tests are considered to be more appropriate then there is only one significant result. Either way round, the evidence for selection is hardly overwhelming. Borowsky argued that a one-sided test is appropriate because positive values of $\overline{S}$ indicate the parallel variation which he was searching for. However, it was argued at the beginning of this chapter that negative correlations between species can occur so that, under some circumstances at least, a two-sided test should be used. At any rate, that is the point of view taken here.

In the absence of selection, the 26 tests shown in Table 7.6 should be independent. Therefore Fisher's method can be used to consider whether the test results, taken as a whole, indicate the action of selection. Fisher's method is explained in Section A.15 of the Appendix. If the alternative to the null hypothesis is that $\overline{S}$ values will all tend to depart from zero in the same direction, then it is necessary to calculate

$$X^2 = -2 \sum_{i=1}^{n} \log_e (p_i),$$

and compare this to the chi-square distribution with $2n$ degrees of freedom. Here p_i is the proportion of randomized $\overline{S}$ values larger than the observed value for the ith test, as given by the second to last column in Table 7.6. Thus

$$X^2 = -2\{\log_e (0.252) + \log_e (0.975) + \ldots + \log_e (0.131)\}$$
$$= 76.28,$$

with $2 \times 26 = 52$ degrees of freedom. Since the mean value of a chi-square distribution with 52 degrees of freedom is 52, the question is whether or not 76.28 is significantly different from 52, rather than whether it is significantly large. That is to say, it is a two-sided test that is needed. In fact X^2 is significantly different from the chi-squared mean at the 5% level of significance, so that overall there is evidence of selection.

An alternative, somewhat simpler, way of looking at the overall results is to notice that in the absence of selection, the probability of $\overline{S}$ being positive should be one-half. However, 19 out of 26 $\overline{S}$ values are positive. The sign test (Steel and Torrie, 1980, p. 538) shows that this result is significant at the 5% level. The number of positive results is too large to be attributed to chance alone.

7.3 ALTERNATIVES TO BOROWSKY'S TEST

There are two obvious alternatives to Borowsky's test that can be mentioned at this point. The first of these is Mantel's test, which was discussed in Section 6.8. A matrix of distances between colonies can be constructed based upon the distribution variables for species A. Another matrix of distances can be constructed based upon the distribution variables for species B. Mantel's test can then be used to test whether species A distances are correlated with species B distances. A positive matrix correlation indicates that 'close' colonies for species A also tend to be 'close' for species B, possibly because these 'close' colonies have been affected by natural selection in similar ways. A negative correlation indicates that 'close' colonies for species A tend to be distant for species B, and vice versa. It seems difficult to think of a reason why this might happen. Therefore a one-sided test, looking for positive association only, is appropriate. If a significant result is found then it is necessary to rule out the possibility that species A and species B both exhibit patterns of isolation by distance. Clearly if this is the case then geographically close colonies will tend to have similar distributions for species A, and also for species B. This may then generate a significant test result for Mantel's test which has nothing at all to do with natural selection.

A second alternative to Borowsky's test is a canonical correlation analysis (Harris, 1975, p. 20) using, for example, the BMDP6M computer program (Dixon, 1981). This technique was described briefly in Section 6.7 in terms of relating the distribution of a single species to environmental variables. It can be used just as well to relate the distribution of one species to the distribution of another.

Note that whereas Borowsky's test searches for parallel variation, Mantel's test and canonical correlation analysis search for any association between the distributions of the species being tested. For example, suppose that the proportion of morph type 1 of species A has no correlation with the proportion of the same morph for species B but is correlated with the proportion of morph type 2 of species B. This correlation between different morphs is not particularly relevant to Borowsky's test and should therefore not lead to a significant result. However, the correlation is directly relevant for Mantel's test, or a canonical correlation analysis, so that these tests may well give a significant result.

In practice, sample sizes from different colonies are liable to vary somewhat so that sampling errors in gene or morph proportions, or in estimates of means for quantitative variables, will tend to be larger in some colonies than others. This causes no particular problems with the tests of Borowsky or Mantel because of their non-parametric nature. The only thing is that it may be better to exclude colonies with very small sample sizes from these tests in order to avoid their sampling errors obscuring real effects. However, canonical correlation analysis assumes that the data from all colonies are equally reliable and this needs to be approximately true at least. Luckily, it seems that most of the variation in data is liable to come from colony-to-colony variation rather than from sampling errors, unless sample sizes are very small indeed. The following section indicates how an 'adequate' sample size for each colony can be determined.

7.4 DETERMINING 'ADEQUATE' SAMPLE SIZES FROM COLONIES

A sample taken from a colony will be of adequate size providing that the sampling errors involved in estimating gene or morph proportions, or mean values for quantitative variables, are small in comparison with the variation that exists between colonies.

If it is gene or morph proportions that are being considered then the model of Section 6.4 can be used to assess the size of sampling errors. For example, suppose that there are K morphs and L colonies, and that n_{kl} is the count of morph k for the sample from the lth colony. Then it can be assumed that this count has mean and variance

$$E(n_{kl}) = \pi_k \theta_l \tag{7.5}$$

and

$$\text{var}(n_{kl}) = \pi_k \theta_l \{1 + \alpha \theta_l\} \tag{7.6}$$

where π_k is the mean proportion of this morph over all colonies and θ_l is the expected value of m_l, the total sample size from colony l. The parameter α reflects variation from colony to colony. If $\alpha = 0$ then n_{kl} behaves like a simple Poisson count from a population with the same morph proportions in all colonies. If α is large then the proportions of different morphs vary greatly from colony to colony.

A method for estimating α is described in Section 6.4 and illustrated in Example 6.2. It amounts to replacing the observed counts n_{kl} by adjusted values

$$a_{kl} = n_{kl}/(1 + \alpha m_l), \tag{7.7}$$

and calculating a chi-square value for the table of adjusted morph frequencies for K morphs at L colonies. The value of α can be chosen so as to make the calculated chi-square equal to its expected value, $(L-1) \times (K-1)$. The idea here is that with the correct choice of α the adjustment will make the samples from different colonies look like repeated samples from a population with constant morph proportions.

Having estimated α, the relative importance of sampling variation and colony to colony variation can easily be determined. According to equations (6.8) the proportion of morph k at colony l, $\hat{\pi}_{kl} = n_{kl}/m_l$, has variance

$$\text{var}(\hat{\pi}_{kl}) \simeq \pi_k(1 - \pi_k)/m_l + \alpha \pi_k(1 - \pi_k). \tag{7.8}$$

The first term on the right-hand side of this equation is the sampling variance of $\hat{\pi}_{kl}$; the second term is the variance due to variation between colonies. It follows, therefore, that if sampling errors are to be relatively unimportant then it is necessary that $1/m_l$ should be much less than α. A reasonable rule might be that the variance due to variation between colonies should be at least twice the sampling variance, so that $\alpha \geqslant 2/m_l$, or

$$m_l \geqslant 2/\alpha. \tag{7.9}$$

Equations (7.5) to (7.9) have been developed in terms of morph proportions. However, they can be used equally well with gene proportions.

Suppose next that the distribution of an animal in the lth colony is summarized by the mean value $\bar{X}_l$ of a quantitative variable X, based upon a sample of size m_l. Let the within-colony variance of X be V_W and the variance between colonies be V_B. The values of V_W and V_B can be estimated by a one-factor analysis of variance as shown, for example, by Steel and Torrie (1980, p. 149; the random model). Then the variance of $\bar{X}_l$ is

$$\operatorname{var}(\bar{X}_l) = V_B + V_W/m_l,$$

where V_W/n_l is the component of the variance due to sampling errors. A reasonable requirement might then be $V_B \geqslant 2V_W/m_l$ or

$$m_l > 2V_W/V_B. \tag{7.10}$$

Example 7.2   Association between *Cepaea* species

Table 7.7 shows some of the data published by Clarke (1960, 1962a; with a different classification of morphs) for mixed colonies of the snails *Cepaea nemoralis* and *C. hortensis* in Southern England. The selection was on the basis of a minimum sample size of 15 for each species, and a grassland habitat. This is a case where two species can be classified using the same morph system. There is a good deal of evidence, for both *C. nemoralis* and *C. hortensis*, that the distribution of morphs is related to the nature of the habitat where they live, possibly because of the activity of predators hunting by sight. Clarke (1962a) suggested that this may be a situation where predators practise apostatic selection, so that in a series of mixed colonies from roughly similar habitats, there should be a negative relationship between the proportions of similar morphs for *C. hortensis* and *C. nemoralis*.

Consider the *C. hortensis* frequencies only. These provide a 20×5 table of morph frequencies for which the ordinary contingency table chi-square statistic is 1126.7. With $(20-1) \times (5-1) = 76$ degrees of freedom, this is very significantly large so that there seems to be considerable variation in morph proportions from colony to colony. The transformation of equation (7.7) with $\alpha = 0.11$ reduces the chi-square value down to its expected value of 76, so equation (7.9) suggests that *C. hortensis* sample sizes need to be $2/0.11 \simeq 18$ or more in order that the variance of sampling errors in morph proportions should be half or less of the variance due to differences between colonies.

For *C. nemoralis* the chi-square value for the 20×5 table of sample morph frequencies is 481.1. The transformation of equation (7.7) with $\alpha = 0.16$ reduces the chi-squared value to its expected value of 76. Hence, according to equation (7.9), samples sizes of $2/0.16 \simeq 13$ or more are adequate for this species.

From these calculations it seems fair to regard a sample size of 15 or more as

Table 7.7 Morph distributions for *Cepaea nemoralis* and *C. hortensis* in 20 mixed colonies with open habitat in Southern England.

Colony	*C. hortensis* morph frequencies					*C. nemoralis* morph frequencies				
	YFB	YOB	YUB	P	B	YFB	YOB	YUB	P	B
Shippon	26	0	40	13	0	54	5	3	82	20
Headington Wick	33	0	28	0	1	5	6	0	34	0
Shippon Fen 2	68	0	59	26	0	12	3	1	24	2
Shepherd's Rest 1	175	18	34	0	0	16	35	0	18	0
Shepherd's Rest 2	19	0	0	0	0	13	17	0	2	0
Stanford on Vale	44	0	14	0	0	5	0	0	10	0
Chisledon	173	0	121	0	0	6	0	2	10	1
Farington	99	0	84	0	0	18	9	0	39	5
The Ham	20	0	16	0	0	8	3	1	2	9
Wanborough Plain	50	3	0	0	0	2	0	0	28	0
Charlbury Hill	9	3	7	0	0	2	5	1	12	5
White Horse Hill 1	77	6	2	0	0	4	17	0	14	3
White Horse Hill 2	24	2	0	0	0	6	16	0	0	0
White Horse Hill 3	102	14	0	0	0	7	21	0	18	0
Dragons Hill 1	76	5	342	0	0	2	9	0	31	6
West Down 1	200	0	199	0	0	0	6	3	0	9
West Down 2	52	0	44	0	0	0	5	3	2	5
Sparsholt Down	5	0	14	0	0	13	22	0	0	0
Little Hinton	13	0	2	0	0	5	6	0	10	0
Dragons Hill 4	36	9	32	0	0	1	4	0	11	1

being adequate for both species of snail. Thus all the data shown in Table 7.7 can be used for tests on species associations.

For Borowsky's test, the statistic $\bar{S}$ of equation (7.3) comes out at 0.016, which is certainly no indication of a negative relationship. Using the computer program BOROWSKY provided in the Appendix it was found that 562 out of 1000 values from the randomized distribution of $\bar{S}$ were less than 0.016. Clearly this test gives no evidence here of either positive or negative association between the two species of *Cepaea*.

A matrix of distances between the 20 colonies can be constructed on the basis of the *C. hortensis* morph proportions using the distance function (7.1), where x_{ij} is the proportion of morph *j* in colony *i*. Another distance matrix can be constructed using *C. nemoralis* morph proportions. The two distance matrices can then be compared using Mantel's test. The result is that the *g* test statistic of equation (6.13) is 0.06, which is not at all significant. The matrix correlation of equation (6.16) is only $r = 0.05$. There is no indication at all that colonies which are similar for *C. hortensis* frequencies are also similar for *C. nemoralis* frequencies.

A canonical correlation analysis to relate the proportions of YOB, YUB, P and B morphs of *C. hortensis* to proportions of the same morphs for *C. nemoralis* also failed to provide a significant result.

It seems that the data of Table 7.7 provide no evidence of any relationships, positive or negative, between *C. hortensis* and *C. nemoralis* frequencies. This contradicts Clarke's (1962a) conclusion that there is a negative relationship between certain morph proportions for the two species, but there are several explanations for this. Firstly, Clarke used data from more colonies than have been used here because he did not insist on a minimum sample size of 15 for each species in each colony. Secondly, he used a weighting system based upon sample sizes for calculating correlation coefficients. Thirdly, he considered the correlations between individual morphs rather than for all morphs taken together.

Clarke's method of weighting data according to sample sizes is not appropriate since, as has been shown above, sampling errors that are dependent on sample sizes are not the most important source of variation for samples of 15 or more snails. If weighting is to be used, then a more complicated weighting system is required.

If individual morphs are considered one at a time, then it is hardly surprising to find some correlations between proportions for different species. For the data of Table 7.7 there are $5 \times 5 = 25$ correlation coefficients that can be calculated between *C. hortensis* and *C. nemoralis* morphs and it can be expected that one or two of these will be significant at the 5% level. It is much better to compare all morph proportions of *C. hortensis* to all morph proportions for *C. nemoralis* at the same time, using Borowsky's test, Mantel's test, or canonical correlations, when looking for evidence of associations.

Carter (1967) was not able to verify Clarke's negative correlations when he collected more data from mixed colonies of *C. hortensis* and *C. nemoralis*, and he criticized Clarke's conclusion that there is a negative relationship between the species. However, Clarke (1969) countered Carter's arguments and reaffirmed his belief in the existence of negative correlations.

Example 7.3   Association between *Hydrobia* species

An example of data on a quantitative variable is provided by Fenchel's (1975a,b) study of the mud snails *Hydrobia ventrosa* and *H. ulvae* in northern Jutland. Table 7.8 shows average lengths for the two species for 15 sympatric and 17 allopatric colonies. Considering the differences between the allopatric and sympatric colonies, this seems to be a case of character displacement as shown in Fig. 7.1(c), or else something that mimics this phenomenon.

For the present example, only the data for sympatric colonies will be considered. It is interesting to ask whether there is a significant correlation in

Table 7.8   Average lengths (mm) of *Hydrobia ulvae* and *H. ventrosa* from 15 sympatric and 17 allopatric colonies in northern Jutland. About 100 snails were measured in each colony. The colony numbers are as used by Fenchel (1975b). These data were read from Figures 2 and 3 of Fenchel's paper. This may have introduced a few small errors.

	Sympatric colonies H. ulvae H. ventrosa			Allopatric colonies H. ulvae H. ventrosa	
Colony	X	Y	Colony	X	Y
7	3.8	2.6	57	3.2	
6	4.5	2.8	58	2.9	
12	3.8	2.9	63	3.4	
23A	4.4	3.5	69	3.0	
9	4.0	2.6	65	3.3	
10	3.5	2.9	18	3.3	
10A	4.4	2.4	23	2.8	
19	4.0	2.5	3	3.5	
I	4.4	2.9	2	3.5	
II	4.2	3.4	37		3.0
III	4.1	3.3	66		3.5
IV	4.3	3.1	43		3.1
V	3.6	2.7	70A		3.1
GVI	5.0	3.5	70B		3.4
70D	4.5	2.7	70E		3.6
			13		2.7
			GI		3.1
Means	4.17	2.92		3.21	3.19
Std. dev.	0.40	0.36		0.26	0.29

these colonies between the average length of *H. ulvae* (X) and the average length of *H. ventrosa* (Y). The answer is no: the observed correlation, $r = 0.29$, is not significantly different from zero at the 5% level.

With only a single quantitative variable there is no point in extending the analysis beyond the calculation of a correlation coefficient. Within sympatric colonies there is no evidence of an interaction between the two species. Of course, this could just be due to the small number of colonies. Only correlations outside the range -0.59 to $+0.59$ are significant with 15 colonies.

7.5 COMPARISON BETWEEN ALLOPATRIC AND SYMPATRIC COLONIES

Judging from Examples 7.1 to 7.3, one might conclude that it is fairly difficult to find evidence of association between species by just looking at sympatric colonies. The problem is that even when associations exist, they are likely only to have quite mild effects on distributions. These will not show up as significant unless many colonies are examined.

On the other hand, it seems relatively easy to find a difference between sympatric and allopatric colonies. In a recent review of the evolutionary consequences of interspecific competition, Arthur (1982a) has discussed a number of examples. The only problem is the one that has been noted above: it is always possible to explain the difference between the two types of colony in terms of some unknown underlying environmental variable. That is to say, species B may be absent from some colonies because an environmental factor is missing. The lack of this factor may not stop species A being present but it might influence the morph distribution of A. Then species A will show a difference between allopatric and sympatric colonies that has nothing at all to do with species B.

Of course, if all that is being looked for is evidence of selection, it does not matter why there is a difference between allopatric and sympatric colonies, providing that non-selective explanations can be ruled out. The main requirement then is that the allopatric and sympatric colonies should be interspersed over a wide geographical area so that each colony provides more or less independent data and the relative geographical locations of different colonies can be ignored. Given these conditions, several tests are possible.

For data on gene or morph proportions, one of the tests given in Sections 6.3 and 6.4 can be used to see whether the distribution of species A varies between allopatric and sympatric colonies. There are then two 'habitats', corresponding to the absence and presence of species B. A similar comparison between allopatric and sympatric colonies can be made for species B. Alternatively, a matrix of distances between colonies can be constructed based upon the distribution of one of the species, and this can be correlated with an allopatric/sympatric distance matrix using Mantel's test given in Section 6.8. The allopatric/sympatric matrix can have a distance of 0 between two allopatric colonies or between two sympatric colonies, and a distance of 1 between an allopatric and a sympatric colony. Mantel's test will then see whether this form of grouping produces colonies that are relatively similar when compared with colonies as a whole. Mantel's test can also be used with quantitative variables since these variables can be used to construct distance matrices just as well as gene or morph proportions.

In some cases it may be desirable to test the hypothesis that the variance of a single quantitative variable is constant in allopatric and sympatric colonies. This can be done using Bartlett's test. See, for example, Steel and Torrie (1980, p. 471). Assuming that the colony-to-colony variance is the same for sympatric and allopatric colonies, ordinary analysis of variance can be used to compare mean values for a single quantitative variable.

When there are several quantitative variables, the λ_1 statistic of Section 3.4 can be used to test for a constant covariance matrix for colony-to-colony variation. Assuming that this test is not significant, so that the variation between sympatric colonies seems to be the same as the variation between allopatric ones, the test statistic λ_2 can be used to test for equal means for the two types of colony. See Schindel and Gould (1977) for an example of how character displacement for several quantitative variables can be studied using standard multivariate techniques.

Example 7.4   Allopatric and sympatric colonies of *Cepaea hortensis*

Clarke (1960, 1962a) provided data on morph distributions for allopatric colonies of *Cepaea hortensis* as well as for sympatric colonies with *Cepaea nemoralis*. Table 7.7 gives his results for the samples from sympatric colonies, with open habitats, where there were 15 or more specimens of both species of snail. Table 7.9 gives the rest of Clarke's data for colonies with open habitats where there are samples of 15 or more *C. hortensis*.

Altogether there are 33 allopatric colonies and 27 sympatric colonies. We can ask whether there is any evidence of a difference between these. According to the chi-square randomization test of Section 6.3 the answer is a clear 'no'. Treating sympatric and allopatric colonies as two different 'habitats' for *C. hortensis*, the within-habitat total chi-square value is 3183.0 with 232 degrees of freedom. This is very significantly large, which shows, of course, that there is a good deal of variation within the two categories 'sympatric' and 'allopatric'. However, 1000 random allocations of the 60 colonies as 33 allopatric and 27 sympatric yielded randomization chi-square values ranging from a low of 2742.7 to a high of 3833.7, with a mean of 3352.5. The observed value is not significantly low at the 5 % level. There is therefore no evidence of a difference between the allopatric and sympatric colonies.

Arthur (1978, 1980) studied a number of different sets of data involving allopatric and sympatric colonies of *Cepaea* and found significant differences

Table 7.9 Morph distributions of *Cepaea hortensis* in colonies with open habitats in Southern England. These are additional to the results given in Table 7.7.

	Colony	YFB	YOB	YUB	P	B
		Morph frequencies				
Allopatric	Marcham	9	0	17	8	20
colonies	Cumnor	44	1	7	0	0
	Wooton Rivers	17	0	0	4	1
	Marston-Elsfield	26	0	0	0	0
	Stowood Crossroads	26	3	8	2	2
	Dare Hill A	133	65	14	0	0
	Derry Hill A	32	0	9	3	0
	Rockley, Berk	17	3	9	4	0
	Wooton Abington B	142	26	52	10	0
	Swerford	46	1	12	0	0
	Christmas Common	9	0	38	0	0
	Fiddler's Hill	174	0	39	16	0
	Ashbury Hill	25	3	0	0	0
	Derry Hill B	6	0	7	12	0
	Silbury Hill A	159	2	70	0	29
	Silbury Hill B	83	0	32	0	7
	Silbury Hill C	46	0	41	2	2
	Silbury Hill D	53	0	23	0	2
	Dragons Hill	60	4	259	0	0
	Walkers Hill	7	0	22	0	1
	Morgans Hill	14	4	9	10	0
	Three Barrows	19	0	55	0	5
	The Ball, Pewsey	29	0	32	4	0
	Chisledon-Ogbourne	165	6	142	7	5
	Wheatley	80	1	48	0	4
	Cowley-Chislehmptn	21	0	0	5	4
	Wooton-Abington A	52	2	35	8	0
	Woodperry Corner	43	0	28	0	0
	Wheatley Holton	51	5	32	0	0
	Kingstone Coombes	41	2	1	0	0
	Etchilhampton	18	0	30	0	0
	Kingston Bagpuize	11	1	11	1	0
	Forest Hill-Stanton	8	0	13	0	0
Sympatric	Cothill Fen	8	0	5	6	0
colonies	Wooton	14	0	13	9	0
shared	Watchfield	4	0	12	0	0
with	Hill Barn Tumulus	15	1	0	0	0
C. nemoralis	Little Hinton	48	1	12	0	0
	White Horse Hill 4	31	2	0	0	0
	White Horse Hill 5	18	0	0	0	0

between the two types of colony in many cases. He suggested that the most likely explanation is competitive selection. Later (Arthur, 1982b) he made a detailed survey of one area where a significant difference is found and concluded that climatic selection could be the real explanation. Whatever the true situation, it does seem that in some cases, some form of selection is responsible for a difference between allopatric and sympatric colonies of *Cepaea*, although no difference is apparent with Clarke's data.

Example 7.5   Allopatric and sympatric colonies of *Hydrobia*

Table 7.8 shows Fenchel's (1975b) data on average lengths of *Hydrobia ulvae* and *H. ventrosa* in sympatric and allopatric colonies. It has already been shown, in Example 7.3, that there is no significant correlation between the sizes of the two species in sympatric colonies.

Bartlett's test can be used to test the hypothesis that the variance of lengths is the same for both species, in both sympatric and allopatric colonies. The variance estimates for differences between colonies are $S_1^2 = 0.1600$ for sympatric *H. ulvae*, $S_2^2 = 0.1296$ for sympatric *H. ventrosa*, $S_3^2 = 0.0676$ for allopatric *H. ulvae*, and $S_4^2 = 0.0841$ for allopatric *H. ventrosa*, with degrees of freedom 14, 14, 8 and 7, respectively. From these values Bartlett's test statistic (Steel and Torrie, 1980, p. 471) is 2.10 with 3 degrees of freedom which is certainly not significantly large when compared with the chi-square distribution. There is therefore no real evidence against the hypothesis that the variance is constant, even although the sample variances are larger for sympatric than allopatric colonies.

A two-factor analysis of variance on colony mean lengths produces the results shown in Table 7.10. There are highly significant differences between species, and between allopatric and sympatric colonies. The highly significant interaction reflects the fact that the two species diverge in sympatric colonies.

Table 7.10 Two-factor analysis of variance on colony mean lengths for *Hydrobia ulvae* and *H. ventrosa*.

Source of variation	Sum of squares	d.f.	Mean square	F ratio
Species	4.37	1	4.37	36.3*
Type of colony	1.28	1	1.28	10.6†
Interaction	4.05	1	4.05	33.6*
Residual	5.17	43	0.12	

* Significantly large at the 0.1 % level
† Significantly large at the 1 % level

Arthur (1982a) suggests that these data provide the clearest evidence yet obtained for character displacement in a natural population.

7.6 OVERALL GENE FREQUENCY COMPARISONS BETWEEN SPECIES

Ayala and Tracey (1974) surveyed genetic variation at 28 genetic loci in six Caribbean populations of four sympatric species of *Drosophila*. They found that within any one species, gene frequencies show geographical variation but are nevertheless fairly constant. However, between species the pattern of differences seemed rather remarkable. For any pair of species, nearly half the genetic loci had essentially identical gene distributions while nearly all the other half had completely different distributions. Later studies confirmed this to be the general pattern for *Drosophila*, and Ayala and Gilpin (1974) argued that this is not what can be expected from isolation between different species if there is no selection. Instead, they suggest that stabilizing selection maintains quasi-stable equilibria at many genetic loci but environmental effects can cause rapid changes to quite different equilibria.

Nei and Tateno (1975) have disputed this conclusion. They suggest that the mathematical model used by Ayala and Gilpin involves unrealistic assumptions, such as no mutation. When more realistic assumptions are made and incorporated into a computer simulation, the pattern found by Ayala and his colleagues can be generated without the action of natural selection.

On the whole, it seems that searching for evidence of selection in overall gene frequency differences between species is not going to be successful. It is far better to use a colony-by-colony approach, as discussed in the earlier sections of this chapter.

8 Gene frequency changes at a single genetic locus

Up to the present point, genetic considerations have been of minor importance to the topics that have been discussed. This is true even though in some instances it has been the analysis of gene frequencies that have been of interest. However, the present chapter brings a change of emphasis. Now the genetics of selection becomes the main concern, particularly gene frequency changes from generation to generation at a single genetic locus.

The chapter opens with a review of the genetics of selection at an autosomal locus. Section 8.5 then describes a test for selection. This is followed by six sections that consider various special cases of the estimation of selective values for different genotypes. There is then a cautionary note on the bias and variance of estimates. Estimation of selective values in general, particularly the use of the method of maximum likelihood, is discussed at some length in Sections 8.13 to 8.15.

Finally, the chapter concludes with short discussions on the estimation of effective population sizes, selection related to sex, and the analysis of phenotype frequencies in place of the analysis of genotype frequencies.

8.1 THE INTRODUCTION OF GENETIC CONSIDERATIONS

The present chapter concentrates on the study of gene frequency changes over time at a single genetic locus. Many of the methods that are discussed were originally developed for analysing karyotype (chromosome) frequencies rather than gene frequencies. However, this is no problem since the algebra of selection is essentially the same for chromosomes and genes.

A problem with all studies on single loci is that it is very difficult to prove that selection is acting directly on the particular locus being considered (Clarke, 1975; McDonald, 1983). It might be possible to show that gene frequencies are changing at the locus. However, this may be due to selection on a second locus that is in genetic disequilibrium with the first one, so that genetic 'hitch-hiking' may be occurring. This is discussed further in Section 10.2. Here it will merely be noted that the problem is particularly liable to occur with laboratory populations. Mukai and Yamazaki (1980) have argued that the level of selection estimated in the laboratory is often much too high to be associated with one locus and is probably to a large extent the result of genetic disturbances caused when setting up populations.

8.2   TWO ALLELES AT A SINGLE LOCUS

An allele is a particular form of a gene. The simplest possible situation occurs when there are two alleles, A_1 and A_2, say, at a single autosomal locus. Then there are three possible genotypes A_1A_1, A_1A_2 and A_2A_2, of which A_1A_1 and A_2A_2 are homozygotes and A_1A_2 is the heterozygote. An A_1A_1 individual has two A_1 alleles while an A_1A_2 individual has one A_1 and one A_2 allele. The situation is different for sex-linked genes where one sex has only one gene instead of two (see Section 8.17).

An example of an autosomal locus with two alleles is provided by the scarlet tiger moth *Panaxia dominula* in an isolated colony at Cothill, in the Oxford area of England. In this case the two alleles *dominula* and *medionigra* give rise to the three genotypes *dominula*, *medionigra* and *bimacula*: two *dominula* alleles produce the *dominula* genotype; a *dominula* and a *medionigra* allele produce the *medionigra* genotype; two *medionigra* alleles produce the *bimacula* genotype. The forewings of *P. dominula* are black with white and yellow spots and the hind wings are bright red with black spots. *Medionigra* has smaller spots on the forewings and more black on the hind wings than *dominula*. *Bimacula* have much more black on the wings.

Ford (1975) describes how he noticed a number of *medionigra* specimens at Cothill in 1936 and 1938, although he had not seen any in 1921 when he visited the area before. A study of 168 moths from museums that were caught in various years prior to 1929 revealed only four *medionigra*. Ford concluded that the *medionigra* were increasing in frequency so he decided to begin a systematic study of the population. This study continued for many years (Fisher and Ford, 1947; Sheppard, 1951b; Williamson, 1960; Sheppard and Cook, 1962; Ford and Sheppard, 1969), resulting in the data shown in Table 8.1. It appears that the frequency of the *medionigra* gene increased from 1928 until about 1940, declined until about 1955, and then varied somewhat from year to year (Fig. 8.1). It seems likely that some part of the gene frequency changes were due to selection.

8.3   HARDY–WEINBERG EQUILIBRIUM

If a population is very large and the adults mate at random, then the proportions of different types of progeny will be as shown in Table 8.2. Thus for the progeny the genotypes A_1A_1, A_1A_2 and A_2A_2 will have frequencies p^2, $2pq$ and q^2. If these progeny then mate at random and there is no selection, then it is clear that in the next generation the same frequencies will be repeated. This is called Hardy–Weinberg equilibrium: in a large population with random

Table 8.1 Results of sampling the Cothill population of *Panaxia dominula*. There is one generation per year.

Year	Approximate population size	Genotype frequencies			Estimated gene frequency of medionigra $q(\%)$
		dominula	medionigra	bimacula	
pre-1928	?	164	4	0	1.2
1939	?	184	37	2	9.2
1940	?	92	24	1	11.1
1941	2 250	400	59	2	6.8
1942	1 600	183	22	0	5.4
1943	1 000	239	30	0	5.6
1944	5 500	452	43	1	4.5
1945	4 000	326	44	2	6.5
1946	7 000	905	78	3	4.3
1947	6 000	1 244	94	3	3.7
1948	3 200	898	67	1	3.6
1949	1 700	479	29	0	2.9
1950	4 100	1 106	88	0	3.7
1951	2 250	552	29	0	2.5
1952	6 000	1 414	106	1	3.6
1953	8 000	1034	54	1	2.6
1954	11 000	1 097	67	0	2.9
1955	2 000	308	7	0	1.1
1956	11 000	1 231	76	1	3.0
1957	16 000	1 469	138	5	4.6
1958	15 000	1 285	94	4	3.7
1959	7 000	460	19	1	2.2
1960	2 500	182	7	0	1.9
1961	1 400	165	7	0	2.0
1962	216	22	1	0	2.2
1963	470	58	1	0	0.9
1964	272	31	0	0	0.0
1965	625	79	2	0	1.2
1966	315	37	0	0	0.0
1967	406	50	0	0	0.0
1968	978	128	3	0	1.2
1969	5 712	508	38	0	3.5
1970	4 493	444	31	0	3.4
1971	7 084	637	9	0	0.7
1972	3 471	335	5	0	0.7

mating and no selection a stable equilibrium state for two alleles will be reached in one generation such that if the gene frequencies are p and $q = 1 - p$, then the genotype frequencies are p^2, $2pq$ and q^2.

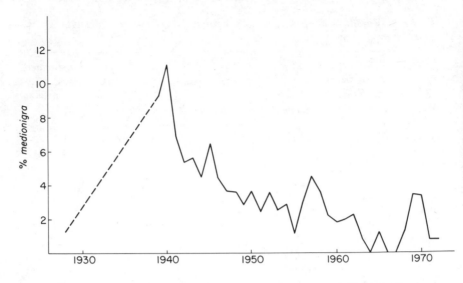

Fig. 8.1 The frequency of the *medionigra* gene at Cothill, 1928–72.

Table 8.2 Frequencies of different progeny from random mating in a population with three morphs A_1A_1, A_1A_2, A_2A_2 in the proportions d, $2h$ and r where $d + 2h + r = 1$.

Type of mating (male × female)	Proportion	Proportion of progeny		
		A_1A_1	A_1A_2	A_2A_2
$A_1A_1 \times A_1A_1$	d^2	d^2	0	0
$A_1A_1 \times A_1A_2$	$2hd$	hd	hd	0
$A_1A_1 \times A_2A_2$	dr	0	dr	0
$A_1A_2 \times A_1A_1$	$2hd$	hd	hd	0
$A_1A_2 \times A_1A_2$	$4h^2$	h^2	$2h^2$	h^2
$A_1A_2 \times A_2A_2$	$2hr$	0	hr	hr
$A_2A_2 \times A_1A_1$	rd	0	rd	0
$A_2A_2 \times A_1A_2$	$2hr$	0	hr	hr
$A_2A_2 \times A_2A_2$	r^2	0	0	r^2
Totals	1	$(d+h)^2$ $= p^2$	$2(d+h)(r+h)$ $= 2pq$	$(r+h)^2$ $= q^2$

Note: In matings between homozygotes such as $A_1A_1 \times A_2A_2$, only one type of progeny can be produced (A_1A_2). However, in a mating involving one heterozygote such as $A_1A_1 \times A_1A_2$, the outcomes A_1A_1 and A_1A_2 can both occur with equal probability. In the case of two heterozygotes, $A_1A_2 \times A_1A_2$, the outcomes A_1A_1, A_1A_2 and A_2A_2 have probabilities $\frac{1}{4}$, $\frac{1}{2}$ and $\frac{1}{4}$, respectively.

8.4 SELECTION ON A RANDOMLY MATING POPULATION

Suppose that a large population is subject to selection from the time that the zygotes are formed until the time that they mate at random. Let w_1, w_2 and w_3 denote the relative survival rates for the genotypes A_1A_1, A_1A_2 and A_2A_2, respectively. The situation is then as shown in Table 8.3.

Table 8.3 Selection operating on a large randomly mating population between the time of formation of the zygotes A_1A_1, A_1A_2 and A_2A_2 until the time that these zygotes mate.

	Genotype relative frequencies			Gene relative frequencies	
	A_1A_1	A_1A_2	A_2A_2	A_1	A_2
In parents	–	–	–	p	$q = 1 - p$
Progeny before selection	p^2	$2pq$	q^2	p	q
Progeny after selection	w_1p^2	$2w_2pq$	w_3q^2	$w_1p^2 + w_2pq$	$w_3q^2 + w_2pq$

At the time the parents mate the proportion of A_1 genes is $p = 1 - q$. Through random mating the proportions of progeny A_1A_1, A_1A_2 and A_2A_2 are expected to be p^2, $2pq$ and q^2. Thus, before selection the proportion of A_1 genes is still p for the progeny. After selection, at the adult stage, the progeny genotypes are in the ratios $w_1p^2 : 2w_2pq : w_3q^2$. Hence the proportion of A_1 genes among the adult progeny is

$$p' = (w_1p^2 + w_2pq)/(w_1p^2 + 2w_2pq + w_3q^2),$$

and the change in this proportion over the generation is

$$\Delta p = p' - p = (p/\bar{w})(w_1p + w_2q - \bar{w}), \tag{8.1}$$

where

$$\bar{w} = w_1p^2 + 2w_2pq + w_3q^2. \tag{8.2}$$

The equivalent expression for the change in the proportion of A_2 genes is

$$\Delta q = (q/\bar{w})(w_2p + w_3q - \bar{w}), \tag{8.3}$$

where, of course, $\Delta p = -\Delta q$.

The relative survival rates w_1, w_2 and w_3 are called *selective values* or *fitnesses*, while $\bar{w}$ is the *mean fitness*. Note that equations (8.1) and (8.3) are unaffected if w_1, w_2 and w_3 are all multiplied by the same constant. For this reason it is usual to consider the selective values of two of the genotypes relative to the other one. For example,

$$\theta_2 = w_2/w_1 \qquad \text{and} \qquad \theta_3 = w_3/w_1 \qquad (8.4)$$

are the selective values of $A_1 A_2$ and $A_2 A_2$ relative to $A_1 A_1$. Also, for reasons explained in Section 1.5, it is sometimes better to consider logarithms of these values

$$\gamma_2 = \log_e(w_2/w_1) \qquad \text{and} \qquad \gamma_3 = \log_e(w_3/w_1). \qquad (8.5)$$

If there is no selection then $\theta_2 = \theta_3 = 1$ so that $\gamma_2 = \gamma_3 = 0$. One advantage of using logarithms comes from the fact that $\log_e(w_1/w_2) = -\log_e(w_2/w_1)$. Hence the amount of selection does not depend upon which genotype is taken as the standard.

A number of special cases are recognized as being important for the relationship between w_1, w_2 and w_3. For example, suppose that A_1 is dominant to A_2 so that genotypes $A_1 A_1$ and $A_1 A_2$ have the same phenotype. If $A_2 A_2$ is the fittest genotype it is then possible to take $w_1 = w_2 = 1 - s$ and $w_3 = 1$, where $0 \leqslant s \leqslant 1$. Table 8.4, case (iii), shows how the proportion of A_1 genes changes over one generation for this type of selection. Four other special cases are also shown in this table.

Heterosis occurs when the heterozygote $A_1 A_2$ is fitter than both homozy-

Table 8.4   The change in one generation in the proportion of A_1 genes with different kinds of selection. Selective values (relative survival rates) are expressed in terms of selective coefficients s. In all cases it is assumed that the initial proportion of A_1 genes is p, where $q = 1 - p$.

		Selective value of			Change in	Equilibrium
Case	Condition	$A_1 A_1$	$A_1 A_2$	$A_2 A_2$	p (Δp)	when p is
(i)	Complete dominance selection against $A_2 A_2$	1	1	$1-s$	$\dfrac{sq^2 p}{1 - sq^2}$	0 or 1
(ii)	Complete dominance selection against A_2	1	$1-s$	$1-s$	$\dfrac{sqp^2}{1 - 2sq + sq^2}$	0 or 1
(iii)	Complete dominance selection against A_1	$1-s$	$1-s$	1	$\dfrac{-spq^2}{1 - 2sp + sp^2}$	0 or 1
(iv)	Heterozygote intermediate	1	$1-\frac{1}{2}s$	$1-s$	$\dfrac{\frac{1}{2}spq}{1 - sq}$	0 or 1
(v)	Heterosis	$1-s_1$	1	$1-s_2$	$\dfrac{-pq(s_1 p - s_2 q)}{1 - s_1 p^2 - s_2 q^2}$	$0, \dfrac{s_2}{s_1 + s_2}$ or 1

gotes. This is case (v) of Table 8.4. It is of particular interest because it is the only case shown where there is an equilibrium gene frequency that is not equal to zero or one. Equilibrium occurs when $\Delta p = 0$ in equation (8.1).

As discussed in Section 1.4, selection may act at other times in the life cycle of an organism as well as between the formation of genotypes and their development to adults. At reproduction the morphs may differ in their ability to produce offspring. It is also possible that gene frequencies are altered by non-random mating behaviour. However, for the analysis of data the models shown in Table 8.4 are often sufficient. Or, to be more accurate, these models may be all that it is practical to use. At any rate, for the next few sections of this chapter these complications will be ignored. They are discussed further in Section 8.13.

8.5   A TEST FOR SELECTION

In their original paper on the Cothill population of *Panaxia dominula*, Fisher and Ford (1947) proposed a test for whether selection is operating with two alleles at one locus. This test has subsequently been extended by Schaffer *et al.* (1977) and Wilson (1980). See also Yardley *et al.* (1977).

The experimental design with the test proposed by Schaffer *et al.* involves a population being sampled in the manner shown in Fig. 8.2. That is to say, in generation 0 there is an initial population of adults for which the frequency of a marker gene is p. From this population is drawn a random sample of $n_1/2$ individuals for analysis and a random sample of $N_1/2$ individuals to form the parents for generation 1. This process then continues for a number of generations.

If it is assumed that no selection takes place then the gene frequency p will still change from one generation to the next because the population sizes of parents, $N_1/2, N_2/2, \ldots$, are finite. Also there will be sampling errors in estimates of the gene frequencies because the samples are of finite sizes $n_1/2$, $n_2/2, \ldots$ Furthermore, the gene frequencies from generation to generation will be correlated because if the $N_i/2$ individuals chosen to form the ith generation happen to have a large proportion of the marker gene then all of the following generations will also tend to have high values.

Let

$$\hat{p}_i = r_i/n_i$$

where r_i is the number of marker genes in the sample of $n_i/2$ individuals from the ith generation. This is a binomial proportion of the total number of n_i genes carried by the individuals. Schaffer *et al.* (1977) followed Fisher and Ford (1947) in proposing the use of the arcsine transformation

$$y_i = 2\sin^{-1}(\hat{p}_i^{1/2}) \tag{8.6}$$

to produce a variate with a variance that does not depend upon p_i. Here y_i is in

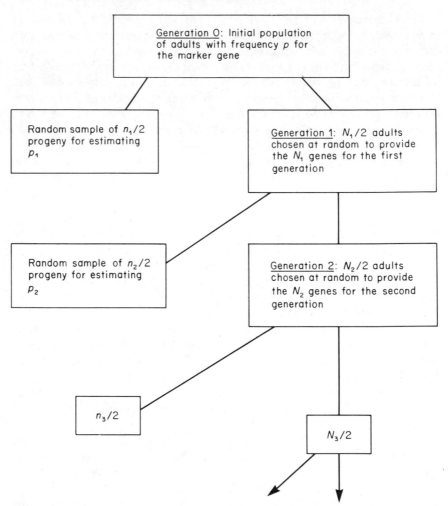

Fig. 8.2   Schaffer *et al.*'s (1977) model for sampling a population in a test for selection. This leads to equations (8.7) for variances and covariances.

radians. Then, with no selection, the expected value of y_i is

$$\mu = E(y_i) \simeq 2\sin^{-1}(p^{1/2})$$

where p is the initial gene frequency. Also,

$$\left.\begin{array}{l} \operatorname{var}(y_i) \simeq \dfrac{1}{n_i} \prod_{j=1}^{i-1} (1 - 1/N_j) + \sum_{k=1}^{i-1} \dfrac{1}{N_k} \prod_{j=1}^{k-1} (1 - 1/N_j), \\[2ex] \operatorname{cov}(y_i, y_r) \simeq \prod_{k=1}^{i-1} \dfrac{1}{N_k} \prod_{j=1}^{k-1} (1 - 1/N_j), \end{array}\right\} \quad (8.7)$$

for $i < r$. A sum with an upper limit less than the lower limit is defined to be zero, while a product with an upper limit less than a lower limit is defined to be unity.

In Fisher and Ford's original paper the experimental arrangement was different from the one shown in Figure 8.2. They were concerned particularly with the *Panaxia dominula* population at Cothill where the samples each year were taken from the flying adult population, with replacement. Thus the sample of $n_i/2$ individuals from the ith generation were included in the $N_i/2$ that provided the genes for the next generation. This suggests the alternative sampling scheme shown in Fig. 8.3. In this case equations (8.7) become

and

$$\left.\begin{array}{l} \text{var}(y_i) \simeq \sum_{k=2}^{i} \frac{1}{N_k} \prod_{j=k+1}^{i} (1 - 1/N_j) + \frac{1}{n_i} - \frac{1}{N_i} \\[2em] \text{cov}(y_i, y_r) \simeq \sum_{k=2}^{i} \frac{1}{N_k} \prod_{j=k+1}^{i} (1 - 1/N_j), \end{array}\right\} \qquad (8.8)$$

where $i < r$. It can be seen that equations (8.7) and (8.8) all simplify considerably if the population sizes are large so that $(1 - 1/N_i) \simeq 1$.

If a population is followed for a number of generations and sampled for some or all of these, then the variances and covariances of the y_i values can be calculated using either equations (8.7) or (8.8), depending upon which sampling scheme is most appropriate (Figs 8.2 and 8.3). Hence a covariance matrix may be constructed. Suppose that the observed y values are $y_{(1)}, y_{(2)}, \ldots, y_{(s)}$, where $y_{(i)}$ is used instead of y_i to emphasize that the ith observation may not be taken in the ith generation. Let the calculated $s \times s$ covariance matrix be $\mathbf{V}$. Then, on the hypothesis of no selection, the minimum variance estimator of μ is

$$\hat{\mu} = \sum_{i=1}^{s} \sum_{j=1}^{s} h_{ij} y_{(i)} \bigg/ \sum_{i=1}^{s} \sum_{j=1}^{s} h_{ij} \qquad (8.9)$$

where h_{ij} is the element in the ith row and jth column of the inverse matrix $\mathbf{V}^{-1}$ (Appendix, Section A.6). Furthermore, the test statistic

$$T = (\mathbf{Y} - \hat{\mu})' \mathbf{V}^{-1} (\mathbf{Y} - \hat{\mu}) \qquad (8.10)$$

will approximately have a chi-square distribution with $s - 1$ degrees of freedom, where $\mathbf{Y}' = (y_{(1)}, y_{(2)}, \ldots, y_{(s)})$ and $\hat{\mu}' = (\hat{\mu}, \hat{\mu}, \ldots, \hat{\mu})$. If selection is occurring then T will tend to be large.

A problem here is that T gives a completely general test for selection and does not take into account the fact that if selection occurs then it may be directional. To allow for this, Schaffer *et al.* proposed the linear regression model $E(y_i) = \beta_0 + \beta_1 i$. The generalized least-squares estimators of β_0 and β_1 are then given by

$$\hat{\mathbf{B}} = \begin{bmatrix} \hat{\beta}_0 \\ \hat{\beta}_1 \end{bmatrix} = (\mathbf{X}' \mathbf{V}^{-1} \mathbf{X})^{-1} \mathbf{X}' \mathbf{V}^{-1} \mathbf{Y}$$

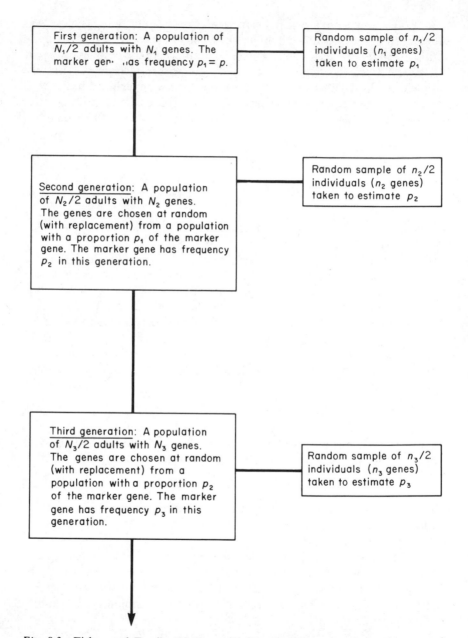

Fig. 8.3 Fisher and Ford's (1947) model for sampling a population in a test for selection. This leads to equations (8.8) for variances and covariances.

where

$$X = \begin{bmatrix} 1 & t_1 \\ 1 & t_2 \\ \vdots & \vdots \\ 1 & t_s \end{bmatrix}$$

(Appendix, Section A.9). Here $t_1, t_2, \ldots, t_s$ are the generation numbers corresponding to the observations $y_{(1)}, y_{(2)}, \ldots, y_{(s)}$.

The T statistic of equation (8.10) can now be split into the two parts $T = T_1 + T_2$, where

$$T_1 = \hat{\mathbf{B}}'\mathbf{X}'\mathbf{V}^{-1}\mathbf{X}\hat{\mathbf{B}} - \hat{\mu}^2 \sum_{i=1}^{s} \sum_{j=1}^{s} h_{ij}, \qquad (8.11)$$

with 1 degree of freedom, can be used to test for a linear trend in y_i values, and

$$T_2 = T - T_1, \qquad (8.12)$$

with $s - 2$ degrees of freedom, can be used to test for non-linear trends.

Wilson (1980) has extended this test, assuming that there are several populations generated from a single genetic source. She considered the possibility of polynomial trends in the y_i values and proposed a method for obtaining bounds on effective population sizes using data from several populations.

The greatest problem with using the test with a natural population is undoubtedly the determination of the population sizes N_i, which need to be effective population sizes of genes in terms of breeding adults. These sizes may be far less than twice the number of adults in the population (Falconer, 1981, Ch. 4). (See Section 8.16 for a further discussion of this matter.)

Example 8.1   Testing for selection on
Panaxia dominula at Cothill

The test for selection can be applied to the data on *Panaxia dominula* at Cothill that was discussed in Section 8.2. Genotype frequencies are shown in Table 8.1. Figure 8.1 shows a plot of the estimated percentages of *medionigra* genes in this colony from 1928 to 1972. A 1928 estimate of 1.2% is probably an overestimate since it is based upon museum specimens, and collectors tend to concentrate on unusual varieties. It seems that from 1928 to 1940 the percentage of *medionigra* increased up to about 10%. There was then a more or less steady decline to about 2% in 1955. From then on there seem to have been yearly fluctuations from nearly zero up to about 4%. The question to be asked is whether the observed fluctuations can be accounted for solely by random genetic drift from year to year. The steady downward trend from 1940 to 1955 makes this seem unlikely, but it is interesting to see what result there is from a formal test.

The sampling scheme displayed in Figure 8.3 should be a fair approximation for this example. The population sizes shown in Table 8.1 were determined by mark–recapture and other methods. The N_i values of equations (8.8) can be taken to be twice these sizes since they are in terms of genes rather than individuals. Then T_1 and T_2 values can be calculated using equations (8.11) and (8.12).

It is found that $T_1 = 7.36$ with 1 degree of freedom and $T_2 = 125.65$ with 30 degrees of freedom. When compared with critical values for the chi-square distribution, T_1 is significantly large at the 5 % level and T_2 is significantly large at the 0.1 % level. It seems, therefore, that the observed changes in the proportion of *medionigra* genes cannot reasonably be attributed to random genetic drift. The trend can be accounted for partially by a linear term in the arcsine transformed proportions.

This result is not surprising, given the strong downward trend from 1940 to 1955. However, the same conclusion is reached when the periods 1940–55 and 1956–72 are tested separately. For 1940–55, $T_1 = 19.54$ with 1 degree of freedom and $T_2 = 24.32$ with 13 degrees of freedom; for 1956–72, $T_1 = 8.00$ with 1 degree of freedom and $T_2 = 77.59$ with 15 degrees of freedom. These test statistics are all significantly large at the 5 % level at least.

If effective population sizes were smaller than twice the number of adults, then the T_1 and T_2 values just given are not correct. However, it turns out that reducing the assumed population sizes increases the significance of the test results because it has the effect of decreasing the sampling variances of the estimated gene proportions. Halving the assumed population sizes makes $T_1 = 4.42$ and $T_2 = 129.8$ for the full period 1940–72. Quartering the assumed population sizes makes $T_1 = 0.1$ and $T_2 = 186.5$ for the full period. Thus reducing the effective population sizes decreases the significance of the linear trend term but increases the significance of the non-linear trend term.

Fisher and Ford (1947) tested for selection on the *medionigra* gene at Cothill using data collected up to 1946 and essentially the total statistic $T = T_1 + T_2$. They obtained a significant chi-square value and concluded that there was evidence of selection. This conclusion was disputed by Wright (1948) on the grounds of uncertainty about effective population sizes. The collection of data in subsequent years has confirmed Fisher and Ford's conclusion, but nevertheless the points made by Wright do seem to have been valid. In a recent analysis of the data, Wright (1978, p. 171) argues that although selection seems to be responsible for the decline in the *medionigra* gene from 1939 to 1968, other fluctuations are largely due to random genetic changes in the fairly small colony.

8.6   ESTIMATION OF SELECTIVE VALUES: NO DOMINANCE, SAMPLES AFTER SELECTION

Estimation of selective values is simplest when samples of adults are taken from a population after selection in successive generations, providing that all genotypes can be distinguished, and random mating can be assumed. Considering this case, let p_{i-1} denote the proportion of the allele A_1 after selection in generation $i-1$. Random mating will then produce Hardy–Weinberg ratios $p_{i-1}^2 : 2p_{i-1}q_{i-1} : q_{i-1}^2$ for the genotypes $A_1 A_1$, $A_1 A_2$ and $A_2 A_2$ at the start of generation i, where $q_{i-1} = 1 - p_{i-1}$. Selection in the ith generation will change these ratios to $w_{i1}p_{i-1}^2 : 2w_{i2}p_{i-1}q_{i-1} : w_{i3}q_{i-1}^2$, where w_{i1}, w_{i2} and w_{i3} are the selective values for this generation. It follows that if π_{i1}, π_{i2} and π_{i3} are the population proportions of $A_1 A_1$, $A_1 A_2$ and $A_2 A_2$, respectively, after selection in generation i, then

$$
\left.
\begin{aligned}
\pi_{i1} &= c\, w_{i1}\, p_{i-1}^2, \\
\pi_{i2} &= 2c\, w_{i2}\, p_{i-1}\, q_{i-1}, \\
\pi_{i3} &= c\, w_{i3}\, q_{i-1}^2,
\end{aligned}
\right\} \tag{8.13}
$$

and

where c is the constant required to make $\pi_{i1} + \pi_{i2} + \pi_{i3} = 1$.

Since only the relative values of w_{i1}, w_{i2} and w_{i3} are important, it is valid to set $w_{i2} = 1$. Then w_{i1} and w_{i3} become selective values relative to $A_1 A_2$. In that case equations (8.13) produce

$$
\left.
\begin{aligned}
w_{i1} &= 2\,q_{i-1}\,\pi_{i1}/(p_{i-1}\,\pi_{i2}), \\
w_{i3} &= 2\,p_{i-1}\,\pi_{i3}/(q_{i-1}\,\pi_{i2}).
\end{aligned}
\right\} \tag{8.14}
$$

and

These equations can be used to estimate the selective values. If a random sample of n_i individuals is taken from the population at the end of generation i then the sample proportions $\hat{\pi}_{i1}$, $\hat{\pi}_{i2}$ and $\hat{\pi}_{i3}$ of the genotypes $A_1 A_1$, $A_1 A_2$ and $A_2 A_2$ are unbiased estimators of π_{i1}, π_{i2} and π_{i3}, with multinomial variances and covariances

$$
\left.
\begin{aligned}
\operatorname{var}(\hat{\pi}_{ij}) &= \pi_{ij}(1 - \pi_{ij})/n_i, \\
\operatorname{cov}(\hat{\pi}_{ij}, \hat{\pi}_{ik}) &= -\pi_{ij}\pi_{ik}/n_i, \quad j \neq k.
\end{aligned}
\right\} \tag{8.15}
$$

and

A similar random sample of n_{i-1} individuals taken at the end of generation $i-1$ can be thought of as a sample of $2n_{i-1}$ genes. The sample proportion $\hat{p}_{i-1}$ of A_1 genes is then an unbiased estimator of p_{i-1} with the binomial variance

$$
\operatorname{var}(\hat{p}_{i-1}) = p_{i-1}(1 - p_{i-1})/(2n_{i-1}). \tag{8.16}
$$

Substituting the estimates $\hat{\pi}_{ij}$ and $\hat{p}_{i-1} = 1 - \hat{q}_{i-1}$ into equations (8.14) gives

$$
\left.
\begin{aligned}
\hat{w}_{i1} &= 2\,\hat{q}_{i-1}\,\hat{\pi}_{i1}/(\hat{p}_{i-1}\,\hat{\pi}_{i2}), \\
\hat{w}_{i3} &= 2\,\hat{p}_{i-1}\,\hat{\pi}_{i3}/(\hat{q}_{i-1}\,\hat{\pi}_{i2}).
\end{aligned}
\right\} \tag{8.17}
$$

and

For finding biases, variances and the covariances it is easier to work with the logarithms $\hat{v}_{i1} = \log_e(\hat{w}_{i1})$ and $\hat{v}_{i3} = \log_e(\hat{w}_{i3})$ of these estimators. Taylor series approximation show that

$$\left. \begin{array}{c} \text{bias}(\hat{v}_{i1}) \simeq (1/p_{i-1} - 1/q_{i-1})/(4n_{i-1}) + (1/\pi_{i2} - 1/\pi_{i1})/(2n_i), \\[2mm] \text{bias}(\hat{v}_{i3}) \simeq (1/q_{i-1} - 1/p_{i-1})/(4n_{i-1}) + (1/\pi_{i2} - 1/\pi_{i3})/(2n_i), \end{array} \right\} \quad (8.18)$$

$$\left. \begin{array}{c} \text{var}(\hat{v}_{i1}) \simeq 1/(2\,p_{i-1}q_{i-1}n_{i-1}) + 1/(\pi_{i1}n_i) + 1/(\pi_{i2}n_i), \\[2mm] \text{var}(\hat{v}_{i3}) \simeq 1/(2\,p_{i-1}q_{i-1}n_{i-1}) + 1/(\pi_{i2}n_i) + 1/(\pi_{i3}n_i), \end{array} \right\} \quad (8.19)$$

and

$$\text{cov}(v_{i1}, \hat{v}_{i3}) \simeq 1/(\pi_{i2}n_i) - 1/(2\,p_{i-1}q_{i-1}n_{i-1}). \quad (8.20)$$

There will be some correlation between estimates from successive generations. That is to say, $\hat{v}_{i-1\,1}$ and $\hat{v}_{i-1\,3}$ will be correlated with $\hat{v}_{i1}$ and $\hat{v}_{i3}$ since both pairs of estimates use the sample from generation $i-1$. However, there will be no correlation between estimates for generations further apart.

The assumption that samples are taken after all selection has been completed is very important for the estimation equations that have been presented in this section. Prout (1965) has demonstrated the serious errors that can arise if these equations are used when the assumption is not true. This matter is discussed further in Section 8.13.

Example 8.2   Selective values for *Panaxia dominula* at Cothill

For *Panaxia dominula* at Cothill it is probably a fair assumption that a sample of flying adults provides a sample after selection. Therefore the estimation equations given in the previous section can be applied to the data shown in Table 8.1. However, in most years there are so few individuals of the genotype $A_2 A_2$ (*bimacula*) that it is only really possible to estimate selective values for $A_1 A_1$ (*dominula*) relative to $A_1 A_2$ (*medionigra*). Table 8.5 shows estimates of $v_{i1} = \log_e(w_{i1})$ calculated using equation (8.17), with biases calculated using equations (8.18) and standard errors calculated using equations (8.19). No estimates are available for 1964 to 1968 because of the absence of *medionigra* from yearly samples. The weighted averages shown at the foot of Table 8.5 are minimum variance values obtained using equation (A28) of the Appendix. The chi-square values are for test (a) of the same section of the Appendix (Section A.6). They are significantly large, which indicates that selective values did not remain constant over time, possibly because of stochastic variation in the population gene frequencies. Odd and even years have been averaged separately because estimates more than one year apart are uncorrelated.

Table 8.5 Estimates of logarithms of selective values of *dominula* relative to *medionigra* for *Panaxia dominula* at Cothill, based upon the data in Table 8.1.

Year	Sample size n_i	Genotype proportions $\hat{\pi}_i{}^1$	$\hat{\pi}_{i2}$	Proportion of A_1 genes $\hat{p}_i$	$\hat{v}_i$	Bias	Standard error
1939	223	0.825	0.166	0.908	–	–	–
40	117	0.786	0.205	0.889	−0.25	0.03	0.28
41	461	0.868	0.128	0.932	0.53*	−0.01	0.25
42	205	0.893	0.107	0.946	0.20	0.01	0.26
43	269	0.888	0.112	0.944	−0.10	−0.01	0.29
44	469	0.911	0.086	0.955	0.23	−0.01	0.24
45	372	0.876	0.118	0.935	−0.36	0.00	0.22
46	986	0.918	0.079	0.957	−0.48*	−0.01	0.19
47	1341	0.928	0.070	0.963	0.18	0.00	0.16
48	966	0.930	0.069	0.964	0.04	0.00	0.16
49	508	0.943	0.057	0.971	0.21	0.01	0.23
50	1194	0.926	0.074	0.963	−0.29	−0.01	0.22
51	581	0.950	0.050	0.975	0.38	0.01	0.22
52	1521	0.930	0.070	0.964	−0.38*	−0.01	0.21
53	1089	0.949	0.050	0.974	0.35*	0.00	0.17
54	1164	0.942	0.058	0.971	−0.14	0.00	0.19
55	315	0.978	0.022	0.989	0.98*	0.06	0.40
56	1308	0.941	0.058	0.970	−1.02*	−0.07	0.40
57	1612	0.911	0.086	0.954	−0.42*	0.00	0.15
58	1383	0.929	0.068	0.963	0.28*	0.00	0.14
59	480	0.958	0.040	0.978	0.61*	0.02	0.26
60	189	0.963	0.037	0.981	0.16	0.05	0.45
61	172	0.959	0.041	0.980	−0.10	0.00	0.54
62	23	0.957	0.043	0.978	−0.10	0.42	1.09
63	59	0.983	0.017	0.991	0.96	0.01	1.42
.	.	.	.	.	.	.	.
68	131	0.977	0.023	0.988	–	–	–
69	546	0.930	0.070	0.965	−1.13	−0.15	0.59
70	475	0.935	0.065	0.966	0.04	0.00	0.25
71	646	0.986	0.014	0.993	1.60*	0.04	0.38
72	340	0.985	0.015	0.993	−0.08	0.04	0.56

Weighted average of $\hat{v}_i$ for even years: $\bar{v} = 0.04$, SE$(\bar{v}) = 0.06$, chi-square is 24.97 with 13 degrees of freedom.
Weighted average of $\hat{v}_i$ for odd years: $\bar{v} = 0.14$, SE$(\bar{v}) = 0.06$, chi-square is 53.59 with 13 degrees of freedom.

* More than two standard errors from zero, and therefore significantly different at the 5% level on an approximate test. These values are evidence of selection, but not in a constant direction.

8.7 ESTIMATION OF SELECTIVE VALUES: NO DOMINANCE, SAMPLES BEFORE SELECTION

The method of estimation discussed in Section 8.6 was based upon a comparison between genotype frequencies after selection with the frequencies expected before selection, after random mating. Such a comparison can no longer be made when a population is only sampled before selection in each generation. Instead, selective values have to be estimated from gene frequency changes from generation to generation. For this situation some equations used by Dobzhansky and Levene (1951) are appropriate.

Suppose that at the start of generation i, before selection, the alleles A_1 and A_2 have proportions p_i and $q_i = 1 - p_i$ in a randomly mating population. Then equation (8.1) shows that at the start of the next generation the expected proportion of allele A_1 is

$$p_{i+1} = (w_1 p_i^2 + w_2 p_i q_i)/(w_1 p_i^2 + 2w_2 p_i q_i + w_3 q_i^2).$$

Here w_2 can be arbitrarily set at 1 so that w_1 and w_3 are selective values for the $A_1 A_1$ and $A_2 A_2$, respectively, relative to the heterozygote $A_1 A_2$. If $u_i = p_i/q_i$ then the last equation provides the relationship

$$p_{i+1}/q_{i+1} = (w_1 p_i^2 + p_i q_i)/(p_i q_i + w_3 q_i^2),$$

or

$$u_{i+1} = (w_1 u_i^2 + u_i)/(u_i + w_3),$$

so that

$$w_3 = (w_1 u_i + 1)u_i/u_{i+1} - u_i. \tag{8.21}$$

Assuming that w_1 and w_3 remain constant, a similar equation,

$$w_3 = (w_1 u_{i+1} + 1)u_{i+1}/u_{i+2} - u_{i+1},$$

holds for the transition from the start of generation $i+1$ to the start of generation $i+2$. Solving the last two equations for w_1 then gives

$$w_1 = \{u_i u_{i+2}(1 - u_{i+1}) - u_{i+1}^2(1 - u_{i+2})\}/(u_{i+1}^3 - u_i^2 u_{i+2}). \tag{8.22}$$

Given estimates $\hat{p}_i$, $\hat{p}_{i+1}$ and $\hat{p}_{i+2}$ of the proportion of allele A_1 in three consecutive generations, these may be used to estimate u_i, u_{i+1} and u_{i+2} values using the obvious equation

$$\hat{u}_i = \hat{p}_i/(1 - \hat{p}_i). \tag{8.23}$$

Substitution into equation (8.22) then provides an estimate of w_1 and equation (8.21) provides an estimate of w_3.

No explicit formulae seem to have been published for the variances, covariances and biases of estimators obtained in this way. However, numerical approximations can be calculated using equations (A25) to (A27) of the

Appendix. In particular,

$$\text{var}(\hat{w}_j) \simeq \Delta_{ji}^2 + \Delta_{ji+1}^2 + \Delta_{ji+2}^2, \tag{8.24}$$

where Δ_{ji} is the change in $\hat{w}_j$ that occurs when $\hat{p}_i$ is changed by one standard deviation. Similarly,

$$\text{cov}(\hat{w}_1, \hat{w}_3) \simeq \Delta_{1i}\Delta_{3i} + \Delta_{1\,i+1}\Delta_{3\,i+1} + \Delta_{1\,i+2}\Delta_{3\,i+2}. \tag{8.25}$$

If $\hat{p}_i$ is a simple binomial proportion then, of course, its standard deviation is just $\text{SD}(\hat{p}_i) = \sqrt{\{p_i(1 - p_i)/n_i\}}$ for a sample of size n_i.

Example 8.3   Selection on laboratory populations of *Drosophila*

Fig. 8.4 shows the results of an experiment on *Drosophila pseudoobscura* that was reported by Dobzhansky and Pavlovsky (1953). Four experimental populations were set up in which initially 80% of chromosomes were chiricahua (CH) and the remaining 20% were standard (ST). The populations were then sampled at irregular times for one year, during which the percentage of CH chromosomes dropped to about 18%. Since the changes were similar for all four populations they are treated as a single population for the present example.

Table 8.6 shows details of the estimation of selective values for these data. The first three chromosome proportions can be used to estimate one pair of fitnesses. The last three chromosome proportions can be used to estimate a second, independent pair. In carrying out the calculations there is a problem because the sample times do not match generation times, which Dobzhansky and Pavlovsky took to be 25 days. The first step in estimation has therefore involved using linear interpolation to determine chromosome proportions at 25, 50, 120 + 25 = 145 and 120 + 50 = 170 days. Obviously this has increased the errors of estimation to some extent.

Equations (8.24) and (8.25) can be used to determine variances and covariances. Table 8.7 shows details of the calculations. Estimates, plus and minus one standard error, are as follows: for the first three samples, $\hat{w}_1 = 0.385 \pm 0.057$ and $\hat{w}_3 = 0.705 \pm 0.208$; for the last three samples, $\hat{w}_1 = 1.12 \pm 1.69$ and $\hat{w}_3 = 1.30 \pm 0.98$. Thus the samples up to 80 days appear to give fairly good estimates of the selective values of genotypes $A_1 A_1$ and $A_2 A_2$ relative to $A_1 A_2$. It seems that $A_1 A_1$ had very low fitness, while $A_2 A_2$ had about the same fitness as $A_1 A_2$. On the other hand, the samples at 120, 160 and 200 days have provided estimates that are so unreliable as to be quite worthless. It seems clear that the estimates given by the method of Section 8.7 will be of very little value unless: (a) population changes, as measured by p_i, p_{i+1} and p_{i+2}, are substantial, and (b) sampling errors are small in comparison to these changes.

One disturbing aspect of the estimation is the extremely high correlation between $\hat{w}_1$ and $\hat{w}_3$. For the early estimates this is $r = 0.964$; it becomes

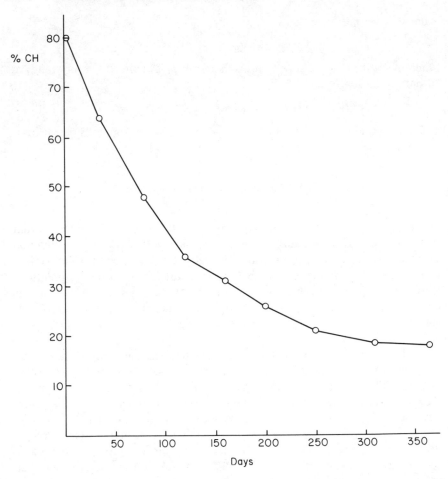

Fig. 8.4   Changes in the percentage of CH (chiricahua) chromosomes in experimental populations of *Drosophila pseudoobscura*. Sample points are indicated by the open circles.

$r = 0.998$ for the later estimates. This indicates that $\hat{w}_1$ and $\hat{w}_3$ are almost linearly related. Clearly the data are not really sufficient for estimating two selective values.

8.8   ESTIMATION OF SELECTIVE VALUES: DOMINANCE, SAMPLES BEFORE SELECTION

Suppose that A_1 is a dominant gene. Then the genotypes $A_1 A_1$ and $A_1 A_2$ give rise to a single phenotype $A_1 A_-$, where A_- indicates either A_1 or A_2. Hence,

Table 8.6   Estimation of w_1 and w_3 for *Drosophila pseudoobscura*. Linear interpolation has been used to determine proportions of CH chromosomes at 25-day intervals after day 0 and day 120. All sample sizes were 1200 chromosomes.

	Sample chromosome frequencies		Proportion	Interpolated values		$\hat{u} = \dfrac{\hat{p}}{(1-\hat{p})}$		
Day	CH	ST	CH($\hat{p}$)	Day	$\hat{p}$		$\hat{w}_1$	$\hat{w}_3$
0	–	–	0.8000	0	0.8000	4.0000		
35	764	436	0.6367	25	0.6834	2.1581	0.3846	0.7051
80	559	641	0.4658	50	0.5797	1.3794		
120	433	767	0.3608	120	0.3608	0.5645		
160	375	825	0.3125	145	0.3306	0.4939	1.1239	1.3036
200	312	888	0.2600	170	0.2994	0.4273		

Note: $\hat{w}_1$ is calculated from equation (8.22) and $\hat{w}_3$ is then calculated from equation (8.21).

given a sample from a population there is no way of estimating the proportion p of A_1 genes without making assumptions about the relative genotype frequencies. However, if there is random mating then the newly formed zygotes $A_1 A_1$, $A_1 A_2$ and $A_2 A_2$ will be in the Hardy–Weinberg proportions $p^2 : 2pq : q^2$, where $q = 1 - p$. Therefore if a random sample of n individuals is taken from a population before selection p and q can be estimated as

$$\hat{q} = \sqrt{(n_2/n)} \quad \text{and} \quad \hat{p} = 1 - \hat{q}, \tag{8.26}$$

where n_2 is the number of $A_2 A_2$ individuals in the sample. These estimators can be shown by a Taylor series approximation to have variances

$$\text{var}(\hat{p}) = \text{var}(\hat{q}) \simeq (1 - q^2)/(4n), \tag{8.27}$$

assuming that n_2/n is a simple binomial proportion.

Since there is dominance, it is possible to assume that $w_1 = w_2 = 1 - s$ and $w_3 = 1$. For this case, Table 8.4 shows that the change in the proportion of A_1 genes over one generation is

$$\Delta p = - spq^2/(1 - 2sp + sp^2), \tag{8.28}$$

starting with the proportion p. Rearranging this equation gives

$$s = \Delta q/((1 - q)\{q^2 + \Delta q(1 + q)\}), \tag{8.29}$$

where $\Delta q = - \Delta p$ is the change in the frequency of the A_2 gene. This suggests the estimator of s,

$$\hat{s}_i = (\hat{q}_{i+1} - \hat{q}_i)/((1 - \hat{q}_i)\{\hat{q}_i^2 + (\hat{q}_{i+1} - \hat{q}_i)(1 + \hat{q}_i)\}), \tag{8.30}$$

Table 8.7 Calculation of variances and covariances for selective values using equations (8.24) and (8.25). In these equations Δ_{ji} is the change in $\hat{w}_j$ that occurs when $\hat{p}_i$ is increased by one estimated standard deviation, $\text{SD}(\hat{p}_i) \simeq \sqrt{\{\hat{p}_i(1-\hat{p}_i)/1200\}}$.

(a) Calculations for samples at 0, 35 and 80 days
In this case $p_1 = 0.80$ is known exactly, without sampling errors. Hence only sampling errors in $\hat{p}_2$ and $\hat{p}_3$ have to be considered; Δ_{11} and Δ_{31} are zero.

		Interpolation					
Day	$\hat{p}$	Day	$\hat{p}$	$\hat{w}_1$	$\hat{w}_3$	Δ_{1i}	Δ_{3i}
0	0.8000	0	0.8000				
35	0.6506*	25	0.6933	0.440	0.887	$0.440 - 0.385$	$0.887 - 0.705$
80	0.4658	50	0.5890			$= 0.055$	$= 0.182$
0	0.8000	0	0.8000				
35	0.6367	25	0.6834	0.371	0.604	$0.371 - 0.385$	$0.604 - 0.705$
80	0.4802*	50	0.5845			$= -0.014$	$= -0.101$

* increased by one standard deviation

$\text{var}(\hat{w}_1) \simeq (0.055)^2 + (-0.014)^2 = 0.0032 \quad \text{SE}(\hat{w}_1) \simeq 0.057$

$\text{var}(\hat{w}_3) \simeq (0.182)^2 + (-0.101)^2 = 0.0433 \quad \text{SE}(\hat{w}_3) \simeq 0.208$

$\text{cov}(\hat{w}_1, \hat{w}_3) \simeq (0.055)(0.182) + (-0.014)(-0.101) = 0.0114$

$r = \text{cov}(\hat{w}_1, \hat{w}_3)/\{\text{SE}(\hat{w}_1)\text{SE}(\hat{w}_3)\} \simeq 0.964$

(b) Calculations for samples at 120, 160 and 200 days
Here there are three proportions with sampling errors.

		Interpolation					
Day	$\hat{p}$	Day	$\hat{p}$	$\hat{w}_1$	$\hat{w}_3$	Δ_{1i}	Δ_{3i}
120	0.3746*	120	0.3746				
160	0.3125	145	0.3358	0.819	1.167	$0.819 - 1.124$	$1.167 - 1.304$
200	0.2600	170	0.2994			$= -0.305$	$= -0.137$
120	0.3608	120	0.3608				
160	0.3259*	145	0.3390	2.756	2.249	$2.756 - 1.124$	$2.249 - 1.304$
200	0.2600	170	0.3094			$= 1.632$	$= 0.945$
120	0.3608	120	0.3608				
160	0.3125	145	0.3306	0.799	1.094	$0.799 - 1.124$	$1.094 - 1.304$
200	0.2727*	170	0.3026			$= -0.325$	$= -0.210$

* increased by one standard deviation

$\text{var}(\hat{w}_1) \simeq (-0.305)^2 + (1.632)^2 + (-0.325)^2 = 2.862 \quad \text{SE}(\hat{w}_1) \simeq 1.69$

$\text{var}(\hat{w}_3) \simeq (-0.137)^2 + (0.945)^2 + (-0.210)^2 = 0.956 \quad \text{SE}(\hat{w}_3) \simeq 0.98$

$\text{cov}(\hat{w}_1, \hat{w}_3) \simeq (-0.305)(-0.137) + (1.632)(0.945) + (-0.325)(-0.210) = 1.6522$

$r = \text{cov}(\hat{w}_1, \hat{w}_3)/\{\text{SE}(\hat{w}_1)\text{SE}(\hat{w}_3)\} \simeq 0.998$

based on the change in q from the start of generation i to the start of generation $i+1$.

An approximate variance for $\hat{s}_i$ can be determined using equation (A25) of the Appendix, Section A.5. That is,

$$\mathrm{var}(\hat{s}_i) \simeq \Delta_{ii}^2 + \Delta_{i\,i+1}^2, \tag{8.31}$$

where Δ_{ij} is the change in $\hat{s}_i$ that results from a change of one standard deviation in $\hat{q}_j$. If a series of estimates are calculated then $\hat{s}_i$ and $\hat{s}_{i+1}$ will both depend upon $\hat{q}_{i+1}$. Equation (A26) of the Appendix then gives

$$\mathrm{cov}(\hat{s}_i, \hat{s}_{i+1}) \simeq \Delta_{i\,i+1}\Delta_{i+1\,i+1}. \tag{8.32}$$

An approximate bias can be calculated using equation (A27).

8.9 ESTIMATION FROM SAMPLES SEVERAL GENERATIONS APART

Suppose that a population is sampled, before selection, in two generations that are not consecutive. Suppose also that there is dominance so that the genotypes A_1A_1 and A_1A_2 have equal and constant selective values relative to the recessive homozygote A_2A_2, and equation (8.28) gives the gene frequency change per generation. An estimator of the selective coefficient s was developed for this situation by Clarke and Murray (1962a, b) in order to compare samples of the snail *Cepaea nemoralis* taken from the Berrow sand dunes in Somerset in 1926 with samples from the same places taken in 1959–60.

Clarke and Murray argued as follows. First, it can be noted that the change in the proportion of A_2 genes per generation is

$$\Delta q = sq^2(1-q)/\{1 - s(1 - q^2)\}. \tag{8.33}$$

This equation can be approximated by the differential equation

$$\frac{dq}{dt} = sq^2(1-q)/\{1 - s(1 - q^2)\}, \tag{8.34}$$

since $\Delta q/\Delta t \simeq dq/dt$, taking $\Delta t = 1$ generation of time. The solution of the differential equation is

$$s = \frac{\log_e\left\{\dfrac{q_t(1-q_0)}{q_0(1-q_t)}\right\} - \dfrac{1}{q_t} + \dfrac{1}{q_0}}{t + \log_e\left\{\dfrac{q_t}{q_0}\right\} - \dfrac{1}{q_t} + \dfrac{1}{q_0}}$$

where q_0 is the gene frequency at time 0 and q_t is the gene frequency at time t. Clearly, if estimates $\hat{q}_0$ and $\hat{q}_t$ of q_0 and q_t are available, then these can be substituted into the equation. The result is Clarke and Murray's estimator of s,

$$\hat{s} = \frac{\log_e\left\{\dfrac{\hat{q}_t(1-\hat{q}_0)}{\hat{q}_0(1-\hat{q}_t)}\right\} - \dfrac{1}{\hat{q}_t} + \dfrac{1}{\hat{q}_0}}{t + \log_e\left\{\dfrac{\hat{q}_t}{\hat{q}_0}\right\} - \dfrac{1}{\hat{q}_t} + \dfrac{1}{\hat{q}_0}}. \tag{8.35}$$

The variance of this estimator can be found, using the Taylor series method, to be

$$\text{var}(\hat{s}) \simeq \left[\left\{\frac{1}{1-q_0} + \frac{1-s}{q_0}\left(1+\frac{1}{q_0}\right)\right\}^2 \text{var}(\hat{q}_0)\right.$$
$$\left. + \left\{\frac{1}{1-q_t} + \frac{1-s}{q_t}\left(1+\frac{1}{q_t}\right)\right\}^2 \text{var}(\hat{q}_t)\right] \Big/$$
$$\left[t + \log_e\left(\frac{q_t}{q_0}\right) - \frac{1}{q_t} + \frac{1}{q_0}\right]^2. \tag{8.36}$$

If a random sample of n_i individuals is taken from a population before selection in generation $i+1$, and this yields n_{i2} of the recessive genotype A_2A_2, then from equations (8.26) and (8.27) $\hat{q}_i = \sqrt{(n_{i2}/n_i)}$ and $\text{var}(\hat{q}_i) \simeq (1-q_i^2)/(4n_i)$. Clarke and Murray (1962b) give an equation for the variance of $\hat{s}$ which appears to be based upon the incorrect assumption that q_i can simply be estimated by counting the A_2 genes in the sample of n_i genes. Wall et al. (1980) provide another variance equation which agrees with equation (8.36) when $s = 0$.

Clarke and Murray's estimator should give reasonable results providing that the following three conditions are satisfied:

(1) The true value of s should be small, so that Δq, the change in q per generation, is small.
(2) The total change in q, $q_t - q_0$, should be large relative to the sampling errors in $\hat{q}_0$ and $\hat{q}_t$.
(3) Neither q_0 nor q_t should be close to zero or one.

Condition (3) is needed in order to avoid infinite values of $\hat{s}$. It will be seen, for example, that if $\hat{q}_0 = 0.95$ and $\hat{q}_t = 1.0$, then $\hat{s} = +\infty$. This occurs even though the maximum possible value for s is $+1.0$, which happens when the three genotypes A_1A_1, A_1A_2 and A_2A_2 have fitnesses of 0.0, 0.0 and 1.0, respectively.

When s is not small, Clarke and Murray's estimator will not produce the value of $\hat{s}$ such that repeated application of equation (8.33) gives the change from $\hat{q}_0$ to $\hat{q}_t$ in t generations. For example, suppose that $\hat{q}_0 = 0.05$ and $\hat{q}_5 = 0.15$. Substituting these estimates into equation (8.35), with $t = 5$, gives $\hat{s} = 0.748$. However, if equation (8.33) is applied five times, starting with $q_0 = 0.05$, it gives $q_1 = q_0 + \Delta q_0 = 0.057$, $q_2 = q_1 + \Delta q_1 = 0.065, \ldots,$

$q_5 = q_4 + \Delta q_4 = 0.110$. Therefore, $s = 0.748$ is not large enough to change q from 0.05 to 0.15 in five generations.

This bias in Clarke and Murray's estimator can be corrected using the results shown in Table 8.8. The table shows the values produced by equation

Table 8.8   Value of s given by equation (8.35) as a percentage of the value of s that gives the correct change in q from generation 0 to generation t.

	Initial gene frequency (%)							Initial gene frequency (%)						
	1	10	30	50	70	90	99	1	10	30	50	70	90	99
t			$s = 0.10$							$s = -0.10$				
1	100	99	98	99	101	103	105	100	101	102	101	100	97	95
5	100	99	98	99	101	104	105	100	101	102	102	100	97	95
10	100	99	98	99	101	104	105	100	101	102	102	100	98	95
25	100	99	99	100	103	104	105	100	101	102	102	101	99	96
50	100	99	100	102	104	105	*	100	101	101	102	101	100	97
t			$s = 0.30$							$s = -0.45$				
1	100	98	96	97	103	112	118	100	104	108	107	101	90	83
5	100	97	96	100	107	115	118	100	104	108	108	105	94	84
10	100	97	97	104	111	116	119	100	104	107	108	106	99	87
25	100	96	106	112	*	*	*	100	103	106	107	107	104	97
50	100	103	*	*	*	*	*	100	103	105	105	105	104	101
t			$s = 0.50$							$s = -1.0$				
1	100	96	93	97	108	126	137	101	109	122	121	107	83	71
5	100	95	97	109	121	133	138	101	109	119	122	117	99	76
10	99	95	108	120	129	136	*	101	108	117	120	119	109	90
25	99	107	*	*	*	*	*	101	106	112	114	114	111	103
50	99	*	*	*	*	*	*	101	105	108	110	110	109	104
t			$s = 0.70$							$s = -2.50$				
1	99	95	93	101	120	150	170	103	127	186	200	145	89	53
5	99	93	110	132	150	165	*	102	124	164	186	179	131	78
10	99	102	*	*	*	*	*	102	121	149	164	166	144	103
25	99	*	*	*	*	*	*	102	116	130	137	139	133	116
50	98	*	*	*	*	*	*	102	111	119	122	124	122	114
t			$s = 0.80$							$s = -4.00$				
1	99	94	94	106	132	173	198	104	152	398	645	260	80	40
5	99	95	128	154	175	*	*	104	144	261	387	379	198	87
10	99	117	*	*	*	*	*	104	137	204	257	272	209	121
25	98	*	*	*	*	*	*	104	126	153	169	175	165	133
50	*	*	*	*	*	*	*	103	118	131	138	140	138	125

* In these cases the gene frequency after selection exceeds 99.99%.

(8.35) as a percentage of true values of s. Interpolation in the table shows that when $s = 0.75$, $q_0 = 0.05$, and $t = 5$, then $\hat{s}$ as produced by equation (8.35) is about 97 % of what it should be. Therefore, the estimate of $\hat{s} = 0.748$ obtained from $\hat{q}_0 = 0.05$ and $\hat{q}_5 = 0.15$ should be corrected to $\hat{s} = 0.748/0.97 = 0.77$. This corrected estimate is actually still slightly low, since the value of s that changes q from 0.05 to 0.15 in five generations is 0.78 to two decimal places.

In cases where the correction required for the estimate is small, equation (8.36) can still be used for the variance. When a substantial correction is required it may be better to use equation (A25) of the Appendix and take

$$\text{var}(\hat{s}) \simeq \Delta_0^2 + \Delta_t^2, \tag{8.37}$$

where Δ_i is the change in $\hat{s}$ that results from a change of one standard deviation in $\hat{q}_i$. Bias can be approximated using equation (A27).

Suppose that a series of samples are taken from a population at times $t_1, t_2, \ldots, t_m$. Then it is possible to use equation (8.35) to estimate a selective coefficient for each pair of consecutive samples. Let $\hat{s}_i$ denote the estimate based upon the samples at time t_i and t_{i+1}. Clearly, $\hat{s}_i$ and $\hat{s}_{i+1}$ will be correlated to some extent since they both use the sample at time t_{i+1}. The covariance can be approximated by using equation (A26) of the Appendix, which provides

$$\text{cov}(\hat{s}_i, \hat{s}_{i+1}) \simeq \Delta_{i\,i+1}\Delta_{i+1\,i+1} \tag{8.38}$$

where Δ_{ij} is the change in $\hat{s}_i$ that results from a one standard deviation change in $\hat{q}_{t_j}$, the estimated gene frequency at time t_j.

Clarke and Murray's approach to estimation is easily modified when different assumptions are made about selective values. For example, suppose that the selective values of $A_1 A_1$, $A_1 A_2$ and $A_2 A_2$ are taken to be $1 - s$, $1 - \frac{1}{2}s$ and 1, respectively. An approximate differential equation for the process is then

$$\frac{dq}{dt} = \tfrac{1}{2}sq(1-q)/\{1 - s(1-q)\},$$

with the solution

$$2\log_e\left\{\frac{q_t(1-q_0)}{q_0(1-q_t)}\right\} - 2s\log_e\left(\frac{q_t}{q_0}\right) = st,$$

where q_0 is the proportion q at time 0 and q_t is the proportion at time t. Substituting estimates of q_0 and q_t into the last equation produces

$$\hat{s} = 2\log_e\left(\frac{\hat{q}_t(1-\hat{q}_0)}{\hat{q}_0(1-\hat{q}_t)}\right)\bigg/\left\{t + 2\log_e\left(\frac{\hat{q}_t}{\hat{q}_0}\right)\right\}. \tag{8.39}$$

This estimator seems to have been first published by Curtis *et al.* (1978). The

Taylor series approximate variance is

$$\text{var}(\hat{s}) \simeq 4\left(\left\{\frac{1}{1-q_0} + \frac{1-s}{q_0}\right\}^2 \text{var}(\hat{q}_0) + \left\{\frac{1}{1-q_t} + \frac{1-s}{q_t}\right\}^2 \text{var}(\hat{q}_t)\right) \bigg/$$
$$\left\{t + 2\log_e\left(\frac{q_t}{q_0}\right)\right\}^2.$$

It would be straightforward to produce a table of corrections for this estimator of a similar form to Table 8.8.

8.10   ESTIMATION OF SELECTIVE VALUES: DOMINANCE, SAMPLES AFTER SELECTION

Suppose that at the start of a generation, after random mating, a population contains a proportion p of allele A_1 and a proportion $q = 1 - p$ of the alternative allele A_2. Then the genotypes should be in Hardy–Weinberg ratios $p^2 : 2pq : q^2$. If selection on the genotypes during the generation gives selective values w_1, w_2 and w_3, then at the end of the generation these ratios will be changed to $w_1 p^2 : 2w_2 pq : w_3 q^2$. Hence a random sample taken at the end of the generation is expected to contain a proportion

$$\pi = w_3 q^2 / (w_1 p^2 + 2w_2 pq + w_3 q^2) \tag{8.40}$$

of the genotype $A_2 A_2$. Rearranging this equation produces a quadratic equation in q, for which the only valid solution is

$$q = \frac{\pi(w_2 - w_1) + \sqrt{\{\pi^2 w_2^2 + \pi(1 - \pi)w_1 w_3\}}}{w_3(1 - \pi) + \pi(2w_2 - w_1)}. \tag{8.41}$$

This equation, in conjunction with equations already provided, allows selective values to be estimated from samples taken after selection. An iterative approach is required.

For example, assuming dominance, let $w_1 = w_2 = 1 - s$ and $w_3 = 1$. Then from equation (8.29),

$$s = (q_{i+1} - q_i)/[(1 - q_i)\{q_i^2 + (q_{i+1} - q_i)(1 + q_i)\}]. \tag{8.42}$$

Also, equation (8.41) becomes

$$q_i = \sqrt{\{\pi_i(1 - s)/(1 - s\pi_i)\}}. \tag{8.43}$$

If estimates $\hat{\pi}_i$ and $\hat{\pi}_{i+1}$ are available of the proportion of $A_2 A_2$ individuals at the end of generations i and $i + 1$, then equations (8.42) and (8.43) can be solved iteratively to find the corresponding estimate of s. An example is provided in

Table 8.9 Estimation of a selective coefficient s from two generations of data by solving equations (8.42) and (8.43) iteratively. The data are assumed to be from two samples taken after selection, as follows:

Generation (i)	A_1A_-	A_2A_2	n_i	$\hat{\pi}_i$
1	152	98	250	0.392
2	144	106	250	0.424

Iteration	Initial value of $\hat{s}$	Generation (i)	q_i	Final value of $\hat{s}$
1	0.000	1	0.6261	0.155
		2	0.6512	
2	0.155	1	0.5938	0.159
		2	0.6192	
3	0.159	1	0.5929	0.159
		2	0.6183	

Note: The steps in the iteration consist of: (a) estimating q_1 and q_2 using equation (8.43) with the initial value for $\hat{s}$; and (b) calculating a new value for $\hat{s}$ using equation (8.42). In the present example, three iterations have produced $\hat{s} = 0.16$ to two decimal places.

Table 8.9. An approximate variance is given by

$$\text{var}(\hat{s}) \simeq \Delta_i^2 + \Delta_{i+1}^2,$$

where Δ_i is the change in $\hat{s}$ resulting from a one standard deviation change in $\hat{\pi}_i$.

8.11 ESTIMATION FROM SAMPLES AFTER SELECTION SEVERAL GENERATIONS APART

If two samples are taken from a population after selection, several generations apart, then Clarke and Murray's estimator of equation (8.35) can be used in conjunction with equation (8.43) to estimate a selective coefficient s, assuming that the genotypes A_1A_1, A_1A_2 and A_2A_2 have selective values of $1-s$, $1-s$ and 1, respectively. Thus, suppose that $\hat{\pi}_0$ is the estimated proportion of A_2A_2 after selection in generation 0, while $\hat{\pi}_t$ is the corresponding proportion in generation t. Then the following three equations that need to be solved to find an estimate of s:

$$\hat{q}_i = \sqrt{\{\hat{\pi}_i(1-\hat{s})/(1-\hat{s}\hat{\pi}_i)\}}, \qquad i = 0, t, \tag{8.44}$$

and

$$\hat{s} = \frac{\log_e \{\hat{q}_t(1 - \hat{q}_0)/\hat{q}_0(1 - \hat{q}_t)\} - 1/\hat{q}_t + 1/\hat{q}_0}{t + \log_e (\hat{q}_t/\hat{q}_0) - 1/\hat{q}_t + 1/\hat{q}_0}. \tag{8.45}$$

Starting with $\hat{s} = 0$, $\hat{q}_0$ and $\hat{q}_t$ can be evaluated using equation (8.44). These estimates of q can then be substituted into equation (8.45) to get a better value for $\hat{s}$. This can be corrected for error using Table 8.8 in the manner that has been described in Section 8.9. The new value of $\hat{s}$ can then be used in equation (8.44) to get improved values for q_0 and q_t, and so on. The iterative process can continue until stable values are obtained.

The same approach can be used with different assumptions about selective values, providing that the selective values can still be expressed in terms of a single parameter. For instance, if $A_1 A_1$, $A_1 A_2$ and $A_2 A_2$ have selective values of $1 - s$, $1 - \frac{1}{2}s$ and 1, respectively, then equation (8.44) needs to be modified and equation (8.45) needs to be replaced by equation (8.39).

Whatever assumptions are made about selective values, the error in the estimate of s will depend upon the errors in $\hat{\pi}_0$ and $\hat{\pi}_t$. Consequently, equation (8.37) provides a numerical approximation for the variance of $\hat{s}$ where Δ_i is the change in $\hat{s}$ that results from a one standard deviation change in $\hat{\pi}_i$. If a sequence of estimates of s are available from a single population then there will be some correlation between consecutive estimates. Equation (8.38) can be used to determine a numerical approximation for $\text{cov}(\hat{s}_i, \hat{s}_{i+1})$ where Δ_{ij} is defined as the change in $\hat{s}_i$ that is brought about by a one standard deviation change in $\hat{\pi}_{t_j}$.

8.12   A NOTE OF CAUTION

Caution is needed when using the variance formulae provided above. They relate only to sampling errors in selective values and selective coefficients. However, in reality population values will show some stochastic variation from generation to generation because population sizes are finite. Also, there will be environmental changes with time that will change selective forces. Without a knowledge of population size there is no way of predicting how large the stochastic variation will be on a particular population. It is for this reason that evidence for selection can usually only be found by considering data either from a single population over a long period of time or, alternatively, data from a number of independent populations with some general relationship. An example of this latter situation occurs when several isolated populations in the same geographical area have approximately the same estimated selective values.

Caution is also needed because of the possibility of large biases in estimators based upon small samples. This has been emphasized recently by calculations carried out by Anxolabhere et al. (1982). They show that for sample sizes that are often used (100, or less) biases may not be negligible and, what is more, may result in the false conclusion that selective values are frequency dependent. Some assessment of bias can be made using Taylor series approximations, as mentioned in the preceding sections. However, if possible, it is best to use simulation to determine the distribution of the estimators used in any study.

Example 8.4 Selective values for *Cepaea nemoralis* at Berrow

Table 8.10 shows sample frequencies of dark brown and other colours of *Cepaea nemoralis* at the locations shown on Fig. 8.5 on the sand dunes at Berrow in Somerset, for the years 1926, 1959/60 and 1975. These are only part of a somewhat larger set of data (Clarke and Murray, 1962a, b; Murray and Clarke, 1978). The shell colour of *C. nemoralis* is determined by several alleles at a single locus. The allele A_1 for dark brown is dominant to the alleles for other colours that are collectively taken as A_2 for the purposes of this example.

Table 8.11 shows estimates of selective coefficients s, based upon the assumption that the genotypes A_1A_1 and A_1A_2 have the selective value $1 - s$ relative to A_2A_2. Following Clarke and Murray, a generation time of three years has been assumed, although there is some doubt about whether this is correct. However, this is not a serious problem since Clarke and Murray (1962b) have shown that there is a simple approximate relationship between the generation time and $\hat{s}$: the selective coefficient per year is constant. Halving the generation time results in a doubling of $\hat{s}$.

At locations D19, D29 and D30, the 1959/60 sample consisted entirely of non–dark brown (A_2A_2) snails. For all three of these locations this has meant

Table 8.10 Frequencies of dark brown (n_1) and other colours (n_2) of *Cepaea nemoralis* at Berrow, Somerset, from samples taken in 1926, 1959/60 and 1975.

Location	1926			1959/60			1975		
	n_1	n_2	$\hat{\pi}_2$	n_1	n_2	$\hat{\pi}_2$	n_1	n_2	$\hat{\pi}_2$
D19	5	137	0.965	0	132	1.000	–	–	
D29	1	47	0.979	0	342	1.000	4	163	0.976
D30	2	228	0.991	0	220	1.000	3	61	0.953
D40	5	175	0.972	3	103	0.972	1	92	0.989
D46	7	161	0.958	1	58	0.983	–	–	
D48	63	406	0.866	9	80	0.899	7	160	0.958
D50	22	111	0.835	20	125	0.862	21	222	0.914
D52	95	332	0.778	41	316	0.885	4	46	0.920
D54	18	231	0.928	11	569	0.981	2	109	0.982
D56	14	101	0.878	18	286	0.941	21	110	0.840

$\hat{\pi}_2 = n_2/(n_1 + n_2)$ = the estimated proportion of A_2A_2 snails.

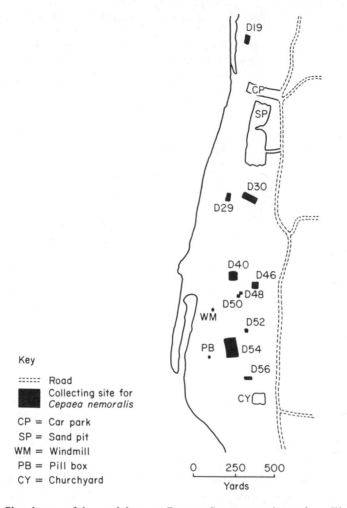

Fig. 8.5   Sketch map of the sand dunes at Berrow, Somerset, redrawn from Figure 1 of Clark and Murray (1962a). Collecting sites are drawn to scale.

that $\hat{s}_1$ is plus infinity. For locations D29 and D30, $\hat{s}_2$ is minus infinity. The infinite $\hat{s}_1$ values have been changed to 1.0 since this is the maximum possible value of s. However, the minus infinity values for $\hat{s}_2$ have been left alone on the grounds that such values are possible. They imply that the genotype A_2A_2 has zero fitness relative to the other two genotypes.

It was pointed out in Section 8.9 that equation (8.35) may not give the correct value of $\hat{s}$. 'Correct' here means the value of $\hat{s}$ that gives the change from $\hat{q}_0$ to $\hat{q}_t$ by repeated application of equation (8.33) for a single generation

Table 8.11   Estimates of selective values from the data in Table 8.10 assuming samples before selection.

Location	$\hat{s}_1$	Formula SE	Numerical SE	$\hat{s}_2$	Formula SE	Numerical SE	Numerical r
D19	1.000	–	–	–	–	–	–
D29	1.000	–	–	$-\infty$	–	–	–
D30	1.000	–	–	$-\infty$	–	–	–
D40	−0.002	0.067	0.054	0.197	0.197	0.168	−0.43
D46	0.082	0.098	0.069	–	–	–	–
D48	0.029	0.034	0.030	0.190	0.104	0.090	−0.63
D50	0.020	0.031	0.029	0.107	0.067	0.062	−0.53
D52	0.072	0.018	0.018	0.082	0.111	0.094	−0.26
D54	0.124	0.035	0.031	0.010	0.160	0.122	−0.34
D56	0.071	0.034	0.030	−0.234	0.074	0.069	−0.45

Note: The 'formula' standard errors have been calculated from equation (8.36) with s, q_0 and q_t replaced with $\hat{s}$, $\hat{q}_0$ and $\hat{q}_t$. The 'numerical' standard error comes from equation (8.37) with Δ_i defined as the change in $\hat{s}$ resulting from a one standard deviation reduction in $\hat{q}_i$. A reduction was used rather than an increase since some $\hat{q}$ values are very close to 1.0. The numerical value of r uses equation (8.38) to determine $\text{cov}(\hat{s}_1, \hat{s}_2)$.

change Δq in q. Table 8.8 can be used to adjust the estimate. The adjustments made on this basis have only very minor effects on the non-infinite estimates.

It is unnecessary to use the numerical equation (8.37) to calculate standard errors in this example since an explicit formula is available. The reason for providing the numerically calculated values in Table 8.11 was simply to show how these compare with values obtained directly from the formula. On the whole the agreement is good, with the numerically calculated values tending to be slightly lower than those calculated from the formula.

Since Clarke and Murray sampled adult snails, it is not really reasonable to assume that samples were taken before selection. However, if it is assumed that samples were taken after selection, estimates hardly change. The largest effect is on the estimate of s_2 for location D56. This is -0.234 assuming samples before selection and -0.231 assuming samples after selection, the sample 'after-selection' value being found by solving equations (8.44) and (8.45). In reality, Clarke and Murray's samples were no doubt taken after some, but not all, selection had taken place. Thus the estimates of selective coefficients should fall somewhere between the 'sample before selection' estimates and the 'sample after selection' estimates. In the present case these two extremes are so close together that the 'sample before selection' estimates can be accepted as they stand.

Taking the estimates of s_1 for locations D40 to D56, equation (A28) of the Appendix, Section A.6, gives a weighted mean of $\bar{s}_1 = 0.063$, with standard error 0.012. The chi-square statistic of test (a) of Section 6 of the Appendix is 7.24

with 7 degrees of freedom, which is not significantly large. For these locations there seems to have been a more or less constant amount of selection from 1926 to 1959/60. The non-infinite estimates of s_2 give a weighted mean of $\bar{s}_2 = 0.017$, with standard error 0.039. The chi-square statistic is 17.24 with 6 degrees of freedom. This is significantly large and it seems therefore that for the period 1959 to 1975, selective coefficients varied from location to location, averaging out at about zero. The results for the second time interval 1959/60–1975 obviously bring into question one of the basic assumptions of Clarke and Murray's method of estimation. It seems doubtful that s remained constant at any of the locations for the 1959–1975 period.

The other basic assumption of Clarke and Murray's method, that the fitness is equal for $A_1 A_1$ and $A_1 A_2$ snails, is not important in this example. Since the proportion of A_2 alleles is high, it follows that there must have been a very low proportion of the $A_1 A_1$ dark brown homozygotes at all locations. Thus estimated selective coefficients are based almost entirely on data for the two genotypes $A_1 A_2$ and $A_2 A_2$. Consequently, all that has really been estimated is the selective coefficient for the heterozygotes $A_1 A_2$ relative to the homozygotes $A_2 A_2$.

It is unfortunate that the data from colonies D19, D29 and D30 cannot be used to help estimate average values for s. Clarke and Murray overcame this problem by 'inventing' a single dark brown snail for the samples that did not contain one. However, this approach seems somewhat arbitrary since it means that the q value estimated depends on the sample size that happens to be taken. A better method involves using full maximum likelihood estimation with results from several samples from one or more locations. This is discussed further in Example 8.5 below.

8.13 ESTIMATION OF SELECTIVE VALUES IN GENERAL

When constant selective values are to be estimated using data from more than three generations, there are no simple equations that will necessarily give efficient estimates. The general problem of estimation seems to have been first considered by Wright and Dobzhansky (1946), who proposed a least squares method for determining selective values for three karyotypes (chromosome types) $A_1 A_1$, $A_1 A_2$ and $A_2 A_2$. They also considered the extension of their method for the case of six karyotypes $A_1 A_1$, $A_1 A_2$, $A_1 A_3$, $A_2 A_2$, $A_2 A_3$, $A_3 A_3$, each with a different fitness. Levene et al. (1954) noted certain difficulties with the least squares method because observations are not given proper weighting. They proposed instead a minimum chi-square method which they claimed is approximately equivalent to the maximum likelihood method. They used this method to estimate the fitnesses of six karyotypes.

Cavalli (1950) developed a method for estimating selective values that involves approximating gene frequency changes using a differential equation in a somewhat similar way to Clarke and Murray (1962b). Cavalli only specifically considered the case where there is heterosis, so that A_1A_2 individuals are fitter than A_1A_1 and A_2A_2. In principle, the approach can be extended to other cases with two alleles.

The special case of heterosis is important because it involves an equilibrium gene frequency that is not trivial. If the selective values for A_1A_1, A_1A_2 and A_2A_2 are taken as $1 - s_1$, 1 and $1 - s_2$, then, as shown in Table 8.4, the proportion of A_2 alleles has a stable equilibrium when $q = s_1/(s_1 + s_2)$. That is to say, the change in q per generation, Δq, is equal to zero for this value of q. For this reason it has sometimes been suggested that gene frequency changes Δq be plotted against q and a linear regression line fitted. Then the point where $\Delta q = 0$ corresponds to $s_1/(s_1 + s_2)$ and the regression coefficient of q may be equated to the theoretical value of $d(\Delta q)/dq$ (Wright and Dobzhansky, 1946). This method is rather rough and can have considerable bias (Kirby, 1974, Whittam, 1981).

Prout (1965) emphasized that it is not possible to obtain valid estimates of selective values using genotype frequencies from two consecutive generations unless genotypes are counted after all selection has taken place. Since one important component of selection may be differential fertility, it might not be sufficient in this respect to sample mature adults. For example, the analysis of genotype frequencies of *Panaxia dominula* at Cothill that has been given in Example 8.2 above can be criticized because it involves the implicit assumption that flying adults mate at random.

Prout's analysis is based upon a two-allele model. Assume that a population starts generation 0 with a proportion p of allele A_1 and a proportion $q = 1 - p$ of allele A_2. Through random mating the genotypes A_1A_1, A_1A_2 and A_2A_2 will then be in the Hardy–Weinberg ratios

$$p^2 : 2pq : q^2.$$

If a sample is taken after some selection has taken place then these ratios will be changed to

$$E_1 p^2 : 2pq : E_3 q^2,$$

where E_1 and E_2 are the 'early' components of the selective values of A_1A_1 and A_2A_2, relative to A_1A_2. By the end of generation 0 more selection will have occurred, including some related to differential fertility, so that the genotype frequencies will be in ratios

$$E_1 L_1 p^2 : 2pq : E_3 L_3 q^2,$$

where L_1 and L_3 are the 'late' components of the selective values. Net selective values for A_1A_1 and A_2A_2 relative to A_1A_2 are therefore $w_1 = E_1 L_1$ and $w_3 = E_3 L_3$, respectively. At this point the frequency of A_1 alleles will be proportional to $E_1 L_1 p^2 + pq$ and the frequency of A_2 alleles will be proportional to $E_3 L_3 q^2 + pq$. Random mating will therefore produce genotypes in the ratios

$$(E_1 L_1 p^2 + pq)^2 : 2(E_1 L_1 p^2 + pq)(E_3 L_3 + pq) : (E_3 L_3 q^2 + pq)^2$$

at the start of generation 1. From 'early' selection these ratios become

$$E_1(E_1 L_1 p^2 + pq)^2 : 2(E_1 L_1 p^2 + pq)(E_3 L_3 + pq) : E_3(E_3 L_3 q^2 + pq)^2$$

by the time that a sample is taken from generation 1.

Thus the changes in genotype proportions are complicated functions of the E and L components of the selective values w_1 and w_3. The equations given in Section 8.6 for samples after selection are certainly not valid for estimating w_1 and w_3. Indeed, the estimates produced by equations (8.17) are functions of p. If these equations are used on data collected after partial selection then estimated selective values will indicate spurious frequency-dependent selection.

In a later paper, Prout (1969) showed that at least four generations of data are needed to estimate E_1, E_3, L_1 and L_3. However, maximum likelihood estimation with four generations produces disappointing estimates, with large standard errors, occasional negative values, and convergence difficulties for the iterative method of calculation. These are all characteristics of a model with too many parameters for the available data.

Two ways round these difficulties have been adopted with laboratory populations. Firstly, separate measurements of the different components of fitness have been made. A comparison has then been made between predicted changes based on these components and actual population changes (Prout, 1971a, b). A problem with this approach is recreating the actual population conditions in the separately conducted fitness assessment experiments. Secondly, an experimental regime has been set up so that it is possible to count genotypes at several different stages in the life cycle. This was done by Bungaard and Christiansen (1972) but their approach is only applicable to very special populations where the females only mate once and can be isolated so that their offspring can be counted.

De Benedictis (1977, 1978) has argued that although Prout's algebra is quite correct, its importance may have been overemphasized in recent years, since estimates of selective values obtained in the classical manner, ignoring the possibility of early and late selection, are still of value for describing and predicting gene frequency changes. Significant values are evidence for selection, although the mechanism for selection may be oversimplified. This point of view seems to have some validity. For example, with the population of *Panaxia dominula* at Cothill (Example 8.2) it is surely of value to know what magnitude of selective values is required to describe the observed changes in genotype frequencies.

Du Mouchel and Anderson (1968) seem to have provided the first true maximum likelihood model for estimating selective values for any number of alleles at a single locus. This model is only approximate in that it assumes that the population being sampled is effectively infinite, so that all gene frequency changes are attributed to selection and none to random genetic drift. However, this is not too serious a problem when it comes to the analysis of data from natural populations of unknown size. Even with this simplifying assumption, selective values have to be estimated by iterative calculations. Wilson (1970) extended the model to cases where selection is sex dependent, and also to cases where samples are taken after partial selection in each generation.

Du Mouchel and Anderson (1968) and Wilson (1970) were somewhat pessimistic about the estimation of selective values in natural populations. Du Mouchel and Anderson noted the unrealistically large sample sizes that are needed for accurate estimation when four or five alleles are present. Wilson observed that estimation is virtually impossible when there are complications due to sex-dependent or partial selection.

Templeton (1974) reconsidered maximum likelihood estimation using a finite population model. This model is different from that of Du Mouchel and Anderson because it recognizes that some part of gene frequency changes is due to random genetic drift. Unfortunately, however, it does not allow for sample data. Rather, the model is based upon complete counts of genes either before or after all selection in a sequence of consecutive generations. Further work on finite population, complete census models of this nature will be found in papers by Watterson (1979, 1982) and Watterson and Anderson (1980).

In practice, estimation of selective values can be done using one of the standard computer programs for maximum likelihood. The available data will usually consist of counts of either phenotypes or genes in several generations. The expected values of these counts can be written in terms of initial gene frequencies, selective values, and sample size parameters. The distribution of the counts, conditional on the overall total count, can be regarded as multinomial. In that case, Reed and Schull's (1968) FORTRAN program MAXLIK, for which the code is provided in Section A.16 of the Appendix, can be used for estimation. As explained in Section A.2 of the Appendix, the multinomial assumption is reasonable if all the sample phenotype or genotype counts are independent Poisson variables. This may often be true for samples from natural populations, since sample sizes cannot be fixed in advance. However, even if sample sizes are fixed in advance this will not effect the estimation of selection parameters. Thus MAXLIK can be used with either fixed or random sample sizes.

White and White (1981) have published an ALGOL computer program for the estimation of selective values and say that a FORTRAN version is available on request. This program takes a series of estimates of the proportion of a dominant phenotype in different generations and determines selective values by the principle of least squares. Phenotypes are assumed to be counted at the same stage in each generation after some, but not all, selection has occurred. A particularly useful facility is that different sets of selective values can apply at different times.

As an example of the use of their program, White and White describe a situation where a laboratory population of the Australian sheep blowfly *Lucilia cuprina* was started with 90% of individuals resistant to the insecticide diazinon and 10% of individuals susceptible. Resistance is conferred by a single allele, which is almost completely dominant. The population was maintained for 38 generations, during which time the proportion of resistant individuals declined to about 10%. White and White estimated selective values on the assumption that resistant individuals were initially at a severe disadvantage, but this disadvantage declined with time. This was done by assigning one set of selective values to generations 1 and 2 and another set to generations 13 to 38. Generations 3 to 12 were given selective values midway between the two sets. Other applications of White and White's program are described by Partridge (1978) (studying warfarin resistance in rats) and Muggleton (1983) (studying melathion resistance in beetles).

8.14   MAXIMUM LIKELIHOOD ESTIMATION WITH THE DOMINANCE SELECTION MODEL

In Example 8.4, Clarke and Murray's data on *Cepaea nemoralis* populations on the Berrow sand dunes (Table 8.10) were analysed by estimating a selective coefficient for each of several different locations and then averaging these. A problem with this approach is that no use can be made of the data from locations D19, D29 and D30 because there were no dark brown snails in the samples taken in 1959/60. Maximum likelihood estimation using the program MAXLIK overcomes this problem. It can be assumed, for one time period, that selection was the same at all the locations, and a common selective value can be estimated.

Assume that the dominance selection model is appropriate. That is to say, the genotypes $A_1 A_1$ and $A_1 A_2$ are indistinguishable and have selective values of $1 - s$ relative to the recessive genotype $A_2 A_2$. In that case it has been shown before that if q_i is the proportion of A_2 alleles immediately after random mating at the start of generation i, then the one-generation change in q_i is given by the equation

$$q_{i+1} = q_i + s q_i^2 (1 - q_i)/\{1 - s(1 - q_i^2)\}. \qquad (8.46)$$

Suppose that there are L locations. Let q_{0l} denote the proportion of A_2 alleles at location l at the start of generation zero. Then repeated use of equation (8.46) will generate the sequence q_{1l}, q_{2l}, q_{3l}, and so on. Thus q_{tl} can be determined for any integer value of t. With an organism with overlapping generations such as *C. nemoralis* it will be a fair approximation to determine q_{tl} for non-integer values of t by linear interpolation. For example, $q_{4.5l} = \frac{1}{2}(q_{4l} + q_{5l})$.

If a sample is taken at time zero, and another sample at time t, for each of the

Table 8.12 Data for the maximum likelihood estimation of a common selective value for L locations.

Location	Time 0			Time t		
	A_1A_-	A_2A_2	Total	A_1A_-	A_2A_2	Total
1	n_{101}	n_{201}	n_{01}	n_{1t1}	n_{2t1}	n_{t1}
2	n_{102}	n_{202}	n_{02}	n_{1t2}	n_{2t2}	n_{t2}
$\vdots$	$\vdots$	$\vdots$	$\vdots$	$\vdots$	$\vdots$	$\vdots$
L	n_{10L}	n_{20L}	n_{0L}	n_{1tL}	n_{2tL}	n_{tL}

L locations, then there will be sample frequencies as shown in Table 8.12. Given the sample sizes n_{0l} and n_{tl} at the lth location, the expected values of morph counts are then

$$E(n_{1il}) = (1 - q_{il}^2)\,n_{il}$$

and
$$\left. \right\} \quad (8.47)$$

$$E(n_{2il}) = q_{il}^2 n_{il},$$

Details of how these equations can be used with the computer program MAXLIK are provided in Section A.17 of the Appendix.

The selective values of A_1A_1 and A_1A_2 relative to A_2A_2 are $1 - s$, which cannot be negative. To ensure that a negative value does not occur during the course of the maximum likelihood iterations, it may be useful to take $1 - s = \exp(u)$, so that $u = \log_e(1 - s)$, and estimate u instead of s. The usual Taylor series approximation then provides

$$\text{var}(\hat{s}) \simeq \exp(2u)\,\text{var}(\hat{u}). \quad (8.48)$$

If samples are regarded as having been taken after selection, then only slight modifications are needed for maximum likelihood estimation. After selection in generation i the genotypes A_1A_1, A_1A_2 and A_2A_2 are expected to have frequencies in the ratios $(1 - s)p_{il}^2 : 2(1 - s)p_{il}q_{il} : q_{il}^2$ at the lth location. Therefore equations (8.47) must be changed to

$$E(n_{1il}) = (1 - s)(1 - q_{il}^2)n_{il}/\{1 - s(1 - q_{il}^2)\}$$

and
$$\left. \right\} \quad (8.49)$$

$$E(n_{2il}) = q_{il}^2 n_{il}/\{1 - s(1 - q_{il}^2)\},$$

for $i = 0$ and t. No other changes are required.

Another application of maximum likelihood with the dominance selection model involves estimating a selective value on the basis of a series of samples taken at different times from a single population, assuming that the selective value remains constant. In this case, equation (8.46) can still be used to generate a series $q_1, q_2, q_3, \ldots$ of gene frequencies, starting from an initial value q_0. Let n_{1i} denote the number of A_1A_- individuals found in a sample of size n_i taken before selection in the ith generation. Let n_{2i} denote the number of A_2A_2 individuals in the sample. These frequencies have expected values

$$E(n_{1i}) = (1 - q_i^2) n_i$$

and

$$E(n_{2i}) = q_i^2 n_i,$$

$$\left.\vphantom{\begin{array}{c}a\\a\\a\end{array}}\right\}\qquad (8.50)$$

for maximum likelihood estimation with MAXLIK (see Section A.18 of the Appendix for more details). Taking $u = \log_e(1 - s)$ as the selection parameter will avoid negative selective values of $A_1 A_-$ relative to $A_2 A_2$.

If samples are taken after selection then the only change that needs to be made is to replace equations (8.50) with

$$E(n_{1i}) = (1 - s)(1 - q_i^2) n_i / \{1 - s(1 - q_i^2)\},$$

and

$$E(n_{2i}) = q_i^2 n_i / \{1 - s(1 - q_i^2)\}.$$

$$\left.\vphantom{\begin{array}{c}a\\a\\a\end{array}}\right\}\qquad (8.51)$$

Example 8.5   Maximum likelihood estimation with *Cepaea nemoralis* at Berrow

Consider again the data shown in Table 8.10 on the frequency of dark brown *Cepaea nemoralis* on the Berrow sand dunes. There are ten colonies for which samples are available in both 1926 and 1959/60. Clarke and Murray argued that the generation time for this snail is about three years, so the two samples are about 11.3 generations apart.

Assuming the same selection at all ten locations, and samples before selection, MAXLIK was used to find the maximum likelihood estimate of $u = \log_e(1 - s)$ for the dominance selection model. This estimate is $\hat{u} = -0.069$, with standard error 0.012, corresponding to an estimated selection coefficient of $\hat{s} = 1 - \exp(\hat{u}) = 0.067$. Equation (8.48) provides $\mathrm{SE}(\hat{s}) \simeq \exp(\hat{u}) \, \mathrm{SE}(\hat{u}) = 0.011$. The chi-square goodness of fit statistic output by MAXLIK is 14.21 with 9 degrees of freedom. This is not significantly large at the 5% level. Therefore the model of a constant selective value fits the data quite well. This is remarkable in a way, since the estimation and the test of goodness of fit involve no allowance for the stochastic variation that can be expected from location to location because of finite populations sizes.

If samples are assumed to have been taken after selection then equations (8.47) are replaced by equations (8.49). This makes virtually no difference at all to the estimated selective value or its standard error.

When the same calculations are carried out on the eight locations for which samples are available in both 1959/60 and 1975, a somewhat different result is obtained. The estimate of $u = \log_e(1 - s)$ is $\hat{u} = 0.027$ with standard error 0.038, for samples assumed to be either before or after selection. This provides the estimate $s = -0.027$, with a standard error of approximately 0.039. The chi-square goodness of fit statistic is 35.77 with 7 degrees of freedom, which is significantly large at the 0.1 % level. Thus from 1959/60 to 1975 the locations do not seem to have had the same selective coefficients. Quite apart from this, the low estimate of $s = -0.027$, which is less than two standard errors from zero, gives no indication at all of selection in any consistent direction.

The conclusions from this maximum likelihood analysis of Clarke and Murray's data are almost the same as the conclusions that were reached in Example 8.4 with rather simpler calculations.

A somewhat different application of maximum likelihood involves estimating a selective value for a single population on the basis of a series of samples taken at different times. This can be illustrated by some further data from the *Cepaea nemoralis* populations on the Berrow sand dunes that are provided in Table 8.13 (Clarke and Murray, 1962a, b; Murray and Clarke, 1978). If unbanded and dark brown snails are ignored, then the allele for the mid-banded 00300 condition, A_1 say, is dominant to the alleles for other banding types that can be collectively denoted by A_2. Table 8.13 shows sample frequencies of mid-banded $(A_1 A_-)$ and other banded $(A_2 A_2)$ snails for the years 1926, 1959/60, 1963, 1969 and 1975, for yellow, pink and pale brown snails, for six of the localities studied by Clarke and Murray.

Assuming that the selective value of mid-banded snails relative to other banded snails $(1 - s)$ remained constant from 1926 to 1975, and samples were taken before selection, expected phenotype sample frequencies are given by equations (8.50). Using these with MAXLIK provides the estimates shown in Table 8.14. Estimates based upon equations (8.51) for samples after selection are almost exactly the same. For all locations the constant selection model fits the data fairly well. The estimated value of s is more than two standard errors from zero for locations D48, D52 and D54. Overall it appears that selection was taking place, generally with s being negative so that the mid-banded snails $(A_1 A_-)$ had higher fitness than the other banded $(A_2 A_2)$ snails. The total of

Table 8.13 Data from five colonies of *Cepaea nemoralis* at Berrow in Somerset. Only yellow, pink and pale brown shells have been counted.

Location	1926		1959/60		1963		1969		1975	
	n_1	n_2	n_1	n_2	n_1	n_2	n_1	n_2	n_1	n_2
D40	10	165	13	90	5	50	10	88	11	81
D48	60	345	19	61	11	38	43	148	28	131
D50	14	97	22	103	4	44	28	204	30	192
D52	33	299	41	274	17	82	19	51	8	38
D54	41	189	159	408	61	103	52	140	33	74
D56	19	82	47	239	23	106	21	129	18	92

n_1 = sample frequency of mid-banded (00300) snails
n_2 = sample frequency of other banded snails

Table 8.14 Maximum likelihood estimation of the selective coefficient of mid-banded *Cepaea nemoralis* relative to other band types using the data of Table 8.13.

Location	$\hat{u}$	$SE(\hat{u})$	$\hat{s}$	$SE(\hat{s})$	χ_3^2
D40	0.050	0.026	−0.051	0.027	1.08
D48	0.028	0.013	−0.028	0.013	3.05
D50	0.001	0.021	−0.001	0.021	3.39
D52	0.052	0.018	−0.053	0.019	6.16
D54	0.051	0.015	−0.052	0.016	6.94
D56	−0.016	0.021	0.016	0.021	6.90
				Total	27.52

Note: Almost identical estimates are found assuming samples either before or after selection.

the chi-square goodness of fit values is 27.52 with 18 degrees of freedom. This is nearly significantly large at the 5 % level. Hence, overall, the dominance fitness model gives a fair, but not particularly good fit to the data.

The minimum-variance weighted average of the estimates of s from all six locations is given by equation (A28) of Section A.6 of the Appendix to be $\bar{s} = -0.030$, with standard error 0.0074. The chi-square test statistic (a) of that section is 10.69 with 5 degrees of freedom, which is not quite significantly large at the 5 % level. It seems therefore that the hypothesis that all locations had the same selection coefficient cannot be rejected.

8.15 MAXIMUM LIKELIHOOD ESTIMATION WITH SEVERAL ALLELES

Suppose that there are k alleles $A_1, A_2, \ldots, A_k$ at a single genetic locus. There are then $k(k+1)/2$ genotypes, consisting of k homozygotes $A_1 A_1$, $A_2 A_2, \ldots, A_k A_k$ and $k(k-1)/2$ heterozygotes $A_1 A_2, A_1 A_3, \ldots, A_{k-1} A_k$. Let $p_i(g)$ denote the proportion of genes that are A_i at the start of generation g. Then, assuming random mating, the frequencies of the genotypes will be as shown in Table 8.15. It is expected that a proportion $p_i(g)^2$ of genotypes will be $A_i A_i$. Since $A_i A_j$ and $A_j A_i$ are equivalent, it is expected that the proportion of $A_i A_j$ is $2p_j(g)p_j(g)$, for $i \neq j$.

If the selective value of $A_i A_j$ is w_{ij}, then after selection in the gth generation the relative frequency of $A_i A_i$ will change to $w_{ii}p_i(g)^2$ while the relative frequency of $A_i A_j$ will be $2w_{ij}p_i(g)p_j(g)$. Hence, counting two A_i genes for each $A_i A_i$ individual and one A_i gene for each $A_i A_j$ individual, it follows that the frequency of A_i after selection will be proportional to

$$2w_{ii}p_i(g)^2 + \sum_{j \neq i} 2w_{ij}p_i(g)p_j(g) = 2p_i(g) \sum_{j=1}^{k} w_{ij}p_j(g).$$

Table 8.15 Expected genotype proportions through random mating, which results in a random pairing of two genes.

First gene	Second gene				Total
	A_1	A_2	$\ldots$	A_k	
A_1	$p_1(g)^2$	$p_1(g)p_2(g)$	$\cdots$	$p_1(g)p_k(g)$	$p_1(g)$
A_2	$p_2(g)p_1(g)$	$p_2(g)^2$	$\cdots$	$p_2(g)p_k(g)$	$p_2(g)$
$\vdots$	$\vdots$	$\vdots$		$\vdots$	$\vdots$
A_k	$p_k(g)p_1(g)$	$p_k(g)p_2(g)$	$\cdots$	$p_k(g)^2$	$p_k(g)$
Total	$p_1(g)$	$p_2(g)$	$\cdots$	$p_k(g)$	1

Therefore the expected proportion of A_i genes at the start of generation $g + 1$ is

$$p_i(g+1) = p_i(g) \sum_{j=1}^{k} w_{ij} p_j(g) \bigg/ \sum_{r=1}^{k} p_r(g) \sum_{j=1}^{k} w_{rj} p_j(g). \qquad (8.52)$$

This last equation forms the basis of Du Mouchel and Anderson's (1968) maximum likelihood method for estimating the selective values w_{ij}. Given initial gene frequencies $p_1(0), p_2(0), \ldots, p_k(0)$, and selective values w_{11}, $w_{12}, \ldots, w_{kk}$, it is possible to calculate expected gene frequencies after any number of generations. These can then be used to calculate expected sample frequencies for the different alleles.

Suppose that a population is sampled at times $t_1, t_2, \ldots, t_m$. Then equation (8.52) can be used to generate corresponding gene frequencies. Let the sample at time t_j consist of n_j genes, and assume that it is taken before selection. Let n_{ij} denote the sample number of A_i genes. Then the expected value of n_{ij}, given n_i, is

$$E(n_{ij}) = p_i(t_j)n_j.$$

This is the equation needed for maximum likelihood estimation with MAXLIK.

One very real problem with the estimation of selective values when there are more than two alleles is that the estimates obtained can be very highly correlated. In practical terms this means that the maximum of the likelihood function may be rather badly defined, so that many different combinations of selective values give essentially the same maximum value. As a result, the iterative process used by MAXLIK to determine estimates may not converge. Du Mouchel and Anderson (1968) noted that non-convergence is common with data on four or more alleles, and they used an alternative steepest-ascent maximizing procedure when the normal approach did not work. Another practical problem encountered by Du Mouchel and Anderson was that sometimes the likelihood function was maximized with some selective values being negative. When this happened they searched for a set of non-negative selective values that gave virtually the same value for the likelihood function. This was possible because of the ill-defined maximum of the function.

When a likelihood function has an ill-defined maximum this is a clear indication that the model being estimated is over-parameterized, which suggests that some form of reparameterization, with fewer parameters, will be beneficial. Working in terms of logarithms of selective values has the advantage of not allowing negative selective values; one possibility is to take

$$\log_e(w_{ij}) = \begin{cases} 2u_i, & j = i \\ u_i + u_j + h_{ij}, & j \neq i, \end{cases}$$

where $u_k = 0$. Here u_i is the additive effect of the allele A_i on the logarithm of fitness, while h_{ij} is a heterozygote effect, with $h_{ji} = h_{ij}$. Additive effects on logarithms of selective values are, of course, equivalent to multiplicative effects on selective values themselves. Setting $u_k = 0$ has the effect of making $w_{kk} = 1$ so that all selective values are defined relative to the genotype $A_k A_k$. Parameters can be reduced or eliminated by taking h_{ij} constant or zero.

When mating is at random, the proportion of allele A_i after selection in generation g is expected to be the same as the proportion before selection in generation time $g + 1$. Therefore, as far as genes are concerned, samples can be either before or after selection. However, the situation is obviously not the same when genotypes are involved. Thus, suppose that a population is sampled after selection in generations $t_1, t_2, \ldots, t_m$ and genotype frequencies are recorded. Let the sample at time t_r consist of n_r individuals (not genes), of which n_{ijr} are of genotype $A_i A_j$. Then homozygote $A_i A_i$ has an expected frequency proportional to $w_{ii} p_i(t_r)^2$, while a heterozygote $A_i A_j$ has an expected frequency proportional to $2 w_{ij} p_i(t_r) p_j(t_r)$. It follows that the expected value of n_{ijr} is

$$E(n_{ijr}) = \begin{cases} w_{ii} p_i(t_r)^2 n_r / C_r, & j = i \\ 2 w_{ij} p_i(t_r) p_j(t_r) n_r / C_r, & j \neq i \end{cases}$$

where

$$C_r = \sum_{i=1}^{k} \sum_{j=1}^{k} w_{ij} p_i(t_r) p_j(t_r).$$

This equation provides the expected frequencies that are needed for maximum likelihood estimation using MAXLIK.

8.16   ESTIMATING THE EFFECTIVE POPULATION SIZE

An novel approach for testing for selection was proposed by Krimbas and Tsakas (1971). They noted that if p_{i0} is a sample proportion of allele A_i in generation 0, and p_{it} is a sample proportion in generation t, then, in the absence of selection, the expected value of $(\hat{p}_{it} - \hat{p}_{io})^2$ can be expressed as

$$E\{(\hat{p}_{it} - \hat{p}_{io})^2\} = \text{var}(p_{it}|p_{io}) + \text{var}(\hat{p}_{io}|p_{io}) + \text{var}(\hat{p}_{it}|p_{io}),$$

where p_{io} and p_{it} are the population proportions. Here $\text{var}(p_{it}|p_{io})$ is the variance in p_{it} due to random genetic drift, while $\text{var}(\hat{p}_{ij}|p_{io})$ is the variance for the sampling error involved in estimating $\hat{p}_{ij}$. There are then the approximations

$$\text{var}(p_{it}|p_{io}) \simeq p_{io}(1 - p_{io})\{1 - (1 - 1/N_e)^t\} \simeq p_{io}(1 - p_{io})t/N_e,$$

where N_e is the effective population size in terms of genes. Also, assuming binomial sampling errors, $\text{var}(\hat{p}_{ij}|p_{io}) = p_{io}(1 - p_{io})/n_j$, where n_j is the number of genes sampled in generation j. It follows that

$$E\{(\hat{p}_{it} - \hat{p}_{io})^2\} \simeq p_{io}(1 - p_{io})(t/N_e + 1/n_0 + 1/n_t),$$

so that

$$E\{(\hat{p}_{it} - \hat{p}_{io})^2\}/\{p_{io}(1 - p_{io})\} \simeq t/N_e + 1/n_0 + 1/n_t.$$

If there are k alleles, this gives

$$\frac{1}{k}\sum_{i=1}^{k} E\{\hat{p}_{it} - \hat{p}_{io})^2\}/\{p_{io}(1 - p_{io})\} \simeq t/N_e + 1/n_0 + 1/n_t.$$

Rearranging this equation and replacing unknown values by their estimates then provides

$$\hat{N}_e = t \left/ \left[\frac{1}{k}\sum_{i=1}^{k} (\hat{p}_{it} - \hat{p}_{io})^2/\{p_{io}(1 - \hat{p}_{io})\} - 1/n_0 - 1/n_t \right] \right. \qquad (8.53)$$

as an estimator of the effective population size.

Krimbas and Tsakas used this method to estimate the effective size of a population of the olive fruit fly *Dacus oleae*. They applied it with two different genes that are both highly polymorphic, and obtained almost identical results. This, they suggested, is because changes at both loci were due to genetic drift only. If $\hat{N}_e$ is very different for two loci then this implies that they are being effected by selection in different ways. Krimbas and Tsakas's estimates of N_e were only about one-sixth of the number of genes in the population, but this was considered to be realistic.

Pamilo and Varvio-Aho (1980) examined the validity of equation (8.53) and concluded that it is subject to serious errors. However, this conclusion depends upon the particular assumptions that were made about the way genes are sampled (Nei and Tajima, 1981). According to Pollack (1983), a good estimator of N_e is

$$\hat{N}_e = t \left/ \left\{ \sum_{i=1}^{k} (\hat{p}_{it} - \hat{p}_{io})^2/\bar{p}_i - 1/n_0 - 1/n_t \right\} \right. ,$$

where $\bar{p}_i = (\hat{p}_{it} + \hat{p}_{io})/2$. See Pollack's paper for more details of its derivation, and an equation for variance. See also Tajima and Nei (1984). Pollack provides results that indicate that $\hat{N}_e$ is not strongly affected by selection. This is good as far as estimating the real effective population size is concerned. However, it does of course also mean that Krimbas and Tsakas's idea of detecting selection by variation in $\hat{N}_e$ from locus to locus is unlikely to be successful unless selection is very intense.

8.17   DIFFERENT SELECTIVE EFFECTS ON MALES AND FEMALES

So far in this chapter it has been assumed that selective values are the same for males and females. Clearly this may not always be the case, so it is appropriate to mention that Christiansen, Bungaard and Barker (1977) have considered in some detail the mathematics of situations where there is differential selection on the two sexes. Using Prout's (1965) model of early and late selection, they have reconfirmed his finding that estimating fitness values from samples taken before selection is complete in each generation will give the false impression of frequency-dependent selection.

The most obvious case of a sex difference occurs when the genetic locus being considered is sex linked. Then one of the chromosomes involved determines the sex, often through a system where one sex is homogametic, XX, and the other heterogametic, XY. If there are two alleles A_1 and A_2 at a locus on the X chromosome then in the homogametic sex there are three genotypes A_1A_1, A_1A_2 and A_2A_2; but in the heterogametic sex there are only two, A_1Y and A_2Y. The general dynamics of selection on a sex-linked locus is described by Cook (1971, Ch. 2). Equations for estimating selective values have been provided by Alvarez *et al.* (1983).

8.18   ANALYSIS OF PHENOTYPE FREQUENCIES

There is perhaps no need to be unduly concerned about estimating selective values when analysing data on phenotype frequencies. For example, with Clarke and Murray's (1962a, b) data on dark brown *Cepaea nemoralis* on the Berrow sand dunes, the main interest was in seeing whether there is evidence of time changes that are consistent in different locations. This can be investigated simply by treating the proportion of dark-brown snails as the dependent variable and carrying out a logit regression using GLIM or an equivalent computer program. This will be just as valid as an analysis based on estimating selective values assuming a genetic model that may not be correct. The analysis of *C. nemoralis* morph proportions carried out by Murray and Clarke (1978) was in effect this type of logit regression.

9 Equilibrium gene frequencies at a single locus

Chapter 8 was concerned with how gene frequency changes with time at a single locus can provide evidence of selection. The present chapter is still concerned with analyses based upon a single locus. Now, however, interest turns to the distribution of genes at a single point in time. One motivation for some of the methods that will be discussed is the idea that selection may have quite small effects over the limited periods of time that populations can be studied. Hence trying to detect selection by the methods of Chapter 8 may fail. Nevertheless, small selective effects operating over many generations may result in gene frequency distributions that are inconsistent with a hypothesis of no selection.

After a short introduction, the general problem of estimating gene frequencies from phenotype frequencies is considered. The simplest efficient approach is through gene counting. After gene frequencies have been estimated it is sometimes possible to carry out a goodness of fit test of the genetic model to the data. A significant result on such a test may then indicate selection.

In recent years there has been considerable interest in the extent to which gene frequency distributions of natural populations can be accounted for by the infinite alleles, neutral mutation model for evolution. A test of this model is discussed in Sections 9.8 and 9.9. One problem with the test is that it is not, strictly speaking, applicable to data obtained by the standard technique of electrophoresis. For this reason 'charge-state' models of neutrality have been proposed as the basis for alternative tests. A test of a charge-state model is described in Section 9.11.

Many of the problems associated with studying selection on natural populations can be overcome by sampling mother–offspring combinations at the breeding time for monogamous populations with discrete generations. The chapter concludes with a discussion of this type of data.

9.1 INTRODUCTION

The development of new techniques for detecting genetic variation, particularly electrophoretic methods, has meant that in the last few years it has become much easier to generate large sets of data on present-time gene distributions for many species. This is one reason for the upsurge of interest that there has been in testing whether distributions seem to have been effected by selection. However, that is not the whole story. Along with the development of the new laboratory techniques came the discovery that natural populations are polymorphic at large numbers of loci and that there is much more genetic variation in nature than had been previously thought. As has been discussed in Chapter 1, this led some geneticists to put forward the hypothesis that most genes are selectively neutral. As a result there has been even more interest in testing whether or not observed gene frequency distributions agree with this hypothesis, although, unfortunately, most tests based on electrophoretic data have turned out to be of somewhat doubtful validity.

The simplest approach for looking for evidence of selection involves testing whether a random sample from a population gives phenotype frequencies that are in Hardy–Weinberg ratios. It may be recalled from Section 8.3 that in a large, randomly mating population the frequencies of newly formed zygotes should be in Hardy–Weinberg ratios. In the absence of selection these ratios should be retained to maturity. A test based on this principle requires the estimation of gene frequencies. This will therefore be the first topic discussed in this chapter.

9.2 THE ESTIMATION OF GENE FREQUENCIES

The estimation of gene frequencies is straightforward when all the possible genotypes can be distinguished from each other. In that case, a random sample of n individuals can be thought of as a random sample of $2n$ genes. If this sample contains r_i of allele i, then the sample proportion

$$\hat{p}_i = r_i/2n \tag{9.1}$$

is an unbiased estimator of the population proportion with the usual multinomial variance

$$\text{var}(\hat{p}_i) = p_i(1 - p_i)/2n. \tag{9.2}$$

Also, the multinomial covariance of p_i and p_j is

$$\text{cov}(\hat{p}_i, \hat{p}_j) = -\hat{p}_i \hat{p}_j/2n, \qquad i \neq j. \tag{9.3}$$

When dominance exists, this introduces complications. The first step is to write down the relationships that exist between phenotype and gene proportions. An *ad hoc* approach to estimation then involves equating observed and expected phenotype proportions and solving the resulting equations. The following example will show the general idea.

Example 9.1   *Ad hoc* estimation with the ABO blood grouping system

In the simplest model for the human ABO blood grouping system there are three alleles *a*, *b* and *o* at a single locus. These will be assumed to be in proportions p, q and $r = 1 - p - q$ in a population. Standard testing methods detect the *a* and *b* alleles but cannot distinguish *ao* individuals from *aa*, or *bo* individuals from *bb*, so that *a* and *b* are dominant to *o*. The possible phenotypes and their probabilities are as shown in Table 9.1, based on the assumption of random mating. It follows from this table that for a random individual taken from the population the following equations give the probabilities of having different blood groups:

$$\left.\begin{aligned}
P(\text{A}) &= P(aa) + P(ao) = p^2 + 2pr, \\
P(\text{B}) &= P(bb) + P(bo) = q^2 + 2qr, \\
P(\text{AB}) &= P(ab) \qquad\quad = 2pq,
\end{aligned}\right\} \qquad (9.4)$$

and

$$P(\text{O}) = P(oo) \qquad\quad = r^2,$$

where $P(\text{A}) + P(\text{B}) + P(\text{AB}) + P(\text{O}) = 1$.

Table 9.1 Phenotype–genotype relationships and probabilities for the ABO blood grouping system. The genes *a*, *b* and *o* are assumed to be in proportions p, q and $r = 1 - p - q$ in both male and female adults. The table shows the possible pairs of genes, with their probabilities, and the resulting blood groups.

			Second gene		
			a	*b*	*o*
	a	Probability	p^2	pq	pr
		Blood group	A	AB	A
First	*b*	Probability	pq	q^2	qr
gene		Blood group	AB	B	B
	o	Probability	pr	qr	r^2
		Blood group	A	B	O

The *ad hoc* method for estimating p, q and r involves equating these probabilities to sample proportions of the different blood groups. There are then three independent equations and two unknown independent gene frequencies. Two of the equations can be chosen to be satisfied exactly to determine the unknowns.

This can be illustrated on some data provided by Fisher and Taylor (1940) for Scotland. A sample of 10 969 individuals consisted of 3755 of group A, 1144 of group B, 364 of group AB and 5706 of group O. Substituting these as proportions into equations (9.4) taking $3755/10\,969 = 0.3423$, and so on, gives

$$P(A) = p^2 + 2pr \simeq 0.3423,$$
$$P(B) = q^2 + 2qr \simeq 0.1043,$$
$$P(AB) = 2pq \qquad \simeq 0.0332,$$

and

$$P(O) = r^2 \qquad \simeq 0.5202.$$

(9.5)

From the last equation, $r \simeq \sqrt{0.5202} = 0.7212$. Substituting this into the first equation then gives the quadratic

$$p^2 + 1.4425p - 0.3423 \simeq 0,$$

with two solutions $p \simeq -1.6500$ or 0.2074. Taking the positive root provides $q = 1 - p - r \simeq 1 - 0.2074 - 0.7212 = 0.0714$. The *ad hoc* estimates are therefore $\hat{p} = 0.2074$, $\hat{q} = 0.0714$ and $\hat{r} = 0.7212$. Taking a different pair of equations provides slightly different values. For example, the second and last equations produce $\hat{p} = 0.2098$, $\hat{q} = 0.0690$ and $\hat{r} = 0.7212$.

9.3   MAXIMUM LIKELIHOOD ESTIMATION BY GENE COUNTING

When there are more phenotypes than alleles, as in the ABO model, *ad hoc* estimates will have no general optimum properties. The only good thing about them is that they may be easy to calculate. However, in cases where the number of alleles is equal to the number of phenotypes there will be no choice in the equations used for estimation, and the *ad hoc* estimates will in fact be maximum likelihood estimates.

If necessary, *ad hoc* estimates can be changed to maximum likelihood estimates by making use of an iterative gene counting method that has been developed by Ceppellini *et al.* (1955) and Yasuda and Kimura (1968). The principle behind this method is the allocation of phenotype frequencies to genotypes using the best available gene frequency estimates. The process is illustrated in Example 9.2 below.

Alternatively, maximum likelihood estimates can always be found using the computer program *MAXLIK which is listed in the Appendix. All that is necessary is to write a FREQ subroutine to express phenotype proportions in terms of gene frequencies as in equations (9.4).

Example 9.2   Gene counting with the ABO system

For the ABO blood grouping system, the A group individuals have either genotype *aa*, for which the probability is p^2, or genotype *ao*, for which the probability is $2pr$. Hence a proportion $p^2/(p^2 + 2pr) = p/(p + 2r)$ of the A

Table 9.2 Estimation of gene frequencies for the ABO blood grouping system by gene counting for the Scottish data.

| | | Gene frequencies | | | Allocated genotype frequencies | | | | | | AB | O | Gene counts | | | |
| | | | | | A | | | B | | | | | | | | |
Iteration		p	q	r	aa	ao	Total	bb	bo	Total	ab	oo	a	b	o	Total
Improvement	1	0.207 40	0.071 40	0.721 20	472	3283	3755	54	1090	1144	364	5706	4591	1562	15 785	21 938
of ad hoc	2	0.209 27	0.071 20	0.719 53	477	3278	3755	54	1090	1144	364	5706	4596	1562	15 780	21 938
estimates	3	0.209 50	0.071 20	0.719 30	477	3278	3755	54	1090	1144	364	5706	4596	1562	15 780	21 938
Variance	1	0.209 50	0.071 20	0.719 30	485	3331	3816*	54	1090	1144	364	5706	4665	1562	15 833	22 060
calculations	2	0.211 47	0.070 81	0.717 72	490	3326	3816	54	1090	1144	364	5706	4670	1562	15 828	22 060
	3	0.211 70	0.070 81	0.717 50	491	3325	3816	54	1090	1144	364	5706	4671	1562	15 827	22 060
	4	0.211 74	0.070 81	0.717 45	491	3325	3816	54	1090	1144	364	5706	4671	1562	15 827	22 060
	1	0.209 50	0.071 20	0.719 30	477	3278	3755	56	1122	1178*	364	5706	4596	1598	15 812	22 006
	2	0.208 85	0.072 62	0.718 53	476	3279	3755	57	1121	1178	364	5706	4595	1599	15 812	22 006
	3	0.208 81	0.072 66	0.718 53	476	3279	3755	57	1121	1178	364	5706	4595	1599	15 812	22 006
	1	0.209 50	0.071 20	0.719 30	477	3278	3755	54	1090	1144	383*	5706	4615	1581	15 780	21 976
	2	0.210 00	0.071 94	0.718 06	479	3276	3755	55	1089	1144	383	5706	4617	1582	15 777	21 976
	3	0.210 09	0.071 99	0.717 92	479	3276	3755	55	1089	1144	383	5706	4617	1582	15 777	21 976
	1	0.209 50	0.071 20	0.719 30	477	3278	3755	54	1090	1144	364	5782*	4596	1562	15 932	22 090
	2	0.208 06	0.070 71	0.721 23	473	3282	3755	53	1091	1144	364	5782	4592	1561	15 937	22 090
	3	0.207 88	0.070 67	0.721 46	473	3282	3755	53	1091	1144	364	5782	4592	1561	15 937	22 090

* Phenotype frequency increased by one standard deviation for the calculation of variances and covariances.

group individuals can be assigned to genotype *aa* and the remainder can be assigned to *ao*. Similarly, the B group individuals are either *bb*, for which the probability is q^2, or *bo*, for which the probability is $2qr$. Hence a proportion $q^2/(q^2 + 2qr) = q/(q + 2r)$ of the B group individuals can be assigned to genotype *bb* and the remainder to *bo*. The other two blood groups, AB and O, can be assigned directly to genotypes. Having assigned all individuals to genotypes, a count of genes can be made and hence gene proportions can be determined. The procedure is iterative. To begin with, any values for p, q and r can be used to allocate the phenotype frequencies to genotypes. Improved values are then found by gene counting. These new values can then be used in a further round of improvement. This process can be continued until convergence is obtained.

The first three rows of Table 9.2 show the calculations for Fisher and Taylor's (1940) Scottish data from Example 9.1. (The calculations in the lower rows of the table are concerned with determining variances and they can be ignored for the present.) The initial values taken for p, q and r are 0.2074, 0.0714 and 0.7212, which are the *ad hoc* values obtained previously by solving the first and last of equations (9.5). Using these values, the 3755 A group individuals are allocated as $3755 \times 0.2074/(0.2074 + 2 \times 0.7212) = 472$ *aa* and 3283 *ao*, while the 1144 B group individuals are allocated as $1144 \times 0.0714/(0.0714 + 2 \times 0.7212) = 54$ *bb* and 1090 *bo*. Counting genes then gives 4591 of *a*, 1562 of *b* and 15 785 of *o*, where these add to 21 938, which is twice the number of individuals. Improved values for p, q and r are therefore $4591/21\,938 = 0.209\,27$, $1562/21\,938 = 0.071\,20$ and $15\,785/21\,938 = 0.719\,53$. A further iteration changes these to 0.209 50, 0.071 20 and 0.719 30, which give the same genotype allocations to the nearest integer. This seems to be sufficient accuracy for all practical purposes.

9.4   VARIANCES AND COVARIANCES FOR GENE FREQUENCY ESTIMATORS

Variances and covariances for estimators of population gene frequencies can be determined conveniently using equations (A25) and (A26) of the Appendix, Section A.5, with sample phenotype counts being regarded as having independent Poisson distributions. Thus suppose that $\hat{p}_1, \hat{p}_2, \ldots, \hat{p}_n$ are estimators of the proportions of n alleles that are functions of m phenotype counts $a_1, a_2, \ldots, a_m$. Then

$$\text{var}(\hat{p}_i) \simeq \Delta_{i1}^2 + \Delta_{i2}^2 + \ldots + \Delta_{im}^2, \tag{9.6}$$

and

$$\text{cov}(\hat{p}_i, \hat{p}_j) \simeq \Delta_{i1}\Delta_{j1} + \Delta_{i2}\Delta_{j2} + \ldots + \Delta_{im}\Delta_{jm}, \tag{9.7}$$

where Δ_{ir} is the change in $\hat{p}_i$ resulting from a one standard deviation change in a_r. On the Poisson assumption, a one standard deviation change in a_r is approximately $\sqrt{a_r}$, taking a_r itself as an estimate of its expected value.

(Remember that the variance of a Poisson variate is equal to its expected value.) For reasons that are discussed in Section 2 of the Appendix, these variances and covariances should be close to the values that apply when the phenotype counts are regarded as being multinomially distributed with a fixed total. In other words, the Poisson assumption is not crucial here. It is merely a convenient 'trick' for calculating numerical variances and covariances. On the other hand, the Poisson assumption may be realistic with many sets of data where the total sample size is not fixed before sampling begins.

The use of equations (9.6) and (9.7) can be illustrated on the Scottish ABO blood group data. The relevant calculations are shown in the bottom part of Table 9.2. First, the observed A group frequency is increased from 3755 to $3755 + \sqrt{3755} = 3816$, and new gene proportions are determined by a new round of gene counting. Second, the B group frequency is increased from 1144 to $1144 + \sqrt{1144} = 1178$, and new gene proportions are determined. Then the AB frequency and finally the O frequency are increased. This leads to the following array of Δ_{ij} values where, for example $\Delta_{11} = 0.211\,74 - 0.209\,50 = 0.002\,24$ while $\Delta_{23} = 0.071\,99 - 0.071\,20 = 0.000\,79$:

$$
\begin{bmatrix}
\Delta_{11} & \Delta_{21} & \Delta_{31} \\
\Delta_{12} & \Delta_{22} & \Delta_{32} \\
\Delta_{13} & \Delta_{23} & \Delta_{33} \\
\Delta_{14} & \Delta_{24} & \Delta_{34}
\end{bmatrix}
=
\begin{bmatrix}
0.002\,24 & -0.000\,39 & -0.001\,85 \\
-0.000\,69 & 0.001\,46 & -0.000\,77 \\
0.000\,59 & 0.000\,79 & -0.001\,38 \\
-0.001\,62 & -0.000\,53 & 0.002\,16
\end{bmatrix}
$$

Hence from equation (9.6) $\mathrm{var}(\hat{p}) \simeq 8.47 \times 10^{-6}$ (standard error $\simeq 0.0029$), $\mathrm{var}(\hat{q}) \simeq 3.19 \times 10^{-6}$ (standard error $\simeq 0.0018$), and $\mathrm{var}(\hat{r}) \simeq 10.59 \times 10^{-6}$ (standard error $\simeq 0.0033$). Covariances can also be found using equations (9.7). For example, $\mathrm{cov}(\hat{p}, \hat{q}) \simeq -0.56 \times 10^{-6}$.

Actually, it is unnecessary to determine variances and covariances in this numerical manner for maximum likelihood estimators of the gene frequencies for the ABO model. Explicit variance and covariance equations are given, for example, by Yasuda and Kimura (1968). The advantage of the purely numerical approach is its generality. In the present case, Yasuda and Kimura's formulae produces virtually the same results as equations (9.6) and (9.7).

Equation (A27) of the Appendix, Section 5, can be used to approximate biases of gene frequency estimators if this is required.

9.5 TESTS FOR HARDY–WEINBERG GENE FREQUENCY RATIOS

The phenotype probabilities of equations (9.4) for the ABO blood group model are based upon the assumption that genotypes are in Hardy–Weinberg ratios. This assumption can be tested by seeing whether the observed phenotype frequencies are in agreement with the expected frequencies that are obtained by substituting estimated gene proportions into these equations. A test is possible because there are more phenotypes than alleles. In cases where the number of phenotypes is equal to the number of alleles, no test is possible because the estimation process will ensure exact agreement between observed and expected phenotype frequencies. If there are more alleles than phenotypes, it is not even possible to estimate the gene proportions.

The general testing procedure involves three stages. Thus suppose that there are n alleles at a single locus, giving rise to m phenotypes, with $m > n$. To begin with, the gene proportions $p_1, p_2, \ldots, p_n$ are estimated. The estimates must be maximum likelihood ones, or their equivalents. *Ad hoc* estimates obtained by equating some of the observed phenotype frequencies to their expected values may not be sufficiently accurate for this purpose. The second stage in the test is the calculation of expected phenotype frequencies $E_1, E_2, \ldots, E_m$ to compare with the observed ones $O_1, O_2, \ldots, O_m$. The third part of the test involves calculating the chi-square statistic

$$\chi^2 = \sum_{i=1}^{m} (O_i - E_i)^2/E_i,$$

with $m - n - 2$ degrees of freedom. The number of degrees of freedom is the usual 'number of frequencies $- 1 -$ number of estimated parameters'. There are $n - 1$ estimated parameters since the last gene frequency can be written as $p_n = 1 - p_1 - p_2 - \ldots - p_{n-1}$. A significantly large value of χ^2 is evidence against the hypothesis that the gene frequencies are in Hardy–Weinberg ratios. For the chi-square approximation to be valid, the expected frequencies should not be very small. Preferably they should all be five or more.

The expected phenotype proportions are related to genotype proportions in the manner illustrated in Table 9.1 for the ABO blood grouping system. In general, if there are n alleles $A_1, A_2, \ldots, A_n$ with proportions $p_1, p_2, \ldots, p_n$, then the Hardy–Weinberg condition says that the proportion of the genotype $A_i A_i$ is p_i^2, while the proportion of the genotype $A_i A_j$ is $2p_i p_j$, $i \neq j$. Phenotype proportions are obtained by adding up the Hardy–Weinberg genotype proportions, taking into account any dominance relationships.

There are a number of possible explanations for genotype frequencies not being in Hardy–Weinberg ratios. An obvious one is non-random mating. Other explanations are migration, differential survival of different genotypes, misclassification of phenotypes, or a small population size. If data from several different subpopulations are added together then there will tend to be an apparent excess of heterozygotes. This is called the Wahlund effect (Spiess, 1977, p. 351; Wallace, 1981, p. 165).

Example 9.3   Hardy–Weinberg test with the ABO blood group

As an illustration, consider again Fisher and Taylor's (1940) Scottish ABO blood group data. The observed numbers of A, B, AB and O individuals are

$n_A = 2755$, $n_B = 1144$, $n_{AB} = 364$ and $n_O = 5706$. Maximum likelihood estimates of the frequencies of the a, b and o alleles have been found by gene counting to be $\hat{p} = 0.2095$, $\hat{q} = 0.0712$ and $\hat{r} = 0.7192$ on the basis of these data. These estimates may be substituted into equations (9.4) to get expected proportions in the different blood groups. Multiplication by the total sample size of 10 969 then gives expected frequencies to compare with the observed frequencies. Thus, with an obvious notation

$$E_A = 10969 \times (\hat{p}^2 + 2\hat{p}\hat{r}) = 3786.9,$$
$$E_B = 10969 \times (\hat{q}^2 + 2\hat{q}\hat{r}) = 1179.0,$$
$$E_{AB} = 10969 \times (2\hat{p}\hat{q}) \qquad = 327.1,$$

and

$$E_O = 10696 \times (\hat{r}^2) \qquad = 5673.7.$$

The chi-square value is then

$$\chi^2 = \frac{(2755 - 3786.9)^2}{3786.9} + \frac{(1144 - 1179.0)^2}{1179.0} + \frac{(364 - 327.2)^2}{327.2} +$$

$$+ \frac{(5706 - 5673.7)^2}{5673.7} = 5.63,$$

with 1 degree of freedom.

This test statistic is significant at the 5 % level but not the 1 % level. There is therefore some evidence of a disturbance from the Hardy–Weinberg frequencies for these data. Fisher and Taylor remark: 'Systematic errors, not all of which are yet understood, do undoubtedly affect the rarest of the four blood groups (AB).' Subsequently it has been discovered that there is more than one version of the A allele and this may well have contributed to some misclassifications. The significant result was not seriously considered by Fisher and Taylor to be due to selection.

9.6 HARDY–WEINBERG TESTS WITH TWO ALLELES

Hardy–Weinberg tests with two alleles have received particular attention. Thus suppose that there are two alleles A_1 and A_2 at a single locus, so that according to the Hardy–Weinberg principle the three genotypes $A_1 A_1$, $A_1 A_2$ and $A_2 A_2$ are in ratios $p^2 : 2pq : q^2$, where $q = 1 - p$ is the population proportion of allele A_1. All genotypes must be distinguishable.

The ordinary chi-square test of Section 9.5 can be used to compare sample frequencies with the Hardy–Weinberg expectations. The first step is the estimation of the gene frequency p. The estimate can then be used to calculate expected frequencies to compare with the observed ones. For example, suppose that a sample gives 188 of genotype $A_1 A_1$, 29 of genotype $A_1 A_2$, and 6 of genotype $A_2 A_2$. The total of 223 individuals have 446 genes, of which $2 \times 188 + 29 = 405$ are A_1 and the remaining 41 are A_2. Therefore $\hat{p} = 405/446 = 0.908$. This provides the following expected genotype frequencies in the sample: $A_1 A_1$, $223 \, \hat{p}^2 = 183.9$; $A_1 A_2$, $223 \times 2\hat{p}(1 - \hat{p}) = 37.3$; and $A_2 A_2$, $223(1 - \hat{p})^2 = 1.9$. The value of chi-square is then

$$\chi^2 = (188 - 183.9)^2/183.9 + (29 - 37.3)^2/37.3 + (6 - 1.9)^2/1.9 = 10.79,$$

with 1 degree of freedom, which corresponds to a significance level of about 0.1%. There is very strong evidence against the Hardy–Weinberg hypothesis.

The large χ^2 value is mainly due to the difference between the observed and expected frequency of $A_2 A_2$. There is, of course, a certain amount of doubt about the validity of the usual chi-square test with an expected frequency as small as 1.9. For small-sample situations like this, various alternatives to this simple chi-square test have been investigated. Elston and Forthofer (1977) consider six different possibilities, including an exact test that is due to Haldane (1954b). It seems that the usual chi-square test will give approximately the correct significance level providing that the total number of individuals in the sample is more than about 150. For sample sizes up to 200, Vithayasai (1973) has tabulated exact significance levels.

For sample sizes of 20 or more, Elston and Forthofer describe a fairly simple procedure for obtaining a good approximation to the exact significance level without the need for extensive calculations. Without going into details, it can be said that this procedure gives a significance level of 0.2% on the data that have just been considered. According to Haldane (1954b), the exact significance level for these data is 0.5%. For this example the usual chi-square test without any adjustments has given essentially the same result as Elston and Forthofer's procedure.

It is often assumed that in the absence of selection the probability of the observed number of heterozygotes exceeding the expected number is 1/2. Given a series of independent samples from p different populations, this suggests that a test for selection can be made by seeing whether the number of cases of heterozygote excess is significantly different from $p/2$. Majumder and Chakraborty (1981) have examined this idea. They show that for small samples the probability of heterozygote excess can be somewhat different from 1/2 and that this 'obvious' test may therefore not be valid. However, they provide a table of probabilities of heterozygote excess, and show how a test can be carried out taking true probabilities into account. See their paper for more details.

9.7 LIMITATIONS OF HARDY–WEINBERG TESTS

Although tests for Hardy–Weinberg frequencies may be of value for detecting selection, they are of little use when it comes to showing the direction of effects.

An example will illustrate the difficulties. It will be supposed that a large population of adults has a proportion $p = \frac{1}{2}$ of A_1 alleles and mates at random. The progeny are then initially in Hardy–Weinberg proportions, but this is altered by differential survival as follows:

	$A_1 A_1$	$A_1 A_2$	$A_2 A_2$	p
Initial number of progeny	100	200	100	0.50
Survival rate	0.5	0.8	1.0	–
Survivors	50	160	100	0.42

Taking the p value for survivors and working out Hardy–Weinberg frequencies gives these to be 54.5 for $A_1 A_1$, 151.0 for $A_1 A_2$, and 104.5 for $A_2 A_2$. The chi-square value is then

$$\frac{(50 - 54.5)^2}{54.5} + \frac{(160 - 151.0)^2}{151.0} + \frac{(100 - 104.5)^2}{104.5} = 1.10,$$

which is nowhere near significant at the 5 % level, even though strong selection has taken place. Furthermore, the difference between the observed and expected frequencies for $A_2 A_2$ indicate a lack of this genotype even though it had the highest survival rate of all.

It was Lewontin and Cockerham (1959) who first explained why the testing of sample frequencies can give this type of result. They showed that if the survival rates of $A_1 A_1$, $A_1 A_2$ and $A_2 A_2$ individuals are w_1, w_2 and w_3, respectively, from the time of formation until the time of sampling, then there will be an apparent excess of heterozygotes if $w_1 w_3 < w_2^2$ and an apparent lack if $w_1 w_3 > w_2^2$. If it happens that $w_1 w_3 = w_2^2$ then the genotype frequencies will be in Hardy–Weinberg proportions no matter how strong the selection is. Lewontin and Cockerham concluded that tests for Hardy–Weinberg frequencies are mainly of value in detecting situations where the heterozygote is at an advantage or at a disadvantage compared with both homozygotes. They show, for example, that if $w_1 = w_3 = 0.5$ while $w_2 = 1.0$, then a sample size of 95 will have an 0.9 probability of providing a 5 % significant chi-square test. However, if $w_1 = 0.5$, $w_2 = 1.0$ and $w_3 = 1.6$ then a sample of about 4000 is required for the same power.

9.8   THE EWENS–WATTERSON TEST

According to the infinite alleles model for neutral mutations, every gene mutation gives rise to a completely new allele which is functionally equivalent to all other alleles (Kimura and Crow, 1964). Thus, at any particular locus there are an infinite variety of possible allelic types and the current distribution is purely the result of chance events that have made some more frequent than others. Using this model, Ewens (1972) considered the information that can be gained from a random sample of $n/2$ individuals, with n genes, where k alleles are observed to have frequencies $n_1, n_2, \ldots, n_k$, so that $\Sigma n_i = n$. He showed that the probability function for the sample frequencies, conditional upon the observed values of n and k, is

$$p(n_1, n_2, \ldots, n_k) = 1 \left/ \left\{ B(k, n) \prod_{i=1}^{k} n_i \right\} \right. \tag{9.8}$$

providing that sufficient time has elapsed for the population to reach equilibrium. In this equation,

$$B(k, n) = \sum 1 \left/ \left(\prod_{i=1}^{k} n_i \right) \right. \tag{9.9}$$

where the summation is over the possible values of the n_i, subject to the restrictions that $n_i > 0$ for all i and $\Sigma n_i = n$.

The Ewens–Watterson test is based upon this distribution. The idea is to summarize the data in terms of the sample 'homozygosity',

$$\hat{F} = \sum_{i=1}^{k} n_i^2/n^2, \tag{9.10}$$

and see whether the observed value is within the likely range for the infinite alleles model. Note that this is equivalent to what is usually called 'Simpson's index' in population biology.

The smallest possible value of $\hat{F}$ is $F_{\min} = 1/k$ which occurs when all alleles have the same frequency $n_i = n/k$. Small values of $\hat{F}$ indicate that allele frequencies are too 'even', which suggests that there is a tendency for selection in favour of heterozygotes in the population. The maximum possible value of $\hat{F}$ occurs when all except one allele are observed once only, so that $n_1 = n - k + 1$, $n_2 = n_3 = \ldots = n_k = 1$. This largest value is $F_{\max} = \{(n-k+1)^2 + (k-1)^2\}/n^2$, which must be less than or equal to one. It equals one if $k = 1$. High values of $\hat{F}$ indicate that heterozygotes are selected against in the population.

Since the distribution of $\hat{F}$ is difficult to calculate exactly, it has been a common practice to determine it by simulation. A large number of sample values of $n_1, n_2, \ldots, n_k$ are generated on a computer from the distribution (9.8) and $\hat{F}$ values determined. The empirical percentage points of the $\hat{F}$ distribution are then used in place of the exact ones for determining significance.

Watterson (1978a) analysed some data originally due to Singh et al. (1976) using this approach. In this study a total of 27 alleles were found in 146 genes from the xanthine dehydrogenase locus of Drosophila pseudoobscura using four electrophoretic conditions. The alleles had the following frequency distribution using Watterson's notation:

$$1^{10}, 2^3, 3^7, 5^2, 6^2, 8^1, 11^1, 68^1.$$

This means that ten alleles were seen once each, three alleles were seen twice each, and so on, up to one allele which was seen 68 times. The sample homozygosity is calculated from equation (9.10) to be

$$\hat{F} = (10 \times 1^2 + 3 \times 2^2 + 7 \times 3^2 + 2 \times 5^2 + 2 \times 6^2 + 8^2 + 11^2 + 68^2)/146^2$$

$$= 0.2353.$$

Watterson generated 2000 samples with $n = 146$ and $k = 27$ and found only eight of these (0.4 %) to have $\hat{F}$ values as large as 0.2353. Thus on a two-sided test the observed value is significantly far from the centre of the distribution at about the 0.8 % level. This is not a value that can be expected from the infinite alleles, neutral mutation model: there is clear evidence against it. Selection against heterozygotes is indicated.

There is a certain difficulty with high values of $\hat{F}$ that has been noted by Ewens (1977a, b). This is that the most likely distributions of allele frequencies under the distribution (9.8) may happen to give very high $\hat{F}$ values. Ewens quotes the case where $n = 500$ and $k = 7$ where the highest probability occurs for $n_1 = n_2 = n_3 = 1, n_4 = n_5 = 2, n_6 = 3$ and $n_7 = 490$. This looks very much like a case where the seventh allele has a selective advantage in comparison with the others. For this reason he suggests that the test can only be used successfully to test for low $\hat{F}$ values. However, this argument does not invalidate a test of significance which is only based upon the idea that a value which is significantly large at the α % level will only occur α % of the time for the distribution (9.8). Selection must change this frequency, although it may not change it very much.

What is more of a problem is the assumption of an equilibrium distribution for allele numbers. It is known that equilibrium is approached at a very slow rate. For a population of size N it will take something in the order of N generations for this to be reached, starting from a situation with a single allele at a locus (Ewens and Gillespie, 1974). 'Bottlenecks' in population numbers can also strongly affect allele distributions (Ewens, 1977a, b). It seems clear that for natural populations the assumption of equilibrium will usually be something of an act of faith.

Another even more serious problem is the assumption that all alleles can be identified. This is not generally true for data obtained by electrophoresis for reasons that are explained in Section 9.10 below. In the example quoted above for the xanthine dehydrogenase locus of D. pseudoobscura, Singh et al. (1976) found 27 alleles for 146 genes using four electrophoresis conditions. An additional heat-sensitivity test revealed that in fact at least 37 alleles were present. Using a different statistic from that of equation (9.10), Ewens and Gillespie (1974) concluded that non-identification of alleles has only moderate effects on the sample homozygosity. The main effect is to increase the evenness of allele frequencies. Therefore high values of F are still evidence against the hypothesis of neutrality. However, this conclusion is based upon the 'charge-state' model for electrophoretic data, which is thought nowadays to be of doubtful validity. Clearly, the results of applying the Ewens–Watterson test to electrophoretic data must be viewed with some scepticism. A significant result could disappear if all the alleles could be identified.

These problems may make it seem that the Ewens–Watterson test has little value at all. This is not completely the case, as will be seen in Example 9.4. Furthermore, the distribution (9.8) does have two important positive qualities (Ewens, 1979a). Firstly, it is remarkably robust in that it arises almost unavoidably for any model of neutral mutations of the infinite allele type. This is important since the 'true' model is not known. Secondly, it is not strongly affected by the existence of some deleterious alleles that are unimportant to evolution. The point here is that it is generally accepted that at least some deleterious alleles exist but that these are eliminated soon after they arise, and therefore are not usually seen. The hypothesis that people really want to test is the one of generalized neutrality: an allele is either selectively neutral or it is deleterious. Ewens (1979a) found that, if anything, the effect of deleterious alleles is to make a test based upon the $\hat{F}$ statistic somewhat conservative.

It is, of course, rather inconvenient to have to simulate many samples in order to determine percentage points for the distribution of $\hat{F}$. Unfortunately, this seems to be the simplest way to test for significance. Watterson (1977, p. 808) gives formulae for the moments of the distribution but these are not easy to calculate. Watterson (1978b) has also provided a table of simulated percentage points for values of k up to 10 and n up to 2000. An extended version of this table is given here as Table 9.3. Ewens (1979b, Appendix C) provides 2.5%, 5% and 97.5% points for $\hat{F}$ for k up to 30.

If resort is made to simulating the distribution of $\hat{F}$ it is important to have an efficient method of doing this. A FORTRAN computer program based on an algorithm proposed by Stewart (1977) is provided in Section A.19 of the Appendix.

9.9 THE EWENS–WATTERSON TEST ON A SUBDIVIDED POPULATION

Using computer simulation, Ewens and Gillespie (1974) have examined the effect of population subdivision on the distribution of sample homozygosities. They considered a population of 250 individuals divided into five subpopulations of 50. It was found that if 2% or more of the genes in a subpopulation in each generation come from the other subpopulations, then the population can be regarded as being effectively the same as a single population of 250 for the purpose of testing for neutrality.

Slatkin (1982) also used simulation to consider the effect of population

Table 9.3 Percentage points for the sample homozygosity (multiplied by 100) as determined from 1000 simulated samples.

k	n	Min.	1	2½	5	10	20	30	40	50	60	70	80	90	95	97½	99	Max.
2	50	50	50	50	51	52	60	68	76	82	89	92	96	96	96	96	96	96
	100	50	50	50	51	53	62	72	80	87	92	96	98	98	98	98	98	98
	200	50	50	50	51	54	62	73	83	90	94	96	98	99	99	99	99	99
	500	50	50	50	51	55	67	79	88	94	96	98	99	100	100	100	100	100
	1000	50	50	51	51	56	68	81	91	95	98	99	99	100	100	100	100	100
	2000	50	50	51	53	60	74	86	93	97	99	99	100	100	100	100	100	100
3	50	33	34	35	39	43	48	51	56	61	67	73	78	89	92	92	92	92
	100	33	34	36	39	45	49	54	60	66	75	82	89	92	94	96	96	96
	200	33	35	38	41	45	50	55	64	71	79	85	90	95	97	97	98	98
	500	33	35	39	43	48	52	59	67	75	84	90	95	98	98	99	99	99
	1000	33	36	41	45	50	54	63	72	80	88	92	96	98	99	99	100	100
	2000	33	37	43	48	50	57	65	76	85	91	96	98	99	100	100	100	100
4	50	25	28	31	32	34	38	41	45	48	53	60	67	75	81	85	88	88
	100	25	28	31	34	37	42	46	49	54	59	67	75	85	89	90	94	94
	200	25	28	31	34	37	43	47	51	57	65	73	81	89	92	95	96	97
	500	25	31	34	36	40	46	51	56	63	71	79	87	93	96	97	98	99
	1000	25	32	34	37	42	48	52	59	66	74	83	89	94	97	98	99	99
	2000	25	33	36	39	44	50	55	62	70	79	86	92	96	98	99	100	100

k	n																	
5	50	20	24	25	26	28	31	34	37	40	44	48	54	64	71	75	81	85
	100	20	25	27	29	32	35	39	43	46	50	56	63	73	80	85	87	92
	200	20	27	28	31	34	38	42	46	50	56	62	70	80	88	91	93	96
	500	20	28	31	33	36	41	45	50	55	62	69	78	87	91	94	96	98
	1000	20	29	30	34	38	44	48	52	58	65	73	82	90	94	96	98	99
	2000	20	29	32	34	38	45	49	55	63	70	78	85	93	96	98	99	100
7	50	14	18	19	20	22	24	26	28	30	32	36	40	49	55	60	65	78
	100	14	20	21	22	24	27	30	32	34	37	41	46	56	65	70	78	88
	200	14	20	22	24	26	29	32	35	40	43	48	55	65	73	78	83	94
	500	14	22	24	25	28	33	36	40	44	48	55	63	74	80	86	89	98
	1000	14	22	24	26	29	33	37	42	46	51	59	68	78	84	89	92	99
	2000	14	24	26	28	32	36	40	44	49	54	62	70	81	89	93	96	99
10	50	10	13	14	14	15	17	18	19	21	22	24	28	32	35	39	45	68
	100	10	15	16	17	18	20	22	24	25	27	30	34	38	43	48	56	83
	200	10	16	17	18	19	22	24	26	28	31	34	39	47	55	61	69	91
	500	10	17	18	19	21	24	27	29	32	35	40	45	55	65	72	78	96
	1000	10	17	19	21	23	26	29	32	35	39	44	50	61	70	75	84	98
	2000	10	17	19	22	25	29	32	35	39	42	48	56	68	76	83	90	99

k = number of alleles n = sample size (genes)

In some cases the percentage points are equal for different percentages because of repeated values. For example, with $k = 2$ and $n = 50$ the maximum possible homozygosity of 0.96 was obtained in 225 of the 1000 simulated samples. This means that the percentage points from 78% to 100% are all equal to 0.96. The percentage point for $P\%$ is defined as the value such that $P\%$ of the sample values are less than or equal to it.

subdivision. However, he considered much lower migration rates than those used by Ewens and Gillespie. He noted that mutation and migration are similar in some ways and that consequently it may be useful to regard a subdivided population with little migration among local populations as a collection of independent populations, each with an effective mutation rate representing the introduction of new alleles by both mutation and migration. On this basis the Ewens–Watterson test can be carried out separately on each subpopulation and the test results combined in some way to determine the overall significance.

Slatkin's computer simulations involved migration rates of the same order of magnitude as mutation rates. There were ten subpopulations with 128 individuals in each, with the number of immigrant genes in a subpopulation ranging from about 0.1 % to 0.4 % per generation. The subpopulations were set out either along a line or as a 5×2 array. In all cases, treating the subpopulations as being independent was a good approximation.

Taking Ewens and Gillespie's (1974) results together with Slatkin's (1982) results suggests that, in general, data from different subpopulations can be pooled providing that a moderate or high amount of migration has taken place in the past. However, if migration has been low, so that for a particular subpopulation this has not been much more important than mutation for the introduction of new alleles, then the subpopulations can be treated as being independent. Of course, determining migration rates may not be easy. Slatkin (1981) has proposed a method for estimating them based upon allele distributions in subpopulations.

Slatkin (1982) also considered the power of the Ewens–Watterson test to detect selection with three models for mutant genes. He found that with his subdivided populations the Ewens–Watterson test has reasonable power to detect some types of selection but not others.

The method of combining test results that was used by Slatkin (1982) involved classifying the probability levels of sample homozygosities into the ranges (0, 0.05), (0.05, 0.1), (0.1, 0.5) and (0.5, 1.0) and then seeing whether the frequencies in the intervals are significantly different from their expected values, assuming no selection. In the following examples, Fisher's method (Appendix, Section A.15) is used instead. Intuitively, Fisher's method seems likely to be more powerful than Slatkin's method.

Example 9.4   Testing inversion polymorphisms
in *Drosophila pseudoobscura*

An interesting application of the Ewens–Watterson test is described by Haymer and Hartl (1981). Following Hartl *et al.* (1980), they avoided the problem of the identification of alleles by working with chromosome data from samples of *Drosophila pseudoobscura*. Chromosome inversions were treated as 'alleles', this use of the Ewens–Watterson test being justified on two grounds. Firstly, chromosomes segregate during meiosis in the usual Mendelian fashion and males and females heterozygous for chromosome inversions produce gametes bearing each type in an approximately 1:1 ratio. Secondly, recombination between inversions is extremely rare so that chromosomes retain their integrity except for occasional 'mutations'.

The observed distributions used by Haymer and Hartl were for third-chromosome polymorphisms. The data were originally reported by Dobzhansky and Epling (1944). Table 9.4 shows the results for Keen Camp, California. Here there were five collecting sites, with samples taken in 1939 and 1940. An ordinary chi-square test for differences between chromosome proportions at different locations gives a test statistic of 23.77 with 12 degrees of freedom for 1939. This is significantly large at the 5% level. There is therefore some evidence of site differences in that year. However, for 1940 the same test (ignoring the small number of Santa Cruz chromosomes) gives a chi-square value of 13.34 with 12 degrees of freedom. This is far from being significantly large so that there is no evidence of site differences in that year.

Sample homozygosities are shown in Table 9.4, together with their

Table 9.4 Third-chromosome karyotype frequencies for *Drosophila pseudoobscura* at five collecting sites at Keen Camp, on Mount San Jacinto, California.

	Site	ST	AR	CH	TL	SC	Sample size	Homozygosity $\hat{F}$	Corresponding percentage point*
1939	A	140	137	147	21	0	445	0.305	1.0%
	B	143	119	174	17	0	453	0.465	2.0%
	C	90	125	174	11	0	400	0.338	3.5%
	D	94	120	141	15	0	370	0.317	2.5%
	E	85	102	125	6	0	318	0.329	2.5%
	Total	552	603	761	70	0	1986		
1940	A	198	132	220	15	3	568	0.326	5.0%
	B	118	87	162	15	0	382	0.329	2.5%
	C	114	87	168	15	0	384	0.332	2.5%
	D	123	108	167	19	1	418	0.315	4.0%
	E	199	124	281	25	1	630	0.339	5.0%
	Total	752	538	998	89	5	2382		

Chromosome types are standard (ST), arrowhead (AR), chiricahua (CH), tree line (TL), and Santa Cruz (SC).
* Approximate probability of obtaining a homozygosity this low, or lower, as found by interpolating in Table 9.3.

percentage points found by interpolating in Table 9.3. All except one of the homozygosities are significantly low at about the 5% level.

An overall assessment of the significance of the 1939 results can be made in two ways. If it is assumed that there was a good deal of population interchange between the five sites, then the samples can be pooled to give a total sample of size $n = 1986$ with homozygosity $\hat{F} = 0.318$ and $k = 4$ alleles. From Table 9.3 this is significant at about the 1% level on a two-sided test.

Alternatively, if it is assumed that the five sites were very isolated, then each site can be treated as giving independent data. The site results can then be combined using Fisher's method (Appendix, Section A.15). The general idea here is that if n independent significance tests result in percentage points $p_1, p_2, \ldots$, and p_n for a test statistic, then the overall significance of the tests can be determined by treating

$$X^2 = -2 \sum_{i=1}^{n} \log_e(p_i) \qquad (9.11)$$

as a chi-square variate with $2n$ degrees of freedom. If the p_i values tend to be low, then X^2 will be high; if the p_i values tend to be high, then X^2 will be low. A two-sided test for X^2 is appropriate if high or low p_i values are considered to be evidence of selection.

In the present case, Table 9.4 shows that for 1939 the probabilities of sample homozygosities as low as those observed for sites A to E were $p_1 \simeq 0.010$, $p_2 \simeq 0.020$, $p_3 \simeq 0.035$, $p_4 \simeq 0.025$ and $p_5 \simeq 0.025$. Hence

$$X^2 = -2(\log_e 0.010 + \log_e 0.020 + \log_e 0.035 + \log_e 0.025 + \log_e 0.025)$$

$$= 38.49.$$

Regarded as a chi-square variate with 10 degrees of freedom, this is significant at the 0.1% level on a two-sided test.

It seems that for the 1939 data, regarding the five collecting sites as being from a single population results in a sample homozygosity that is significant at the 1% level. On the other hand, regarding the sites as being independent results in overall significance at the 0.1% level.

For 1940, pooling the data from the five sites gives $\hat{F} = 0.328$ with $n = 2382$ and $k = 5$. The homozygosity is then significant at about the 5% level on a two-sided test. Alternatively, treating the sites as being independent gives $X^2 = 33.18$ with 10 degrees of freedom, which is significant at the 0.1% level on a two-sided test. Hence for 1940 the overall results are significant at something between the 5% and 0.1% levels.

In this particular example it seems clear that it is best to treat the five collecting sites as providing a sample from a single population. Presumably these sites were quite close together in the general area of Keen Camp so that there must have been a good deal of migration between them. Indeed, as mentioned above, the chromosome proportions for the different sites are not significantly different in 1940. Therefore the overall significance of the results is best taken as being at the 1% level in 1939 and 5% in 1940. The low homozygosities that have been found suggests that there is a tendency for selection in favour of heterozygotes in the population.

In their analysis of the Keen Camp data, together with data from other locations, Haymer and Hartl (1981) combined the results for 1939 and 1940 for an overall test of significance. This does not seem to be justifiable since nobody has investigated the effect of pooling samples taken from a single population at different times. Thus all that can be said with regard to the two years is that they have produced consistent results.

Example 9.5 Testing allele distributions
for *Plethodon dorsalis*

The example used by Slatkin (1982) to illustrate his method for testing for neutrality in a subdivided population involved 14 loci from populations of the salamander *Plethodon dorsalis*. The data were originally published by Larson and Highton (1978). Slatkin notes that the electrophoretic techniques used to obtain them probably do not distinguish all alleles and that therefore no definite conclusions are possible from the Ewens–Watterson test. Still the example is of some interest in demonstrating how calculations are carried out.

Here it will be sufficient to consider data from just two of the 14 loci. Table 9.5 shows Larson and Highton's results for the esterase locus (excluding those for samples of less than 50 genes). Slatkin found evidence of selection in

Table 9.5 Allele distributions for the esterase locus for samples from different populations of the salamander *Plethodon dorsalis* in the Southern Blue Ridge Mountains.

Sample	*Observed number of alleles* a	c	e	g	h	i	k	*Samples size* n	*Homozygosity* $\hat{F}$	*Approx. corresponding percentage point*[*]
5	65		5					70	0.867	55%
6		54	10					64	0.736	35%
8		24		28		8		60	0.396	6.0%
12			50	10				60	0.722	33%
14				12			48	60	0.680	28%
15	26	15	19					60	0.351	2.5%
16	53	1						54	0.964	80%
18			15		40	23		78	0.387	5.0%
20					37	31		68	0.504	2.5%
21					3	47		50	0.887	60%

[*] Found by interpolation in Table 9.3.

Table 9.6 As for Table 9.5 except that the distributions are for the 6-phosphogluconate dehydrogenase (6-Pgd) locus.

| | Observed number of alleles | | | | Sample size | Homozygosity | Approx. correspondinging percentage |
Sample	a	b	c	e	n	$\hat{F}$	point
5	68		2		70	0.944	70%
6	38		26		64	0.518	10%
8		11	49		60	0.701	30%
12			40	20	60	0.556	15%
14			20	40	60	0.556	15%
15			30	30	60	0.500	2.5%
16				54	54	–	–
18			78		78	–	–
20			68		68	–	–
21		2	40	8	50	0.667	60%

this case. Table 9.6 shows sample results for the 6-Pgd locus, where Slatkin found no evidence of selection.

The populations sampled cover a fairly wide geographical area. According to Slatkin (1981), migration is only at a low level. Consequently, it is appropriate to treat each sample as being independent for the purpose of analysis. On this basis, equation (9.11) can be used to test the overall significance of the data. For the esterase locus alone, the chi-square test statistic is 35.90 with 20 degrees of freedom. This is significantly large at the 5 % level on a two-sided test. For the 6-Pgd locus, the chi-square test statistic is 23.71 with 14 degrees of freedom, which is just significant at the 5 % level.

A test for no selection at both the esterase and the 6-Pgd loci is given by using the total of the chi-square values for each locus separately. This is 35.90 + 23.71 = 59.61 with 20 + 14 = 34 degrees of freedom, which is significant at the 1 % level. The results from the two loci have reinforced each other because in both cases the sample homozygosities have tended to be rather low. Of course, care must be taken with this type of combination of tests to ensure that significantly low chi-square values from certain loci (indicating high homozygosities) and significantly high chi-squared values from other loci (indicating low homozygosities) do not cancel each other out. The test is only sensitive if all homozygosities tend to be high or all homozygosities tend to be low.

As mentioned before, no firm conclusions can be reached with the present data. However it seems that sample homozygosities are low for both the esterase and the 6-Pgd loci, indicating selection in favour of heterozygotes. It may be recalled from the discussion in Section 9.8 that the likely effect of not distinguishing all alleles is to give a low homozygosity. Consequently, this may be the true explanation for the significant test results.

9.10   THE NATURE OF ELECTROPHORETIC DATA

It has been pointed out several times that a major problem with using the Ewens–Watterson test is that it relies upon the assumption that all alleles can be identified, although this is probably not true for most data obtained by electrophoresis. This has led to considerable interest in the development of models that take into account the special nature of these data.

There is no need here to describe the details of electrophoretic techniques. An elementary treatment is given, for example, by Wallace (1981, Ch. 4). For the present purpose it is sufficient to note that the results obtained from testing a single individual are of the form shown in Fig. 9.1. Thus different alleles at a

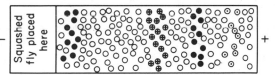

Protein molecules move different distances towards the positive pole according to their electrical charges.

'Developers' can be used to detect particular enzyme genes. The fly is a homozygote for this gene, with two ⊕ alleles.

For this gene the fly is a heterozygote having two ● alleles with different charges.

For this gene the fly is a homozygote having two ⊙ alleles with the same charge.

Fig. 9.1   A simplified explanation of the electrophoretic technique. A squashed fly is placed in an electric field on a gell material moistened slightly with an alkaline buffer solution. After several minutes some water-soluble protein molecules will have moved towards the positive pole. Enzyme genes can be detected as bands by suitably processing the gel.

Table 9.7 Charge-state distributions for the Lap-5 and Odh-1 loci of *Drosophila willistoni* in Caribbean populations in 1972.

Locus	Electro-phoresis distance (mm)	Charge state	Santiago	Santo Domingo	Maya-güez	Barran-quitas	Yungue	St. Kitts
Lap-5	94	6	0	0	0	0	2	0
	96	5	0	2	2	3	0	1
	98	4	14	19	18	20	7	3
	100	3	191	166	304	459	399	295
	103	2	108	144	129	139	88	62
	105	1	7	5	5	3	2	3
			320	336	458	624	498	364
Odh-1	86	7	0	0	0	0	1	0
	94	5	6	5	0	0	2	0
	96	4	0	0	1	0	1	0
	100	3	179	155	107	364	321	111
	104	2	2	4	1	6	8	1
	108	1	1	2	1	2	1	0
			188	166	110	372	334	112

locus give rise to bands at different positions on a gel. Testing large numbers of individuals results in a distribution of these bands. As an example, Table 9.7 shows a small part of the data collected by Ayala and Tracey (1974) for six Caribbean populations of *Drosophila willistoni*. It will be seen from this table that in Santiago, Dominican Republic, 160 individuals (320 genes) were tested at the Lap-5 locus. This produced four band distances. The most frequent distance was arbitrarily taken as being 100 mm on the gel and the other three observed distances were then at 98, 103 and 105 mm. Different alleles do not necessarily result in different band distances and therefore there may well have been more than four alleles present.

The observed bands at the Lap-5 locus are 2–3 mm apart and can be thought of as representing equally spaced 'electromorphs' with values 1, 2, 3, 4, 5 and 6. For the Odh-1 locus, the situation is similar but with electromorphs from 1 to 7 with a distance between states of about 4 mm. It seems possible that an electromorph of 6 can occur for this locus, corresponding to a gel distance of about 90 mm, but it has not been observed.

9.11 CHARGE-STATE MODELS

Ohta and Kimura (1973) were the first to set up a mathematical model specifically for electrophoretic data. They assumed that the electromorphs correspond to proteins with different net electric charges, that there is no selection associated with different charge states, and that mutations change the state either by $+1$ or -1 units. The model was later extended by Brown *et al.* (1975) and Wehrhahn (1975) to allow for two-step mutation changes.

The two-step model is defined as follows. Suppose that a population is such that a proportion p_i of alleles have a charge state i, for $i = 0, \pm 1, \pm 2, \ldots$, where $\sum p_i = 1$. A proportion β of all mutations increase the charge of an allele by one unit and a proportion γ of mutations increase the charge by two units. Further proportions β and γ result in charge changes of -1 and -2 units, respectively. The remaining proportion $\mu = 1 - 2\beta - 2\gamma$ of mutations cause no charge changes and are therefore not detected by electrophoresis.

Moran (1975) has shown that with this model the proportions p_i do not tend towards fixed values as time increases. Nevertheless, the existing charge states in a population should remain bunched together. This bunched set 'wanders' over the possible levels. The wandering makes it difficult to test whether the model fits data, but progress is possible by considering the distribution of heterozygosity for individuals. Thus, suppose that a random sample of m individuals is taken from a population and results in m_{ij} individuals with one allele at charge level i and one allele at charge level j. The data can then be written in the form of an upper triangular array, as shown in Table 9.8. The diagonal sum

$$D_j = \sum_{i=1}^{r-j} m_{i\,i+j}$$

is the number of individuals with two alleles that are j charge steps apart. It has

Table 9.8 Classification of a random sample of m individuals according to the charge states of their two alleles.

Charge state for lowest allele	Charge state for highest allele				Diagonal sums
	1	2	3	... r	
1	m_{11}	m_{12}	$m_{13} \ldots m_{1r}$		$m_{1r} = D_{r-1}$
2		m_{22}	$m_{23} \ldots m_{2r}$		$m_{1r-1} + m_{2r} = D_{r-2}$
3			$m_{33} \ldots m_{3r}$		$m_{1r-2} + m_{2r-1} + m_{3r} = D_{r-3}$
.				.	.
.					.
.					.
r				m_{rr}	$\sum m_{ii} = D_0$

been shown by Brown *et al.* (1975) and Weir *et al.* (1976) that under the assumption of random mating in an isolated population, the expected value of D_0 (the number of charge state homozygotes) is of the form

$$E(D_0) = \frac{(1 - \lambda_1)(1 - \lambda_2)(1 + \lambda_1 \lambda_2)}{(1 + \lambda_1)(1 + \lambda_2)(1 - \lambda_1 \lambda_2)} m \qquad (9.12)$$

while

$$E(D_j) = \frac{2 E(D_0)}{(1 + \lambda_1 \lambda_2)(\lambda_1 - \lambda_2)} \{(1 - \lambda_2^2)\lambda_1^{j+1} - (1 - \lambda_1^2)\lambda_2^{j+1}\}, \qquad (9.13)$$

$j = 1, 2, 3, \ldots$ For a sensible distribution it is necessary that $-1 < \lambda_2 \leqslant 0 < \lambda_1 < 1$. These expected values apply after sufficient time has elapsed for the population to reach a stationary state. For a population of effective size N_e this requires of the order of $8 N_e$ generations from the time of some disturbance.

The two parameters λ_1 and λ_2 in these equations are related to the mutation rates β, γ and μ that were defined above, and also to the effective population size. Writing

$$u = 4N_e \mu \beta \qquad \text{and} \qquad v = 4 N_e \mu \gamma,$$

there are relationships

$$u = (\lambda_1 + \lambda_2)(1 + \lambda_1 \lambda_2)/\{(1 - \lambda_1)(1 - \lambda_2)\}^2 \qquad (9.14)$$

and

$$v = -\lambda_1 \lambda_2/\{(1 - \lambda_1)(1 - \lambda_2)\}^2. \qquad (9.15)$$

The ratio $u/v = \beta/\gamma$ indicates the relative frequencies of one- and two-step mutations. For $u \geqslant 0$ it is necessary that $\lambda_1 + \lambda_2 \geqslant 0$, or $\lambda_1 \geqslant -\lambda_2$. If $\lambda_1 = -\lambda_2$ then it follows from equations (9.13) that $E(D_1) = E(D_3) = E(D_5) = \ldots = 0$ since one-step mutations do not occur.

Weir *et al.* (1976) have used these results as a basis for a test of the hypothesis that electrophoretically detectable protein polymorphisms are selectively neutral. The idea is that the parameters λ_1 and λ_2 can be estimated by minimizing the chi-square statistic

$$x^2 = \sum \{D_j - E(D_j)\}^2/E(D_j). \qquad (9.16)$$

Then x^2 can be tested to see whether it is significantly large. If it is significantly large then the charge-state model seems incorrect, possibly because of selection taking place.

Actually, Weir *et al.* did not express their test in quite this way. They did not begin with a data matrix of the form shown in Table 9.8. Instead, they assumed that only the proportions of alleles with different charge states are known. Assuming Hardy–Weinberg equilibrium these can then be allocated to genotypes, and estimated D_j values calculated. Thus suppose that n_i alleles with charge state i are observed in a sample of n alleles, for $i = 1, 2, \ldots, r$. This is a proportion n_i/n so it is expected that through random mating a proportion $(n_i/n)^2$ of individuals will be homozygotes for charge state i. Thus the sample of n alleles is expected to consist of $m = n/2$ individuals of which $\hat{m}_{ii} = m(n/n_i)^2$ are homozygotes for this charge state. This argument gives an estimated D_0 value of

$$\hat{D}_0 = \sum_{i=1}^{r} \hat{m}_{ii} = \left(\sum_{i=1}^{r} n_i^2 \right) \Big/ 2n \qquad (9.17)$$

which can be used in place of the unknown true values in equation (9.16). In a similar way, the Hardy–Weinberg proportion of individuals with one allele of charge state i and another of charge state k is $2(n_i/n)(n_k/n)$, this being twice the product of the allele proportions. Therefore a sample of m individuals is estimated to contain $\hat{m}_{ik} = 2m(n_i/n)(n_k/n)$ with this genotype. Consequently, there is an estimate of D_j,

$$\hat{D}_j = \sum_{i=1}^{r-j} \hat{m}_{i\,i+j} = \sum_{i=1}^{r-j} n_i n_{i+j}/n, \qquad (9.18)$$

$$j \neq 0.$$

Weir et al. proposed that the D_j values should be estimated in this way and then the pseudo-chi-square statistic

$$X_1^2 = \sum \{\hat{D}_j - E(D_j)\}^2 / E(D_j)$$

should be minimized to determine λ_1 and λ_2 of equations (9.12) and (9.13). The minimum value can then be tested to see if it is significantly large. This seems a reasonable procedure because X_1^2 will presumably be close to the genuine chi-square statistic of equation (9.16). Early simulations indicated that a test based on X_1^2 is conservative, in the sense that if the charge state model is correct then significantly large values occur less often than they should (Weir et al., 1976). However, more recent work has cast some doubts on the chi-square approximation (Brown et al., 1981).

The method for minimizing X_1^2 that was proposed by Weir et al. needs a special computer program. Certain rules are used to determine how many terms to include in the chi-square sum, and different combinations of λ_1 and λ_2 within the region $0 < \lambda_1 < 1$, $-1 < \lambda_2 < 0$ are tried to find which values are best. A somewhat simpler approach involves first choosing λ_1 and λ_2 so that the mean and standard deviation of the D_j distribution are fitted exactly.

The basis of the method of moments is the following equations. These can be used either with the actual D_j values, as defined in Table 9.8, or the estimated values of equations (9.17) and (9.18). For convenience, the actual D_j values are shown in the equations. The observed mean of the j values is

$$\bar{j} = \sum_{j=0}^{r-1} j D_j/m$$

while the observed variance is

$$s^2 = \sum j^2 D_j/m - (\bar{j})^2.$$

From equations (9.12) and (9.13) it is not difficult to find that the expected values of these moments are

$$\mu = \frac{2}{(\lambda_1 - \lambda_2)(1 - \lambda_1 \lambda_2)} \left\{ \frac{\lambda_1^2(1 - \lambda_2)^2}{(1 - \lambda_1)(1 + \lambda_1)} - \frac{\lambda_2^2(1 - \lambda_1)^2}{(1 - \lambda_2)(1 + \lambda_2)} \right\}, \quad (9.19)$$

and

$$\sigma^2 = \frac{2}{(\lambda_1 - \lambda_2)(1 - \lambda_1 \lambda_2)} \left\{ \frac{\lambda_1^2(1 - \lambda_2)^2}{(1 - \lambda_1)^2} - \frac{\lambda_2^2(1 - \lambda_1)^2}{(1 - \lambda_2)^2} \right\} - \mu^2. \quad (9.20)$$

Fig. 9.2 shows values of the mean μ corresponding to different values for λ_1 and λ_2. It will be seen that when λ_1 is greater than about 0.6, the mean is almost unaffected by the value of λ_2. It will also be seen that the curves for λ_2 values are made broken when $\lambda_2 = -\lambda_1$. It was pointed out above that if $\lambda_1 < -\lambda_2$ then the mutation parameter u is negative. This is impossible so the condition $\lambda_1 \geqslant -\lambda_2$ is necessary for a biologically meaningful model.

Fig. 9.3 shows values for the standard deviation σ for different values of λ_1 and λ_2. The standard deviation can be seen to be almost unaffected by the value of λ_2, providing λ_1 is greater than 0.3. The standard deviation is equal to the mean for λ_1 of 0.7 or more.

Figs 9.2 and 9.3 may be used to estimate λ_1 and λ_2. For example, suppose that $\bar{j} = 0.17$ and $s = 0.54$, which are the values for the Odh-1 locus in Santiago using the data given in Table 9.7. First of all, take $\lambda_2 = 0$. From Fig. 9.2, it can be seen that a mean of 0.17 corresponds to $\lambda_1 \simeq 0.09$. Reference to Fig. 9.3 shows that $\lambda_2 = 0$ and $\lambda_1 = 0.09$ gives a standard deviation of $\sigma \simeq 0.43$, which is smaller than the observed standard deviation of 0.54. A non-zero λ_2 value will give a higher standard deviation so the next step is to try, say, $\lambda_2 = -0.2$. Fig. 9.2 then shows that the mean of 0.17 corresponds to $\lambda_1 \simeq -0.2$, while Fig. 9.3 shows that $\lambda_1 = 0.20$ and $\lambda_2 = -0.2$ gives a standard deviation of about 0.57. This is now larger than the observed standard deviation of 0.54 which indicates that λ_2 needs to be slightly closer to zero. Continuing in this way shows that the correct mean and standard deviation are given by $\hat{\lambda}_1 = 0.18$ and $\hat{\lambda}_2 = -0.15$. Having obtained moment estimates of λ_1 and λ_2 by trial and error, these can be improved, if necessary, by changing them slightly and seeing whether this results in a smaller value for a chi-square statistic.

With many sets of data the best possible fit is found by taking $\lambda_2 = 0$. In this case equations (9.19) and (9.20) reduce to

$$\mu = 2\lambda_1 / \{(1 - \lambda_1)(1 + \lambda_1)\}, \quad (9.21)$$

and

$$\sigma^2 = 2\lambda_1(1 + \lambda_1^2) / \{(1 - \lambda_1)(1 + \lambda_1)\}^2 \quad (9.22)$$

A moment estimator $\hat{\lambda}_1$ of λ_1 is found by equating the observed mean $\bar{j}$ to μ to give

$$\bar{j} = 2\hat{\lambda}_1 / \{(1 - \hat{\lambda}_1)(1 + \hat{\lambda}_1)\},$$

or

$$\hat{\lambda}_1 = 0.5\bar{j}(1 - \hat{\lambda}_1)(1 + \hat{\lambda}_1). \quad (9.23)$$

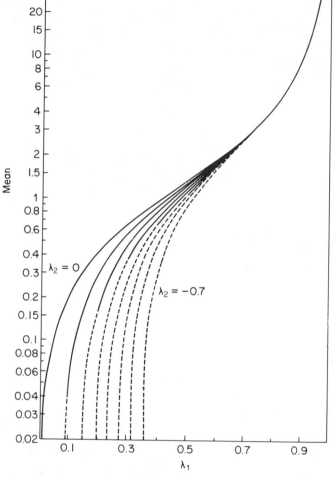

$\lambda_2 = 0$

$\lambda_2 = -0.7$

Fig. 9.2 The mean charge levels for the one- and two-step charge-state model. The broken curves are for biologically impossible combinations of λ_1 and λ_2. Curves are given for λ_2 in steps of 0.1.

This last equation is also obtained by using the principle of maximum likelihood with the D_j values being treated as Poisson variates (Weir *et al.*, 1976, p. 656).

Equation (9.23) is useful for obtaining $\hat{\lambda}_1$ by iteration. Substituting an approximate value for $\hat{\lambda}_1$ in the right-hand side will give an improved value. For example, suppose $\bar{j} = 1.0$, and take $\hat{\lambda}_1 = 0$ as an initial approximation for $\hat{\lambda}_1$. Putting this in the right-hand side gives $\hat{\lambda}_1 = 0.5$ as second approximation.

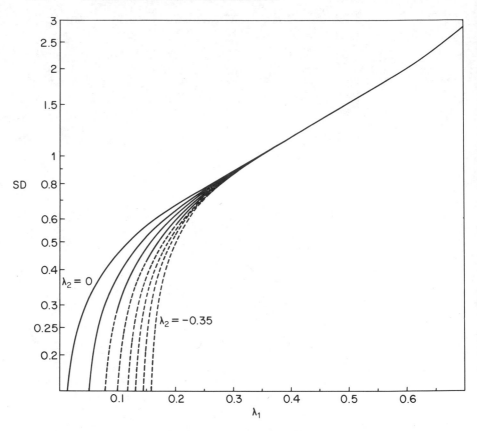

Fig. 9.3   The standard deviation of charge levels for the one- and two-step charge-state model. The broken curves are for biologically impossible combinations of λ_1 and λ_2. Curves are given for λ_2 in steps of 0.05.

Putting this in the right-hand side gives the third approximation $\hat{\lambda}_1 = 0.375$. Continuing in this way yields $\hat{\lambda}_1 = 0.414$, as the solution of equation (9.23).

   Wehrhahn's (1975) approach for fitting the charge-state model to data was different from that of Weir *et al.* It is less satisfactory to the extent that it relies upon the assumption that the probability of a two-step mutation is one-tenth of the probability of a one-step mutation. See also Cobbs (1979).

   The value of the charge-state model has been widely debated (Johnson 1974, 1977; Cobbs and Prakash, 1977; Ramshaw and Eanes, 1978; Ramshaw *et al.*, 1979; Fuerst and Ferrell, 1980; Ayala, 1982; McCommas, 1983; Brown *et al.*, 1981). Different authors have somewhat different views. What does seem to be clear is that any conclusions reached on the basis of this model must be taken with some reservations. The general consensus of opinion seems to be that the 'true' neutral mutation situation is a mixture of the infinite alleles model and the charge-state model, perhaps along the lines suggested by Li (1976). The best hope for progress lies in improving electrophoretic and related techniques so that almost all alleles can be detected and the Ewens–Watterson test can be used with confidence.

There is one fact that makes the situation at the present time slightly better than it would otherwise be. If the true situation is a mixture between the infinite alleles model and the charge-state model then the observed variation in allelic mobilities will be higher than what is expected from the charge-state model alone. Thus if allelic mobilities consistently show a significantly low amount of variation for the charge-state model, then this is perhaps some evidence of selection. On the other hand, low variation could be due to there not having been enough time elapsed for the population to have reached equilibrium.

Example 9.6   The charge-state model with
Drosophila willistoni

Weir *et al.* (1976) fitted the charge-state model to all of the data published by Ayala and Tracey (1974) on *Drosophila willistoni*. These data were for 27 loci in six Carribbean populations sampled in 1972. A good fit was obtained for 70 % of loci. In the cases where the model did not fit there tended to be too few alleles with extreme charge levels. Table 9.7 shows the data for two of the 27 loci. Lap-5 is an example of a locus where the charge-state model does not fit. In contrast, the fit is very good for the Odh-1 locus.

Consider the Lap-5 results for Santiago. The numbers of alleles with charge states 1 to 4 are $n_1 = 7, n_2 = 108, n_3 = 191$ and $n_4 = 14$, with a total of $n = 320$ genes and $m = 160$ individuals. From equation (9.17) the estimated number of individuals in the sample with two alleles in the same charge state is $\hat{D}_0 = 75.6$. From equation (9.18) the estimated number of individuals with two alleles j steps apart, for $j = 1, 2$ and 3 are $\hat{D}_1 = 75.2, \hat{D}_2 = 8.9$, and $\hat{D}_3 = 0.3$. The mean value of j is therefore estimated as $\bar{j} = \Sigma j D_j/m = 0.59$ while the standard deviation is estimated as $s = \{\Sigma j^2 D_j/m - (\bar{j})^2\} = 0.60$.

Fig. 9.2 shows that if $\lambda_2 = 0$ then a mean of 0.59 corresponds to $\lambda_1 \simeq 0.27$. In that case Fig. 9.3 shows that the standard deviation should be about 0.82. Making λ_2 non-zero has the effect of increasing the standard deviation for a given mean value and hence it must be concluded that $\lambda_2 = 0$ is appropriate for these data, with $\lambda_1 \simeq 0.27$. An improved estimate of λ_1 can be found by solving equation (9.23) by iteration. Starting with $\lambda_1 = 0.27$, the estimate $\hat{\lambda}_1 = 0.273$ is found in three iterations.

When $\lambda_2 = 0$ equations (9.12) and (9.13) reduce to

$$E(D_0) = m(1 - \lambda_1)/(1 + \lambda_1)$$

and

$$E(D_j) = 2E(D_0)\lambda_1^j.$$

Using these equations, the estimate $\hat{\lambda}_1 = 0.273$ gives expected values $E(D_0) = 91.4$, $E(D_1) = 49.9$, $E(D_2) = 13.6$, $E(D_3) = 3.7$, $E(D_4) = 1.0$, $E(D_5) = 0.3$, and $E(D_6) = 0.1$.

Weir *et al.*'s procedure for calculating a chi-square value involves including a term $\{D_j - E(D_j)\}^2/E(D_j)$ if $E(D_j)$ is greater than one or if $\hat{D}_j$ is greater than zero. In the present instance this amounts more or less to pooling the last three frequencies. The test statistic is then

$$X_1^2 = \frac{(75.6 - 91.4)^2}{91.4} + \frac{(75.2 - 49.9)^2}{49.9} + \frac{(8.9 - 13.6)^2}{13.6} + \frac{(0.3 - 3.7)^2}{3.7}$$

$$+ \frac{(0 - 1.4)^2}{1.4}$$

$$= 21.7.$$

The number of degrees of freedom is the number of classes after pooling, which is five, minus one because the observed and expected frequencies add to 160, and minus one because λ_1 has been estimated from the data. Thus there are three degrees of freedom and the test statistic is significantly large at the 0.1 % level. There is clear evidence against the model. Using their slightly different numerical method Weir *et al.* found $X_1^2 = 20.1$.

The data for Lap-5 in all six populations show low standard deviations. Consequently, for all populations λ_2 has been taken to be zero. For five of the populations a significantly high X_1^2 value is obtained. A summary of the results is given in Table 9.9. The meaning of $\lambda_2 = 0$ is that the two-step mutation parameter v is zero so that two-step charge changes do not occur. However, this observation is only relevant in cases where the charge-state model fits the data. For Lap-5 all that can be said is that even taking $\lambda_2 = 0$, the standard deviation of the charge levels is too low for the model to be correct, which may be because of selection against alleles with extreme charge states.

The charge level distributions are of quite a different nature for the Odh-1 locus. In this case the standard deviations are large relative to the means with the result that for five of the six populations, non-zero λ_2 values are needed to account for the variation.

For the Santiago population the allele frequencies at charge states 1 to 5 (Table 9.7) are $n_1 = 1, n_2 = 2, n_3 = 179, n_4 = 0$ and $n_5 = 6$, with a total of 188. From equations (9.17) and (9.18) these provide the estimates $\hat{D}_0 = 85.32$, $\hat{D}_1 = 1.91, \hat{D}_2 = 6.66, \hat{D}_3 = 0.06$ and $\hat{D}_4 = 0.03$. The mean of this distribution is $\bar{j} = 0.17$ while the standard deviation is $s = 0.54$. It has been shown in Section 9.11 that this mean and standard deviation correspond to moment estimates of $\hat{\lambda}_1 = 0.18$ and $\hat{\lambda}_2 = -0.15$.

Substituting these moment estimates into equations (9.12) and (9.13) gives the expected values $E(D_0) = 83.73$, $E(D_1) = 5.16$, $E(D_2) = 4.68$, $E(D_3) = 0.28$ and $E(D_4) = 0.13$. Weir *et al.*'s rule is that there is a contribution to the chi-

Table 9.9 Summary of the results from fitting the charge-state model to data on the Lap-5 and Odh-1 loci of six Caribbean populations of *Drosophila willistoni*.

	Number of individuals tested (m)	Number of charge states (r)	Mean		Standard deviation		$\hat{\lambda}_1$	$\hat{\lambda}_2$	$\hat{u}$	$\hat{v}$	X_1^2	d.f.
			Observed ($\bar{J}$)	Expected (μ)	Observed (s)	Expected (σ)						
Lap-5 Santiago	160	4	0.59	0.59	0.60	0.82	0.273	0	0.517	0	21.7*	3
Santo Domingo	168	5	0.65	0.65	0.65	0.89	0.299	0	0.608	0	24.4*	3
Mayagüez	229	5	0.53	0.53	0.59	0.77	0.247	0	0.436	0	23.9*	3
Barranquitas	312	5	0.44	0.44	0.57	0.70	0.212	0	0.341	0	21.6*	3
Yunque	249	5	0.36	0.36	0.55	0.61	0.173	0	0.252	0	10.1†	3
St Kitts	182	5	0.37	0.37	0.50	0.62	0.177	0	0.262	0	4.8	3
Odh-1 Santiago	94	4	0.17	0.17	0.54	0.57	0.20	−0.19	0.011	0.040	0.2	2
Santo Domingo	83	4	0.21	0.20	0.58	0.60	0.21	−0.19	0.022	0.044	0.6	2
Mayagüez	55	4	0.07	0.08	0.33	0.34	0.11	−0.09	0.021	0.011	0.1	1
Barranquitas	186	3	0.05	0.06	0.26	0.27	0.08	−0.06	0.021	0.005	0.3	1
Yunque	167	6	0.11	0.12	0.41	0.43	0.14	−0.11	0.032	0.017	3.3	2
St Kitts	56	2	0.02	0.02	0.13	0.13	0.01	0	0.009	0	0	0

Note: When $\hat{\lambda}_2 = 0$ the estimate of $\hat{\lambda}_1$ is a moment estimate which is also a pseudo-maximum likelihood estimate. When $\lambda_2 > 0$ the estimates have been chosen to minimize X_1^2. There are some disagreements between this table and Table 5 of Weir *et al.* (1976) because of slightly different numerical methods being used. However, in all cases these are of a minor nature and do not effect any conclusions. The X_1^2 values marked with asterisks are significantly large at the 5% (†) or 0.1% level (*).

squared statistic if either $E(D_j)$ is greater than one or $\hat{D}_j$ is greater than zero. Thus

$$X_1^2 = \frac{(85.32 - 83.73)^2}{83.73} + \frac{(1.91 - 5.16)^2}{5.16} + \frac{(6.66 - 4.68)^2}{4.68}$$

$$+ \frac{(0.06 - 0.28)^2}{0.28} + \frac{(0.03 - 0.13)^2}{0.13} = 3.18,$$

with 2 degrees of freedom.

This is not the minimum value that can be found for X_1^2. Table 9.10 shows that a minimum is found when $\hat{\lambda}_1 = 0.20$ and $\hat{\lambda}_2 = -0.19$, in which case $X_1^2 = 0.25$. It is clear that the charge-state model gives a good fit to these data. A similar good fit is found for the Odh-1 locus in the other populations (Table 9.9). There is certainly no evidence of selection at this locus.

Table 9.10   Values of X_1^2 for different combinations of λ_1 and λ_2 for the Odh-1 locus at Santiago.

λ_1	λ_2						
	-0.14	-0.15	-0.16	-0.17	-0.18	-0.19	-0.20
0.16	2.74	1.53	∞ [b]	$-$	$-$	$-$	$-$
0.17	3.77	1.99	0.90	∞	$-$	$-$	$-$
0.18	5.03	3.18 [a]	1.49	0.50	∞	$-$	$-$
0.19	6.45	4.60	2.81	1.20	0.29	∞	$- -$
0.20	7.98	6.16	4.37	2.64	1.09	0.25 [c]	∞
0.21	9.63	7.84	6.06	4.31	2.63	1.14	0.34
0.22	11.39	9.62	7.85	6.10	4.39	2.76	1.31

Notes: a The chi-squared value for the moment estimates $\hat{\lambda}_1 = 0.18$ and $\hat{\lambda}_2 = -0.15$ is 3.18.
b If $\lambda_1 + \lambda_2 = 0$ then $E(D_1) = E(D_3) = 0$. Since $\hat{D}_1$ and $\hat{D}_3$ are non-zero this results in $X_1^2 = \infty$. It is biologically impossible for $\lambda_1 + \lambda_2$ to be negative.
c The minimum X_1^2 value is about 0.25 when $\lambda_1 = 0.20$ and $\lambda_2 = -0.19$.

9.12   POPULATION SAMPLES INCLUDING MOTHER–OFFSPRING COMBINATIONS

In a series of papers, Christiansen and his colleagues have developed methods for the analysis of population samples including mother–offspring combinations (Christiansen and Frydenberg, 1973, 1976; Christiansen et al., 1973, 1974, 1977; Christiansen, 1977, 1980; Østergaard and Christiansen, 1981). The motivation for this was the realization of the extent to which simple samples of adults from a series of consecutive generations fall short of what is required to study properly the different components of selection that can occur with a single genetic locus.

Christiansen and Frydenberg (1973) noted that an ideal sampling system for a simple monogamous population with discrete generations involves:

(a) A sample of the zygote population in one generation.
(b) A sample of the adult population produced from the zygotes, separated into males and females.
(c) A sample of the breeding population, recording the males, the females and the mating pairs.
(d) Determination of the number of zygotes produced by the different pairs.
(e) A sample of the new zygote population.

This ideal system allows an analysis of: (1) zygotic selection by a comparison of (a) and (b); (2) sexual selection by a comparison of (b) and (c); (3) mating patterns by a study of (c); (4) fecundity selection by a study of (d); and (5) gametic selection by a comparison of (e) with (c) and (d). Unfortunately, such comprehensive data will be very difficult to obtain for natural populations. However, Christiansen and Frydenberg suggested that a practical system that is nearly as good can be designed by exploiting the possibility of sampling mother–offspring combinations. If a sample is taken from an adult population when the females are pregnant then the individuals sampled can be scored for their genotypes and classified as males, sterile females and fertile females. One randomly chosen foetus from each pregnant female can also be scored for its genotype to provide information on breeding males. Table 9.11 shows the situation when there are two alleles and hence three genotypes. The generalization of the form of data when there are three or more alleles is fairly straightforward (Østergaard and Christiansen, 1981). It is assumed that the sampled population is sufficiently large so that stochastic variation within the population is negligible in comparison with sampling errors.

Using data as shown in Table 9.11, a series of six selection hypotheses can be tested in turn. The idea is that each successive hypothesis implies more restrictions on the expected sample frequencies than the preceeding hypo-

Table 9.11   The format of data including mother–offspring combinations.

Genotype	Number of fertile females	Offspring genotype frequencies A_1A_1	A_1A_2	A_2A_2	Number of sterile females	Number of adult males
A_1A_1	f_1	c_{11}	c_{12}	–	s_1	m_1
A_1A_2	f_2	c_{21}	c_{22}	c_{23}	s_2	m_2
A_2A_2	f_3	–	c_{32}	c_{33}	s_3	m_3
Total	n_f		n_f		n_s	n_m

thesis. Likelihood ratio methods can be used to test the hypotheses against chi-square distributions. Christiansen and his colleagues used a different testing procedure that will give approximately the same numerical values, except possibly for small samples. The important result as far as likelihood ratio testing is concerned is that if sample frequencies $O_1, O_2, \ldots, O_r$ follow a multinomial distribution with expected frequencies $E_1, E_2, \ldots, E_r$, then the statistic

$$G = 2 \sum_{i=1}^{r} O_i \log_e (O_i/E_i) \qquad (9.24)$$

will approximately be a chi-square variate with the degrees of freedom being $r - 1$ minus one for every parameter that has to be estimated from the sample in order to determine the expected frequencies. If data from several independent multinomial distributions are available then the G values and their degrees of freedom can be added up and the total treated as a chi-square variate. Large values of G indicate that the model used to determine the expected frequencies is not correct.

For the present application it is convenient to view the offspring frequencies $c_{11}, c_{12}, \ldots, c_{33}$ as being a sample from one multinomial distribution, the sterile female frequencies s_1, s_2, and s_3 as a sample from a second independent multinomial distribution, and the adult male frequencies m_1, m_2 and m_3 as a sample from a third independent multinomial distribution. The expected frequencies can then be parameterized as

$$E(c_{ij}) = \gamma_{ij} n_f, \qquad (9.25)$$

and

$$E(s_i) = \beta_i n_s, \qquad (9.26)$$

$$E(m_i) = \alpha_i n_m, \qquad (9.27)$$

where $\gamma_{11} + \gamma_{12} + \gamma_{21} + \gamma_{22} + \gamma_{23} + \gamma_{32} + \gamma_{33} = 1$, $\beta_1 + \beta_2 + \beta_3 = 1$, and $\alpha_1 + \alpha_2 + \alpha_3 = 1$ (Table 9.12).

If no further restrictions are put on the γ, β and α parameters then their

Table 9.12   Class probabilities for the unrestricted model 0.

Genotype	Offspring genotypes			Sterile females	Adult males
	$A_1 A_1$	$A_1 A_2$	$A_2 A_2$		
$A_1 A_1$	γ_{11}	γ_{12}	–	β_1	α_1
$A_1 A_2$	γ_{21}	γ_{22}	γ_{23}	β_2	α_2
$A_2 A_2$	–	γ_{32}	γ_{33}	β_3	α_3
Total		1		1	1

maximum likelihood estimators are

$$\hat{\gamma}_{ij} = c_{ij}/n_f, \qquad (9.28)$$

$$\hat{\beta}_i = s_i/n_s, \qquad (9.29)$$

and

$$\hat{\alpha}_i = m_i/n_m. \qquad (9.30)$$

With these parameter values, the observed and expected sample frequencies are equal and the likelihood ratio goodness of fit statistic is $G_0 = 0$, with zero degrees of freedom. This unrestricted model makes no assumptions about relationships in the data and can conveniently be called *model 0*.

If there is no gametic selection in females and no foetal survival differences between the genotypes A_1A_1, A_1A_2 and A_2A_2, then half of the offspring of A_1A_2 mothers should have an A_1 maternal gene and half should have an A_2 maternal gene. Whatever the male genotype proportions are, it is expected that half the offspring from A_1A_2 mothers will be A_1A_2. This implies that

$$\gamma_{22} = \gamma_{21} + \gamma_{23}, \qquad (9.31)$$

so that γ_{22} is no longer an independent parameter. In this case, equations (9.29) and (9.30) still provide maximum likelihood estimators of β_i and α_i. Maximum likelihood estimators of the γ parameters are also still provided by equation (9.28) for $i \neq 2$. However, for A_1A_2 mothers,

$$\hat{\gamma}_{21} = f_2 c_{21}/\{2n_f (c_{21} + c_{23})\},$$
$$\hat{\gamma}_{23} = f_2 c_{23}/\{2n_f (c_{21} + c_{23})\}, \qquad (9.32)$$

and, $\hat{\gamma}_{22} = \hat{\gamma}_{21} + \hat{\gamma}_{23}$. This model will be called *model 1*. The log-likelihood goodness of fit statistic of equation (9.24) receives non-zero contributions from c_{21}, c_{22} and c_{23} so that it becomes

$$G_1 = 2 \sum_{j=1}^{3} c_{2j} \log_e \{c_{2j}/E(c_{2j})\}. \qquad (9.33)$$

It has one degree of freedom which comes from the single restriction (9.31).

Assume next that for mating males the proportion of genes that are A_1 is μ_1 and the proportion that are A_2 is $\mu_2 = 1 - \mu_1$. Assume also that there is random mating in the breeding population. Then if the female genotypes A_1A_1, A_1A_2 and A_2A_2 are in proportions ϕ_1, ϕ_2 and $\phi_3 = 1 - \phi_1 - \phi_2$, it follows that

$$\begin{aligned}
\gamma_{11} &= \mu_1 \phi_1, & \gamma_{12} &= \mu_2 \phi_1, \\
\gamma_{21} &= \tfrac{1}{2}\mu_1 \phi_2, & \gamma_{22} &= \tfrac{1}{2}\phi_2, & \gamma_{23} &= \tfrac{1}{2}\mu_2 \phi_2, \\
\gamma_{32} &= \mu_1 \phi_3, & \text{and} & \gamma_{33} &= \mu_2 \phi_3.
\end{aligned} \qquad (9.34)$$

For example, the offspring of an A_1A_1 mother will receive an A_1 paternal gene with probability μ_1, in which case it will be of genotype A_1A_1. The probability of the mother–offspring combination $A_1 A_1 – A_1 A_1$ is therefore the probability of the mother (ϕ_1) times the probability of the offspring given the mother (μ_1), i.e., $\gamma_{11} = \mu_1 \phi_1$.

The model with relationships (9.34) will be called *model 2*. It is a special case of *model 1*. Equations (9.29) and (9.30) still provide maximum likelihood estimators of β_i and α_i. The other parameters have maximum likelihood estimators

$$\hat{\phi}_i = f_i/n_f, \tag{9.35}$$

and

$$\hat{\mu}_i = (c_{11} + c_{21} + c_{32})/(n_f - c_{22}), \tag{9.36}$$

where f_i is the number of fertile females of genotype i (Table 9.11). The log-likelihood goodness of fit statistic receives non-zero contributions from all the c_{ij} frequencies and becomes

$$G_2 = 2 \sum_{i,j} c_{ij} \log_e \{c_{ij}/E(c_{ij})\}. \tag{9.37}$$

Model 2 has seven independent parameters α_1, α_2, β_1, β_2, ϕ_1, ϕ_2 and μ_1, compared to ten parameters for the perfectly fitting model 0 and nine parameters for model 1. Therefore G_2 can be compared with the chi-square distribution with three degrees of freedom to test for the goodness of fit of model 2 while $G_2 - G_1$ can be compared with the chi-square distribution with one degree of freedom to compare model 2 with model 1.

The next model to be considered, *model 3*, involves the assumption that all males mate equally successfully, so that the proportion of A_1 alleles in mating males is equal to the proportion for adult males in general. This means that

$$\mu_1 = \alpha_1 + \tfrac{1}{2}\alpha_2 \quad \text{and} \quad \mu_2 = \tfrac{1}{2}\alpha_2 + \alpha_1 \tag{9.38}$$

in equations (9.34). Model 3 has six independent parameters α_1, α_2, β_1, β_2, ϕ_1 and ϕ_2. Maximum likelihood estimators of β_i and ϕ_i are still provided by equations (9.29) and (9.35). Estimation of the male genotype proportions is not so straightforward. It can be done by a genotype counting procedure which is illustrated in the example that follows. This is more convenient than the Newton–Raphson method suggested by Christiansen and Frydenberg (1973) when calculations are done on a hand calculator.

The goodness of fit statistic for model 3 receives non-zero contributions from offspring and adult male sample frequencies and can be written as

$$G_3 = 2 \sum_{i,j} c_{ij} \log_e \{c_{ij}/E(c_{ij})\} + 2 \sum_i m_i \log_e \{m_i/E(m_i)\}. \tag{9.39}$$

It has four degrees of freedom. The difference in the goodness of fit of model 2 and model 3 can be examined by $G_3 - G_2$ with one degree of freedom.

If there is no differential female mating success, then the genotype proportions should be the same for fertile and sterile females, so that $\beta_i = \phi_i$.

Assuming that this is the case in addition to all the assumptions of model 3 gives *model 4*. This has independent parameters α_1, α_2, ϕ_1 and ϕ_2. The maximum likelihood estimates of the α parameters are the same as for model 3. The maximum likelihood estimate of ϕ_i is

$$\hat{\phi}_i = (f_i + s_i)/(n_f + n_s). \tag{9.40}$$

The goodness of fit statistic involves all the sample frequencies and therefore takes the form

$$G_4 = 2\sum_{i,j} c_{ij} \log_e\{c_{ij}/E(c_{ij})\} + 2\sum_i s_i \log_e\{s_i/E(s_i)\}$$
$$+ 2\sum_i m_i \log_e\{m_i/E(m_i)\}. \tag{9.41}$$

It has six degrees of freedom. The difference $G_4 - G_3$ has two degrees of freedom.

If zygotic selection is the same for males and females then adult genotype proportions should be the same for both sexes, so that $\alpha_i = \phi_i$. Assuming that this is the case, in addition to all the assumptions of model 4, gives *model 5*. This has only the two independent parameters α_1 and α_2. Explicit maximum likelihood estimates of these parameters are not available but they can be obtained by the same genotype counting process as used for fitting model 3. The goodness of fit statistic, G_5, involves all the sample frequencies and has eight degrees of freedom. The difference $G_5 - G_4$ has two degrees of freedom.

Finally, we come to *model 6*, which says that the adult genotype frequencies are in Hardy–Weinberg equilibrium for males and females, in addition to all the assumptions of model 5. This means that $\alpha_1 = \mu_1^2$, $\alpha_2 = 2\mu_1\mu_2$ and $\alpha_3 = \mu_2^2$, where $\mu_1 = 1 - \mu_2$ is the proportion of A_1 genes. The maximum likelihood estimator of μ_1 is

$$\hat{\mu}_1 = \frac{2f_1 + f_2 + c_{11} + c_{21} + c_{32} + 2s_1 + s_2 + 2m_1 + m_2}{3n_f - c_{22} + 2n_s + 2n_m}. \tag{9.42}$$

The goodness of fit statistic, G_6, has nine degrees of freedom. The difference $G_6 - G_5$ has one degree of freedom.

Christiansen and his colleagues proposed that models 1 to 6 should be fitted one by one to data. If model i has been accepted then a significant difference $G_{i+1} - G_i$ indicates that selection is occurring in connection with the extra assumptions of model $i+1$ over model i. See Christiansen and Frydenberg (1973) and Christiansen et al. (1973) for a more detailed discussion of the reasons why significant results can be obtained at different stages. The first of these two papers also contains information on the important question of the power of the tests to detect selection.

There are two assumptions that are implicit in all of the six models. The first of these is that the population was in gene frequency equilibrium during the period of time covering all generations represented in the sampled adults. The second assumption is that the adult population was homogenous over age classes within each sex. The first assumption can be tested by considering samples from successive years. The second assumption can be tested by dividing adults into age classes. Appropriate analyses are summarized by Christiansen (1980). The generalizations of the six models to allowing for more than two alleles at any number of loci have been provided by Østergaard and Christiansen (1981). Christiansen (1980) also considers the analysis of data involving more than one offspring from each mother when each brood is sired by one male.

Example 9.7   Adult and mother–offspring data on *Zoarces viviparus*

Christiansen and his co-workers have illustrated their methods of analyses for mother–offspring data with extensive data on the eelpout *Zoarces viviparus* at Kaloe Cove, Jutland. The present example will largely follow the analysis of Christiansen *et al.* (1973) of samples for 1969 and 1970 which were scored for the two-allele esterase polymorphism Est III. In these samples the adults were not aged. From 1971 to 1974 yearly samples of aged adults were taken. Christiansen (1977) and Christiansen *et al.* (1977) discuss the analysis of the full set of data from 1969 to 1974 at some length. They conclude that the only clear component of selection occurs between the formation of zygotes and maturity, heterozygotes being less fit than homozygotes.

Mother–offspring data provide no information on the fecundity of females with different genotypes. This can be studied using data on brood sizes. Brood size data collected from 1969 to 1971 indicate no evidence of selective fecundity for *Z. viviparus* (Christiansen, 1977).

Table 9.13 shows the mother–offspring and adult data for 1969 and 1970. To begin with, we can consider the analysis of the 1969 results in some detail. Thus the six selection models of Section 9.12 can be fitted in turn to the sample frequencies in order to assess the evidence for various types of selection. These models can be thought of as developing from the trivial model 0 which has all the expected sample frequencies equal to their observed values. Table 9.14 shows the expected frequencies for the six models, with corresponding goodness of fit statistics.

Model 1 assumes no female gametic selection, so that half of the offspring from $A_1 A_2$ mothers should be $A_1 A_2$. There are 357 mother–offspring combinations with $A_1 A_2$ mothers. The expected number of $A_1 A_2 - A_1 A_2$ combinations is therefore $E(c_{22}) = 357/2 = 178.5$. Equations (9.32) give the estimated probabilities associated with $A_1 A_2 - A_1 A_1$ and $A_1 A_2 - A_2 A_2$ combinations to be

$$\hat{\gamma}_{21} = 357 \times 65/\{2 \times 782(65 + 119)\} = 0.0806,$$

and

$$\hat{\gamma}_{23} = 357 \times 119/\{2 \times 782(65 + 119)\} = 0.1476.$$

Table 9.13   Observed genotypic distributions for pregnant females, mother–offspring combinations, sterile females and adult males for the Est III polymorphism for *Zoarces viviparus* at Kaloe Cove, 1969 and 1970.

Year	Genotype	Fertile females	Offspring			Sterile females	Adult males
			A_1A_1	A_1A_2	A_2A_2		
1969	A_1A_1	111	41	70	–	8	54
	A_1A_2	357	65	173	119	32	200
	A_2A_2	314	–	127	187	29	177
	Total	782		782		69	431
1970	A_1A_1	54	19	35	–	8	32
	A_1A_2	158	33	82	43	15	84
	A_2A_2	165	–	62	103	11	93
	Total	377		377		34	209

Multiplying these probabilities by the total mother–offspring frequency of 782 gives the expected values $E(c_{21}) = 63.1$ and $E(c_{23}) = 115.4$. All other expected frequencies are equal to their observed values. The goodness of fit statistic is $G_1 = 0.340$ with one degree of freedom. There is no evidence of female gametic selection.

Model 2 says that the proportion of paternal A_1 alleles, μ_1, is the same for all maternal genotypes. This proportion is estimated from equation (9.36) to be

$$\hat{\mu}_1 = (41 + 65 + 127)/(782 - 173) = 0.3826.$$

The proportion of paternal A_2 genes is then estimated as $\hat{\mu}_2 = 1 - \hat{\mu}_1 = 0.6174$. The proportions of different genotypes of fertile females are estimated at their observed values $\hat{\phi}_1 = 111/782 = 0.1419$, $\hat{\phi}_2 = 357/782 = 0.4565$ and $\hat{\phi}_3 = 314/782 = 0.4015$. Substituting these estimates into equations (9.34) gives mother–offspring probabilities. Multiplication by 782 then gives the expected mother–offspring frequencies. The expected frequencies for sterile females and adult males are equal to their observed values. The goodness of fit statistic is $G_2 = 1.730$ with three degrees of freedom. There is no evidence of non-random mating. The increase in lack of fit in moving from model 1 to model 2 is $G_2 - G_1 = 1.390$ with two degrees of freedom. There is no evidence against the additional assumptions made in model 2 as compared to model 1.

Model 3 says that all adult males mate equally successfully, so that the proportion of paternal A_1 genes is equal to the proportion of this gene in the adult male population. Thus the mother–offspring probabilites are given by

Table 9.14 Expected sample frequencies for models fitted to the 1969 data on *Zoarces viviparus* at Kaloe Cove.

Model (i)	Genotype	Mother–offspring combinations			Sterile females	Adult males	Goodness of fit statistic	
		A_1A_1	A_1A_2	A_2A_2			G_i^*	d.f.
0	A_1A_1	41†	70†	–	8†	54†	0	0
	A_1A_2	65†	173†	119†	32†	200†		
	A_2A_2	–	127†	187†	29†	177†		
1	A_1A_1	41†	70†	–	8†	54†	0.340	1
	A_1A_2	63.1	178.5	115.4	32†	200†		
	A_2A_2	–	127†	187†	29†	177†		
2	A_1A_1	42.5	68.5	–	8†	54†	1.730	3
	A_1A_2	68.3	178.5	110.2	32†	200†		
	A_2A_2	–	120.1	193.9	29†	177†		
3	A_1A_1	40.8	70.2	–	8†	57.3	2.872	4
	A_1A_2	65.6	178.5	112.9	32†	202.4		
	A_2A_2	–	115.4	198.5	29†	171.3		
4	A_1A_1	40.2	69.1	–	9.6	57.3	3.098	6
	A_1A_2	65.7	178.7	113.0	31.5	202.4		
	A_2A_2	–	115.9	199.3	27.8	171.3		
5	A_1A_1	39.6	67.9	–	9.5	59.3	3.320	8
	A_1A_2	66.4	180.3	114.0	31.8	198.8		
	A_2A_2	–	115.5	198.3	27.7	173.0		
6	A_1A_1	39.0	67.0	–	9.3	58.4	3.434	9
	A_1A_2	67.0	181.9	114.9	32.1	200.5		
	A_2A_2	–	114.9	197.3	27.6	172.1		

* Goodness of fit statistics need to be calculated from expected frequencies to two decimal places.
† Expected frequency = observed frequency.

equations (9.38) and (9.34). They are functions of the fertile female genotype proportions ϕ_1, ϕ_2 and ϕ_3 and the adult male genotype proportions α_1, α_2 and α_3. The estimates of the female proportions are the same as for model 2. Estimates of the male proportions can be calculated by genotype counting.

Genotype counting is just like the gene counting that was described in Section 9.3. The principle involved is that all relevant observed sample frequencies have to be allocated to male genotypes in the correct proportions. For example, the mother–offspring combination $A_1 A_1 – A_1 A_1$ has probability $\gamma_{11} = (\alpha_1 + \frac{1}{2}\alpha_2)\phi_1$ according to equations (9.34) and (9.38). This combination can occur in two ways. There can be an $A_1 A_1$ father, which contributes $\alpha_1 \phi_1$ to the probability γ_{11}, or an $A_1 A_2$ father, which contributes $\frac{1}{2}\alpha_2 \phi_1$ to γ_{11}. An $A_2 A_2$ father is impossible. The total number of $A_1 A_1 – A_1 A_1$ mother–offspring combinations observed is c_{11}. These can be allocated to paternal genotypes $A_1 A_1$ and $A_1 A_2$ in the ratio $\alpha_1 \phi_1 : \frac{1}{2}\alpha_2 \phi_1$. Thus $c_{11}\alpha_1 /(\alpha_1 + \frac{1}{2}\alpha_2)$ should be allocated to $A_1 A_1$ fathers and $\frac{1}{2}c_{11}\alpha_2 /(\alpha_1 + \frac{1}{2}\alpha_2)$ to $A_1 A_2$ fathers. As a second example, consider the mother–offspring combination $A_1 A_2 – A_2 A_2$ which has probability $\gamma_{23} = \frac{1}{2}(\frac{1}{2}\alpha_2 + \alpha_3)\phi_2$ according to equations (9.34) and (9.38). The sample frequency c_{23} now needs to be allocated to $A_1 A_2$ and $A_2 A_2$ fathers in the ratio $\frac{1}{4}\alpha_2 \phi_2 : \frac{1}{2}\alpha_3 \phi_2$ or $\frac{1}{2}\alpha_2 : \alpha_3$. In other words, $\frac{1}{2}c_{23}\alpha_2 /(\frac{1}{2}\alpha_2 + \alpha_3)$ should be allocated to $A_1 A_2$ males and the remainder to $A_2 A_2$ males. Applying similar arguments to all the mother–offspring combinations shows that the allocation of these to male genotypes amounts to: (a) taking the sample total $c_{11} + c_{21} + c_{32}$ and allocating a fraction $\alpha_1 /(\alpha_1 + \frac{1}{2}\alpha_2)$ of this to $A_1 A_1$ males and the remainder to $A_1 A_2$ males; and (b) taking the sample total $c_{12} + c_{23} + c_{33}$ and allocating a fraction $\frac{1}{2}\alpha_2 /(\frac{1}{2}\alpha_2 + \alpha_3)$ of this to $A_1 A_2$ males and the remainder to $A_2 A_2$ males. The mother–offspring combination $A_1 A_2 – A_1 A_2$, with observed frequency c_{22}, need not be allocated since the number in this class is shown by equations (9.34) not to be a function of the male genotype proportions. Likewise, the sterile female proportions are irrelevant. However, adult male frequencies need to be included in the genotype counts.

To begin the genotype counting process, the proportions α_1, α_2 and α_3 are approximated by their observed values in the sample of adult males. For 1969 this gives $\hat{\alpha}_1 \simeq 54/431 = 0.1253$, $\hat{\alpha}_2 \simeq 200/431 = 0.4640$ and $\hat{\alpha}_3 \simeq 177/431 = 0.4107$. These values are then used to allocate the sample frequencies $c_{11} + c_{21} + c_{32}$ and $c_{12} + c_{23} + c_{33}$ to male genotypes as described above. The male genotypes are then counted, including adult males, and the proportions found to be $A_1 A_2$, $A_1 A_2$ and $A_2 A_2$ give new estimates of $\hat{\alpha}_1$, $\hat{\alpha}_2$ and $\hat{\alpha}_3$ that are better than the initial values. If the process is repeated five times, stable estimates $\hat{\alpha}_1 = 0.1329$, $\hat{\alpha}_2 = 0.4696$ and $\hat{\alpha}_3 = 0.3975$ are obtained (Table 9.15).

From equations (9.38) the adult male allele proportions are estimated as $\hat{\mu}_1 = \hat{\alpha}_1 + \frac{1}{2}\hat{\alpha}_2 = 0.3677$ and $\hat{\mu}_2 = 1 - \hat{\mu}_1 = 0.6323$. From model 2, $\hat{\phi}_1 = 0.1419$, $\hat{\phi}_2 = 0.4565$ and $\hat{\phi}_3 = 0.4015$. Using these estimates, mother–offspring probabilities can be calculated from equations (9.34). Multiplication by 782 then gives expected mother–offspring frequencies. Sterile female expected frequencies are still equal to the observed frequencies. Adult male expected frequencies are calculated as $E(m_i) = 431\hat{\alpha}_i$. The goodness of fit statistic of equation (9.39) is then $G_3 = 2.872$ with four degrees of freedom. The difference $G_3 - G_2 = 1.142$ with one degree of freedom. There is no evidence of differential male mating success.

Model 4 assumes no differential mating success for females. This means that the genotype proportions are the same for mothers and sterile females, with these proportions being estimated using equation (9.40) as

$$\hat{\phi}_1 = (f_1 + s_1)/(n_f + n_s) = (111 + 8)/(782 + 69) = 0.1398,$$

$$\hat{\phi}_2 = (f_2 + s_2)/(n_f + n_s) = (357 + 32)/(782 + 69) = 0.4571,$$

and $\hat{\phi}_3 = 1 - \hat{\phi}_1 - \hat{\phi}_2 = 0.4031$. Substituting these values, together with the model 3 estimates of male gene proportions $\hat{\mu}_1 = 0.3677$ and $\hat{\mu}_2 = 0.6323$, into equations (9.34) gives expected mother–offspring proportions. These are then converted into the expected frequencies by multiplying by 782. Sterile female expected frequencies are given by $E(s_i) = 69\hat{\phi}_i$. Adult male expected frequencies are the same as for model 3. The goodness of fit statistic for model 4 is then $G_4 = 3.098$ with six degrees of freedom. The increase over model 3 is $G_4 - G_3 = 0.226$ with two degrees of freedom. There is no evidence of differential mating success for females.

Model 5 assumed that if zygotic selection exists then it is the same for males and females. In other words, adult genotype proportions are equal for males and females so that $\phi_i = \alpha_i$. There are therefore only three parameters in this model, the common genotype proportions α_1, α_2 and α_3. These can be estimated by genotype counting using the same method as shown in Table 9.15, except that all adult genotypes need to be counted instead of only adult male genotypes. To this end the mothers, the sterile females and the adult males can be regarded as a single group of 1282 individuals consisting of 173 A_1A_1, 589 A_1A_2 and 520 A_2A_2. However, the offspring have to be allocated to their father's genotypes using the estimates of α_1, α_2 and α_3. Thus a fraction $\alpha_1/(\alpha_1 + \frac{1}{2}\alpha_2)$ of the total $c_{11} + c_{21} + c_{32} = 233$ can be counted as A_1A_1 fathers and the remainder as A_1A_2. A fraction $\frac{1}{2}\alpha_2/(\frac{1}{2}\alpha_2 + \alpha_3)$ of the total $c_{12} + c_{23} + c_{33} = 376$ can be counted as A_1A_2 fathers and the remainder as A_2A_2. The mother–offspring combination A_1A_2–A_1A_2 could be allocated to A_1A_1, A_1A_2 and A_2A_2 fathers in the ratio $\alpha_1 : \alpha_2 : \alpha_3$, but this would have no effect on the final estimates of genotype proportions so the frequency c_{22} can be ignored.

The genotype counting is shown in Table 9.16. Final estimates of $\hat{\alpha}_1 = 0.1375$, $\hat{\alpha}_2 = 0.4612$ and $\hat{\alpha}_3 = 0.4013$ are found after three iterations, starting with the genotype proportions in adults as initial values. The adult gene proportions are then estimated as $\hat{\mu}_1 = \hat{\alpha}_1 + \frac{1}{2}\hat{\alpha}_2 = 0.3681$ for A_1 and $\hat{\mu}_2 = 1 - \hat{\mu}_1 = 0.6319$ for A_2. Taking $\hat{\phi}_i = \hat{\alpha}_i$ in equations (9.34) gives estimated mother–offspring probabilities which convert to expected frequencies on multiplication by 782. Expected frequencies for sterile females and adult males are given by $E(s_i) = 69\hat{\alpha}_i$ and $E(m_i) = 431\hat{\alpha}_i$. The goodness of fit statistic is then $G_5 = 3.320$ with eight degrees of freedom. The increase in moving from model 4 to model 5 is $G_3 - G_4 = 0.222$ with two degrees of freedom. If zygotic selection occurs then it appears to have similar effects on males and females.

Model 6 says that there is no zygotic selection, so that adult genotype

Table 9.15 Genotype counting to determine male genotype frequencies for model 3.

Iteration	Estimated male genotype proportions			Male genotype	Allocation of $c_{11} + c_{21} + c_{32}$	Allocation of $c_{12} + c_{23} + c_{33}$	Adult males	Count of genotypes
	$\hat{\alpha}_1$	$\hat{\alpha}_2$	$\hat{\alpha}_3$					
1	0.1253	0.4640	0.4107	A_1A_1	81.7	–	54	135.7
				A_1A_2	151.3	135.7	200	487.0
				A_2A_2	–	240.3	177	417.3
					233	376	431	1040
2	0.1305	0.4683	0.4013	A_1A_1	83.4	–	54	137.4
				A_1A_2	149.6	138.5	200	488.1
				A_2A_2	–	237.5	177	414.5
					233	376	431	1040
3	0.1321	0.4693	0.3986	A_1A_1	83.9	–	54	137.9
				A_1A_2	149.1	139.3	200	488.4
				A_2A_2	–	236.7	177	413.7
					233	376	431	1040
4	0.1326	0.4696	0.3978	A_1A_1	84.1	–	54	138.1
				A_1A_2	148.9	139.6	200	488.5
				A_2A_2	–	236.4	177	413.4
					233	376	431	1040
5	0.1328	0.4697	0.3975	A_1A_1	84.2	–	54	138.2
				A_1A_2	148.8	139.6	200	488.4
				A_2A_2	–	236.4	177	413.4
					233	376	431	1040
	0.1329	0.4696	0.3975		(final estimates)			

Note: At each iteration a fraction $\alpha_1/(\alpha_1 + \frac{1}{2}\alpha_2)$ of the $c_{11} + c_{21} + c_{32}$ mother–offspring combinations are allocated to A_1A_1 and the remainder to A_1A_2 fathers. Also, a fraction $\frac{1}{2}\alpha_2/(\frac{1}{2}\alpha_2 + \alpha_3)$ of the $c_{12} + c_{23} + c_{33}$ are allocated to A_1A_2 and the remainder to A_2A_2 fathers. New estimates of α_1, α_2 and α_3 are found from the total genotype counts. For example, at the end of iteration 1, $\hat{\alpha}_1 = 135.7/1040 = 0.1305$.

Table 9.16 Genotype counting to determine adult genotype frequencies for model 5.

Iteration	Estimated adult genotype proportions			Adult genotype	Allocation of $c_{11} + c_{21} + c_{32}$	Allocation of $c_{12} + c_{23} + c_{33}$	All sampled adults	Count of genotypes
	$\hat{\alpha}_1$	$\hat{\alpha}_2$	$\hat{\alpha}_3$					
1	0.1249	0.4594	0.4056	A_1A_1	86.2	–	173	259.2
				A_1A_2	146.8	135.9	589	871.7
				A_2A_2	–	240.1	520	760.1
					233	376	1282	1891
2	0.1371	0.4610	0.4020	A_1A_1	86.9	–	173	259.9
				A_1A_2	146.1	137.0	589	872.1
				A_2A_2	–	239.0	520	759.0
					233	376	1282	1891
3	0.1374	0.4612	0.4014	A_1A_1	87.0	–	173	260.0
				A_1A_2	146.0	137.2	589	872.2
				A_2A_2	–	238.8	520	758.8
					233	376	1282	1891
	0.1375	0.4612	0.4013		(final estimates)			

frequencies are in Hardy–Weinberg equilibrium. Thus $\alpha_1 = \mu_1^2$, $\alpha_2 = 2\mu_1\mu_2$ and $\alpha_3 = \mu_2^2$, where μ_i is the population proportion of the allele A_i. From equation (9.42),

$$\hat{\mu}_1 = \frac{2(111) + 357 + 41 + 65 + 127 + 2(8) + 32 + 2(54) + 200}{3(782) - 173 + 2(69) + 2(431)}$$

$$= 0.3681.$$

Estimates of genotype proportions are therefore $\hat{\alpha}_1 = \hat{\mu}_1^2 = 0.1355$, $\hat{\alpha}_2 = 2\hat{\mu}_1\hat{\mu}_2 = 0.4652$ and $\hat{\alpha}_3 = \hat{\mu}_2^2 = 0.3993$, where $\hat{\mu}_2 = 1 - \hat{\mu}_1 = 0.6319$. Mother–offspring probabilities are found by substituting these estimates into equations (9.34), taking $\hat{\phi}_i = \hat{\alpha}_i$. Multiplication by 782 then gives expected mother–offspring frequencies. Expected frequencies for sterile females and adult males are found as $E(s_i) = 69\hat{\alpha}_i$ and $E(m_i) = 431\hat{\alpha}_i$. The goodness of fit statistic is $G_6 = 3.434$ with nine degrees of freedom. The increase in moving from model 5 to model 6 is $G_6 - G_5 = 0.144$ with one degree of freedom. Neither of these values is at all significant in comparison with the chi-square distribution so there is no evidence of any selection at all with the 1969 data.

For the 1970 data, Christiansen *et al.* (1973) provide the test statistics shown in Table 9.17. In this case there is a significant chi-square value for zygotic selection. The goodness of fit statistic increases significantly in moving from model 5 to model 6 although the total fit of model 6 is quite acceptable. One significant test statistic in so many tests could easily occur by chance. However, subsequent samples have confirmed that zygotic selection does take place with *Z. viviparus* at Kaloe Cove (Christiansen, 1977; Christiansen *et al.*, 1977). Some

Table 9.17   Summary of tests on the 1970 *Zoarces viviparus* data.

Model	X^2	d.f.	Difference X^2	d.f.
0 No assumptions	0	0		
			0.28	1
1 No female gametic selection	0.28	1		
			1.11	2
2 + Random mating	1.39	3		
			0.75	1
3 + No male reproductive selection	2.14	4		
			2.27	2
4 + No female sexual selection	4.41	6		
			0.25	2
5 + Equal zygotic selection	4.65	8		
			6.19*	1
6 + No zygotic selection	10.85	9		

* Significantly large at the 5% level.

idea of the magnitude of the zygotic selection can be obtained by noting that in 1969 the genotype numbers for offspring and adults were as follows:

	A_1A_1	A_1A_2	A_2A_2
Offspring	106	370	306
Adults	173	589	520

This suggests selective values for A_1A_1, A_1A_2 and A_2A_2 in the ratios $(173/106):(589/370):(520/306)$ or $1.03:1.00:1.07$. There is apparently a disadvantage in being a heterozygote. Similar calculations for 1970 give selective values of $1.56:1.00:1.28$, which show the same effect but rather stronger. Selective values can be estimated separately for males and females if necessary.

The question of whether the gene frequencies in the population were changing with time can be tested separately for adult males, adult females and offspring, with the tests on males and offspring being independent. For males, the following two-way table can be made up:

	1969	1970	Total
A_1 genes	308	148	456
A_2 genes	554	270	824
Total	862	418	1280

A standard contingency table test then gives a chi-square value of 0.01 with one degree of freedom. There is certainly no evidence of a difference between the two years. Similar insignificant values are found for adult females and offspring.

10 Linkage disequilibrium and selection at two or more loci

The previous two chapters were concerned with analyses involving a single genetic locus. The present chapter extends the discussion of genetic data to situations where two or more loci are considered together.

To begin with, the algebra of random mating is considered for two loci. This introduces the idea of linkage disequilibrium parameters, which measure the extent to which the gene frequencies at two loci are non-randomly associated. Section 10.2 discusses the question of how linkage disequilibrium can provide evidence of selection. This is then followed by three sections on the estimation of linkage disequilibrium parameters under a variety of circumstances, and a section on the measurement of disequilibrium at more than two loci.

In Section 10.7 interest turns to the problem of estimating selective values for two loci considered together. A relatively simple approach based on an assumption of equilibrium is described. The chapter concludes with a brief summary of work that has been done on the estimation of multi-locus selective values in general.

10.1 RANDOM MATING WITH TWO LOCI

We have seen in Section 8.3 that in a large randomly mating population with no selection, the genotypes for a single autosomal (not sex-linked) locus will be in Hardy–Weinberg ratios. Thus if there are two alleles A_1 and A_2 with a proportion p of A_1 for adults, then the offspring will consist of the genotypes A_1A_1, A_1A_2 and A_2A_2 in the ratios $p^2 : 2pq : q^2$, where $q = 1 - p$. In an infinite population these genotype frequencies would be retained indefinitely for successive generations. For a finite population the value of p will gradually change through random drift.

If two autosomal loci are considered together, then Hardy–Weinberg genotype ratios can be expected at each locus separately. Also, the genotypes for the two loci will have independent frequencies. Thus, suppose that locus 1 has alleles A_1 and A_2 with population proportions p_1 and $q_1 = 1 - p_1$, respectively, and locus 2 has alleles B_1 and B_2 with population proportions p_2 and $q_2 = 1 - p_2$, respectively. There are then three genotypes for each locus, and thus $3 \times 3 = 9$ genotypes for the two loci together. With independent association between the loci, the proportion of the genotype $A_iA_jB_kB_l$ will just be the proportion of $A_i A_j$ times the proportion of $B_k B_l$. For example, the proportion of the genotype $A_1 A_1 B_1 B_2$ will be $(p_1^2)(2p_2 q_2)$.

If for some reason the independent association between two loci is disturbed then it takes more than one generation to recover the independence. To see the reason for this it is necessary to consider the proportions of different types of gametes (eggs or sperm) that are produced by different genotypes. Table 10.1 summarizes the different possibilities. The genotypes are written in this table in the form $A_1 B_2/A_2 B_1$ to stress the fact that each individual is derived from an egg and a sperm. Thus $A_1 B_2/A_2 B_1$ can be thought of as a combination of a gamete $A_1 B_2$ with a gamete $A_2 B_1$. Working in terms of gametes like this is necessary because the two loci being considered may be on the same chromosome. If that is the case then the individuals $A_1 B_2/A_2 B_1$ and $A_1 B_1/A_2 B_2$ are not the same in terms of the gametes that they produce. During the formation of gametes each chromosome aligns side by side with its partner, and breakage and reunion can occur, possibly resulting in a recombination of alleles. Without recombination, an $A_1 B_2/A_2 B_1$ individual produces equal numbers of $A_1 B_2$ and $A_2 B_1$ gametes. When a recombination occurs then the gametes $A_1 B_1$ and $A_2 B_2$ can also be produced by this individual. Thus if r is the probability of a recombination then the probabilities for different gametes are $\frac{1}{2}(1-r)$ for $A_1 B_2$, $\frac{1}{2}(1-r)$ for $A_2 B_1$, $\frac{1}{2}r$ for $A_2 B_2$ and $\frac{1}{2}r$ for $A_1 B_1$. On the other hand, for an $A_1 B_1/A_2 B_2$ individual the probabilities are $\frac{1}{2}(1-r)$ for each of the non-recombination gametes $A_1 B_1$ and $A_2 B_2$ and $\frac{1}{2}r$ for each of the recombination gametes $A_1 B_2$ and $A_2 B_1$.

Table 10.1 The proportions of different gametes produced by different genotypes.

Genotype	Population proportion	Proportions of gametes produced			
		$A_1 B_1$	$A_1 B_2$	$A_2 B_1$	$A_2 B_2$
$A_1 B_1/A_1 B_1$	P_{11}^2	1	–	–	–
$A_1 B_1/A_1 B_2$	$2P_{11}P_{12}$	$\frac{1}{2}$	$\frac{1}{2}$	–	–
$A_1 B_1/A_2 B_1$	$2P_{11}P_{21}$	$\frac{1}{2}$	–	$\frac{1}{2}$	–
$A_1 B_1/A_2 B_2$	$2P_{11}P_{22}$	$\frac{1}{2}(1-r)$	$\frac{1}{2}r$	$\frac{1}{2}r$	$\frac{1}{2}(1-r)$
$A_1 B_2/A_1 B_2$	P_{12}^2	–	1	–	–
$A_1 B_2/A_2 B_1$	$2P_{12}P_{21}$	$\frac{1}{2}r$	$\frac{1}{2}(1-r)$	$\frac{1}{2}(1-r)$	$\frac{1}{2}r$
$A_1 B_2/A_2 B_2$	$2P_{12}P_{22}$	–	$\frac{1}{2}$	–	$\frac{1}{2}$
$A_2 B_1/A_2 B_1$	P_{21}^2	–	–	1	–
$A_2 B_1/A_2 B_2$	$2P_{21}P_{22}$	–	–	$\frac{1}{2}$	$\frac{1}{2}$
$A_2 B_2/A_2 B_2$	P_{22}^2	–	–	–	1

Population proportions of gametes produced
$A_1 B_1$: $P_{11}' = P_{11} - r(P_{11}P_{22} - P_{12}P_{21})$
$A_1 A_2$: $P_{12}' = P_{12} + r(P_{11}P_{22} - P_{12}P_{21})$
$A_2 B_1$: $P_{21}' = P_{21} + r(P_{11}P_{22} - P_{12}P_{21})$
$A_2 B_2$: $P_{22}' = P_{22} - r(P_{11}P_{22} - P_{12}P_{21})$

If two loci are on different chromosomes then they are said to be *unlinked* and the probability of recombination must be $r = \frac{1}{2}$. If two loci are on the same chromosome then they are said to be *linked*. If they are close together then the probability of a chromosome break occurring between them is small. Consequently, r will be small. On the other hand, if the two loci are far apart then the probability of a break between them will be large and r may be close to $\frac{1}{2}$. In some cases recombination never occurs for linked loci. For instance, this is the situation with male *Drosophila*.

Having considered the implications of recombination, it is now possible to see why it is that a disturbance to a population which results in an association between two loci may take several or many generations to lose its effect. Thus suppose that in one generation the proportions of gametes that are $A_i B_j$ is P_{ij} among adults, where $P_{11} + P_{12} + P_{21} + P_{22} = 1$. Then, assuming random mating, the genotype proportions will be as shown in the second column of Table 10.1. In offspring, the gamete proportions are then found by adding up the output that can be expected from each genotype. For example, $A_1 B_1$ gametes can only be produced by $A_1 B_1 / A_1 B_1$, $A_1 B_1 / A_1 B_2$, $A_1 B_1 / A_2 B_1$, $A_1 B_1 / A_2 B_2$ and $A_1 B_2 / A_2 B_1$ individuals, with the output proportions shown in the table. The total proportion of $A_1 B_1$ gametes produced is then

$$P'_{11} = (1)P_{11}^2 + (\tfrac{1}{2})\{2P_{11}P_{12}\} + (\tfrac{1}{2})\{2P_{11}P_{21}\} + \tfrac{1}{2}(1-r)\{2P_{11}P_{22}\}$$
$$+ (\tfrac{1}{2}r)\{2P_{12}P_{21}\},$$

$$= P_{11} - r(P_{11}P_{22} - P_{12}P_{21}).$$

Similar expressions for the output of the other gametes are shown at the foot of the table.

The quantity

$$D = P_{11}P_{22} - P_{12}P_{21} \tag{10.1}$$

is called the *linkage disequilibrium parameter*. In terms of this parameter, the gamete proportions among adults in one generation, P_{11}, P_{12}, P_{21} and P_{22}, become

$$\left. \begin{array}{l} P'_{11} = P_{11} - rD \\ P'_{12} = P_{12} + rD \\ P'_{21} = P_{21} + rD \\ P'_{22} = P_{22} - rD \end{array} \right\} \tag{10.2}$$

in the next generation. It follows that the new linkage disequilibrium parameter is

$$D' = P'_{11}P'_{22} - P'_{12}P'_{21}$$

$$= (1-r)D. \tag{10.3}$$

Thus, in the transition from one generation to the next, the linkage disequilibrium is reduced by a factor of $1 - r$. More generally, if D_0 is the linkage disequilibrium in generation 0 then in generation t this will become

$$D_t = (1 - r)^t D_0. \qquad (10.4)$$

Hence the linkage disequilibrium will decay effectively to zero in a few generations for unlinked loci, or for linked loci with a recombination rate close to $\frac{1}{2}$. However, for closely linked genes, with $1 - r \simeq 1$, many generations will be required to achieve the same reduction. When D is zero, two loci are said to be in *linkage equilibrium* because, as shown by equations (10.2), the gamete proportions P_{11}, P_{12}, P_{21} and P_{22} will then remain constant from generation to generation.

Alternative expressions for D involves the gene frequencies. If p_1 is the proportion of A_1 alleles at the first locus and p_2 is the proportion of B_1 alleles at the second locus, then $p_1 = P_{11} + P_{12}$ and $p_2 = P_{11} + P_{21}$. It is not difficult to show that

$$D = P_{11} - p_1 p_2. \qquad (10.5)$$

Thus linkage equilibrium is obtained when the proportion of $A_1 B_1$ gametes is equal to the proportion of A_1 alleles multiplied by the proportion of B_1 alleles. Further expressions for D are

$$D = p_1(1 - p_2) - P_{12} = (1 - p_1)p_2 - P_{21} = P_{22} - (1 - p_1)(1 - p_2). \qquad (10.6)$$

These equations show that in the equilibrium state $P_{12} = p_1(1 - p_2)$, $P_{21} = (1 - p_1)p_2$ and $P_{22} = (1 - p_1)(1 - p_2)$.

Some authors (e.g., Hedrick *et al.*, 1978) have objected to the term 'linkage disequilibrium' on the grounds that linkage is not the only factor that effects the value of D. Indeed, two loci may be in linkage disequilibrium when they are not linked. 'Gametic disequilibrium' is perhaps a more apt expression. Nevertheless, D will be referred to here as the linkage disequilibrium parameter since this follows the common practice.

When there are two loci with several alleles each, a linkage disequilibrium parameter can be calculated for each possible combination of an allele from the first locus with an allele from the second locus. Thus if the first locus has alleles $A_1, A_2, \ldots, A_I$, with proportions $p_{11}, p_{12}, \ldots, p_{1I}$, the second locus has alleles $B_1, B_2, \ldots, B_J$, with proportions $p_{21}, p_{22}, \ldots, p_{2J}$, and the proportion of gametes that are $A_i B_j$ is P_{ij}, then the linkage disequilibrium between A_i and B_j can be defined as

$$D_{ij} = P_{ij} - p_{1i} p_{2j}, \qquad (10.7)$$

for $i = 1, 2, \ldots, I$ and $j = 1, 2, \ldots, J$. This parameter should be zero if A_i and B_j combine at random.

10.2 LINKAGE DISEQUILIBRIUM
AS EVIDENCE OF SELECTION

In a large, randomly mating population, with no selection, the linkage disequilibrium should get closer to zero with each successive generation. For this reason, the presence of linkage disequilibrium is often regarded as being evidence of selection, particularly if the association between loci is consistent between isolated populations (Lewontin, 1974, p. 315). The idea here is that selection will favour certain combinations of alleles and consequently these combinations will remain at high population frequencies.

Unfortunately, in order to ascribe a particular case of linkage disequilibrium to selection at the loci studied, it is necessary to eliminate several other possibilities, particularly: (a) random genetic drift because of a small population size, (b) migration, and (c) genetic 'hitch-hiking'. Hedrick *et al.* (1978) give a review of the effects of these alternatives to selection. The following important points can be noted about each:

(a) In a finite population the gamete frequencies will fluctuate randomly from generation to generation to a certain extent, even when there is no selection. This will cause some random variation in the value of the linkage disequilibrium parameter D. An idea of the magnitude of this variation is provided by considering the standardized parameter

$$R^2 = D^2/\{p_1(1-p_1)p_2(1-p_2)\},$$

where p_1 and p_2 are the population proportions of alleles A_1 and B_1. The value of R^2 can be shown to range from zero to one. A number of authors have found that for populations segregating at both loci,

$$E(R^2) \simeq 1/(1+4Nr),$$

where N is the effective population size and r is the recombination rate. This will tend to be an underestimate of the mean value of R^2 (Hill, 1976), but it gives some idea of the magnitude of random effects. For example, if $N = 100$ and there is tight linkage so that $r = 0.01$, then $E(R^2) \simeq 0.2$. This is large in comparison with the typical values found for natural populations.

(b) Migration between two populations can cause linkage disequilibrium. Continuous migration can serve to maintain a non-zero level, although when the migration is stopped the linkage disequilibrium will decay to zero. One explanation for an observed non-zero value is therefore migration effects in the past that have not had time to disappear. A related cause of disequilibrium is population subdivision. Ohta (1982) has argued that the most impressive example of observed linkage disequilibrium, that among the major histocompatibility complex of man and mouse, may well simply be due to the populations being subdivided.

(c) Genetic 'hitch-hiking' occurs when a neutral gene is closely linked to a selected gene (Hedrick, 1982). Thomson (1977) has shown that selection at one locus can actually generate linkage disequilibrium between two neutral loci. Thus selection may be occurring, but not at the particular loci being investigated. This may not be considered particularly important since the observed disequilibrium is still caused by selection acting on the region of the chromosome being studied. Thomson argued that linkage disequilibrium at the HLA-A and HLA-B loci of the major histocompatibility system in man may well be due to selection on immune response genes within the HLA system and not at the A and B loci themselves.

There are a number of further considerations that need to be taken into account when considering linkage disequilibrium as evidence of selection. Firstly, selection is most likely to cause linkage disequilibrium at closely linked loci. Langley (1977) did indeed find that for *Drosophila melanogaster* there is more disequilibrium for closely linked loci. Unfortunately, this observation is also what is expected for disequilibrium caused by random genetic drift, so its significance is debatable.

Another consideration is the power of tests for linkage disequilibrium. Brown (1975) has shown that detecting a moderate amount of disequilibrium may require sample sizes that are much larger than the typical sizes that have been used in the past. This problem is compounded by the fact that when data are collected by electrophoresis the 'alleles' are more correctly called 'electromorphs' since, as discussed in Section 9.10, they may really be combinations of several alleles. Zouros *et al.* (1977) considered the loss of power to detect linkage disequilibrium that results from this. Subsequently, Weir and Cockerham (1978) pointed out that the statistical methodology used by Zouros *et al.* is not correct, although the general conclusion that pooling of alleles may reduce the power of tests remains true. However, the relationship between pooling and power is not a simple one and the chance of a significant test result may even increase with pooling.

Hedrick *et al.* (1978) have reviewed the evidence for linkage disequilibrium in natural populations. On the whole, disequilibrium has not been widely detected, possibly because of the lack of power of statistical tests.

10.3 ESTIMATION OF LINKAGE DISEQUILIBRIUM FROM GAMETE FREQUENCIES

The estimation of the linkage disequilibrium parameter D is fairly straightforward when gametes can be counted. This involves being able to distinguish the two types of double heterozygotes A_iB_j/A_kB_l and A_iB_l/A_kB_j, which is done in *Drosophila* studies by establishing lines for single wild-type chromosomes. (See, for example, Langley *et al.*, 1974.) Another possibility is to infer genotypes from the progeny of double heterozygotes.

Suppose that a sample of n gametes gives alleles $A_1, A_2, \ldots, A_I$ with proportions $\hat{p}_{11}, \hat{p}_{12}, \ldots, \hat{p}_{1I}$ at one locus, and alleles $B_1, B_2, \ldots, B_J$ with proportions $\hat{p}_{21}, \hat{p}_{22}, \ldots, \hat{p}_{2J}$ at a second locus. Assume also that $\hat{P}_{ij}$ is the proportion of A_iB_j gametes in the sample. Then the linkage disequilibrium between A_i and B_j can be estimated as

$$\hat{D}_{ij} = \hat{P}_{ij} - \hat{p}_{1i}\,\tilde{p}_{2j}. \tag{10.8}$$

This is an almost unbiased estimator of the true population value D_{ij} for all except very small samples (Weir, 1979).

Assuming random mating,

$$\text{var}(\hat{D}_{ij}) \simeq \{p_{1i}(1 - p_{1i})p_{2j}(1 - p_{2j}) + (1 - 2p_{1i})(1 - 2p_{2j})D_{ij} - D_{ij}^2\}/n,$$
(10.9)

where p_{1i} and p_{2j} are population allele frequencies. If $D_{ij} = 0$ then this simplifies to

$$\text{var}(\hat{D}_{ij}) \simeq p_{1i}(1 - p_{1i})p_{2j}(1 - p_{2j})/n.$$
(10.10)

A suitable test for $D_{ij} = 0$ therefore involves treating

$$X_{ij}^2 = n\hat{D}_{ij}^2/\{p_{1i}(1 - p_{1i})p_{2j}(1 - p_{2j})\}$$
(10.11)

as a chi-square statistic with one degree of freedom to see whether it is significantly large. Weir and Cockerham (1978) have shown that a composite test for all the D_{ij} values being zero involves comparing

$$X_T^2 = n \sum_{i=1}^{I} \sum_{j=1}^{J} \hat{D}_{ij}^2/(\hat{p}_{1i}\hat{p}_{2j})$$
(10.12)

with the chi-square distribution with $(I - 1)(J - 1)$ degrees of freedom to see whether it is significantly large.

Table 10.2, which is abridged from Table II of Brown (1975), shows sample sizes that are required in order to have a 0.9 probability of a significant result with the test of equation (10.11) when a 5% level of significance is used, assuming that there are two alleles at each locus. If the proportion of alleles A_1 and A_2 at the first locus are p_1 and q_1, and the proportions of alleles B_1 and B_2 at the second locus are p_2 and q_2, then $D'_{11} = D_{11}/D_{\max}$, where

$$D_{\max} = \begin{cases} \min(p_1 q_2, q_2 p_1), & D \geqslant 0, \\ \min(p_1 q_1, p_2 q_2), & D < 0. \end{cases}$$

is the maximum of the absolute value of D_{11}. The table shows that very large sample sizes are needed if one or both of the loci have extreme allele proportions. For example, if $p_1 = p_2 = 0.1$, and D_{11} is negative at 20% of its maximum value, so that $D'_{11} = -0.2$, then a sample of 19915 gametes will be needed to get a significant test result 90% of the time.

Example 10.1   Linkage disequilibrium for
Drosophila montana in Colorado

An example of linkage disequilibrium that may be generated by selection is provided by a study of *Drosophila montana* in Colorado. Baker (1975) discusses the results of samples taken from one collecting site over a period of four years and single-year samples from two other sites. Four closely linked enzyme loci were studied by electrophoretic methods. The data for two loci

Table 10.2 The sample sizes (number of gametes) required to get a significant result for the test of equation (10.11) with a probability of 0.9, for various values of $D'_{11} = D_{11}/D_{max}$, assuming a 5% level of significance is required.

p_1	p_2	Maximum negative value of D_{11}	D'_{11}						Maximum positive value of D_{11}
			−1.0	−0.6	−0.2	0.2	0.6	1.0	
0.1	0.1	−0.01	518	1874	19915	388	64	29	0.09
	0.2	−0.02	256	867	8946	704	98	41	0.08
	0.3	−0.03	166	531	5288	1102	138	54	0.07
	0.4	−0.04	120	361	3459	1627	189	70	0.06
	0.5	−0.05	90	259	2361	2361	259	90	0.05
0.2	0.2	−0.04	121	396	4010	303	39	15	0.16
	0.3	−0.06	75	239	2364	481	57	21	0.14
	0.4	−0.08	52	160	1541	717	81	18	0.12
	0.5	−0.10	38	113	1047	1047	113	38	0.10
0.3	0.3	−0.09	44	142	1389	274	30	9	0.21
	0.4	−0.12	29	93	902	414	44	14	0.18
	0.5	−0.15	20	64	609	609	64	20	0.15
0.4	0.4	−0.16	18	60	582	262	26	6	0.24
	0.5	−0.20	11	39	390	390	39	11	0.20
0.5	0.5	−0.25	4	25	259	259	25	4	0.25

(numbers 1 and 2 of Baker) for three samples are shown in Table 10.3. The gametes from sampled flies were ascertained by crossing them with laboratory flies with a null allele at each locus.

The chi-square statistics shown in Table 10.3 show very clear evidence that at all three locations the allele frequencies are not independent for the two loci. In every case there are too few (null, null) and (C, F) gametes and too many (C, null) and (null, F) gametes. These chi-square statistics are approximately equal to the X_T^2 values of equation (10.12).

Table 10.3 Gamete frequencies for *Drosophila montana* at three collecting sites in Gunnison County, Colorado, USA. The loci are for esterases that use alpha naphthyl acetate as a preferred substrate in *in vitro* tests. Data on the rare electromorphs 'D' and 'N' have been omitted.

Sample	Locus 1 allele	Locus 2 allele			Total	χ^2 (2 d.f.)*
		Null	E	F		
Gothic, 1973	null	83	12	210	305	
	C	69	1	22	92	68.34
		152	13	232	397	
Horse Ranch Park, 1973	null	39	2	105	146	
	C	33	1	6	40	42.55
		72	3	111	186	
Ohio Creek, 1974	null	62	6	98	166	
	C	38	0	18	56	16.46
		100	6	116	222	

* Chi-square values are for normal contingency table tests for association. For all three samples the value is very highly significant.

Considering the Gothic sample only, the proportion of null and C alleles at the first locus are estimated as $\hat{p}_{11} = 305/397 = 0.7683$ and $\hat{p}_{12} = 92/397 = 0.2317$, while the proportion of null, E and F alleles at the second locus are $\hat{p}_{21} = 152/397 = 0.3829$, $\hat{p}_{22} = 13/397 = 0.0327$ and $232/397 = 0.5844$. Gamete proportions are $\hat{p}_{11} = 83/397 = 0.2091$, $\hat{p}_{12} = 12/397 = 0.0302$, and so on. The linkage disequilibrium between the null alleles at the two loci is then estimated using equation (10.8) as

$$\hat{D}_{11} = 0.2091 - 0.7683 \times 0.3829 = -0.0851.$$

Using equation (10.9) with estimates substituted for population values gives the variance of $\hat{D}_{11}$ to be var $(\hat{D}_{11}) \simeq 1.15 \times 10^{-4}$ so that the standard error is approximately 0.0107. The test statistic X_{11}^2 of equation (10.11) is 68.32, which is very significantly large in comparison with the chi-square distribution with one degree of freedom. There is very clear evidence of disequilibrium between the null alleles at the two loci. Similar calculations for the other pairs of alleles for the Gothic sample and all pairs of alleles for the other samples are given in Table 10.4. There is significant disequilibrium between the two alleles at the first locus and the null and F alleles at the second locus. Note that the estimates $\hat{D}_{ij}$ sum to zero when added over either the subscript i or the subscript j; for instance at Gothic, $\hat{D}_{11} + \hat{D}_{12} + \hat{D}_{13} = -0.085 + 0.005 + 0.080$. This is a property of the definition of the parameter D_{ij}.

Baker (1975) argued from this evidence, plus similar evidence from two further related loci, that the polymorphisms are maintained by selection. He provided evidence that the populations sampled are isolated from each other, so this seems to be an example of consistent results from independent data – Lewontin's (1974, p. 315) sensitive indicator of natural selection. Further samples of D. montana from Utah have confirmed the association between the loci (Baker and Kaeding, 1981). Allendorf (1983) and Baker (1983) have discussed alternative selective models to account for the data in terms of optimum levels of enzyme activity.

Table 10.4   Linkage disequilibrium estimates for *Drosophila montana*.

Sample	Locus 1 allele	Locus 2 allele	$\hat{D}_{ij}$	$SE(\hat{D}_{ij})$	χ^2_{ij} (1 d.f.)	χ^2_T (2 d.f.)
Gothic,	null	null	−0.085	0.011	68.36*	
1973	null	E	0.005	0.003	1.84	
	null	F	0.080	0.011	58.76*	
	C	null	0.085	0.011	68.36*	
	C	E	−0.005	0.003	1.83	
	C	F	−0.080	0.011	58.76*	68.38*
Horse	null	null	−0.094	0.015	41.11*	
Ranch	null	E	−0.002	0.004	0.23	
Park,	null	F	0.096	0.015	42.28*	
1973	C	null	0.094	0.015	41.11*	
	C	E	0.002	0.004	0.23	
	C	F	−0.096	0.015	42.28*	42.49*
Ohio	null	null	−0.058	0.014	15.72*	
Creek,	null	E	0.007	0.003	2.07	
1974	null	F	0.051	0.014	12.12*	
	C	null	0.058	0.014	15.72*	
	C	E	−0.007	0.003	2.07	
	C	F	−0.051	0.014	12.12*	16.45*

* Chi-square values significant at the 0.1 % level; χ^2_{ij} indicates the significance of the association between particular alleles, while χ^2_T indicates the significance of the overall association between loci.

10.4   MAXIMUM LIKELIHOOD ESTIMATION OF LINKAGE DISEQUILIBRIUM FROM GENOTYPE FREQUENCIES

Experimental techniques to determine gametic frequencies may involve considerable labour for some organisms. A common alternative situation is that a sample of genotype frequencies is available. This provides less information than gametic frequencies since the two gametic pairs A_iB_j/A_kB_l and A_iB_l/A_kB_j together form the genotype $A_iA_kB_jB_l$. If there is dominance at one or both of the loci being considered then this brings a further loss of information.

Suppose that there are two alleles at each locus, with no dominance. There are then nine genotypes, and assuming random mating these will have the population proportions shown in Table 10.5. These proportions come from Table 10.1 with the probability of the genotype $A_1A_2B_1B_2$ being the sum of the probabilities for A_1B_1/A_2B_2 and A_1B_2/A_2B_1. If there is dominance at one of the loci, say with B_1 dominant to B_2, then there will only be six distinguishable phenotypes with the classes B_1B_1 and B_1B_2 of Table 10.5 being combined into the single class B_1B_-. If there is dominance at both loci, with A_1 and B_1 being the dominant alleles, then there will be just four distinguishable phenotypes $A_1A_-B_1B_-$, $A_1A_-B_2B_2$, $A_2A_2B_1B_-$ and $A_2A_2B_2B_2$.

Maximum likelihood estimation of the linkage disequilibrium parameter D for the three situations just considered (no dominance, B_1 dominant, A_1 and B_1 dominant) has been discussed by Hill (1974), who gives references to earlier work. Explicit estimators are not available for the first two cases, so Hill proposed that estimates be obtained numerically by 'chromosome counting'. This is the same type of procedure as 'gene counting' (see Section 9.3). Weir and Cockerham (1979) have noted certain difficulties that can arise with chromo-

Table 10.5   Genotypes and their probabilities for two loci with two alleles each and random mating.

A genotype	B genotype			Total
	B_1B_1	B_1B_2	B_2B_2	
A_1A_1	P_{11}^2	$2P_{11}P_{12}$	P_{12}^2	p_1^2
A_1A_2	$2P_{11}P_{21}$	$2(P_{11}P_{22}+P_{12}P_{21})$	$2P_{12}P_{22}$	$2p_1q_1$
A_2A_2	P_{21}^2	$2P_{21}P_{22}$	P_{22}^2	q_1^2
Total	p_2^2	$2p_2q_2$	q_2^2	

$p_1 = 1 - q_1 =$ proportion of allele A_1
$p_2 = 1 - q_2 =$ proportion of allele B_1
$P_{ij} =$ proportion of gamete A_iB_j

some counting on small samples for the case of no dominance. They point out that the maximum likelihood equations have three solutions and that it is possible to reach the wrong one by Hill's method. They proposed instead that all of the solutions to the maximum likelihood equations be obtained numerically so that there is no doubt about whether the true maximum is found for the likelihood function. They also note that estimation is sometimes simplified when there are zero frequencies for certain genotypes.

For the case where A_1 and B_1 are dominant, Table 10.5 shows that the population proportion of the genotype $A_2 A_2 B_2 B_2$ is P_{22}^2. Also, the proportion of $A_2 A_2$ individuals is q_1^2 and the proportion of $B_2 B_2$ individuals is q_2^2. From equations (10.6), one of the definitions of the linkage disequilibrium D between two loci with two alleles each is $D = P_{22} - q_1 q_2$. Therefore an obvious estimator of D is

$$\hat{D} = \hat{P}_{22} - \hat{q}_1 \hat{q}_2,$$

$$= (N_{22}/N)^{1/2} - (N_{2.} N_{.2})^{1/2}/N \tag{10.13}$$

where N_{22} is the observed number of $A_2 A_2 B_2 B_2$ individuals, $N_{2.}$ is the observed number of $A_2 A_2$ individuals ($A_2 A_2 B_1 B_-$ plus $A_2 A_2 B_2 B_2$), and $N_{.2}$ is the observed number of $B_2 B_2$ individuals ($A_1 A_- B_2 B_2$ plus $A_2 A_2 B_2 B_2$) in a random sample of N individuals. This estimator was apparently first considered by Turner (1968). Hill (1974) points out that it is the maximum likelihood estimator. When the population value of D is zero,

$$\operatorname{var}(\hat{D}) \simeq p_1 (2 - p_1) p_2 (2 - p_2)/(4N), \tag{10.14}$$

where p_1 and p_2 are the population proportions of alleles A_1 and B_1, which can be estimated as $\hat{p}_1 = 1 - (N_{2.}/N)^{1/2}$ and $\hat{p}_2 = 1 - (N_{.2}/N)^{1/2}$. A test for $D = 0$ can be based upon comparing the statistic

$$X^2 = \hat{D}^2/\operatorname{var}(\hat{D}) \simeq 4N\hat{D}^2/\{\hat{p}_1 (2 - \hat{p}_1)\hat{p}_2 (2 - \hat{p}_2)\} \tag{10.15}$$

with the chi-square distribution to see whether it is significantly large.

10.5 BURROWS' ESTIMATOR OF LINKAGE DISEQUILIBRIUM

In view of the lengthy calculations required for maximum likelihood estimation of linkage disequilibrium in the case of no dominance, Cockerham and Weir (1977) proposed a simpler estimator for routine use. This was originally suggested to Cockerham and Weir in a private communication from Peter Burrows. In terms of properties, it compares well with the maximum likelihood estimator. It is unbiased when there is random mating, with a similar variance to the maximum likelihood estimator (Weir, 1979). If mating is not at random then it is not clear what exactly Hill's maximum likelihood estimate can be interpreted as. However, Burrows' estimator still has an interpretation in terms of between- and within-individual components of disequilibrium (Cockerham and Weir, 1977).

Suppose that at one locus there are alleles $A_1, A_2, \ldots, A_I$, with population proportions $p_{11}, p_{12}, \ldots, p_{1I}$. At a second locus let there be alleles B_1, $B_2, \ldots, B_J$, with proportions $p_{21}, p_{22}, \ldots, p_{2J}$. Also, let the frequencies of genotypes in a random sample of N individuals be denoted in the following way: $N(A_iA_jB_kB_l)$ is the number of $A_iA_jB_kB_l$; $N(A_iA_j\bar{B}_kB_l)$ is the number of $A_iA_j\bar{B}_kB_l$, where $\bar{B}_k$ indicates any allele other than B_k; $N(A_iB_k/A_jB_l)$ is the number of individuals with gametes A_iB_k and A_jB_l; and so on. Then the sample linkage disequilibrium between alleles A_i and B_j is given by equation (10.8) as

$$\hat{D}_{ij} = \{2N(A_iA_iB_jB_j) + N(A_iA_iB_j\bar{B}_j) + N(A_i\bar{A}_iB_jB_j)$$
$$+ N(A_iB_j/\bar{A}_i\bar{B}_j)\}/(2N) - \hat{p}_{1i}\hat{p}_{2j}.$$

This cannot be evaluated with genotype data because the sample count $N(A_i\bar{A}_iB_j\bar{B}_j)$ cannot be separated into $N(A_iB_j/\bar{A}_i\bar{B}_j)$ and $N(A_i\bar{B}_j/\bar{A}_iB_j)$. However, it can be partitioned into two components

$$\hat{D}_{ij} = \hat{D}_{ij(w)} + \hat{D}_{ij(b)}$$

where

$$\hat{D}_{ij(w)} = \{N(A_iB_j/\bar{A}_i\bar{B}_j) - N(A_i\bar{B}_j/\bar{A}_iB_j)\}/2N$$

represents linkage disequilibrium 'within' $A_i\bar{A}_iB_j\bar{B}_j$ individuals, and

$$\hat{D}_{ij(b)} = \{2N(A_iA_iB_jB_j) + N(A_iA_iB_j\bar{B}_j) + N(A_i\bar{A}_iB_jB_j)$$
$$+ N(A_i\bar{B}_j/\bar{A}_iB_j)\}/2N - \hat{p}_{1i}\hat{p}_{2j}$$

is a measure of the non-random union of gametes (linkage disequilibrium 'between' individuals). The expected value of $D_{ij(b)}$ is zero for a random mating population.

Burrows' estimator of linkage disequilibrium is the composite measure

$$\hat{\Delta}_{ij} = \hat{D}_{ij(w)} + 2\hat{D}_{ij(b)}$$
$$= \{4N(A_iA_iB_jB_j) + 2N(A_iA_iB_j\bar{B}_j) + 2N(A_i\bar{A}_iB_jB_j)$$
$$+ N(A_i\bar{A}_iB_j\bar{B}_j)\}/2N - 2\hat{p}_{1i}\hat{p}_{2j}. \tag{10.16}$$

Assuming random mating, this is just estimating the usual linkage disequilibrium parameter D_{ij}. Cockerham and Weir (1977) show that it is virtually unbiased, an unbiased estimator being $N\hat{\Delta}_{ij}/(N-1)$. Also, with a randomly mating population,

$$\text{var}(\hat{\Delta}_{ij}) \simeq \{p_{1i}(1-p_{1i})p_{2j}(1-p_{2j})$$
$$+ (1-2p_{1i})(1-2p_{2j})\Delta_{ij}/2 + \Delta_{ij}^2/N\}/N. \tag{10.17}$$

Assuming that there is no linkage disequilibrium, only the first term on the right-hand side of this equation is non-zero. Thus the variance can be estimated by $\hat{p}_{1i}(1 - \hat{p}_{1i})\hat{p}_{2j}(1 - \hat{p}_{2j})/N$. A test for $\Delta_{ij} = 0$ therefore involves seeing whether

$$X_{ij}^2 = N \hat{\Delta}_{ij}^2 / \{\hat{p}_{1i}(1 - \hat{p}_{1i})\hat{p}_{2j}(1 - \hat{p}_{2j})\} \tag{10.18}$$

is significantly large in comparison with the chi-square distribution with one degree of freedom. A test for all Δ_{ij} values being zero involves testing

$$X_T^2 = N \sum_{i=1}^{I} \sum_{j=1}^{J} \hat{\Delta}_{ij}^2 / (\hat{p}_{1i}\hat{p}_{2j}) \tag{10.19}$$

against the chi-square distribution with $(I-1)(J-1)$ degrees of freedom (Weir, 1979).

Some idea of the power of a test based on equation (10.18) can be gained from looking at Table 10.6, which is based upon Table IV of Brown (1975). (Actually the test considered by Brown was slightly different from this but the tabulated values should still be about right.) The table relates to a situation where the two loci being considered have two alleles each, with the proportions of A_1 and A_2 being p_1 and q_1 and the proportions of B_1 and B_2 being p_2 and q_2. The parameter D'_{11} is D_{11}/D_{max}, where D_{max} is the maximum of the absolute value of D_{11}. The table shows, for example, that if $p_1 = p_2 = 0.1$, and D_{11} is negative at 20 % of its maximum value so that $D'_{11} = -0.2$, then a sample of 20 535 individuals will be required in order to have a 0.9 probability of getting a significant result with a test at the 5 % level of significance. Table 10.2 shows that the same power is obtained with a sample of 19 915 gametes. Clearly, this type of sample size will usually be quite impractical.

Example 10.2   Linkage between the MN and Ss blood systems in man

Hill (1974) illustrated his maximum likelihood methods on Cleghorn's (1960) data on the MN and Ss blood groups for English blood donors, shown here in Table 10.7. Substitution of the various genotype frequencies into equation (10.16), together with the sample allele proportions $\hat{p}_{11} = 0.5425$ for $A_1 = M$ and $\hat{p}_{21} = 0.3080$ for $B_1 = S$, gives Burrows' estimate of the linkage disequilibrium between the alleles M and S to be

$$\hat{\Delta}_{11} = \frac{4 \times 57 + 2 \times 140 + 2 \times 39 + 224}{2 \times 1000} - 2 \times 0.5425 \times 0.3080$$

$$= 0.0708.$$

This is close to Hill's maximum likelihood estimate of 0.0701. There is no point in calculating the linkage disequilibrium parameters for the other pairs of loci using equation (10.16). It follows from this equation that $\hat{\Delta}_{21} = \hat{\Delta}_{12} = -\hat{\Delta}_{22} = -\hat{\Delta}_{11}$. Thus the single estimate $\hat{\Delta}_{11}$ completely describes the amount of disequilibrium that is observed for all alleles.

Table 10.6 The sample sizes (number of individuals) required to get a significant result for the test of equation (10.18) with a probability of 0.9, for various values of $D'_{11} = D_{11}/D_{max}$, assuming that a 5% level of significance is used.

Allele proportions		Maximum negative value of D_{11}	D'_{11}						Maximum positive value of D_{11}
p_1	p_2		−1.0	−0.6	−0.2	0.2	0.6	1.0	
0.1	0.1	−0.01	453	2046	20535	316	45	20	0.09
	0.2	−0.02	218	909	9159	632	76	29	0.08
	0.3	−0.03	138	533	5368	1037	114	40	0.07
	0.4	−0.04	97	347	3472	1578	165	54	0.06
	0.5	−0.05	72	238	2335	2335	238	72	0.05
0.2	0.2	−0.04	102	402	4081	272	31	12	0.16
	0.3	−0.06	62	235	2388	453	48	16	0.14
	0.4	−0.08	43	152	1541	695	70	22	0.12
	0.5	−0.10	31	103	1034	1034	103	31	0.10
0.3	0.3	−0.09	37	136	1394	258	25	8	0.21
	0.4	−0.12	24	87	897	401	39	12	0.18
	0.5	−0.15	17	58	599	599	58	17	0.15
0.4	0.4	−0.16	15	55	575	253	23	6	0.24
	0.5	−0.20	10	35	382	382	35	10	0.20
0.5	0.5	−0.25	4	22	251	251	22	4	0.25

Table 10.7   MN and Ss blood group frequencies for a sample from England

Genotype at first locus	Genotype at second locus			
	$B_1 B_1 (SS)$	$B_1 B_2 (Ss)$	$B_2 B_2 (ss)$	Total
$A_1 A_1$ (*MM*)	57	140	101	298
$A_1 A_2$ (*MN*)	39	224	226	489
$A_2 A_2$ (*NN*)	3	54	156	213
Total	99	418	483	1000

Sample proportion of $A_1 = \hat{p}_{11} = (2 \times 298 + 489)/2000 = 0.5425$
Sample proportion of $B_1 = \hat{p}_{21} = (2 \times 99 + 418)/2000 = 0.3080$

Assuming random mating, the variance of $\hat{\Delta}_{11}$ can be estimated from equation (10.17), with Δ_{11}, p_{11} and p_{21} replaced with their estimates, as $\mathrm{var}(\hat{\Delta}_{11}) \simeq 3.664 \times 10^{-5}$. The standard error of $\hat{\Delta}_{11}$ is therefore approximately 0.0061. This compares with Hill's numerically determined standard error of 0.0062 for the maximum likelihood estimator. The chi-square test statistic of equation (10.18) is $X^2_{11} = 94.93$ with one degree of freedom. There is overwhelming evidence of linkage disequilibrium for the MN and Ss loci.

The explanation for the linkage disequilibrium in the present case is very close linkage, and therefore a small recombination fraction r. Presumably some disequilibrium in the English population has occurred through migration and there has not been enough time for it to reduce to zero.

The assumption of random mating can be tested by seeing whether the genotypes are in Hardy–Weinberg frequencies at each locus separately. The appropriate test is described in Section 9.6. Hardy–Weinberg frequencies do seem to hold at both loci.

Example 10.3   Linkage disequilibrium for *Drosophila melanogaster* in North Carolina

Langley *et al.* (1978) have analysed collections of wild *Drosophila melanogaster* taken from various locations throughout North Carolina and the east coast of the United States during the years 1970 to 1973. Eight polymorphic enzyme loci were considered. These were αGphd, Mdh and Adh on chromosome II, and Est-6, Pgm, Odh, Est-C, and Acph on chromosome III. All these loci have one predominant allele and several rare alternatives. Langley *et al.* pooled the rarer alternatives so as to be able to carry out a two-'allele' analysis. They argued that a multi-allele analysis would have been considerably more complicated and would have given little increase in information. In this connection it can be noted that since the data were obtained by electrophoresis, even the most common 'allele' may be a combination of several alleles with the same mobility.

The Burrows estimator of equation (10.16) was used to measure linkage disequilibrium. As noted in the previous example, with two alleles at each locus, $\hat{\Delta}_{11} = -\hat{\Delta}_{12} = -\hat{\Delta}_{21} = \hat{\Delta}_{22}$. Therefore the single value $\hat{\Delta}_{11}$ completely describes the observed association between two loci. Langley *et al.* used the standardized statistic

$$\hat{R} = \tfrac{1}{2}\hat{\Delta}_{11}/\{\hat{p}_{11}(1-\hat{p}_{11})\hat{p}_{21}(1-\hat{p}_{21})\}^{1/2} \qquad (10.20)$$

to compare different populations, since this is less dependent than $\hat{\Delta}_{11}$ on the sample gene proportions $\hat{p}_{11}$ and $\hat{p}_{21}$. From equation (10.17), if $\Delta_{11} = 0$ then

$$\mathrm{var}(\hat{R}) \simeq 1/4N \qquad (10.21)$$

when $\hat{R}$ is estimated from a sample of N individuals.

Suppose that for a particular pair of loci there are estimates $\hat{R}_1, \hat{R}_2, \ldots, \hat{R}_K$ available from K independent samples. Then in the absence of linkage disequilibrium,

$$X_k^2 = \hat{R}_k^2/\mathrm{var}(\hat{R}_k) = 4N_k\,\hat{R}_k^2 \qquad (10.22)$$

is approximately a chi-square variate with one degree of freedom, where N_k is the kth sample size. An overall test for linkage disequilibrium at the locus then involves treating ΣX_k^2 as a chi-square variate with K degrees of freedom, where the summation is over the K samples. Langley *et al.* noted that a significantly large value of ΣX_k^2 may be caused by a constant non-zero R value in the different subpopulations sampled, or by variation in the R values for the different subpopulations. To tell the difference between these situations, they carried out weighted analyses of variance on the $\hat{R}_k$ values, with the weight given to $\hat{R}_k$ being $4N_k$, the reciprocal of its sampling variance.

Langley *et al.*'s paper should be consulted for details of the analysis of variance calculations. Here it can be noted that:

(a) The minimum variance weighted mean of the R_k values from K samples is

$$\bar{R} = \sum_{k=1}^{K} N_k \hat{R}_k / \sum_{k=1}^{K} N_k,$$

with variance

$$\mathrm{var}(\bar{R}) = \sum_{k=1}^{K} N_k^2 \,\mathrm{var}(\hat{R}_k) \Bigg/ \left(\sum_{k=1}^{K} N_k \right)^2 = 1 \Bigg/ \left(4 \sum_{k=1}^{K} N_k \right).$$

(Appendix, Section A.6). A test for $\bar{R}$ being significantly different from zero involves seeing whether

$$\bar{R}^2/\mathrm{var}(\bar{R}) = 4 \sum_k N_k \bar{R}^2 \qquad (10.23)$$

is significantly large in comparison with the chi-square distribution with one degree of freedom.

(b) Some of Langley *et al.*'s samples were taken in the same geographical area within the space of one month. These can be thought of as repeated samples from the same 'region'. On this basis the variance between $\hat{R}_k$ values from different samples could be split into a between-region component σ_R^2 and a within-region component σ_{RS}^2. In other words, a nested analysis of variance was done.

Table 10.8 summarizes Langley *et al.*'s results for North Carolina. For genes on the same chromosome there is one $\bar{R}$ value (for Est-6 × Pgm) that is significantly different from zero. For three other pairs of loci there are significant differences between $\hat{R}$ values for repeated samples from the same area ($\sigma_{RS} > 0$). The situation for the unlinked loci is somewhat similar. Considering the number of tests, these results indicate a consistency over the whole state ($\sigma_R = 0$) and possible regional heterogeneity ($\sigma_{RS} > 0$).

From these and other data, Langley *et al.* concluded that the magnitude of linkage disequilibrium between enzyme loci is small in natural populations of *D. melanogaster*, although linked loci do perhaps display somewhat more linkage disequilibrium than unlinked loci. Adding the test statistic (10.23) across loci gives a total chi-square of 26.5 with 13 degrees of freedom for the linked loci for the North Carolina data. This is significantly large at the 5% level. By comparison with this, the total chi-square for unlinked loci is 14.1 with 15 degrees of freedom, which is just less than the expected value due to sampling errors only. Too much should not be read into this result because the main reason for the difference between linked and unlinked loci is the highly significant result for Est-6 × Pgm.

These results for natural populations are somewhat different from what Langley *et al.* found with cage populations. In cage populations, significant linkage disequilibrium is often found with linked genes, possibly because of small population sizes.

Langley *et al.* concluded their paper by noting that the pattern of results that they found is quite consistent with any of the usual models for the maintenance of genetic variation, selective or otherwise.

10.6 MULTI-LOCUS DISEQUILIBRIUM

So far, attention has been restricted to measuring the association between two genetic loci only. More generally, data may be available for any number of loci and there may be interest in considering these more than two at a time. As with the two-locus case, appropriate methods of analysis depend upon whether the data consists of gamete or genotype frequencies.

Smouse (1974) noted that one approach for gametic data involves fitting log-linear models. For example, suppose that there are three loci being considered, with I alleles at locus A, J alleles at locus B and K alleles at locus C. Then with a log-linear model it is assumed that the sample count of $A_i B_j C_k$ gametes, n_{ijk},

Table 10.8 Results of analyses on North Carolina populations of *Drosophila melanogaster.*

Locus × locus	Number of samples	Number of regions	Linked genes			
			$4\bar{N}$	$\bar{R}$	$\hat{\sigma}_R$	$\hat{\sigma}_{RS}$
αGpdh × Mdh	101	72	498	+ 0.0051	0.0	0.0497*
αGpdh × Adh	102	74	501	+ 0.0073	0.0267	0.0
Mdh × Adh	105	75	509	+ 0.0043	0.0	0.0277
Est-6 × Pgm	70	60	451	− 0.0209*	0.0383	0.0
Est-6 × Est-C	97	68	463	+ 0.0033	0.0	0.0449*
Est-6 × Odh	101	73	490	− 0.0013	0.0030	0.0299
Est-6 × Acph	103	74	495	+ 0.0085	0.0	0.0255*
Pgm × Est-C	65	55	409	+ 0.0018	0.0	0.0294
Pgm × Odh	70	61	448	+ 0.0048	0.0125	0.0159
Pgm × Acph	71	62	447	+ 0.0039	0.0	0.0077
Est-C × Odh	93	66	461	− 0.0068	0.0	0.0242
Est-C × Acph	95	67	463	+ 0.0004	0.0175	0.0
Odh × Acph	101	75	487	+ 0.0012	0.0	0.0366

Locus × locus	Number of samples	Number of regions	Unlinked genes			
			$4\bar{N}$	$\bar{R}$	$\hat{\sigma}_R$	$\hat{\sigma}_{RS}$
αGpdh × Est-6	101	72	493	+ 0.0053	0.0	0.0387*
αGpdh × Pgm	71	62	456	+ 0.0053	0.0	0.0276
αGpdh × Est-C	93	64	460	+ 0.0012	0.0284	0.0
αGpdh × Odh	98	73	489	− 0.0064	0.0045	0.0150
αGpdh × Acph	101	74	488	− 0.0036	0.0	0.0421†
Mdh × Est-6	103	72	504	− 0.0005	0.0	0.0290
Mdh × Pgm	70	60	458	+ 0.0018	0.0266	0.0
Mdh × Est-C	96	66	649	+ 0.0005	0.0	0.0224
Mdh × Odh	101	73	497	+ 0.0050	0.0	0.0412‡
Mdh × Acph	104	75	496	− 0.0099‡	0.0	0.0252
Adh × Est-6	104	74	506	+ 0.0050	0.0	0.0225
Adh × Pgm	71	62	461	+ 0.0025	0.0	0.0390
Adh × Est-C	97	68	472	− 0.0007	0.0	0.0320
Adh × Odh	102	75	497	+ 0.0040	0.0110	0.0318*
Adh × Acph	105	77	499	− 0.0027	0.0	0.0310

* Significant at the 0.1 % level
† Significant at the 1 % level
‡ Significant at the 5 % level

say, has a Poisson distribution with a mean value

$$E(n_{ijk}) = \exp(\mu + \alpha_i + \beta_j + \gamma_k + \alpha\beta_{ij} + \alpha\gamma_{ik} + \beta\gamma_{jk} + \alpha\beta\gamma_{ijk}). \quad (10.24)$$

Here μ, α_i, β_j, $\ldots$, $\alpha\beta\gamma_{ijk}$ are parameters which determine the relationship between the three loci. If $\alpha\beta_{ij} = \alpha\gamma_{ik} = \beta\gamma_{jk} = \alpha\beta\gamma_{ijk} = 0$, for all i, j and k, then the three loci are independent since the expected frequency of n_{ijk} can be written as a constant, $\exp(\mu)$, multiplied by factors for the three loci:

$$\begin{aligned} E(n_{ijk}) &= \exp(\mu + \alpha_i + \beta_j + \gamma_k) \\ &= \exp(\mu) \times \exp(\alpha_i) \times \exp(\beta_j) \times \exp(\gamma_k). \end{aligned}$$

This is equivalent to the expected total sample size, multiplied by the probability of allele A_i, multiplied by the probability of allele B_j, multiplied by the probability of allele C_k. In other words, this model represents the state of no linkage disequilibrium. On the other hand, if $\alpha\beta_{ij} \neq 0$ but $\alpha\gamma_{ik} = \beta\gamma_{jk} = \alpha\beta\gamma_{ijk} = 0$, for all i, j and k, so that

$$E(n_{ijk}) = \exp(\mu + \alpha_i + \beta_j + \gamma_k + \alpha\beta_{ij}),$$

then there is linkage disequilibrium between loci A and B, but not between A and C or between B and C. Thus the parameters $\alpha\beta_{ij}$, $\alpha\gamma_{ik}$ and $\beta\gamma_{jk}$ allow for linkage disequilibria between pairs of loci. The three-locus parameter $\alpha\beta\gamma_{ijk}$ allows for an association between all three loci. This would be needed, for example, if the allele pair $A_i B_j$ had a higher probability of occurring with the allele C_k than with the allele C_l.

The log-linear formulation is very convenient because it allows an analysis using a standard computer program: GLIM (Appendix, Section A.10) or BMDP4F (Dixon, 1981) are two programs that can be used. Chi-square tests are available to compare the various possible models for data to look for evidence of different types of association between loci. However, Karlin and Piazza (1981) point out that this type of analysis is not suitable for detecting particular pairwise disequilibrium relationships. They discuss a number of linkage disequilibrium measures for three loci, particularly in terms of detecting disequilibrium at the HLA-A, B and C loci for the histocompatibility system in man. See also Hill (1976) for a review of measures of linkage disequilibrium at three and four loci.

Hill (1975) noted the usefulness of the type of model given by equation (10.24) and discussed the use of such models with data on genotype frequencies rather than gamete frequencies for three loci. In this case the computations cannot be done with a standard computer program for log-linear models.

For handling genotype data on large numbers of loci, Smouse and Neel (1977) adopted a scoring technique. Thus, consider a locus with two alleles A_1 and A_2. A variable Y_1 can then be defined, where

$$Y_1 = \begin{cases} 1, & \text{for an } A_1 A_1 \text{ individual,} \\ \tfrac{1}{2}, & \text{for an } A_1 A_2 \text{ individual, and} \\ 0, & \text{for an } A_2 A_2 \text{ individual.} \end{cases}$$

Likewise, if there is a second locus with two alleles, then a variable Y_2 can be defined, where

$$Y_2 = \begin{cases} 1, \text{ for a } B_1 B_1 \text{ individual,} \\ \tfrac{1}{2}, \text{ for a } B_1 B_2 \text{ individual, and} \\ 0, \text{ for a } B_2 B_2 \text{ individual.} \end{cases}$$

Any number of two-allele loci can be handled in this way. Every individual will then have a vector $\mathbf{Y}$ of Y values. For example, with four loci an individual with genotype $A_1 A_1 B_2 B_2 C_1 C_2 D_1 D_1$ would have the vector $\mathbf{Y}' = (1, 0, \tfrac{1}{2}, 1)$. The reason for introducing this scoring technique is so as to be able to analyse data using standard multivariate methods.

Smouse et al. (1983) have shown that with L loci the population mean for the vector $\mathbf{Y}'$ is $\boldsymbol{\mu}' = (p_1, p_2, \ldots, p_L)'$, where p_i is the population proportion of the first allele for the ith locus. This is unaffected by non-random mating. However, non-random mating does affect the covariance matrix which has the general form $\boldsymbol{\Sigma} = \mathbf{A}' \mathbf{W} \mathbf{A}$, where

$$\mathbf{A} = \begin{bmatrix} (1+\theta_1)^{1/2} & 0 & 0 & \cdots & 0 \\ 0 & (1+\theta_2)^{1/2} & 0 & \cdots & 0 \\ 0 & 0 & (1+\theta_3)^{1/2} & \cdots & 0 \\ \vdots & \vdots & \vdots & & \vdots \\ 0 & 0 & 0 & \cdots & (1+\theta_L)^{1/2} \end{bmatrix},$$

and

$$\mathbf{W} = \tfrac{1}{2} \begin{bmatrix} p_1(1-p_1) & \Delta_{12} & \Delta_{13} & \cdots & \Delta_{1L} \\ \Delta_{21} & p_2(1-p_2) & \Delta_{23} & \cdots & \Delta_{2L} \\ \Delta_{31} & \Delta_{32} & p_3(1-p_3) \cdots & & \Delta_{3L} \\ \vdots & \vdots & \vdots & & \\ \Delta_{L1} & \Delta_{L2} & \Delta_{L3} & \cdots & p_L(1-p_L) \end{bmatrix}.$$

Here θ_i is a measure of non-random mating at the ith loci, which is zero with random mating. Also, Δ_{ij} is equivalent to the population mean of Burrows' linkage disequilibrium parameter of equation (10.16). Testing for linkage disequilibrium and non-random mating is now reduced to tests on $\mathbf{A}$ and $\mathbf{W}$. Smouse and his colleagues suggest these tests be based on the assumption that $\mathbf{Y}$ has a multivariate normal distribution, although they recognize that this gives rather approximate tests.

The scoring technique was originally devised by Smouse and Neel (1977) to analyse data for eight loci collected from fifty villages of the Yanomama Indians in South America. It was a question of finding a way of quantifying linkage disequilibrium so as to be able to study variation within and between villages. Considerable linkage disequilibrium was found from village to village, but with no consistent patterns. Smouse and Neel concluded that there was no evidence of natural selection at work. Similar results were found by Smouse et al. (1983) in a study of the Yanomama and three other tribal Indian groups from South America.

Brown et al. (1980) have suggested another way for testing for overall linkage disequilibrium. This is based upon the number of loci that are different (heterozygous) when two random gametes are compared in a population. Suppose that there are L loci, with the population frequency of the ith allele at the lth locus being p_{il}. Then for the lth locus the probability of two random gametes being different is

$$h_l = 1 - \sum_i p_i^2.$$

Still considering a random pair of gametes, let X_l be an indicator variable such that $X_l = 1$ if the two alleles differ at the lth locus and $X_l = 0$ if they are the same. Then the probability that $X_l = 1$ is h_l and the probability that $X_l = 0$ is $1 - h_l$. The total number of heterozygous loci is then

$$K = X_1 + X_2 + \ldots + X_L.$$

It follows immediately that the mean value of K is

$$\mu = \sum_{l=1}^{L} h_l. \tag{10.25}$$

The variance of X_l is easily shown to be $h_l(1 - h_l)$. Therefore if the allele distributions are independent for all L loci, it follows that the variance of K is

$$\sigma^2 = \sum_{l=1}^{L} \text{var}(X_l) = \sum_{l=1}^{L} h_l(1 - h_l). \tag{10.26}$$

On the assumption of independent loci, the third and fourth moments about the mean for K are

$$\mu_3 = \sigma^2 - 2(\Sigma h_i^2 - \Sigma h_i^3) \tag{10.27}$$

and

$$\mu_4 = \sigma^2 - 6(\Sigma h_i^2 + 2\Sigma h_i^3 - \Sigma h_i^4) + 3\sigma_k^4. \tag{10.28}$$

Brown et al. have suggested that any linkage disequilibrium is liable to increase the variance of K above σ^2 as given by equation (10.26). A test for linkage disequilibrium at some or all of the L loci therefore involves seeing whether the observed variance of K for a random sample of n pairs of gametes is significantly larger than σ^2. Of course, h_l will have to be estimated to evaluate σ^2. This can be done using

$$\hat{h}_l = 1 - \Sigma \hat{p}_{il}^2, \tag{10.29}$$

where $\hat{p}_{il}$ is the sample proportion of allele i at the lth locus.

If there is no simple association between pairs of loci then the observed third moment of K may differ from μ_3 of equation (10.27) because of associations between three or more loci. Thus the third moment may be a useful index of complex disequilibria. However, no formal test is yet available based upon this principle.

For testing the variance, Brown *et al.* assumed that the sample variance of K is normally distributed. This seems a somewhat doubtful assumption except for rather large samples. On the other hand, K itself is a linear combination of variables and the central limit theorem suggests that this is likely to be normally distributed. A realistic test may therefore be based upon the chi-square distribution. That is to say, if a random sample of n gamete pairs gives the variance s^2 for K, then the statistic

$$X^2 = (n-1)s^2/\sigma^2 \qquad (10.30)$$

can be tested against the chi-square distribution with $n-1$ degrees of freedom. Just how good this test is depends upon how close the distribution of K is to being normal, assuming no linkage disequilibrium. This can be checked by calculating the skewness (μ_3/σ^3) and kurtosis (μ_4/σ^4) to verify that they are close to the normal distribution values of zero and three, respectively.

Given a sample of N individuals in a randomly mating population, it is straightforward to evaluate K for each individual and hence work out the test statistic X^2 of equation (10.30), and to perform the other calculations just described. Some simulations carried out by Chris Aston (private communication) have indicated that this approach works rather well. It seems possible that if it is used with a population which is thought to be randomly mating, but in reality is not, then a significant test result may occur because of the non-random mating.

If a sample of n gametes is available then Brown *et al.* suggest that one way to determine the sample variance and higher moments is to pair up every gamete with itself and every other gamete to form a 'sample' of n^2 gamete pairs. This 'sample' can then be used to evaluate n^2 values of K and hence the variance and other moments of K. Brown *et al.* treat the 'sample' variance of s^2 obtained in this way like a variance calculated from a sample of n individuals. However, according to Chris Aston's simulations this is not a valid procedure. It is valid to pair up the n gametes randomly into $\frac{1}{2}n$ 'individuals' and then calculate the $\frac{1}{2}n$ corresponding K values, with their variance and so on. Unfortunately, pairing up individuals in this manner obviously results in a loss of information since two random pairings will not give the same K values. Some form of randomization test would be better.

In recent papers (Brown and Feldman, 1981; Brown and Marshall, 1981), Brown and his colleagues have developed this approach further, particularly for studies on multilocus structure of plant populations.

Example 10.4   Linkage disequilibrium
for *Escherichia coli*

Whittam *et al.* (1983) used Brown *et al.*'s (1980) test for linkage disequilibrium in a study of 12 enzyme loci in natural populations of *Escherichia coli*. From an analysis of 1705 clones from various human and animal sources in North

America, 302 electrophoretic types were identified and treated as a random sample of gametes. Equations (10.26) to (10.28) produced the theoretical moments $\sigma^2 = 2.292$, $\mu_3 = -0.151$ and $\mu_4 = 15.187$ for K, the number of heterozygous loci for two randomly chosen gametes. The skewness is therefore $\mu_3/\sigma^3 = -0.044$ and the kurtosis is $\mu_4/\sigma^4 = 2.89$ for the distribution of K, assuming no linkage disequilibrium. Since the skewness and kurtosis are close to the normal distribution values of zero and three, this suggests that K can be treated as being normally distributed. Thus the test statistic X^2 of equation (10.30) can be compared with the chi-square distribution with some confidence.

Whittam *et al.* (1983) followed the suggestion of Brown *et al.* in estimating the sample variance s^2 by working out a K value for every possible pair of gametes and treating the variance of these K values as if it is calculated from a random sample of $n = 302$ individuals. As noted above, this procedure is not valid and it would have been better to have paired the 302 gametes up at random to produce 151 'zygotes', with corresponding K values. However, disregarding this point, Whittam *et al.*'s sample variance of 3.218 will be regarded as based upon a sample of 302 zygotes. The statistic X^2 is then

$$X^2 = 301 \times 3.218/2.292 = 422.6,$$

with 301 degrees of freedom.

Testing X^2 for significance is not straightforward with so many degrees of freedom. However, a standard approximation for testing the significance of a chi-square variate with v degrees of freedom involves comparing

$$Z = (2X^2)^{1/2} - (2v-1)^{1/2}$$

with the standard normal distribution. In the present case this gives

$$Z = (2 \times 422.6)^{1/2} - \sqrt{601} = 4.56,$$

which is significantly large at the 0.1 % level. On this basis there is very strong evidence that the variance of K is larger than the value given by equation (10.26) due, no doubt, to substantial linkage disequilibrium.

Following this test, Whittam *et al.* carried out a principal component analysis to identify four-locus allele combinations showing high variation. Smouse's (1974) log-linear analysis was then used to further study the four loci interactions.

10.7 ESTIMATION OF SELECTIVE VALUES FOR TWO LOCI BY TURNER'S METHOD

The difficulties involved in estimating selective values for a single genetic locus have been discussed in Chapter 8. These are compounded when two or more loci are considered together. Good estimates require very large samples. Complications are introduced if samples are taken before all selection is completed.

For two loci, with two alleles each, random mating leads to the genotype proportions shown in Table 10.5 at the time of the formation of zygotes. Selection from that time until maturity will then change the relative frequencies to those shown in Table 10.9, where w_{ij} is the selective value for a type i gamete combined with a type j gamete. Here the four types of gametes are $A_1 B_1, A_2 B_1, A_1 B_2$ and $A_2 B_2$ so that, for instance, w_{12} is the selective value for $A_1 B_1$ with $A_2 B_1$. There are nine distinguishable genotypes and hence nine selective values, assuming that none of the alleles are dominant.

Table 10.9   Relative genotype frequencies after zygotic selection.

Genotype at locus 1	Genotype at locus 2		
	$B_1 B_1$	$B_1 B_2$	$B_2 B_2$
$A_1 A_1$	$w_{11} P_{11}^2$	$2 w_{13} P_{11} P_{12}$	$w_{33} P_{12}^2$
$A_1 A_2$	$2 w_{12} P_{11} P_{21}$	$2 w_{23}(P_{11} P_{22} + P_{12} P_{21})$	$2 w_{34} P_{12} P_{22}$
$A_2 A_2$	$w_{22} P_{21}^2$	$2 w_{24} P_{21} P_{22}$	$w_{44} P_{22}^2$

Note: The four types of gametes are $A_1 B_1, A_2 B_1, A_1 B_2$ and $A_2 B_2$. The selective value of the ith type with the jth type is denoted by w_{ij}. For example, an $A_1 B_1$ gamete combined with an $A_2 B_1$ gamete gives the genotype $A_1 A_2 B_1 B_1$, with selective value w_{12}. The genotype $A_1 A_2 B_1 B_2$ is formed by either $A_1 B_1 + A_2 B_2$ or $A_1 B_2 + A_2 B_1$ gametes. This means that with the notation used here, $w_{23} = w_{14}$.

A rather simple approach for estimating the selective values for unlinked loci was suggested by Lewontin and White (1960), based on the assumption that the population being studied is in an equilibrium state with no linkage disequilibrium. Turner (1972) argued that better results can be obtained by an alternative method which still assumes an equilibrium state, but allows a constant non-zero linkage disequilibrium.

Turner's method is based upon comparing the genotype frequencies in a sample taken after selection with the frequencies expected before selection from random mating. The before-selection proportions should be as shown in Table 10.5 and the after-selection relative frequencies as shown in Table 10.9. Let $N(A_i A_j B_k B_l)$ denote the after-selection sample frequency of the genotype $A_i A_j B_k B_l$. The sample proportion of this genotype is then $N(A_i A_j B_k B_l)/N$ if the total sample size is N individuals. Fitness estimates are then

$$\left.\begin{aligned}
&\hat{w}_{11} = N(A_1 A_1 B_1 B_1)/(N\hat{P}_{11}^2), \quad \hat{w}_{12} = N(A_1 A_2 B_1 B_1)/(2N\hat{P}_{11}\hat{P}_{21}), \\
&\hat{w}_{22} = N(A_2 A_2 B_1 B_1)/(N\hat{P}_{21}^2), \quad \hat{w}_{13} = N(A_1 A_1 B_1 B_2)/(2N\hat{P}_{11}\hat{P}_{12}), \\
&\hat{w}_{23} = N(A_1 A_2 B_1 B_2)/\{2N(\hat{P}_{11}\hat{P}_{22} + \hat{P}_{12}\hat{P}_{21})\}, \\
&\hat{w}_{24} = N(A_2 A_2 B_1 B_2)/(2N\hat{P}_{12}\hat{P}_{22}), \quad \hat{w}_{33} = N(A_1 A_1 B_2 B_2)/(N\hat{P}_{12}^2), \\
&\hat{w}_{34} = N(A_1 A_2 B_2 B_2)/(2N\hat{P}_{12}\hat{P}_{22}) \quad \text{and} \\
&\hat{w}_{44} = N(A_2 A_2 B_2 B_2)/(N\hat{P}_{22}^2),
\end{aligned}\right\} (10.31)$$

where $\hat{P}_{11}$, $\hat{P}_{21}$, $\hat{P}_{12}$ and $\hat{P}_{22}$ are estimates of the gamete proportions in the sample before selection.

Turner argued that if the population is in equilibrium, then the gamete proportions before selection will be equal to the gamete proportions produced by the adults after selection. Therefore the values $\hat{P}_{ij}$ in equations (10.31) can be estimated from the after-selection sample. Each individual in this sample is expected to produce the same number of gametes if there is random mating, so it is just a question of counting the expected output numbers, taking each individual as producing two gametes: two $A_1 B_1$ gametes for each $A_1 A_1 B_1 B_1$ individual; one $A_1 B_1$ and one $A_1 B_2$ gamete for each $A_1 A_1 B_1 B_2$ individual; and so on. Since unlinked loci are being considered, the $A_1 A_2 B_1 B_2$ individuals should produce $A_1 B_1$, $A_2 B_1$, $A_1 B_2$ and $A_2 B_2$ gametes in equal numbers. On this basis, the gamete proportions output by the population after selection are estimated to be

$$
\begin{aligned}
\hat{P}_{11} &= \{2N(A_1 A_1 B_1 B_1) + N(A_1 A_2 B_1 B_1) + N(A_1 A_1 B_1 B_2) \\
&\quad + \tfrac{1}{2}N(A_1 A_2 B_1 B_2)\}/2N, \\[6pt]
\hat{P}_{21} &= \{2N(A_2 A_2 B_1 B_1) + N(A_1 A_2 B_1 B_1) + N(A_2 A_2 B_1 B_2) \\
&\quad + \tfrac{1}{2}N(A_1 A_2 B_1 B_2)\}/2N, \\[6pt]
\hat{P}_{12} &= \{2N(A_1 A_1 B_2 B_2) + N(A_1 A_2 B_2 B_2) + N(A_1 A_1 B_1 B_2) \\
&\quad + \tfrac{1}{2}N(A_1 A_2 B_1 B_2)\}/2N,
\end{aligned}
\tag{10.32}
$$

and

$$
\begin{aligned}
\hat{P}_{22} &= \{2N(A_2 A_2 B_2 B_2) + N(A_1 A_2 B_2 B_2) + N(A_2 A_2 B_1 B_2) \\
&\quad + \tfrac{1}{2}N(A_1 A_2 B_1 B_2)\}/2N,
\end{aligned}
$$

for $A_1 B_1$, $A_2 B_1$, $A_1 B_2$ and $A_2 B_2$, respectively.

There are nine selective values being estimated here from nine genotype frequencies. Just temporarily it is convenient to denote the genotype frequencies as $f_1, f_2, \ldots, f_9$, in any order, and the estimated selective values corresponding to these as $\hat{w}_1, \hat{w}_2, \ldots, \hat{w}_9$. Then, using equation (A25) of the Appendix, Section A.5,

$$
\operatorname{var}(\hat{w}_i) \simeq \sum_{j=1}^{9} \Delta_{ij}^2,
\tag{10.33}
$$

where Δ_{ij} is the change in $\hat{w}_i$ resulting from a one standard deviation increase $(+\sqrt{f_j})$ in the jth genotype frequency. Also, from equation (A26),

$$
\operatorname{cov}(\hat{w}_i, \hat{w}_k) \simeq \sum_{j=1}^{9} \Delta_{ij} \Delta_{kj}.
\tag{10.34}
$$

Biases can be approximated using equation (A27) if this is considered necessary.

Turner's method can be generalized to cases where there are more than two alleles at one or both of the loci being studied. The only problem is that the number of possible genotypes increases very quickly with the number of alleles. The result is then likely to be many very small genotype frequencies with correspondingly poor estimates of selective values.

It seems necessary to stress the importance of the equilibrium assumption as far as the estimates are concerned. An example should be sufficient to demonstrate the errors that can arise when this does not hold. Suppose that a certain population has equal numbers of the gametes $A_1 B_1$, $A_2 B_1$, $A_1 B_2$ and $A_2 B_2$ among adults, who mate at random. Then $P_{11} = P_{21} = P_{12} = P_{22} = 0.25$ and the genotype proportions in the progeny are as follows:

	$B_1 B_1$	$B_1 B_2$	$B_2 B_2$
$A_1 A_1$	0.063	0.125	0.063
$A_1 A_2$	0.125	0.250	0.125
$A_2 A_2$	0.063	0.125	0.063

Let the array of selective values for the genotypes, in the same order, be

$$\left.\begin{array}{ccc} 0.416 & 0.666 & 0.832 \\ 0.666 & 1.065 & 1.332 \\ 0.832 & 1.332 & 1.665 \end{array}\right\} \qquad (10.35)$$

where these values are scaled so as to make the mean selective value equal to one. Then after selection the relative genotype frequencies become

$$\left.\begin{array}{ccc} 0.026 & 0.083 & 0.052 \\ 0.083 & 0.266 & 0.167 \\ 0.052 & 0.167 & 0.105 \end{array}\right\} \qquad (10.36)$$

For example, the relative frequency of $A_1 A_1 B_1 B_1$ becomes the original proportion multiplied by the selective value: $0.063 \times 0.416 = 0.026$.

Suppose now that Turner's method of estimation is applied to a sample of 1000 adults from this population. The expected sample genotype frequencies are then the values given in the array (10.36), multiplied by 1000. Applied to these frequencies, Turner's method gives the following array of estimated selective values:

$$\begin{array}{ccc} 0.84 & 0.97 & 0.88 \\ 0.97 & 1.12 & 1.01 \\ 0.88 & 1.01 & 0.91. \end{array}$$

These bear no resemblance at all to the true values which are as given in the array (10.35). The point is that the population being considered is not in equilibrium. Therefore Turner's estimates of selective values are invalid.

Example 10.5   Selective values for *Keyacris scurra* in New South Wales

In New South Wales, Australia, the flightless grasshopper *Keyacris scurra* (formerly *Moraba scurra*; Hedrick *et al.*, 1978) shows inversion polymorphisms in two different chromosome pairs which do not seem to combine at random (White, 1957; Lewontin and White, 1960; White *et al.*, 1963). This was the example used by Turner (1972) to develop his method for estimating selective values.

For the present example, interest will centre on some data collected in 1955 and 1956 and recorded in Table 1 of White (1957). These data are provided here in Table 10.10, where it will be seen that two colonies (Hall and Royalla A) were sampled in 1955 and four colonies (Wombat, Hall, Royalla B and Williamsdale) were sampled in 1956. White noted that these colonies are isolated relics of a much more widespread and continuous population which existed about 120 years ago. Migration between the colonies is hardly possible at all at the present time because of the changes to the habitat caused by grazing sheep. Only samples of adult males are available.

The two 'loci' being considered are not really loci at all. Rather, they are chromosome types. The CD chromosomes in samples can be classified as Standard (St) or Blundell (Bl) while the EF chromosomes can be classified as Standard or Tidbinbilla (Tid). In practice the chromosome types can be treated as 'alleles' on two unlinked loci and the selective values of different 'genotypes' can be estimated using Turner's (1972) method. However, before this can be done it is necessary to have some evidence that the different colonies are in states of equilibrium with regard to the distribution of CD and EF chromosomes.

Table 10.11 shows estimates $\hat{\Delta}_{11}$ of linkage disequilibrium from Burrows's formula (10.16), values of Langley *et al.*'s standardized statistic $\hat{R}$ of equation (10.20), standard errors for the $\hat{R}$ values from equation (10.21), and chi-square test statistics $X^2 = 4N\hat{R}^2$ from equation (10.22). All of the colonies appear to have positive values for Δ_{11}, although the X^2 statistic is only significant for the Hall colony in 1956. The sum of the X^2 values, 13.48, can be compared with the chi-square distribution with six degrees of freedom to give an overall test for linkage disequilibrium. The sum is significant at the 5 % level, so there is some evidence of linkage disequilibrium for the colonies taken together.

In general, if there are K independent estimates $\hat{R}_1, \hat{R}_2, \ldots, \hat{R}_K$, based on sample sizes $N_1, N_2, \ldots, N_K$, then the total of the X^2 values can be partitioned as

$$\Sigma X_k^2 = 4\Sigma N_k \hat{R}_k^2 = 4\Sigma N_k (\hat{R}_k - \overline{R})^2 + 4\overline{R}^2 \Sigma N_k,$$

Table 10.10   The distribution of chromosome pairs in adult male *Keyacris scurra* from isolated colonies in New South Wales.

Chromosome EF	Colony	Year	Chromosome CD			
			St/St	St/Bl	Bl/Bl	Totals
St/St	Wombat	1956	14	197	509	720
	Hall	1955	52	256	250	558
	Hall	1956	46	221	201	468
	Royalla A	1955	42	184	167	393
	Royalla B	1956	37	186	156	379
	Williamsdale	1956	39	78	45	162
St/Tid	Wombat	1956	5	61	195	261
	Hall	1955	22	87	98	207
	Hall	1956	7	50	63	120
	Royalla A	1955	25	117	136	278
	Royalla B	1956	16	96	86	198
	Williamsdale	1956	21	56	46	123
Tid/Tid	Wombat	1956	0	8	11	19
	Hall	1955	0	2	10	12
	Hall	1956	0	5	7	12
	Royalla A	1955	3	12	14	29
	Royalla B	1956	2	8	13	23
	Williamsdale	1956	3	7	5	15
Totals	Wombat	1956	19	266	715	1000
	Hall	1955	74	234	358	777
	Hall	1956	53	276	271	600
	Royalla A	1955	70	313	317	700
	Royalla B	1956	55	290	255	600
	Williamsdale	1956	63	141	96	300

where $\bar{R} = \Sigma N_k R_k / \Sigma N_k$ is the pooled estimate of R with minimum variance (Appendix, Section A.6). The term $4\bar{R}^2 \Sigma N_k$ can be compared with the chi-square distribution with one degree of freedom; a significantly large value indicates that the mean value of R for all populations is non-zero. The term $4\Sigma N_k (\hat{R}_k - \bar{R})^2$ can be compared with the chi-square distribution with $K - 1$ degrees of freedom; a significantly large value indicates that the value of R varies from population to population.

For the present data, $\bar{R} = 0.0254$ and $\Sigma N_k = 3977$. Therefore $4\bar{R}^2 \Sigma N_k = 10.26$, with one degree of freedom. This is significantly large at the 1 % level. There is clear evidence that the average linkage disequilibrium is not zero.

Table 10.11   Linkage disequilibrium in colonies of *Keyacris scurra* as determined from the data given in Table 10.10.

Colony	Sample size N_k	Burrows's $\hat{\Delta}_{11}$	$\hat{R}_k$	$SE(\hat{R}_k)$	Chi-square statistic X_k^2 (1 d.f.)
Wombat, 1956	1000	0.0020	0.008	0.016	0.23
Hall, 1955	777	0.0075	0.023	0.018	1.59
Hall, 1956	600	0.0150	0.049	0.020	5.87*
Royalla A, 1955	700	0.0103	0.026	0.019	1.85
Royalla B, 1956	600	0.0089	0.023	0.020	1.31
Williamsdale, 1956	300	0.0203	0.047	0.029	2.63

Total X^2 (6 d.f.) = 13.48*

* Significant at the 5% level.

From Table 10.11, the total chi-square value of $4\Sigma N_k \hat{R}_k^2$ is 13.48. Therefore the heterogeneity chi-square value is

$$4\Sigma N_k (\hat{R}_k - \overline{R})^2 = 4\Sigma N_k \hat{R}_k^2 - 4\overline{R}^2 \Sigma N_k$$
$$= 13.48 - 10.26 = 3.22,$$

with five degrees of freedom. This is far from significant, so there is no evidence of R varying from population to population.

The picture that emerges from this analysis is that the colonies of *K. scurra* all display about the same small amount of linkage disequilibrium. This suggests a certain stability in the populations, which is what is required in order for Turner's (1972) method for estimating selective values to be valid. It is only for the Hall colony that samples are available for two times, 1955 and 1956. A chi-square test shows no significant difference between the genotype proportions in the two years, so that there is evidence of stability over time, at least in this location. From an analysis of samples taken at different times from eight colonies of *K. scurra*, White *et al.* (1963) concluded that there is very great stability in genotype frequencies from year to year for this species.

Assuming that equilibrium did exist for the populations, the estimates of selective values shown in Table 10.12 were calculated from equations (10.31) and (10.32), and standard errors were calculated using equation (10.33). From the size of the standard errors it is quite clear that individual estimates are not particularly reliable. However, consistent results between populations suggest that in general the genotypes St/St + St/St, St/St + Bl/Bl, and St/Tid + St/St are less fit than average, the genotype St/Tid + St/Bl has about average fitness, and the genotypes St/St + St/Bl and St/Tid + Bl/Bl have above average fitness. The small number of individuals with a Tid/Tid chromosome has meant that

Table 10.12 Estimated selective values and standard errors for *Keyacris scurra*.

| Genotype | | Wombat 1955 | | Hall 1955 | | Hall 1956 | | Royalla A 1955 | | Royalla B 1956 | | Williamsdale 1955 | | Mean |
EF	CD	$\hat{w}$	SE	$\hat{w}$	SE	$\hat{w}$	SE	$\hat{w}$	SE	$\hat{w}$	SE	$\hat{w}$	SE	$\hat{w}$
St/St	St/St	0.83	0.18	0.89	0.09	0.93	0.09	0.95	0.10	0.85	0.10	1.11	0.10	0.93
	St/Bl	1.05	0.04	1.04	0.04	1.08	0.05	1.03	0.05	1.09	0.05	0.94	0.08	1.04
	Bl/Bl	0.98	0.01	0.96	0.02	0.95	0.03	0.92	0.03	0.94	0.03	0.92	0.08	0.95
St/Tid	St/St	0.88	0.34	1.19	0.18	0.66	0.20	0.98	0.16	0.78	0.16	0.99	0.16	0.91
	St/Bl	0.94	0.08	1.04	0.08	0.95	0.10	1.06	0.26	1.12	0.08	1.00	0.11	1.02
	Bl/Bl	1.06	0.04	1.04	0.06	0.99	0.08	1.14	0.07	0.97	0.07	1.25	0.12	1.08
Tid/Tid	St/St	0.00	–	0.00	–	0.00	–	0.82	0.51	0.83	0.51	0.94	0.43	0.43
	St/Bl	1.44	0.41	0.28	0.19	1.52	0.56	0.71	0.23	0.75	0.23	0.74	0.24	0.91
	Bl/Bl	0.67	0.17	1.16	0.29	1.47	0.42	0.71	0.23	1.11	0.23	0.72	0.26	0.97

the selective values for genotypes that include this chromosome are estimated very poorly.

One of the main concerns of White (1957), Lewontin and White (1960), White *et al.* (1963) and Turner (1972) was to explain the apparent stability of the *K. scurra* populations. They examined this graphically by seeing how the mean population fitness is a function of chromosome frequencies. See their papers for more details.

10.8 MULTIPLE LOCUS ESTIMATION OF SELECTIVE VALUES IN GENERAL

In non-equilibrium situations, Turner's (1972) method for estimating selective values cannot be used. However, the equations (10.31) still apply where $N(A_i A_j B_k B_l)$ is the sample frequency of the genotype $A_i A_j B_k B_l$ after selection in one generation, while $\hat{P}_{11}$, $\hat{P}_{21}$, $\hat{P}_{12}$ and $\hat{P}_{22}$ are estimated gamete proportions produced by randomly mating adults in the previous generation. Thus the selective values w_{ij} can be estimated for a population that is not in equilibrium providing that samples are taken from a population after selection in two consecutive generations. The first sample can then be used to estimate the gamete proportions using equations (10.32). The second sample frequencies are then needed in equations (10.31). Note, however, that this approach only works for unlinked loci since equations (10.32) have been derived on this assumption.

With linked loci the situation is more complicated. Table 10.1 shows that if the gametes $A_1 B_1, A_2 B_1, A_1 B_2$ and $A_2 B_2$ are in proportions P_{11}, P_{21}, P_{12} and P_{22} among randomly mating adults then the gamete proportions among the progeny will be

$$P'_{11} = P_{11} - rD, \qquad P'_{21} = P_{12} + rD,$$

$$P'_{12} = P_{12} + rD \quad \text{and} \quad P'_{22} = P_{22} - rD,$$

where $D = P_{11} P_{22} - P_{12} P_{21}$ and r is the recombination rate. The progeny genotype proportions before selection will then be as shown in Table 10.5 but with P_{ij} replaced by P'_{ij}. After selection, the relative frequency will be as shown in Table 10.9, again with P_{ij} changed to P'_{ij}. Given samples after selection in two consecutive generations, these relationships can be used to estimate the selective values, providing r is known. The first sample can be used to estimate the gametic proportions P'_{ij} in offspring, and hence the expected genotype proportions before selection. The genotype proportions in the second sample can then be compared with these to estimate selective values.

Weir *et al.* (1972) provided equations for the estimation of two locus selective values in organisms with inbreeding, and applied them to data from barley populations. This work was later generalized for three locus selective values by Clegg *et al.* (1978). A number of studies have simplified the estimation of selective values by using recessive lethal genes in laboratory populations. For example, Clarke *et al.* (1981) and Clarke and Feldman (1981) carried out ten-generation selection experiments on laboratory populations of *Drosophila melanogaster* bearing Curly and Plum marked second chromosome inversions. With this system, five of the nine possible genotypes are lethal so that their selective values are known to be zero. Consequently there are only four selective values to be estimated from data. Samples were considered to be taken after partial selection in each generation, so Clarke *et al.* (1981) estimated early and late components of selective values following the one-locus models of Prout (1965, 1969).

A major concern with two-locus studies has been the question of whether there is evidence for epistasis (fitness interactions): is the selective value for a two-locus genotype the product of the selective values for the single loci? The evidence has been reviewed by Hedrick *et al.* (1978). In general, it seems to be a fair comment that one-locus fitness values may have very little relationship with the way that selection operates on that locus.

11 Selection on quantitative variables

Genetic theory is broadly divided into two areas. One area is concerned with specific genetic loci, while the other is concerned with quantitative variables for which the values are determined by several or many loci together. The topics covered in Chapters 8 to 10 have been solely in the first area. It is the second area that is now given prominence.

The chapter starts with a review of the theory of the genetics of quantitative variation. Nonparametric tests for directional selection are then reviewed. There is then a description of a regression model for directional selection that allows for genetic drift. Some models for long-term evolution are described in Sections 11.8 and 11.9. The chapter concludes with a short section on the comparison between the level of variation in marginal and central populations. The idea here is that marginal populations should be less adaptable than central ones due to unequal levels of selection.

11.1 QUANTITATIVE GENETICS

In most cases, phenotypic variation is partly genetic and partly environmental in origin. For example, tall parents tend to have tall children, but height is also determined to some extent by nutrition. In simple cases the genetic and environmental components of variation will be independent, in which case the value X of a character for an individual can be written as

$$X = \mu + G + E$$

where μ is the population mean, from which this individual has a deviation G from genetic causes and E from environmental causes, with G and E being uncorrelated. Then

$$V_P = V_G + V_E,$$

where V_P is the phenotypic variance, V_G is the genotypic variance and V_E is the environmental variance.

The genotypic effect G for an individual is assumed in quantitative genetics to be the result of a large number of genes at different loci all having an effect. Generally the effect of individual genes is not known. Indeed, at the present time little is known even about the way effects are likely to operate. It may be that all genes tend to have effects of about the same size. Alternatively, there may be a few genes with large effects and a large number of genes with small effects. Whatever the situation may be, the genotypic effect can itself be split up into three main components, $G = A + D + I$, where A is the additive effect of an individual's genes according to their average population effect, D is the deviation from this caused by dominance between alleles at the same locus, and I is the effect due to interaction between genes at different loci. The corresponding breakdown for the variance of G is

$$V_{G} = V_{A} + V_{D} + V_{I}.$$

It is important to appreciate that these components of variance are functions of the gene frequencies as well as of the genotypes. They are not necessarily constant over time. Also, different populations are likely to have different components of variance even when the genotypes are the same.

The model $X = \mu + A + D + I + E$ can be made more complicated (and possibly more realistic) by allowing for correlations and interactions between genotypes and environments (Falconer, 1981, p. 121). However, it is sufficient for most purposes.

Heritability is an important parameter of a population. There are two definitions of this in use. The one that is most relevant to natural selection is

$$h^2 = V_{A}/V_{P}, \tag{11.1}$$

the ratio of additive genetic variance to the total phenotypic variance. This is sometimes called 'heritability in the narrow sense'. The alternative definition of heritability is V_{G}/V_{P}, which expresses the extent to which phenotypic values are determined genetically. This is 'heritability in the broad sense'. In the following discussions, heritability will always be defined as in equation (11.1).

The simplest way to estimate heritability in a population is perhaps to rely upon the relationship between parents and progeny. Falconer (1981) shows that if X is the phenotypic value of a single parent and Y is the corresponding mean value for offspring, then the regression equation of Y on X has the form

$$Y = \mu + \tfrac{1}{2}h^2(X - \mu) + e,$$

or

$$Y = \mu(1 - \tfrac{1}{2}h^2) + \tfrac{1}{2}h^2 X + e.$$

Here μ is the mean for the parent population and e reflects environmental effects plus genetic effects, including the effect of the second parent, with a mean of zero. It can be seen that the usual linear regression coefficient provides an estimator of $\tfrac{1}{2}h^2$.

If possible, it is better to use the average X value from both parents (the 'mid-parent' value) in the regression, in which case the theoretical regression equation is

$$Y = \mu + h^2(X - \mu) + e,$$

or

$$Y = \mu(1 - h^2) + h^2 X + e, \tag{11.2}$$

and the regression coefficient is a direct estimator of h^2. These regression procedures remain valid if there is assortative mating so that parents tend to have similar X values. Indeed, this actually increases the precision of the estimation of heritability.

11.2 THE EFFECTS OF DIRECT SELECTION ON A CHARACTER

The relationship shown in equation (11.2) between parental and offspring values of a character makes it possible to predict the effect of selection on a population mean. Consider Fig. 11.1, which shows a hypothetical population in which certain parent–offspring pairs with high character values are selected. This figure shows that if a group of parents with a mid-parent mean $\mu + S$ is chosen to produce the next generation, then the mean for the progeny is

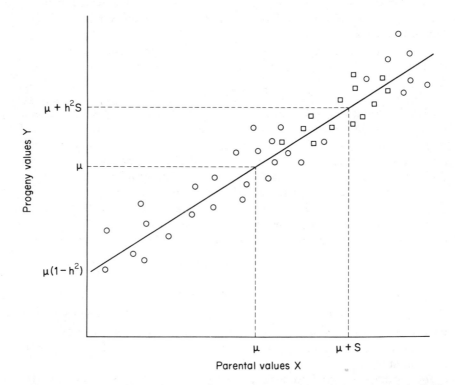

Fig. 11.1 The effects of selection on a population mean. The selected individuals (squares) have a mean of $\mu + S$, where μ is the overall population mean for all individuals, selected and unselected (squares and circles). The expected mean for the progeny of selected individuals is then $\mu + h^2 S$. The theoretical regression line is $Y = \mu(1 - h^2) + h^2 X$, where h^2 is the heritability.

expected to be $\mu + h^2S$. The change in the population mean,

$$R = h^2S, \tag{11.3}$$

is called the *selection response*.

If the heritability h^2 is known, then equation (11.3) enables the effect of selection on the mean to be predicted. The prediction in theory is valid only for one generation because selection can be expected to change h^2. However, this is a problem that is not too important in practice because much more serious deviations from the equation are liable to occur from other causes. It is for this reason that quantitative geneticists differentiate between heritability as defined by the regression in Figure 11.1 and 'realized heritability', which is the h^2 value that needs to be used to explain experimental results.

An illustration of the difficulties involved in using equation (11.3) is provided by artificial selection experiments with a high line and a low line. Each experiment starts with a single population. A high population is generated by selecting individuals with a mean S units higher than average to form the parents for each generation. Thus the response to selection should be an increase in the mean of h^2S per generation for this population. On the other hand, for the low population, which is kept quite separate from the high one, the parents for each generation are chosen to have a mean of S units below average. The response to selection should then be a decrease in the mean of h^2S per generation. Thus the high and low populations should have mean values that move away from the initial population mean at the same rates but in opposite directions. In practice this may not occur.

There are various reasons for an asymmetric response to selection. The essential point to note is that the equation (11.3) can be expected to give a fair approximation for the change in the population mean due to selection until such time as the population approaches some form of genetic limit. However, the h^2 value in the equation may not be equal to the theoretical value given by the regression of progeny values on parent values for one generation.

The effect of selection on the variance of a population is complicated. The effect of artificial selection involving extreme individuals being chosen as parents in each generation should lead in theory to a reduction in the variance of the character being studied. However, this does not often happen: the variance may even increase; often it remains more or less constant. No simple general equation like (11.3) is available to describe what is likely to occur.

11.3 CORRELATED CHARACTERS

So far it has been implicitly assumed that selection is acting directly on the character being considered. This may be the case with artificial selection, but in studying natural selection the best that can be done is to measure characters that are correlated with fitness. There is therefore a need to understand how selection on a character X affects the distribution of a correlated character Y.

In general, the correlation between two characters will be due partly to genetic and partly to environmental causes. The genetic correlation measures the extent to which two measurements reflect the same genes. The environmental correlation represents all other correlations including those due to non-additive genetic effects.

It is the genetic correlation that is relevant to selection. Thus suppose selection is made on the character X so that the individuals that are selected as parents have a mean S above average. Then the response to selection is given by equation (11.3) to be a change of $R_X = h_X^2 S$ in the mean of X, where h_X^2 is the heritability of X. Equations given by Falconer (1981, p. 286) show that the corresponding change in the mean of Y (the correlated response) is expected to be

$$R_Y = h_X h_Y r_A S \sigma_Y / \sigma_X \tag{11.4}$$

where σ_X and σ_Y are the standard deviations of X and Y before selection, h_Y is the heritability of Y, and r_A is the genetic correlation between X and Y. Equation (11.4) can be used to estimate r_A if the other terms are known. Alternatively, this correlation can be estimated using regressions of offspring values on parental values.

11.4   NATURAL SELECTION

Every individual has associated with it a fitness value w, which can for simplicity be assumed to be proportional to the number or progeny produced. With natural populations it will seldom, if ever, be possible to observe w directly. However, it may be the case that fitness is highly correlated with certain other easily measured characters, in which case the operation of selection can be inferred from the effects on these other variables. Even strong selection may have no appreciable effects at all. Equation (11.3) shows that if heritability is zero then selection cannot change the mean, no matter how strong it is.

11.5   FITNESS FUNCTIONS

Fitness functions have been used in earlier chapters to represent the relationship between fitness and one or more quantitative variables. It is appropriate at this point to consider the question of how different types of fitness function affect a population genetically.

Fig. 11.2 shows three possible forms for a fitness function. For function (a) fitness increases with increasing values of a character X. This is directional selection which is trying to increase the population mean. If heritability is zero it cannot do this. There may be a genetic component of fitness but this is all due to dominance deviations and interactions between loci, with no additive genetic variance. The population is in equilibrium, with the distribution of

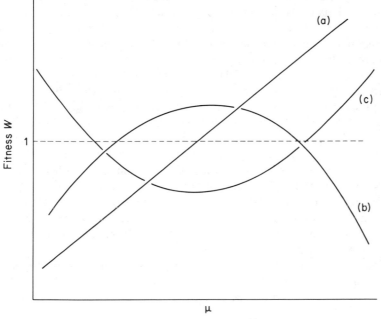

Fig. 11.2   The relationships that are likely between fitness and a quantitative character X (fitness functions). The fitness W is the mean value for all individuals with the value of X. The population mean is μ.

gene frequencies being the best possible under the circumstances. On the other hand, if the heritability is not zero then the population mean should increase from generation to generation. As this increase occurs the heritability should slowly decrease so that eventually an equilibrium is reached. While the population mean is changing, the variance may or may not remain constant. In some cases the measurement being studied is such that an increase in the mean will automatically bring about an increase in the variance. In that case a transformation of the variable may give a constant variance. For example, if the standard deviation is proportional to the mean then a logarithmic transformation will produce a stable variance.

With function (b) of Fig. 11.2 there is stabilizing selection. Individuals with intermediate values of X have the highest fitness. The genetic effects depend upon whether the value of the character is a direct cause of fitness or whether the apparent relationship is only due to correlation with other variables that are related to fitness. In the case of a variable like the clutch size of birds there is a clear causal relationship with fitness. Other things being equal, the birds with the largest clutch size are fittest. However, the number of chicks that can be successfully reared depends upon the food supply. The optimum clutch size is found by a balance between the number and quality of the chicks. A case of spurious relationship is the sternopleural bristle number in *Drosophila melanogaster*. Kearsey and Barnes (1970) show that the survival from egg to adult gives a fitness function like (b), although selection takes place before the flies have developed bristles. Therefore the real criteria for selection is something else which happens to be correlated with bristle number.

With spurious stabilizing selection there is no way of knowing what the genetic effects of selection will be. One possibility is that individuals with intermediate values of X are heterozygous at more loci than extreme individuals and that heterozygotes are fitter than homozygotes. Then the selection will tend to maintain variation in the character (Robertson, 1955). Kearsey and Barnes (1970) argued that this is not the case with sternopleural bristle numbers of *Drosophila melanogaster* because of the shape of the fitness function. True stabilizing selection should tend to reduce both environmental and genetic variance. Both of these effects have been observed in artificial selection experiments (Falconer, 1981, p. 310).

Function (c) of Fig. 11.2 gives disruptive selection. This is the opposite to stabilizing selection, with extreme individuals being favoured. It can be expected to increase both genetic and environmental variance, although in practice this may not occur.

Fig. 1.1 of Chapter 1 shows the expected effects of the different types of fitness function on the distribution of X, if there are any effects at all. Directional selection will move the mean by eliminating one tail of the distribution of X; stabilizing selection will tend to reduce the phenotypic variance; disruptive selection will tend to increase the phenotypic variance.

11.6 NONPARAMETRIC TESTS FOR NATURAL SELECTION

Lande (1977) considered the question of how best to test for natural selection on a quantitative variable given observations on the mean of a single population over a number of generations. He suggested that the sign test or the runs test (Dixon and Massey, 1969) can be applied to the change in the mean from the time of one observation to the next.

These nonparametric tests do not require any particular assumptions about the distribution of the observations. The signs of changes in the population mean are simply recorded; for example:

$$+ + + + + - + + + + - - + +$$

With the sign test it is a question of seeing whether the number of plus signs is significantly different from half of the total number of signs. With the runs test it is a question of whether the number of runs of similar signs (five in the above example) is significantly different from the expected number for a random series. Directional selection should produce a positive or negative trend in the population mean. This will tend to result in either a large or small number of plus signs and a small number of runs.

With these tests it is irrelevant whether the observations are taken at equal points in time. However, the tests require the assumption that successive changes in the population mean are independent, which will not be true unless the sampling errors in estimated population means are negligible. This is easy to see since if the sampling errors are not negligible then the ith observed mean will be of the form

$$\bar{X}_i = \mu_i + e_i,$$

where μ_i is the true population mean and e_i is the sampling error. Hence

$$\Delta \bar{X}_i = \bar{X}_i - \bar{X}_{i-1} = \mu_i - \mu_{i-1} + e_i - e_{i-1}$$

and

$$\Delta \bar{X}_{i+1} = \bar{X}_{i+1} - \bar{X}_i = \mu_{i+1} - \mu_i + e_{i+1} - e_i$$

will be two consecutive observed changes to the mean. Here the actual population changes $\mu_i - \mu_{i-1}$ and $\mu_{i+1} - \mu_i$ will be independent if they are due to random genetic drift only. However, the sampling errors $e_i - e_{i-1}$ and $e_{i+1} - e_i$ are not independent and

$$\text{cov}\,(\Delta \bar{X}_i, \Delta \bar{X}_{i+1}) = -\text{var}\,(e_i).$$

Therefore sampling errors will introduce a negative correlation between successive values of observed changes in the population mean; a positive change will tend to be followed by a negative change, and vice versa. This is the opposite of what can be expected from a trend in the population means (long runs of changes with the same sign). Hence sampling errors will, if anything, tend to hide a significant trend. From this point of view a significant result from a sign test or a runs test is evidence of a trend irrespective of whether or not there are sampling errors in population mean values.

Other tests for randomness of a sequence of means are possible, apart from the runs test and the sign test. Reyment (1982a, b) discusses some of these.

11.7   A REGRESSION MODEL ALLOWING FOR RANDOM GENETIC DRIFT

It might seem that a simple regression of mean values against time would be sufficient to test for directional selection, but this is not the case since some change in a population mean will occur through random genetic drift. A regression model has to take this into account. One approach is as follows.

Consider a population that is subject to natural selection with a constant pressure to change the mean. Let μ_i denote the population mean in the ith generation which is estimated by the mean $\bar{X}_i$ of a random sample of size n_i.

Then

$$\mu_i = \mu_{i-1} + \Delta + \delta_i \qquad (11.5)$$

where Δ is the expected change in the mean per generation due to selection, while δ_i is a deviation caused by random genetic drift. Recursive use of equation (11.5) gives

$$\mu_i = \mu_0 + i\Delta + \sum_{j=1}^{i} \delta_j \qquad (11.6)$$

where μ_0 is the population mean in an initial generation 0. Taking $\overline{X}_i$ as an estimator of μ_i then produces

$$\overline{X}_i = \mu_0 + i\Delta + \sum_{j=1}^{i} \delta_j + e_i$$

where e_i is the sampling error in $\overline{X}_i$. Conditional on the initial population mean of μ_0, $\overline{X}_i$ has variance

$$\mathrm{var}\,(\overline{X}_i) = i\sigma_d^2 + \sigma_p^2/n_i \qquad (11.7)$$

where σ_d^2 is the variance due to random drift and σ_p^2 is the phenotypic variance of X, which is assumed to be constant. The covariance of $\overline{X}_i$ and $\overline{X}_j$ is determined by the number of common δ_j values as

$$\mathrm{cov}\,(\overline{X}_i, \overline{X}_j) = i\sigma_d^2, \qquad 0 \leqslant i < j. \qquad (11.8)$$

What there is here is a linear regression with the expected value of $\overline{X}_i$ being

$$E(\overline{X}_i) = \mu_0 + i\Delta, \qquad (11.9)$$

and the variances and covariances being given by equations (11.7) and (11.8).

In practice, samples may not be taken every generation. Suppose that samples with sizes $n_{(1)}, n_{(2)}, \ldots, n_{(s)}$ are taken at times $t_1, t_2, \ldots, t_s$, these having sample means $\overline{X}_{(1)}, \overline{X}_{(2)}, \ldots, \overline{X}_{(s)}$, and variances $V_{(1)}, V_{(2)}, \ldots, V_{(s)}$. Then immediately there is an estimate

$$\hat{\sigma}_p^2 = \sum_{i=1}^{s} (n_{(i)} - 1) V_{(i)} \bigg/ \sum_{i=1}^{s} (n_{(i)} - 1) \qquad (11.10)$$

of the phenotypic variance σ_p^2, with $\sum (n_{(i)} - 1)$ degrees of freedom. If it happens that $\sigma_d^2 = 0$ then the value of Δ can be estimated by ordinary weighted linear regression as

$$\hat{\Delta} = \sum_{i=1}^{s} n_{(i)} (t_i - \bar{t}) \overline{X}_{(i)} \bigg/ \sum_{i=1}^{s} n_{(i)} (t_i - \bar{t})^2, \qquad (11.11)$$

where $\bar{t} = \Sigma n_{(i)} t_i / \Sigma n_{(i)}$. If $\sigma_d^2 \neq 0$ then $\hat{\Delta}$ is still an unbiased estimator of Δ, with variance

$$\mathrm{var}\,(\hat{\Delta}) = \sigma_p^2/A + B\sigma_d^2/A^2 \qquad (11.12)$$

where

$$A = \sum_{i=1}^{s} n_{(i)}(t_i - \bar{t})^2,$$

$$B = \sum_{i=1}^{s} \sum_{j=1}^{s} n_{(i)} n_{(j)} (t_i - \bar{t})(t_j - \bar{t}) \min(t_i, t_j),$$

and $\min(t_i, t_j)$ is the minimum of the two times t_i and t_j.

The total weighted sum of squares for the regression on the means can be partitioned in the usual way (Steel and Torrie, 1980, p. 270) as $\sum n_{(i)}(\bar{X}_i - \bar{X})^2 = S_1 + S_2$, where $\bar{X} = \sum n_{(i)} \bar{X}_i / \sum n_{(i)}$, $S_1 = A\hat{\Delta}^2$ is the regression sum of squares with one degree of freedom, and S_2 is a residual sum of squares with $s - 2$ degrees of freedom. It is not difficult to show that the expected value of the residual mean square, $M_2 = S_2/(s-2)$, is then as shown in Table 11.1, while the expected value of the regression mean square, $M_1 = S_1$, is $E(M_1) = A\{\Delta^2 + \text{var}(\hat{\Delta})\}$. An F-test comparing the residual mean square (M_2) with the phenotypic variance mean square (M_3) can be used to see if there is evidence that $\sigma_d^2 \neq 0$. If σ_d^2 does seem to be non-zero then its value can be estimated by equating the observed mean squares to their expected values to give

$$\hat{\sigma}_d^2 = (s-2)(M_2 - M_3)/C,$$

where C is as defined in Table 11.1.

The variance of $\hat{\Delta}$ can then be estimated using equation (11.12) and an assessment of whether $\hat{\Delta}$ is significantly different from zero can be made. If σ_d^2 is zero, or close to zero, then the regression mean square (M_1) can be compared with the phenotypic variance mean square (M_3) using an F-test to see if there is evidence that $\Delta \neq 0$.

The constant μ_0 in equation (11.9) can be estimated as

$$\hat{\mu}_0 = \bar{X} - \hat{\Delta}\bar{t} \tag{11.13}$$

if desired.

Example 11.1   The evolution of a horse tooth character

Table 11.2 and Fig. 11.3 show data on a tooth measurement, the paracone height, for four species of horse. These data were first published by Simpson (1953). It is thought that the four species are in direct line of descent, at least to the level of genus. Reasonable guesses for generation times are two years for *Hyracotherium borealis*, three years for *Mosohippus bairdi* and four years for *Merychippus paniensis* and *Neohipparion occidentale*. The equations from the previous section can be used in this case with time measured either by year or by generation. In the following calculations, millions of years are the units used, so that sample times are $t_1 = 0$, $t_2 = 20$, $t_3 = 35$ and $t_4 = 42$.

Table 11.1 Analysis of variance for a weighted regression model.

Source of variation	Sum of squares	Degrees of freedom	Mean square	Expected mean square
Regression on means	$S_1 = A\hat{\Delta}^2$	1	$M_1 = S_1$	$\sigma_p^2 + B\sigma_d^2/A + A\Delta^2$
Deviation of means from regression	$S_2 = \sum n_{(i)}(\bar{X}_i - \bar{X})^2 - S_1$	$s-2$	$M_2 = S_2/(s-2)$	$\sigma_p^2 + C\sigma_d^2/(s-2)$
Phenotypic variance	$S_3 = \sum (n_{(i)}-1)V_{(i)}$	$\sum(n_{(i)}-1)$	$M_3 = \hat{\sigma}_p^2$	σ_p^2

$$A = \sum_{i=1}^{s} n_{(i)}(t_i - \bar{t})^2$$

$$B = \sum_{i=1}^{s}\sum_{j=1}^{s} n_{(i)}n_{(j)}(t_i - \bar{t})(t_j - \bar{t})\min(t_i, t_j)$$

$$C = \sum_{i=1}^{s} n_{(i)}t_i - \sum_{i=1}^{s}\sum_{j=1}^{s} n_{(i)}n_{(j)}\min(t_i, t_j) \left/ \sum_{i=1}^{s} n_{(i)} - B/A \right.$$

Table 11.2   Data on means and standard deviations for a horse tooth character, after transformation to natural logarithms to stabilize the variance. The original measurements were in millimetres.

| Species | Log of paracone height | | Sample size | Age in millions of years |
	Mean	Std. dev.		
Hyracotherium borealis	1.54	0.062	11	50
Mosohippus bairdi	2.12	0.048	14	30
Merychippus paniensis	3.53	0.059	13	15
Neohipparion occidentale	3.96	0.046	5	8

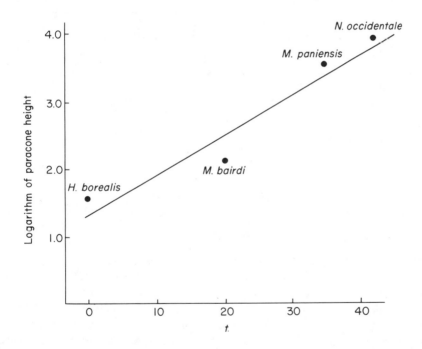

Fig. 11.3   The mean of the logarithm of paracone height for four species of horse that are thought to be in direct line of descent, at least to the level of genus. Time (t) is measured in millions of years since the time of *H. borealis*. The regression line shown, $\mu = 1.32 + 0.059\,t$, is estimated as described in Example 11.1.

From equation (11.10) the pooled estimate for the phenotypic (within-sample) variance is $\sigma_p^2 = 0.003\,04$ with 39 degrees of freedom, so that $\hat{\sigma}_p = 0.055$. The logarithmic transformation has apparently been successful in stabilizing the standard deviation since all the individual sample standard deviations are close to this pooled value.

For the equations of the previous section there are the following values: $\bar{t} = 21.98$, $A = 9576.98$, $\Sigma n_{(i)}(t_i - \bar{t})\bar{X}_{(i)} = 563.14$, $B = 1\,175\,214.91$, $\bar{X} = 2.612$, $\Sigma n_{(i)}(X_i - \bar{X})^2 = 36.071$, and $C = 228.92$. From these, $\hat{\Delta} = 563.14/9576.98 = 0.0588$ and there is the following analysis of variance:

Source	Sum of squares	d.f.	Mean square	Expected mean square
Regression	33.113	1	33.113	$\sigma_p^2 + 122.71\sigma_d^2 + 9576.98\Delta^2$
Deviations	2.958	2	1.479	$\sigma_p^2 + 114.46\sigma_d^2$
Phenotypic variance	0.119	39	0.003	σ_p^2

The F ratio $1.479/0.003 = 493$ is overwhelmingly significant so the observed changes in the mean cannot simply be explained by a constant linear trend. The estimate of the drift variance is $\hat{\sigma}_d^2 = (1.479 - 0.003)/114.46 = 0.0129$. Substituting this and $\hat{\sigma}_p^2 = 0.003$ into equation (11.12) gives var $(\hat{\Delta})$ $\simeq 0.000\,166$, so that the standard error of $\hat{\Delta}$ is estimated as 0.0129. This standard error is clearly not very reliable since $\hat{\sigma}_d^2$ is only based on a deviation from regression mean square with two degrees of freedom. However, the estimate $\hat{\Delta} = 0.0588$ is more than four estimated standard errors from zero so under the circumstances it does seem that there has been a significant linear trend.

From equation (11.13) the estimate of μ_0 is 1.32. Therefore, the mean value of the logarithm of the paracone height can be approximated by the equation

$$\mu = 1.32 + 0.059t$$

over the period studied, where t is millions of years since the time of *Hyracotherium borealis*. The amount of variation through random drift is indicated by the standard deviation of $\sigma_d \simeq 0.11$ per million years.

It hardly seems necessary to point out that these calculations need to be treated with a certain amount of reservation. Apart from anything else, it is questionable whether fossil remains provide random samples of once living populations. (Some further difficulties with fossil material are pointed out at the end of the next section.)

11.8 LANDE'S MODELS FOR LONG-TERM EVOLUTION

Lande (1976) considered the question of how much natural selection is needed to account for observed evolutionary changes over long periods of time. He noted that the form of selection which gives the minimum amount of selective mortality for a given change in the average phenotype is 'truncation' selection, whereby a fixed proportion of the most extreme deviants in one direction do not reproduce. All phenotypes have a fitness of one except those beyond the truncation point, which have a fitness of zero (Fig. 11.4).

If it is assumed that before selection the phenotypic distribution of X is normal with mean μ_0 and standard deviation σ, then it can be shown that after selection the distribution has mean

$$\mu' = \mu - \exp(-\tfrac{1}{2}b^2)/\{(2\pi)^{1/2}\sigma\}.$$

Thus the individuals that reproduce have a mean that is

$$S \simeq \sigma \exp(-\tfrac{1}{2}b^2)/(2\pi)^{1/2}$$

below that for unselected individuals. Equation (11.3) then shows that the selection response in one generation is expected to be

$$R \simeq h^2 \sigma \exp(-\tfrac{1}{2}b^2)/(2\pi)^{1/2}. \tag{11.14}$$

If b is negative, so that truncation is on the left of the mean, then this equation still holds, but the response increases the mean.

Lande argued that it is frequently observed that the phenotypic variance of quantitative characters and the heritability h^2 remain more or less constant during directional selection and that therefore equation (11.14) can be used to find the expected change in the mean through many generations. That is to say,

$$|\mu_t - \mu_0| = h^2 \sigma t \exp(-\tfrac{1}{2}b^2)/(2\pi)^{1/2} \tag{11.15}$$

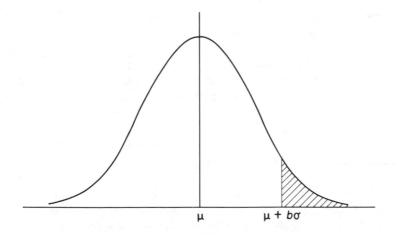

Fig. 11.4   Truncation selection on a normal distribution with mean μ and standard deviation σ for phenotypes. The selection is assumed to occur before reproduction. Individuals with character values greater than $\mu + b\sigma$ do not reproduce.

is the absolute change over t generations, so that

$$\hat{b} = \pm[-2\log_e\{(2\pi)^{1/2}|\mu_t - \mu_0|/(\sigma h^2 t)\}]^{1/2} \qquad (11.16)$$

is an estimate of b.

Truncation is not really a sensible model for natural selection. The value of equation (11.16) is therefore to indicate the minimum of selection that is needed to explain a change in a population mean, on the assumption that all of the change is due to selection and none to random drift.

Suppose now that there is no selection but the mean of a normally distributed population changes because of genetic drift. Lande considered the case where the parents from each generation are N individuals chosen at random from the population, so that N is the effective population size. He shows that in this case the expected value of the population mean in generation t is unchanged, so that $E(\mu_t) = \mu_0$, but the variance of μ_t through genetic drift is

$$\text{var}(\mu_t) = h^2\sigma^2 t/N.$$

Lande used this last equation for a type of test of no selection. He argued that the effective population size, N^*, for which the observed change in the population mean is just significant at the 5% level, is the one for which

$$|\mu_t - \mu_0|/\{\text{var}(\mu_t)\}^{1/2} = 1.96,$$

or

$$(\mu_t - \mu_0)^2/(h^2\sigma^2 t/N^*) = 1.96^2,$$

so that

$$N^* = (1.96)^2 h^2 t\sigma^2/(\mu_t - \mu_0)^2. \qquad (11.17)$$

These calculations suggest that if the true effective population size is bigger than N^*, then the change $\mu_t - \mu_0$ is too large to be accounted for by genetic drift alone. Sampling errors in estimates of μ_t and μ_0 are ignored in the test.

Lande (1976) pointed out the difficulties involved in applying these equations to the fossil material that he was particularly interested in. To begin with, there is no direct way to estimate the heritability h^2. There is a similar problem with determining the number of generations t. Only rather rough values can be determined by comparison with related living populations. Furthermore, as noted in Example 11.1, it is questionable whether fossil remains provide random samples for the estimation of phenotypic means and variances in different generations. It is particularly important to avoid pooling samples from different localities in case this inflates the variance estimate because of differences between populations. Also, geographical variation can very easily appear to be evolutionary change if samples from different generations are taken from different places. For these reasons, calculations based upon fossil material must be regarded with extreme caution.

Leaving these technical difficulties aside, there is perhaps a more important problem in a biased choice of data to analyse. Gould and Eldredge (1977) note

that sequences of mean values from fossil records are usually short and the
probability of a short random sequence showing a unidirectional trend is not
particularly small. For example, if there are four means then the chance of
either an upward or a downward 'trend' by chance alone is 1/4. If only
sequences displaying such 'trends' are analysed then a completely false picture
of evolution will emerge.

Example 11.2   Further analysis of a horse tooth character

The best preserved and most studied characters of fossil mammals are tooth
patterns and dimensions, for which a heritability of about $h^2 = 0.5$ seems
reasonable. Lande (1976) discussed the application of his models to several
examples of this type, including the data on paracone height that have already
been the subject of Example 11.1, and which are recorded in Table 11.2.

The calculation of the minimum amounts of selection needed to account for
the observed changes in the mean are shown in Table 11.3, assuming $h^2 = 0.5$.
It turns out that very weak selection is required. Truncation about 4.6 standard
deviations below the mean is all that is needed, corresponding to the removal
of about one individual in a million by selection in each generation.

Stanley (1979, p. 56) and Gould and Eldredge (1977) have argued that this
type of calculation on horse tooth characters delivers a severe blow to the
gradualistic model of evolution since it is difficult to imagine that selection as
low as one individual in a million could remain constant for up to 50 million
years and not be swamped by other chance processes. They suggest that
selection can only be invoked as the primary source of change if the

Table 11.3 Minimum amounts of selection needed to account for the observed
changes in the mean of the logarithm of the paracone height during the evolution of
horses. Equation (11.16) has been used to estimate the truncation point b (Fig. 11.4)
assuming $h^2 = 0.5$ using the pooled standard deviation estimate $\hat{\sigma} = 0.055$. The
proportions culled are the proportions of individuals more than b below the mean.
Following Lande, generation times are assumed to be two years for *Hyracotherium
borealis*, three years for *Mosohippus bairdi* and four years for *Merychippus paniensis*
and *Neohipparion occidentale*.

Transition	Number of generations (millions)	$\mu_t - \mu_0$	$\hat{b}$	Proportion culled
H. borealis–M. bairdi	10	0.58	−4.79	4×10^{-7}
M. bairdi–M. paniensis	5	1.41	−4.44	2×10^{-6}
M. paniensis–N. occidentale	1.75	0.43	−4.48	2×10^{-6}
H. borealis–N. occidentale	16.75	2.42	−4.59	1×10^{-6}

'punctuated equilibria' theory of evolution is correct. That is to say, if short periods of strong selection are followed by long periods of no change. Lande (1980a) has disagreed with this point of view. In his opinion the low level of selection that is required to explain the evolution of horse tooth characters just illustrates the tremendous power of selection in determining the course of evolution within a lineage. The possible existence of cyclic patterns of species extinctions (Raup and Sepkoski, 1984) adds a new twist to this debate.

One aspect of Lande's calculation that is surely important is the fact that its nature ensures that estimated selection pressures are very low. This can be seen by calculating what would have happened if b had been less extreme than is indicated by Table 11.3. For example, suppose b had been -3.0 rather than -4.6. The proportion culled per generation would still be only about 1 in 1000 but equation (11.15) shows that $\mu_t - \mu_0$ would be 2041, so that $\mu_t = 2043$. This is the logarithm of the paracone height so that the height itself would have a mean of exp (2043) mm! This calculation is clearly ridiculous, and indicates that the effect of selection on evolution has to be exceedingly small for Lande's model to make sense, at least with fossil data. It seems therefore that a low estimated level of selection is inevitable in an example like the one being considered here. The estimate tells us nothing about how selection has really worked. In practice, of course, selection in one direction must proceed towards a limit, and gradually slow down as that limit is approached. A possible model for fossil data is proposed by Hayami (1978).

Equation (11.17) can be used to examine the possibility that selection was by random drift only. The effective population sizes for which the observed changes in means are just significant at the 5% level have been calculated and are shown in Table 11.4. The smallest size is 15 000 and Lande argued that the true effective population size could well have been less than this. Therefore it seems quite conceivable that random drift was largely responsible for the observed changes in means. This conclusion is, of course, open to criticism on the grounds that there is no way at all of knowing what the true effective population size of the horse population was over a period of 42 million years, even to within an order of magnitude.

Table 11.4 Effective population sizes for which the observed changes in the means of the logarithm of paracone height are just significantly large at the 5% level. If the true effective population sizes were larger than the values shown here then the observed changes cannot reasonably be attributed to genetic drift alone.

Transition	$N*$
H. borealis–M. bairdi	1.7×10^5
M. bairdi–M. paniensis	1.5×10^4
M. paniensis–N. occidentale	5.5×10^4
M. borealis–N. occidentale	1.7×10^4

11.9   MULTIVARIATE EXTENSIONS
TO LANDE'S MODELS

Recently Lande (1979, 1980b) has extended his methods for studying selection to include multivariate situations. He considers a set of p characters $\mathbf{X}' = (X_1, X_2, \ldots, X_p)$ with a phenotypic covariance matrix $\mathbf{V}_P$ and a genetic covariance matrix $\mathbf{V}_G$. If these covariance matrices remain approximately constant over time, then a change in the population mean vector from μ_0 in generation 0 to μ_t in generation t can be accounted for in terms of truncation selection on the index

$$I = (\mu_t - \mu_0)' \mathbf{V}_G^{-1} \mathbf{X} = \sum_{i=1}^{p} a_i X_i,$$

which is a linear combination of the original characters. Random samples from the population in generations 0 and t will give sample mean vectors $\overline{\mathbf{X}}_0$ and $\overline{\mathbf{X}}_t$ that are estimates of μ_0 and μ_t, respectively. If an estimate $\hat{\mathbf{V}}_G$ of $\mathbf{V}_G$ is also available then the vector of coefficients a_i in the index can be estimated as

$$\hat{\mathbf{a}} = (\overline{\mathbf{X}}_t - \overline{\mathbf{X}}_0)' \hat{\mathbf{V}}_G^{-1}. \tag{11.18}$$

Index values can then be calculated for generations 0 and t and equation (11.16) can be used to determine the minimum amount of truncation selection needed to explain the change.

Determining an estimate of $\mathbf{V}_G$ is likely to be difficult with fossil material. However, Lande suggests that for many character sets the genetic and phenotypic covariance matrices will be roughly proportional. In that case $\mathbf{V}_G \simeq h^2 \mathbf{V}_P$ where h^2 is an average heritability value. When this is the case, replacing $\hat{\mathbf{V}}_G$ by $\hat{\mathbf{V}}_P$ in equation (11.18) will only mean that I is multiplied by a constant. This will not affect the calculation of the minimum amount of truncation selection. Lande warns, however, that if no information about $\mathbf{V}_G$ is available then it is difficult to draw any conclusions about selective forces from observed evolutionary changes and the phenotypic covariance matrix.

With regard to the question of whether or not changes in a population mean vector can be attributed solely to random genetic drift, Lande notes that if drift is the only factor then the statistic

$$X^2 = (\mu_t - \mu_0)' \mathbf{V}_G^{-1} (\mu_t - \mu_0) N/t$$

will have a chi-square distribution with p degrees of freedom, where N is the effective population size. Setting this statistic equal to the upper 5% point of the chi-squared distribution, say $\chi^2_{0.05}$, and solving for N, gives

$$N^* = t\chi^2_{0.05} / \{ (\mu_t - \mu_0)' \mathbf{V}_G^{-1} (\mu_t - \mu_0) \}$$

as the maximum effective population size consistent with the hypothesis of drift.

For a further discussion of the application of Lande's work to fossil data see the papers by Reyment (1982a, 1983).

11.10 SELECTION IN MARGINAL POPULATIONS

Although natural populations do not have replicates in the usual sense, it may sometimes be possible to study the effect of selection by comparing populations in which selection is not expected to operate at equal levels. This idea was developed by Brakefield (1979) in terms of spot numbers of *Maniola jurtina*. The number of spots on the hind wings of this butterfly behaves as a quantitative variable with heritability of about 0.63 for females and 0.14 for males at 15 °C. The heritability increases with temperature. The ease with which the character can be determined has meant that it has been studied extensively in natural populations (Ford, 1975).

Brakefield notes that it has often been suggested that the genetic structure of populations living on the marginal part of the distribution of a species will be different from that of more central populations. The central populations are likely to be adjusted to a wider range of environments than marginal ones, so that their existence is more secure. The marginal populations will tend to be more resistant to change and should therefore show less variation from year to year than the central ones.

Taking populations in south-east Scotland as being marginal, Brakefield obtained the following results for a one-factor analysis of variance between population spot number means for female *M. jurtina*:

Source of variation	Degrees of freedom	Mean square	F
populations	14	0.2316	10.39
years	43	0.0223	

The F value shows that there are highly significant differences between populations. Taking southern England populations as being central gives the following table, again for females:

Source of variation	Degrees of freedom	Mean square	F
populations	14	0.0907	2.08
years	37	0.0437	

In this case the F ratio is not significantly large at the 5% level.

To test the hypothesis of more year-to-year variation in the central populations, the F ratio $0.0437/0.0223 = 1.96$ can be tested with 37 and 43 degrees of freedom. This is significantly large at the 5% level, so that it does seem that the central populations were more variable (and therefore presumably more adaptable) than the marginal ones.

12 Further analyses using genetic data

This chapter concludes the discussion on tests for selection involving genetic data. Included here are a number of topics that have not fitted easily into the earlier chapters.

To begin with, associations between blood groups and diseases are considered. When found, these associations may provide evidence of selection, although the amount of selection involved is small. The level of selection can only be determined by considering the age at which selection takes place. Section 12.2 is a review of selection on an age-structured population. A somewhat related topic, the index of the opportunity for selection, is discussed in Section 12.3.

Section 12.4 considers the evidence for selection provided by admixture studies. The idea here is that when two populations mix, the frequency of a gene in the hybrid population should fall at an appropriate place between the frequencies in the source populations. If some genes have frequencies that differ substantially from this expectation then this may be evidence of selection on those genes.

Section 12.5 contains a review of a range of global tests for selection that have been proposed using data from many loci at the same time.

The chapter concludes with a discussion of the molecular clock hypothesis. According to this hypothesis, evolutionary change should occur at a constant rate in time, providing that selection is not an important factor. The constancy, or lack of constancy, of the rate of evolution over geological time therefore indicates the importance of selection.

12.1 ASSOCIATIONS BETWEEN BLOOD GROUPS AND DISEASES

It can be argued that by their very nature, human blood group polymorphisms give clear indirect evidence that they are maintained by natural selection. Thus Reed (1975) notes that some blood groups such as MN and rhesus are polymorphic in almost all populations, including small isolated groups. It seems likely that there is some positive mechanism maintaining this universal variation. Furthermore, some blood group systems seem to be very ancient since there are similar, if not identical, polymorphisms in higher primates. It appears to be necessary to invoke selection as the explanation for the maintenance of these polymorphisms for millions of years.

Assuming that selection is involved, it is not clear how it is operating. The genes controlling blood groups have no apparent effect on bodily structure or upon the choice of marriage partners, so they must influence fertility, general viability or susceptibility to diseases. This observation led Ford (1945) to predict that in time the blood groups and other similar human polymorphisms would be found to be associated with specific diseases. When Ford made this prediction there was no clear evidence of any associations for the ABO blood groups, although a few studies had been made, such as that of Buchanan and Higley (1921). Rhesus incompatibility between mother and foetus had been discovered a few years before. However, Ford's prediction proved to be correct. Aird et al. (1953) showed that patients with cancer of the stomach had a higher than average frequency of the A blood group and a lower than average frequency of the O group. Further work by the same workers found that the frequency of the O group is very high in patients suffering from peptic ulcers (Aird et al., 1954). Many other associations of this type have since been uncovered (Mourant et al., 1978).

The statistical analysis of data on the relationship between two blood groups and a disease has been discussed by Woolf (1955), Haldane (1956) and Edwards (1965). They assumed that there are two samples available: patients with the disease in question and a random sample from the general population. Both samples are classified according to their blood group, with the following results:

	Blood group 1	Blood group 2
Patients with disease	a	b
Population sample	c	d

Here a, b, c and d represent counts of the number of cases.

Let p_i denote the population proportion with the disease in blood group i. Let $\phi = p_1/p_2$ be the disease incidence rate for group 1 relative to group 2. Then the question of interest is whether $\phi = 1$, in which case there is no association between the blood groups and the disease. To estimate ϕ we can note that a/c is an estimate of p_1 multiplied by some constant, while b/d is an estimate of p_2 multiplied by the same constant. Therefore the *cross-product ratio*, or *odds ratio*

$$\hat{\phi} = (a/c)/(b/d) = (ad)/(bc)$$

is an estimator of ϕ. It is essentially the same as the selective value estimator $\hat{w}_{ij}$ of equation (2.16).

Woolf (1955) noted that for many purposes $y = \log_e(\hat{\phi})$ is a better measure than $\hat{\phi}$ of the relative incidence of the disease in the two blood groups, the main advantage in taking logarithms being that symmetry is achieved. Thus $\hat{\phi} = 2$ and $\hat{\phi} = \frac{1}{2}$ both indicate a situation where one blood group has double the incidence of the disease in the other group, although they are not equally far from the no-selection value of $\phi = 1$. However, $y = \log_e(2) = 0.69$ and $y = \log_e(\frac{1}{2}) = -0.69$ are symmetric about the no-selection value of $y = 0$.

To avoid infinite estimates when b or c are zero, Haldane (1956) proposed the modified estimator

$$\hat{y} = \log_e[\{(a + \tfrac{1}{2})(d + \tfrac{1}{2})\}/\{(b + \tfrac{1}{2})(c + \tfrac{1}{2})\}]$$

this having variance

$$\text{var}(y) \simeq \frac{1}{a+1} + \frac{1}{b+1} + \frac{1}{c+1} + \frac{1}{d+1}.$$

The modification has very little effect with moderate or large samples. It has already been used in Sections 2.8 and 3.2 for the removal of bias in the estimation of selective values.

Woolf considered the situation where several independent estimates $y_1, y_2, \ldots, y_K$ of $\log_e(\phi)$ are available, for example based upon data collected in different towns. If $w_i = 1/\text{var}(y_i)$ then the weighted mean with minimum variance is the usual $\bar{y} = \Sigma w_i y_i / \Sigma w_i$, with variance $\text{var}(\bar{y}) = 1/\Sigma w_i$ (Appendix, Section A.6). On the assumption that the estimates y_i are normally distributed with zero means, the statistics $X_i^2 = y_i^2 w_i$ will be chi-square variates with one degree of freedom each. Their total $\Sigma y_i^2 w_i$ can then be treated as a chi-square variate with K degrees of freedom, which consists of the sum of $\Sigma(y_i - \bar{y})^2 w_i$ with $K - 1$ degrees of freedom and $\bar{y}^2 \Sigma w_i$ with one degree of freedom. If $\Sigma(y_i - \bar{y})^2 w_i$ is significantly large then this suggests that the y_i values are not all estimating the same quantity, so that the estimates display heterogeneity. If $\bar{y}^2 \Sigma w_i$ is significantly large then this suggests that overall there is evidence that ϕ is not equal to 1, so that there is an association between the blood groups and the disease. See Anderson et al. (1980, p. 122) for a review of alternative approaches to the combination of cross-product ratios.

Mourant et al. (1978, p. 62) note that significant heterogeneity of estimates from different sources is quite common when studying disease associations. They suggest that in many cases the significance is caused by a few 'freaks', perhaps due to technical errors. In other cases the significance may be due to heterogeneity of the disease itself, so that the problem can be overcome by classifying the disease into different types. A danger to guard against is that the significance of a mean $\bar{y}$ value is due to one or two aberrant sets of data in a series where the majority of sets give y values close to zero.

In recent years much of the interest on disease associations has been concerned with the human histocompatibility (HLA) complex (Thomson et al., 1976; Smouse and Williams, 1982; Thomson, 1981). This consists of several closely linked loci, each of which manifests numerous alternative alleles. Many studies have attempted to find associations between the alleles and a particular disease, and many successes have been reported. A few of these have been repeatedly confirmed and seem real. However, most of the reported associations have not withstood closer examination. Smouse and Williams (1982) suggest that the reason for this is quite simple: there are a large number of alleles to be considered, perhaps 30 to 40, and if each of these is tested individually for association with a disease then there is a high probability of finding at least one significant result by chance alone.

The problem of many tests can be overcome by imposing an experimentwise error rate. If there are m alleles to be tested then each test should be made using, say, a $(5/m)\%$ significance level. This will ensure that the probability of getting no significant results will be at least 0.95 when the null hypothesis of no associations is true (Harris, 1975, p. 98).

Using a modified significance level for tests does not overcome all of the problems in testing for associations because the tests for different alleles are not independent. Smouse and Williams (1982) overcame this problem by setting up a scoring system for genotypes of the same type as was used by Smouse and Neel (1977) and Smouse et al. (1983) to study linkage disequilibrium (see Section 10.6). To understand the scoring system, consider a locus with four alleles A_1, A_2, A_3 and A_4. There are then ten possible genotypes, A_1A_1, A_1A_2, A_1A_3, . . . , A_4A_4. Let Y_i denote a random variable which is set equal to one-half of the number of A_i alleles possessed by an individual. Thus, for an A_1A_1 individual $Y_1 = 1$ and $Y_2 = Y_3 = Y_4 = 0$, while for an A_1A_3 individual $Y_1 = Y_3 = \frac{1}{2}$ and $Y_2 = Y_4 = 0$. With this scoring of genotypes, one of the Y_i values is redundant since for all genotypes $Y_1 + Y_2 + Y_3 + Y_4 = 1$. Hence Y_4 can be discarded and Y_1, Y_2 and Y_3 used to describe the genotype. Testing for disease associations then involves seeing whether the multivariate distribution of (Y_1, Y_2, Y_3) has the same mean for individuals with a disease and individuals from the general population. Smouse and Williams (1982) suggest using standard multivariate tests for this purpose. An alternative is to carry out a log-linear model analysis. Khamis and Hinkelman (1984) discuss appropriate procedures for this approach.

In assessing studies of disease associations there are two further points that need emphasizing. The first of these concerns the possibility of associations that are not caused by selection related to the genetic system being studied. One obvious possibility is that a racial subgroup within a population differs from the general population in its genotype distribution and also in its susceptibility to the disease. The racial subgroup will then be over- or underrepresented in the disease group and this in itself may be sufficient to cause a false disease association.

Recently, evidence has been found to suggest that the ABO blood groups are not randomly distributed in different socio-economic groups in England (Beardmore and Karimi-Booshehri, 1983), although this has been disputed (Mascie-Taylor et al., 1984). Since different socio-economic classes are not equally susceptible to various diseases, this could be an important factor in generating disease 'associations' with these classes.

The second point that needs emphasizing about disease associations is that the implied selective effects are very small because most of the diseases involved impose their major effects after the end of the reproductive period. Thus the disease associations do not offer much help in explaining the stability of the genetic polymorphisms.

12.2 SELECTION ON AN AGE-STRUCTURED POPULATION

It has just been remarked that the disease associations that have been observed with blood groups are too small to explain the stability of the blood group polymorphisms. In order to justify this statement it is necessary to consider how selection operates on an age-structured population. A model is needed that takes into account the fact that if selection operates after the reproductive period is over, then it has no long-term effects on a population.

Measuring selection taking into account age structure requires a good deal of information about a population. For this reason calculations have been restricted mainly to determining fitness reductions due to various diseases in man, although an exception is found in O'Donald and Davis's (1975, 1976) study of a population of the Arctic skua (Example 12.1 below). In many applications demographic parameters can only be estimated rather roughly so that the measurements of selective effects can be expected to be of the right order of magnitude at best.

Before considering selection it is necessary to review briefly the usual demographic model for age-structured populations. For convenience it is assumed that the life cycle is divided into a number of discrete age classes of equal length. Let $l(x)$ denote the proportion of individuals surviving from birth until the start of age interval x, for $x = 1, 2, \ldots$. Let $b(x)$ be the average number of offspring per individual during age class x. Then

$$k(x) = l(x)b(x)$$

is the average number of births in age class x per individual born, and

$$R = \sum_x k(\dot{x})$$

is the total number of births per individual born. The mean generation time can be defined as the average age of childbirth,

$$T = \sum_x x\, k(x)/R.$$

Finally, an important parameter is the intrinsic rate of increase, r, which is defined to be the (unique) real solution to the equation

$$\sum_x e^{-rx} k(x) = 1. \tag{12.1}$$

Standard demographic theory shows that if the $k(x)$ values are constant then the population will attain a stable distribution for the relative frequencies in different age classes, with the numbers in each class growing exponentially at the rate r. Thus, if the total population at time zero is N_0, then at time t it will be $e^{rt} N_0$.

Suppose that a population comprises a number of different genotypes. Then $k(x)$ values can be calculated separately for each of these so that $k_i(x)$ can denote the value for the ith genotype while $k(x)$ can still denote the overall population value. Assuming that the population has a stable age structure and is in genetic equilibrium as well, Charlesworth and Charlesworth (1973) show that a reasonable expression for the fitness of the ith genotype is

$$w_i = \sum_x e^{-rx} k_i(x). \tag{12.2}$$

On the other hand, Cavalli-Sforza and Bodmer (1971) argued that a satisfactory measure of selection can be based upon R, the total number of births per individual, providing that all genotypes have the same generation time T. That is to say, the fitness of genotype i relative to genotype 1 (say) can be taken to be

$$w_i = R_i/R_1 = \sum_x k_i(x) / \sum_x k_1(x). \tag{12.3}$$

This is not an appropriate measure when the generation time varies, because if two genotypes have the same value of R then the one with a smaller generation time will have a higher intrinsic rate of increase. If the generation time varies, Cavalli-Sforza and Bodmer suggest that the fitness of genotype i should be taken as proportional to $\exp(r_i \bar{T})$ where r_i is the solution to the equation

$$\sum_x \exp(-r_i x) k_i(x) = 1, \tag{12.4}$$

and $\bar{T}$ is the population mean generation time. Thus the fitness of genotype i relative to genotype 1 can be measured as

$$w_i = \exp(r_i \bar{T}) / \exp(r_1 \bar{T}) = \exp\{(r_i - r_1)\bar{T}\}. \tag{12.5}$$

See Charlesworth (1976) for a more detailed discussion of the difficulties of measuring selection on age-structured populations.

Example 12.1   Selection on Arctic skuas

A colony of Arctic skuas on Fair Isle in Shetland has been studied intensively for many years. Part of the data collected concerns colour phases, and these have been used by O'Donald and Davis (1975, 1976) to estimate fitness values using the equations given above. The colour phases form a genetic polymorphism with dark semi-dominant to pale. There are differences in the mean numbers of chicks fledged by pale, intermediate and dark-phased birds. There are also differences in survival and the age of maturity of the three types.

Table 12.1 gives the data for the calculation of selective values for females only. There is no space here to discuss how the demographic parameters were estimated, but there are two points that need to be noted. Firstly, the x values (age, in years) are not simply the integers 1, 2, 3, ... Birds were first recorded when they started breeding. The first x value for a colour phase is therefore the mean age at first breeding (3.93, 4.56 and 4.55 years for pale, intermediate and dark birds, respectively). Subsequent x values are this initial value $+1$, $+2$, $+3$, and so on. Secondly, the $l(x)$ values were not found by simply counting survivors. The estimation process was more complicated than that and it has resulted in $l(x+1) > l(x)$ in some instances. Since $l(x)$ is supposed to be the proportion surviving to age x this is not really possible. Nevertheless, O'Donald and Davis's estimates have been accepted as they stand because $l(x+1) > l(x)$ causes no particular problems when it comes to estimating fitness values.

Table 12.1 shows that the mean generation time (T) seems to have been different for the three colour phases. Therefore equation (12.3) cannot be used to determine fitness values. Either equation (12.2) or equation (12.5) must be used instead. Consider equation (12.5) first. This requires the solution of equation (12.4) for each genotype.

Taking the pale phase to begin with, and using the information in Table 12.1, the value of r_1 is given by

$$0.125\,e^{-3.93r_1} + 0.223\,e^{-4.93r_1} + \ldots + 0.140\,e^{-16.93r_1} = 1.$$

The solution to this equation can be found graphically, by a numerical method such as root bisection, or using an approximation of O'Donald and Davis (1976). It turns out to be $r_1 = 0.0858$. Similar calculations for the intermediate and dark phases give $r_2 = 0.0582$ and $r_3 = 0.0242$. These are the estimated intrinsic rates of increase for the three phases. The fittest genotype is pale. Relative to this, the fitness values per year of the other genotypes are estimated from equation (12.5) as

$$w_2 = \exp\{(0.0582 - 0.0858)8.997\} = 0.780$$

for intermediate, and

$$w_3 = \exp\{(0.0242 - 0.0858)\,8.997\} = 0.575$$

for dark. The mean generation time $\overline{T} = 8.997$ has been determined by doing the same calculations as shown in Table 12.1, but on the population data pooled for all three genotypes.

When equation (12.1) is solved for the pooled female data, the total

Table 12.1 Estimates of demographic parameters for female Arctic skuas based upon data for the years 1948–62.

Pale phase				Intermediate phase				Dark phase			
x	l(x)	b(x)	k(x)	x	l(x)	b(x)	k(x)	x	l(x)	b(x)	k(x)
3.93	0.428	0.293	0.125	4.56	0.368	0.450	0.166	4.55	0.371	0.479	0.179
4.93	0.363	0.614	0.223	5.56	0.318	0.583	0.185	5.55	0.329	0.605	0.199
5.93	0.354	0.579	0.205	6.56	0.286	0.628	0.180	6.55	0.293	0.633	0.185
6.93	0.340	0.750	0.255	7.56	0.250	0.697	0.174	7.55	0.223	0.650	0.145
7.93	0.346	0.643	0.222	8.56	0.220	0.600	0.132	8.55	0.233	0.722	0.168
8.93	0.292	0.650	0.190	9.56	0.188	0.889	0.167	9.55	0.153	0.600	0.092
9.93	0.281	0.625	0.176	10.56	0.163	0.846	0.138	10.55	0.110	1.000	0.110
10.93	0.216	0.900	0.194	11.56	0.123	0.875	0.108	11.55	0.090	0.750	0.068
11.93	0.161	0.333	0.054	12.56	0.095	1.000	0.095	12.55	0.056	1.000	0.056
12.93	0.066	0.500	0.033	13.56	0.094	0.625	0.059	13.55	0	–	0.000
13.93	0.085	1.000	0.085	14.56	0.091	0.667	0.061				
14.93	0.118	0.500	0.059	15.56	0.084	0.500	0.042		R = 1.201		
15.93	0.172	1.000	0.172	16.56	0.123	1.000	0.123		T = 7.651		
16.93	0.280	0.500	0.140	17.56	0.100	1.000	0.100				
17.93	0	–	0.000	18.56	0	–	0.000				

$R = \sum k(x) = 2.132$
$T = \sum xk(x)/R = 9.441$

$R = 1.728$
$T = 9.840$

population intrinsic rate of increase is $r = 0.0605$ (O'Donald and Davis, 1976, Table 1). Knowing this, alternative estimates of selective values can be determined from equation (12.2). Thus for the pale phase

$$w_1 = 0.125 \exp(-0.0605 \times 3.93) + 0.223 \exp(-0.0605 \times 4.93)$$

$$+ \ldots + 0.140 \exp(-0.0605 \times 17.93)$$

$$= 1.237.$$

The same calculation for the intermediate and dark phases give $w_2 = 0.980$ and $w_3 = 0.764$. Relative to pale, this gives the fitness values of intermediate and dark to be $0.980/1.237 = 0.792$ and $0.764/1.237 = 0.618$. These agree fairly well with the values of 0.780 and 0.575 derived from equation (12.5).

Similar calculations to those just done can be carried out using data from male birds. It turns out that for males the fittest genotype is the dark one. O'Donald and Davis (1977, 1976) discuss the relative importance of survival and reproductive success in the determination of selective values. They also consider the implications of selection as regards the future of the colony. They predict that the pale phase will gradually replace the dark one.

Example 12.2   Selection related to duodenal ulcer

Cavalli-Sforza and Bodmer (1971) have used demographic data to estimate the loss of fitness due to duodenal ulcers for Italian females in 1953. As pointed out by Nei and Yokoyama (1981), their estimate was too small by a factor of about ten because they did not take into account the fact that the group suffering from duodenal ulcers comprises about 10% of the population and that all the deaths from this cause must be assigned to this group.

The left-hand side of Table 12.2 shows the calculations of $k(x)$ values for Italian females in general using the data provided by Cavalli-Sforza and Bodmer (1971, Table 6.6). It will be seen that the number of births per individual is $R = 1.087$. To determine the selective effects of duodenal ulcer it is necessary to determine what reduction this condition causes in R. This is done in the right-hand side of the table. The proportion of females dying of duodenal ulcers in the total population is 178×10^{-6}. Assuming that 10% of the population have a duodenal ulcer at some stage, this means that the proportion dying from duodenal ulcers in the group at risk is ten times this amount, or 178×10^{-5}. The distribution of this proportion over ages is shown in the $q(x)$ column of Table 12.2. The $\Sigma q(x)$ column then indicates the cumulative proportion at age x. For example, by the age of 29 a proportion 0.000 24 of duodenal ulcer sufferers have died. The final column of the table indicates the reduction in births that can be attributed to deaths from duodenal ulcers, assuming that fertility is normal until the age group of death (a questionable assumption). Thus for the age group 25–29 there is a loss of 0.000 08 births, this being the number of births for the population as a whole,

Table 12.2   Estimating the fitness reduction due to duodenal ulcer for Italian females, 1953.

Age group	Total population			q(x)	Losses through duodenal ulcers	
	l(x)	b(x)	k(x)		$\sum q(x)$	$k(x)\sum q(x)$
Under 15	—	0	0	0	0	0
15–19	0.937	0.017	0.016	0.000 06	0.000 06	0.000 00
20–24	0.934	0.192	0.179	0.000 08	0.000 14	0.000 03
25–29	0.930	0.340	0.316	0.000 10	0.000 24	0.000 08
30–34	0.925	0.379	0.351	0.000 16	0.000 40	0.000 14
35–39	0.918	0.147	0.135	0.000 20	0.000 60	0.000 08
40–44	0.909	0.087	0.079	0.000 40	0.001 00	0.000 08
45–49	0.896	0.012	0.011	0.000 78	0.001 78	0.000 02
50 and over	—	0	0	—	—	0
			$R = 1.087$	0.001 78		0.000 43

q(x) = probability of dying from duodenal ulcer in age group x for individuals with a duodenal ulcer

$k(x) = 0.316$, multiplied by the proportion of females dead from duodenal ulcer by age 29, which is 0.000 24.

The total reduction in the average number of offspring is estimated to be 0.000 43. Recalling that the average number of offspring for the population as a whole is $R = 1.087$, this gives $R_1 = 1.087 - 0.000 43 = 1.086 57$ for sufferers from duodenal ulcer. Their fitness compared to the general population can then be estimated using equation (12.3) as $w = R_1/R = 0.999 60$. The proportional reduction in fitness is $(R - R_1)/R = 3.96 \times 10^{-4}$. It can also be noted that if R_2 is the average number of offspring for females without a duodenal ulcer, then $R = 0.9 R_2 + 0.1 R_1$, still taking 10% as the population proportion with the disease. Hence $R_2 = (R - 0.1 R_1)/0.9 = 1.087 05$.

Consider next what implications selection related to duodenal ulcer has with regard to the ABO blood group system, where individuals with blood group O are about 1.4 times more likely to get an ulcer than individuals with other groups. Suppose for the sake of argument that the probability of getting an ulcer is 0.120 for blood group O and 0.085 for other groups. This gives an overall probability of 0.1 for populations in which about half the members are blood group O. Then for blood group O the mean number of offspring will be

$$R_O = 0.120 R_1 + 0.880 R_2 = 1.086 992,$$

while for other blood groups

$$R_{other} = 0.085 R_1 + 0.915 R_2 = 1.087 009,$$

these just being the appropriate weighted averages of the means for the ulcer and the non-ulcer groups. The fitness of O compared to the other groups is therefore $R_O/R_{other} = 0.999 985$.

A selective value for males can be calculated in a similar way and this gives about the same results as for females. It is quite clear that selection associated with duodenal ulcers is going to have little effect on ABO blood group frequencies. This is because deaths from duodenal ulcers are rather rare in people of reproductive age. Therefore the disease has little effect on the average number of children, at least with the assumptions made above.

In this example the relatively simple measure of fitness of equation (12.3) has been used on the assumption that the mean generation time is fairly similar for people with or without a duodenal ulcer. This is reasonable given the small number of deaths from the disease.

12.3 THE INDEX OF OPPORTUNITY FOR SELECTION

Consider a population of N individuals with fitness values $w_1, w_2, \ldots, w_N$, where the fitness of an individual is defined as the expected number of offspring that will reach maturity. Suppose that the offspring of individual i

also have fitness w_i. Then the mean fitness of the N individuals is

$$\bar{w} = \sum_{i=1}^{N} w_i/N,$$

and they are expected to produce Σw_i offspring for which the mean fitness is

$$\bar{w}' = \sum w_i w_i / \sum w_i = \sum w_i^2/(N\bar{w}).$$

Therefore the relative change in mean fitness is

$$(\bar{w}' - \bar{w})/\bar{w} = (\sum w_i^2/N - \bar{w}^2)/\bar{w}^2 = V/\bar{w}^2,$$

where V is the variance of w_i in the parental generation. This is a special case of Fisher's (1930) fundamental theorem of natural selection, which states that 'the rate of increase of fitness of any organism at any time is equal to its genetic variance in fitness at that time'.

Crow (1958) suggested that $V/\bar{w}^2$ can be used as an upper limit to the overall action of natural selection that will apply when all differences in offspring numbers are genetically determined. For any real population, the actual effect of selection may be much lower than this because a great deal of the variation in offspring numbers has no genetic basis. Crow also suggested that $V/\bar{w}^2$ be divided into two components, one due to mortality in the pre-reproductive period and the other due to differential fertility. Thus, suppose that a proportion p_d of individuals die before maturity, and $p_s = 1 - p_d$ survive. Suppose also that a proportion q_i of survivors have i children, for $i = 0, 1, 2, \dots$. Then the population variance of fitness (number of children per individual) is

$$V = p_d(0 - \bar{w})^2 + p_s \sum_i q_i(i - \bar{w})^2$$

$$= p_d\bar{w}^2 + p_s \sum_i q_i(i - \bar{w}_s + \bar{w}_s - \bar{w})^2$$

$$= p_d\bar{w}^2 + p_s \sum_i q_i(i - \bar{w}_s)^2 + p_s(\bar{w}_s - \bar{w})^2$$

$$= V_m + p_s V_f,$$

where

$$\bar{w} = 0 w p_d + p_s \sum_i i q_i$$

is the mean number of offspring per individual born,

$$\bar{w}_s = \sum i q_i = \bar{w}/p_s$$

is the mean number of offspring per mature individual,

$$V_m = p_d\bar{w}^2 + p_s(\bar{w}_s - \bar{w})^2 = \bar{w}^2 p_d/p_s$$

is the variance due to premature mortality, and

$$V_f = \sum q_i(i - \bar{w}_s)^2$$

is the variance due to fertility differences. It then follows that

$$I = V/\bar{w}^2$$
$$= V_m/\bar{w}^2 + p_s V_f/\bar{w}^2$$
$$= p_d/p_s + V_f/(p_s \bar{w}_s^2),$$

or

$$I = I_m + I_f$$

where $I_m = p_d/p_s$ is the index of selection due to mortality, $I_f = V_f/(p_s \bar{w}_s^2)$ is the index of selection due to fertility, and I is the index of the total opportunity for selection. (Crow, 1958, and others have used $p_s I_f$ as the index of selection due to fertility. However, the definition here seems more natural.)

Applications of these indices to human populations have been discussed by several authors. For example, Terrenato et al. (1979) have calculated values for Italy for various periods during the last century. They note that the total index I has reduced by 75% during this time, with the contribution from fertility increasing from 57 to 89%. This is what is expected to happen as a country becomes more industrialized.

Another recent study has been that of Hed and Rasmuson (1981) on two Swedish parishes in the last century. They found that for the small farming parish of Fleninge a proportion $p_s = 0.744$ of children survived to maturity, so that the index of total selection due to mortality was $I_m = p_d/p_s = (1 - 0.744)/0.744 = 0.34$. Also, the mean family size was $\bar{w}_s = 3.34$ children, and the variance was $V_f = 12.71$. Therefore the index of total selection due to fertility was $I_f = 12.71/(0.744 \times 3.34^2) = 1.53$. The index of the total opportunity for selection was then $I = I_m + I_f = 1.87$. Clearly, most of the opportunity for selection can be attributed to fertility differences between women. For the larger parish of Nedertornea, the comparable index values are $I_m = 0.86$, $I_f = 1.47$ and $I = 2.33$. For both parishes the index values are rather high when compared with the results of other studies. Nevertheless, Hed and Rasmuson considered that they did give a good guide to the relative importance of mortality and fertility and that the differences between the two parishes are meaningful. Hed (1984) has continued this work using data on clergymen's wives in Sweden over the period 1600–1849.

Although Crow proposed his indices primarily for studying human populations, they have been used with other species to a limited extent. Crow (1961) gives some examples.

Henneberg (1976) argued that Crow's index of selection due to mortality is unsatisfactory because the fitness of individuals dying during their reproductive period is not estimated as a function of their age at death. He proposed a 'biological state index' for the investigation of natural selection in prehistoric and modern populations of man. This index is $I_{bs} = R_{pot}(1 - p_d)$, where R_{pot} is the reproductive potential remaining after the premenopausal deaths of adults. See his paper for more details.

12.4 ADMIXTURE STUDIES

Suppose that a hybrid population is derived from individuals from two different populations A and B in proportions m and $1 - m$, respectively. Then, in the absence of selection on an allele, its proportion in the hybrid population is expected to be

$$q_H = mq_A + (1 - m)q_B,$$

where q_A and q_B are the proportions for populations A and B. Thus the admixture proportion can be expressed as

$$m = (q_H - q_B)/(q_A - q_B),$$

providing that $q_A \neq q_B$ (Bernstein, 1931). If estimates $\hat{q}_H, \hat{q}_A$ and $\hat{q}_B$ of the gene frequencies are available, then an obvious estimate of m is

$$\hat{m} = (\hat{q}_H - \hat{q}_B)/(\hat{q}_A - \hat{q}_B), \tag{12.6}$$

with variance (Adams and Ward, 1973)

$$\text{var}(\hat{m}) \simeq \{\text{var}(\hat{q}_H) + m^2 \text{var}(\hat{q}_A) + (1 - m)^2 \text{var}(\hat{q}_B)\}/(q_A - q_B)^2.$$

The estimator $\hat{m}$ will be a valid estimator of the amount of population admixture providing that: (a) $\hat{q}_H$, $\hat{q}_A$ and $\hat{q}_B$ are unbiased estimators of population values; (b) mutation and genetic drift have negligible effects; and (c) there is no selection at the locus being considered. Subject to these conditions, estimates of m based upon different loci should differ only because of sampling errors. Thus, if a certain locus provides an estimate of m that is markedly different from the average value for all loci, then this suggests that one or more of the conditions does not hold for this locus. If it can be shown that conditions (a) and (b) are satisfied then it can be argued that the aberrant value of m is due to selection.

The first applications of this idea involved the study of American Negro populations. Matings between American Whites and Negroes must have occurred from the very early days of slavery in America, some 350 years ago. Offspring have almost always been classed as Negroes. Therefore present-day Negro populations are a mixture of their African ancestors and Caucasians. On this basis, Workman *et al.* (1963) and Workman (1968) estimated gene frequencies for 21 loci for Negroes and Whites living in Claxton, Georgia, and also for present-day West African Negroes. They assumed that the gene frequencies for Claxton Whites have been constant since the arrival of Negroes, so that present-day values for Whites can be used for $\hat{q}_A$ in equation (12.6). They similarly assumed that the present-day gene frequencies for West African Negroes are the same as the frequencies that existed during the days of slaving, and used them as $\hat{q}_B$ values. Thus the present-day Claxton Negroes were thought of as a hybrid between present-day Claxton Whites and present-day West African Negroes. Workman *et al.* found that for four alleles the admixture estimates were distinctly higher than they were for other alleles.

They argued that these four alleles may therefore be selectively favoured in the United States, with their frequencies in Claxton Negroes moving towards the Claxton White frequencies at a faster rate than can be explained by the general level of admixture.

Following this early work, Hertzog and Johnston (1968) used the same approach with rhesus blood group frequencies only, and tentatively claimed evidence of selection. However, about the same time, Reed (1969b) argued that there is too much uncertainty about gene frequencies in the African populations from which slaves were taken for selection to be established from admixture estimates.

Blumberg and Hesser (1971) suggested that if two American Negro populations are studied then the admixture estimates can be ranked for each population, and a significant agreement with these rankings is evidence of selection. Adams and Ward (1973) found little evidence of selection from this approach when they used it with five populations. They argued that the few inconsistent significant results that they obtained are probably due to other factors. This conclusion was reinforced by Mandarino and Cadien's (1974) observation that errors in determining African gene frequencies will result in correlations in the rankings of admixture estimates.

Overall it seems fair to accept Reed's (1975) conclusion that the admixture approach is potentially useful for detecting selection in human populations, given good estimates of ancestral gene frequencies. Unfortunately, these are not available for American Negro populations. Despite this caveat, it still seems true that extreme m values for certain loci are at least an indication that selection might be taking place.

The admixture approach can be generalized to situations where a hybrid population is formed from more than two parental populations. Methods for estimating admixture proportions under these circumstances are discussed by Elston (1971). For a Brazilian population with African, Indian and Portugese ancestors, he calculated admixture proportions using ABO blood group frequencies and then separately using rhesus blood group frequencies. There was close agreement, so Elston concluded that there was no evidence of selection at these loci. On the other hand, another study that used the same approach on a Black Carib population on the Atlantic coast of Guatemala suggested aberrant gene frequencies for several loci, possibly due to selection (Crawford et al., 1981).

Only one admixture analysis based upon equation (12.6) seems to have been done on a species other than man (Barker and East, 1980; Barker, 1981). The frequencies of three genes were artificially raised in an Australian population of *Drosophila buzzatii*. The frequencies then declined quite quickly to their original levels. Estimates of admixture proportions indicated that the rate of migration into the population needed to explain the speed of the decline was very high. Estimates also differed significantly according to which gene was used in the calculation. This was taken as evidence for selection either on one or more of the genes studied, or possibly on chromosome regions marked by them.

12.5 OTHER TESTS ON MANY LOCI

A very large number of tests have been proposed that involve comparison of various parameters of allele distributions with the values that can be expected from models of selective neutrality. These can be thought of as global tests because the results from many loci, and possibly many populations, are considered together. It can be argued that such tests are not very informative because most biologists believe that at least some loci are affected by selection. On the other hand, they do give some idea of the overall extent to which gene frequencies can be accounted for without invoking selection. No attempt is made here to do anything more than briefly review the large literature on this topic.

The first test to be mentioned concerns the relationship between the mean and the variance of heterozygosity. The population heterozygosity at a genetic locus is defined as

$$H = 1 - \sum_{i=1}^{k} p_i^2,$$

where the locus has k alleles with proportions $p_1, p_2, \ldots, p_k$. Fuerst et al. (1977) noted that according to the infinite alleles model for neutral mutations, the mean and variance of H for a population that has reached equilibrium are

$$E(H) = \theta/(1 + \theta), \tag{12.7}$$

and

$$\mathrm{var}(H) = 2\theta/\{(1 + \theta)^2 (2 + \theta)(3 + \theta)\}, \tag{12.8}$$

where $\theta = 4N_e v$, with N_e being the effective population size and v the mutation rate for the locus. Fuerst et al. argued that if observed means and variances fit these equations then this is evidence in support of the neutral mutation model. They tested this on a large body of data on protein polymorphisms detected by electrophoresis, involving 125 loci from various groups of mammals, fish and reptiles, and 47 loci from Drosophila.

A problem with using this idea is that an estimate of variance cannot be calculated from a single population value of H. However, if several related species all share the same locus then each species will provide an H value. A mean and variance can be calculated and it can be seen whether these are consistent with the equations. For example, suppose that several species give a mean of 0.09 and a variance of 0.015 for the heterozygosity at a certain locus. From equation (12.7) it is found that

$$\theta = E(H)/\{1 - E(H)\},$$

so that for locus in question $\theta \simeq 0.09/(1 - 0.09) = 0.099$. Hence, from equation (12.8) the variance of H is expected to be var$(H) \simeq 0.025$. Thus the observed variance of 0.015 is smaller than the expected value.

Fuerst *et al.* considered that observed variance-to-mean relationships are generally in agreement with equations (12.7) and (12.8) and that therefore their data support the hypothesis of neutrality. However, there are certain difficulties about accepting this conclusion. Firstly, the value of $\theta = 4N_e v$ will certainly not be constant for different species. The mutation rates v may be more or less constant but the effective population sizes N_e are likely to vary considerably. Generally, this seems likely to increase the observed variance of H above what is expected from equation (12.8). Secondly, the heterozygosity values of two species will be correlated if their divergence time is short. Fuerst *et al.* suggest that this correlation may take as long as 1.6 million years to disappear for some species. Its effect will be to reduce the variance of H. It seems, therefore, that either a high or a low variance for H can be explained easily enough without any need to invoke the idea of selection.

A third difficulty with Fuerst *et al.*'s study concerns the manner in which they decided that equations (12.7) and (12.8) were satisfied. They plotted variances against means for H and compared the plotted points with the curve that is obtained from the equations. This plot shows a wide scatter of data points about the theoretical line. Phillips and Mayo (1981) point out that there is so much scatter that almost any sensible relationship between the mean and variance would fit just as well.

Since the data analysed by Fuerst *et al.* came from electrophoretic studies, it could perhaps be argued that the charge-state mutation model might be more appropriate than the infinite alleles model for determining the variance-to-mean relationship of heterozygosity. However, Fuerst *et al.* showed that both models give very similar predictions for the range of data considered.

A second test is based upon the frequency distribution of different alleles. The idea here is that a random sample of individuals is taken and allele frequencies at L loci are recorded. Each identified allele will have a certain frequency at its locus. The question is whether the distribution of these frequencies is what can be expected from a model of no selection.

It is a standard result of population genetics theory that if the infinite alleles model of neutral mutation holds, then for any particular locus the proportion of alleles with a frequency in the range x_1 to x_2 is given by

$$P(x_1, x_2) = \theta \int_{x_1}^{x_2} (1-x)^{\theta - 1}/x \, dx. \tag{12.9}$$

where θ is as defined for equations (12.7) and (12.8). This result will hold for a collection of alleles obtained from different loci providing that θ is constant. This suggests that θ can be estimated from data on many loci and then observed sample distribution of allele frequencies can be compared with that expected from equation (12.9). This test was used by Ewens in 1969 with 85 *Drosophila* loci (Ewens, 1977b); the observed numbers of alleles with frequencies in the ranges (0.05, 0.1), (0.1, 0.9) and (0.9, 1.0) were 13, 42 and 67, which compare with expected values of 8.2, 43.5 and 63.7. This looks like good agreement but it is not clear how a formal test should be carried out. Maruyama and Yamazaki (1974) made essentially the same test with 400 enzyme loci and claimed that the observed and expected distributions agreed on the basis of a visual comparison. Ohta (1976) considered the test in terms of the charge-state model for neutral mutations and noted that there seems to be a tendency for there to be too many rare alleles.

A problem noted by Ohta (1976) is that equation (12.9) relates to population frequencies rather than sample frequencies. Another problem is that θ values will not be constant for different loci. Both of these problems were investigated by Chakraborty et al. (1980). They showed that for a random sample of n genes, the expected number of alleles with a frequency between x_1 and x_2 is

$$P(x_1, x_2) = \theta \sum_{i = nx_1 + 1}^{nx_2} (1/i) \prod_{j = 1}^{n} (n + 1 - j)/(n + \theta - j) \qquad (12.10)$$

for the infinite alleles model with a constant value of θ. They also provide an alternative expression for the charge-state model. With regard to the problem of θ varying, Nei et al. (1976) have suggested that a reasonable assumption is that the mutation rate v follows a gamma distribution from locus to locus. Equivalent expressions to equations (12.9) and (12.10) can then be determined for either the infinite alleles or the charge-state model. A comparison of the various models with data led Chakraborty et al. (1980) to the conclusion that there are an excess of rare alleles in some populations, but not all. This may be because the populations with an excess have recently experienced a bottleneck in numbers. Alternatively, it may be because these populations have large numbers of slightly deleterious alleles at low frequencies.

For a third type of test, Chakraborty et al. (1980) considered the correlation between the heterozygosity and the number of rare alleles at a locus. Earlier studies had indicated that this correlation tends to be higher than is expected from the infinite alleles model of neutral mutation. However, Chakraborty et al. suggested that these earlier studies were misleading since they did not allow for a varying mutation rate for different loci. They claimed a fair agreement between observed and expected values. Actually, the variation in the data is so large that almost any relationship would fit (Phillips and Mayo, 1981). An equation for the correlation between heterozygosity and the number of alleles in any frequency class has been developed by Chakraborty and Griffiths (1982), assuming a constant mutation rate.

For another test, Fuerst et al. (1977) calculated the distribution of heterozygosity that can be expected according to the infinite neutral alleles model. For 68 species the observed and expected distributions were in agreement according to a Kolmogorov–Smirnov test. Johnson (1972) and Johnson and Feldman (1973) argued that a good measure of the 'evenness' of allele frequencies at a locus is $kF = k(1 - H) = k\Sigma p_i^2$. They argued that this statistic should increase with the number of alleles (k) if the infinite neutral alleles model is correct, but that this does not happen with *Drosophila* data, presumably because of selection. Using more data, Yamazaki and Maruyama (1973) and Kirby and Halliday (1973) concluded that population data are in fact in agreement with the neutral model.

Two tests for neutrality based upon a geographical subdivision of a population have been proposed by Yamazaki and Maruyama (1972, 1974) and Lewontin and Krakauer (1973). Both of these have subsequently been severely criticized (Ewens and Feldman, 1976; Nei and Maruyama, 1975; Robertson, 1975a, b).

Finally, some tests have been based upon a comparison in the gene frequencies of two related populations. Thus, suppose that several loci are sampled from two populations with a common ancestor. The difference between the populations can be measured by Nei's standardized genetic distance, D say, as defined in Section 6.9. For each locus there will be a heterozygosity value $H = 1 - \Sigma p_i^2$ for each population. Let the correlation between these be denoted by r. Then Ohta (1976) has noted that, according to the infinite neutral alleles model for mutations, at time t after divergence

$$D \simeq 2\,vt,$$

and

$$r \simeq \exp\left\{-(4v + 1/N_e)t\right\},$$

where v is the mutation rate and N_e is the harmonic mean of the two effective population sizes. Thus it can be expected that $2D \geqslant -\log_e(r)$. Ohta found that this inequality holds for man but not for *Drosophila*, and he concluded that the *Drosophila* result is due to selection. Chakraborty *et al.* (1978) also studied the observed and expected relationship between D and r, taking into account factors such as the varying mutation rate at different loci. They decided that a rigorous test is not really possible but that data are broadly in agreement with the neutrality model. However, their figures showing the observed relationship between D and r for different populations indicate such a great deal of variation that it is difficult to see how any conclusion can be reached.

It seems fair to sum up the tests described in this section by saying that none of them have been at all conclusive. There are two main reasons for this:

(a) The models used to describe neutrality are of doubtful validity with data obtained by electrophoresis, for reasons that have been discussed in Sections 9.10 and 9.11.
(b) The variation observed in natural populations is so large when data from many loci and many populations are pooled that no simple model can possibly be adequate to account for it.

On the whole, it seems doubtful whether global tests will ever produce unambiguous results.

12.6 THE MOLECULAR CLOCK HYPOTHESIS

It has sometimes been argued that one of the best pieces of evidence in favour of the neutral theory of molecular evolution is the near constancy of evolution rates over geological time. The argument is based upon the following simple observation. Consider a population with effective size N_e and with a mutation rate of v per generation at a certain locus. Then there are $2N_e$ alleles, and $2N_e v$ new alleles are expected to arise each generation through mutations. If all these new alleles are unique, and there is no selection, then it is a standard result (Ewens, 1979b, p. 17) that the probability of a new allele eventually replacing all other alleles is $1/(2N_e)$. Consequently, the long-term average for the number of alleles that are replaced per generation is

$$k = 2N_e v/(2N_e) = v, \tag{12.11}$$

which is independent of the population size. Thus, according to the infinite neutral alleles model for mutations, the number of substitutions of alleles should accumulate at a steady rate over time like a 'molecular clock' (Thorpe, 1982).

The same result is not expected if substitutions are mainly determined by selection. In that case the rate of evolution should depend upon how often and how fast the environment changes. The rate of evolution in living fossils such as the lamprey should be much slower than in a rapidly evolving group such as the primates (Nei, 1975, p. 246).

The molecular clock hypothesis has been studied in terms of substitution rates in amino acid sequences, restriction endonuclease cleavage site maps, and nucleotide sequences. A test of the hypothesis can be made by taking two or more species with a common ancestor and seeing whether the evolutionary rate is the same for all species. In practice it has been found that evolutionary rates are roughly constant over geological time on a yearly basis, but not on a generation basis. Since it is constancy on a generation basis that is predicted by equation (12.11), this has led some authors such as Templeton (1983a) to suggest that the molecular clock hypothesis is best regarded as a reasonable subject to study, with no straightforward explanation at present. On the other hand, Kimura (1979) has argued that constant evolutionary rates either by year or by generation are most easy to explain in terms of a neutral model of evolution.

The first formal statistical test of the molecular clock hypothesis was made by Langley and Fitch (1973, 1974) and Fitch and Langley (1976a, b), using amino acid sequences for several proteins. A certain phylogenetic tree was assumed for a number of vertebrate groups and the minimum possible number of nucleotide substitutions along each branch of the tree was determined for each protein. The substitutions were then assumed to occur with a different constant rate per unit time for each protein, with the number of substitutions along each branch having a Poisson distribution. The substitution rates for proteins, and the branching times for the tree, were then estimated by maximum likelihood. Finally, a goodness of fit test for the model was carried out. The result was a very poor fit.

Unfortunately, there are some technical difficulties with the Langley–Fitch test. Firstly, the assumed phylogenetic tree may be wrong. Secondly, the method for determining the observed number of substitutions along branches of the tree only gives a minimum figure. The bias in this will be higher for long branches than for short branches. (Langley and Fitch made an *ad hoc* correction for this.) Thirdly, and far more important, is the fact that a constant rate of evolution does not necessarily imply a Poisson model for the observed numbers of substitutions (Gillespie and Langley, 1979).

Hudson (1983) considered the Langley–Fitch test in terms of a non-Poisson model with a constant evolutionary rate. By simulating this model he reached the conclusion that Langley and Fitch's result still stands: the molecular clock hypothesis is not supported by data. This was also the view expressed by Czelusniak *et al.* (1982) and Goodman *et al.* (1982) from studying large numbers of amino acid sequences on many species.

Templeton (1983a, b) has argued that at the present time our knowledge of evolution is not sufficient for us to have much confidence in any particular model for evolution. He therefore proposed a nonparametric test for the molecular clock hypothesis. This is perhaps more realistic than Langley and Fitch's parametric test or modifications of their test. Templeton was concerned particularly with the use of restriction endonuclease cleavage site maps. From data on mitochondrial DNA and globin DNA of man and apes, he found strong evidence against the molecular clock hypothesis.

In summary, it seems true to say that formal statistical tests have indicated definite deviations from the molecular clock hypothesis. Nevertheless, in practical terms the hypothesis is sometimes 'nearly' true. Furthermore, the importance of a statistically significant deviation from the hypothesis in terms of a comparison between the neutral and selective theories of evolution is not clear. Apart from these considerations, substitution rates are not simple to estimate. The rate of substitution has been found to vary with the four bases in nucleotide sequences and with positions in codons (Takahata and Kimura, 1981; Gojobori *et al.*, 1982; Kaplan and Risko, 1982). It seems that tests of the neutral mutation model based upon reconstructed phylogenetic trees must be viewed with a certain amount of reservation at the present time.

13 Non-random mating and sexual selection

This chapter concentrates on selection in relationship to the process of reproduction. Particularly, the analysis of data on selective and assortative mating is considered.

After an introductory section, in which the phenomena of female choice and the rare-male advantage are discussed, the analysis of laboratory experiments is considered. These experiments either involve one sex being given a choice of partners of the opposite sex (Section 13.2), or both sexes can have a choice (Section 13.3). A section on the analysis of data from wild populations then follows. Finally, there is a brief mention of Peter O'Donald's models for sexual selection.

13.1 INTRODUCTION

Random mating in a population requires that all possible male–female mating pairs are equally likely to occur. This may never be true in natural populations. Nevertheless, mating may be effectively random with respect to certain phenotypic traits because these traits are distributed randomly in the population and they are irrelevant to the mating process. The situation with man is perhaps a good guide to what happens with other species: mating tends to be distinctly non-random with respect to obvious visible characteristics such as skin colour and height, but appears to be more or less random with respect to 'hidden' characteristics such as blood groups.

When considering the effect of non-random mating on a population, three different aspects need to be considered (Lewontin *et al.*, 1968). Firstly, there may be selective mating, with some individuals having a higher probability of mating than others. Secondly, there can be phenotypic assortative mating which is positive or negative depending upon whether mating pairs tend to have the same phenotype or different phenotypes. Thirdly, there may be inbreeding so that the probability of a male and female mating depends upon the extent to which they share common ancestors. Since an individual that does not mate makes no contribution to the genes in the next generation of a population, it is quite clear that selective mating is liable to change gene frequencies. Assortative mating and inbreeding will not necessarily have any effect on gene frequencies. However, positive assortative mating and inbreeding will promote genetic homozygosity while negative assortative mating will promote heterozygosity.

The present chapter is concerned mainly with the analysis of data on selective and assortative mating. (See Spiess, 1977, for an account of the effects of inbreeding in a population. This aspect of non-random mating is not considered further here.)

In considering data on non-random mating there are two general phenomena that need to be considered. The first of these concerns *female choice*. O'Donald (1980, p. 1) points out that Charles Darwin's original theory of sexual selection involved the idea that males compete with each other for the possession of females, and the females choose with whom they will mate. The male competition part of this theory was accepted quite easily but the female choice part was a good deal more controversial because it was not thought that females of most animal species would have sufficient aesthetic sense to make a choice. However, evidence of female choice has now been found. For example, Petit (1951, 1954, 1958) found that in laboratory mating experiments with *Drosophila melanogaster*, the influence of the genotype of females was relatively unimportant compared with that of the males. In other words, females seemed to choose their mates while males did not. More recently, working with the two-spot ladybird *Adalia bipunctata*, Majerus *et al.* (1982a) have been able to demonstrate that female preference can be inherited.

The second important general phenomenon is the *rare-male advantage*. This was also first demonstrated by Petit with her experiments on *D. melanogaster*. She found that the mating success of mutant and wild flies is a function of the relative frequencies of the two types in culture. Usually the probability of mating for individual males with a certain phenotype increases as the overall proportion of males with that phenotype decreases. The rare-male advantage has been found in other *Drosophila* species and other taxa (Searcy, 1982). Several mechanisms have been suggested to explain it but this is still a matter of some debate (O'Donald, 1980, p. 16; 1983; Searcy, 1982; Spiess, 1982).

Many mating experiments have been concerned with investigating the amount of sexual isolation that there is between geographically separated populations of a species. The speed with which populations can become effectively isolated is of obvious importance to the development of new species. For example, Henderson and Lambert (1982) sampled 29 populations of *D. melanogaster* from around the world and carried out a large number of multiple choice experiments in which males and females from two populations competed for mates. They found no evidence of isolation between populations. The explanation for this finding may be some form of stabilizing selection.

13.2 MALE OR FEMALE CHOICE EXPERIMENTS

A simple way to study mating patterns is by male choice or female choice experiments. In a female choice experiment, males of several types are placed together with females of a single type. After a period of time the number of matings is recorded for each type of male. The different types of male may vary because of their geographical origins, their genotypes, or any other characteristic of interest. Experiments have usually involved only two types of male, one of which is the same as the females. A male choice experiment is carried out in the same way as a female choice experiment. The only difference is that there are two or more types of female placed with a single type of male. For simplicity, the present discussion is worded in terms of female choice experiments only. Reversing the use of the words 'male' and 'female' will cover male choice experiments.

Suppose that an experiment is carried out with females of a single type being placed with K types of male with relative frequencies $A_1, A_2, \ldots, A_K$. It may then be reasonable to assume that at any time after the experiment has begun, the probability that the next male to mate is of type i is

$$P_i = \beta_i A_i \bigg/ \sum_{j=1}^{K} \beta_j A_j, \qquad (13.1)$$

where the positive coefficient β_j reflects the preference of the females for type j males. Thus if β_j is large relative to the other β values, then type j males are at an advantage. On the other hand, if $\beta_j = 0$ then the females will never mate with type j males. Since the scaling of the β_j values does not affect the probabilities P_i, it is convenient to make $\Sigma \beta_i = 1$.

This model is exactly the same as the Chesson–Manly model for competitive survival that was discussed in Section 5.6 if 'survival' is interpreted as meaning non-mating. The maximum likelihood estimator of β_i is

$$\hat{\beta}_i = (d_i/A_i) \bigg/ \sum_{j=1}^{K} (d_j/A_j), \qquad (13.2)$$

where d_i denotes the number of type i males that have been observed to mate by the end of the experiment. Biases, variances and covariances for β_1, $\beta_2, \ldots, \beta_K$ are provided by equations (5.16).

An important assumption in using the above estimator is that the probability of a type i male mating next is given by equation (13.1) throughout an experiment. This will be a reasonable assumption provided that the relative frequencies of available males remain approximately proportional to the initial relative frequencies $A_1, A_2, \ldots, A_K$. If males are only allowed to mate once, then this condition will hold provided that the experiment ends while many males are still unmated. If multiple matings are allowed, then the relative frequencies of available males may remain more or less constant even when most males have mated at least once. However, if males become unavailable after their first mating then the probability of a type i male being chosen first will be

$$P_i = \beta_i n_i \bigg/ \sum_{j=1}^{K} \beta_j n_k,$$

when K types of male are available with frequencies $n_1, n_2, \ldots, n_K$. Thus P_i will change after each mating. Under these conditions an estimate of β_i is given by equation (5.21) to be

$$\hat{\beta}_i = \log_e(r_i/A_i) \bigg/ \sum_{j=1}^{K} \log_e(r_j/A_j) \tag{13.3}$$

where A_j is the number of type j males available at the start of an experiment, of which r_j remain unmated at the end of the experiment.

It may happen that males become unavailable after their first mating and the actual sequence of matings of different types is known. In this case, maximum likelihood estimates of β values can be computed using equation (5.19), with variances and covariances given by equations (5.20).

A common procedure has been to have only $K = 2$ types of male and female and to do two experiments. In one experiment, type 1 females have been placed with type 1 and type 2 males; and then in a second experiment, type 2 females have been given the same choice. Assuming that two experiments have been done like this, it is convenient to use a second subscript on β values to indicate the type of female involved. Thus $\hat{\beta}_{i1}$ is an estimate of the preference coefficient for type i males when type 1 females are used, and $\hat{\beta}_{i2}$ is an estimate for the same type of males but with type 2 females. In that case

$$\hat{b}_1 = \hat{\beta}_{11} - \hat{\beta}_{21} = 2\hat{\beta}_{11} - 1 \tag{13.4}$$

is a measure of the sexual isolation between type 1 females and type 2 males, while

$$\hat{b}_2 = \hat{\beta}_{22} - \hat{\beta}_{12} = 2\hat{\beta}_{22} - 1, \tag{13.5}$$

is a measure of the sexual isolation between type 2 females and type 1 males. Biases and variances for $\hat{b}_1$ and $\hat{b}_2$ can be calculated as

$$\text{bias}(\hat{b}_i) = 2 \text{ bias }(\hat{\beta}_{ii}) - 1, \tag{13.6}$$

$$\text{var}(\hat{b}_i) = 4 \text{ var}(\hat{\beta}_{ii}). \tag{13.7}$$

The indices $\hat{b}_1$ and $\hat{b}_2$ were first proposed by Stalker (1942) with β values determined by equation (13.2). Levene (1949) was the first to use β values determined by equation (13.3).

A joint isolation index based upon $\hat{b}_1$ and $\hat{b}_2$ is the average

$$\hat{b} = (\hat{b}_1 + \hat{b}_2)/2, \tag{13.8}$$

with variance
$$\text{var}(\hat{b}) = \{\text{var}(\hat{b}_1) + \text{var}(\hat{b}_2)\}/4. \tag{13.9}$$

An index of the mating propensity of type 1 females relative to type 2 females is the difference
$$\hat{a} = (\hat{b}_1 - \hat{b}_2)/2, \tag{13.10}$$

which has the same variance as $\hat{b}$. These indices were introduced by Bateman (1949). Levene (1949) suggested that $\hat{a}$ is perhaps better described as an 'excess insemination index' of type 1 females over type 2 females.

Example 13.1   The rare-male advantage with *Drosophila melanogaster*

The influence of the genotype on *Drosophila* sexual behaviour has been the subject of extensive investigation for many years. It has been found that frequency-dependent sexual selection and the rare-male mating advantage are not unusual. A rare-male advantage was first reported by Petit (1954, 1958) working with *D. melanogaster*. However, the phenomenon has become particularly well known through the work of Ehrman and her associates on *D. pseudoobscura* (Ehrman and Probber, 1978). It has been suggested that the phenomenon is only a result of a systematic experimental bias (Kence, 1981; Bryant *et al.*, 1980), but this has been disputed (Leonard and Ehrman, 1983; Knoppien, 1984).

The present example comes from a study by Markow *et al.* (1980) that was designed to test the rare-male advantage with *D. melanogaster*. As part of this study some female choice experiments were carried out using males and females of two strains of flies, Canton-S (CS) and Oregon-R (OR). Details of the experimental results are shown in Table 13.1. It can be seen, for instance, that the first experiment involved 100 CS females being offered 20 CS males and 80 OR males. There were 94 matings, of which 35 were with CS males and 59 with OR males. Obviously some of the CS males mated more than once, so that the β preference parameters should be estimated using equation (13.2) rather than equation (13.3).

Each experimental result shown in Table 13.1 is actually a total from ten experiments. For example, ten independent experiments were carried out in which ten CS females were placed with two CS males and eight OR males for one hour. This resulted in the total of 94 matings that are shown for experiment 1. Adding together experimental results like this does not matter when β values are estimated using equation (13.2). However, this should not be done when β values are estimated using equation (13.3), which takes into account changing proportions as a mating experiment progresses.

Table 13.2 shows estimated β values. For clarity these are denoted as $\hat{\beta}_{CS}$ and $\hat{\beta}_{OR}$ for CS and OR males, respectively, where $\hat{\beta}_{CS} + \hat{\beta}_{OR} = 1$. In any experiment, $\hat{\beta}_{CS} > 0.5$ suggests a preference of the females for CS males,

Table 13.1   Results from experiments on the rare-male advantage with *Drosophila melanogaster*.

Experiment	Females	Males		Males mated		
		CS	OR	CS	OR	Total
1	100 CS	20	80	35	59	94
2	100 CS	50	50	63	37	100
3	100 CS	80	20	73	13	86
4	100 OR	20	80	19	74	93
5	100 OR	50	50	66	34	100
6	100 OR	80	20	63	18	81

Table 13.2   Estimates of preference coefficients β calculated using the data of Table 13.1. Here $\hat{\beta}_{CS}$ and $\hat{\beta}_{OR}$ are the values for CS and OR males, respectively.

Experiment	Females	Percentage CS males	$\hat{\beta}_{CS}$	$\hat{\beta}_{OR}$	Standard error
1	CS	20	0.70	0.30	0.044
2	CS	50	0.63	0.37	0.048
3	CS	80	0.58	0.42	0.073
4	OR	20	0.51	0.49	0.065
5	OR	50	0.66	0.34	0.047
6	OR	80	0.47	0.53	0.067

Biases and standard errors have been calculated using equations (5.16). Since $\hat{\beta}_{OR} = 1 - \hat{\beta}_{CS}$, the standard error is the same for both estimators. Biases are not shown since they are negligible.

$\hat{\beta}_{CS} < 0.5$ suggests a preference for OR males, and $\hat{\beta}_{CS} \simeq 0.5$ suggests no preference. On this basis the rare-male effect appears to exist with CS females. When there were 20% CS males they were strongly preferred to OR males ($\hat{\beta}_{CS} = 0.70$). However, when there were 80% CS males they were only slightly preferred to OR males ($\hat{\beta}_{CS} = 0.58$). There is no evidence of the rare-male effect from the experiments with OR females. They seem to have a strong preference for CS males only when CS and OR males are in equal numbers.

A formal test for frequency-dependent selection could be carried out by doing a regression of $\hat{\beta}_{CS}$ values against the percentage of CS males and seeing whether the regression slope is significant. The situation is rather similar to that described in Example 5.4. The estimated standard errors could be used for a weighted regression which would give a more sensitive analysis (Appendix, Section A.8).

For the experiments with CS females, Stalker's (1942) index of sexual isolation is

$$\hat{b}_{CS} = \hat{\beta}_{CS} - \hat{\beta}_{OR} = 2\hat{\beta}_{CS} - 1.$$

For the experiments with OR females the $\hat{\beta}$ values are reversed to give

$$\hat{b}_{OR} = \hat{\beta}_{OR} - \hat{\beta}_{CS} = 2\hat{\beta}_{OR} - 1.$$

The coefficient of joint isolation is then $\hat{b} = (\hat{b}_{CS} + \hat{b}_{OR})/2$ for a particular percentage of CS males. Similarly, the index of the relative mating propensity of CS to OR females is $\hat{a} = (\hat{b}_{CS} - \hat{b}_{OR})/2$, for a particular percentage of CS males. Variances for $\hat{b}$ and $\hat{a}$ can be calculated as indicated by equation (13.9). This produces the following results:

% CS males	$\hat{b}_{CS}$	SE	$\hat{b}_{OR}$	SE	$\hat{b}$	SE	$\hat{a}$	SE
20	0.40	0.09	−0.02	0.13	0.19	0.08	0.42	0.08
50	0.26	0.09	−0.32	0.09	−0.03	0.06	0.29	0.06
80	0.16	0.15	0.06	0.13	0.10	0.10	0.02	0.10

It can be seen that the isolation between CS females and OR males increases considerably as CS males get more rare. On the other hand, there is little indication of OR females being isolated from CS males. They prefer them to OR males when CS and OR males are equally frequent. Overall, the index $\hat{b}$ of joint isolation differs much from zero only when CS males are rare. The index of relative mating propensity $\hat{a}$ of CS to OR females increases considerably as CS males become more rare.

13.3 MULTIPLE CHOICE EXPERIMENTS

Multiple choice experiments involve placing two or more types of female together with males of the same types. Thus males and females both have a choice of mates. This does not introduce too many problems providing that the proportions of different types of males and females available for mating remain more or less constant throughout an experiment.

To see this, suppose that an experiment is such that there are $A_1, A_2, \ldots, A_K$ of each of K types of male and $B_1, B_2, \ldots, B_K$ of the corresponding females. There are then $A_i B_j$ possible pairs of a type i male with a type j female and it may be reasonable to assume that the probability of a mating of this type is of the form

$$P_{ij} = \beta_{ij} A_i B_j \bigg/ \sum_{r=1}^{K} \sum_{s=1}^{K} \beta_{rs} A_r B_s \qquad (13.11)$$

where the positive value β_{rs} takes into account the likelihood of matings between males of type r and females of type s, relative to other pairings. This is a natural generalization of equation (13.1). The maximum likelihood estimator of β_{ij} comes from generalizing equation (13.2) to

$$\hat{\beta}_{ij} = \{d_{ij}/(A_i B_j)\} \Big/ \left\{ \sum_{r=1}^{K} \sum_{s=1}^{K} d_{rs}/(A_r B_s) \right\}, \qquad (13.12)$$

where d_{rs} is the number of matings observed between type r males and type s females by the end of an experiment. The scaling of the β values is fixed so that

$$\sum_{i=1}^{K} \sum_{j=1}^{K} \beta_{ij} = 1.$$

Biases, variances and covariances for the estimators are found by generalizing equations (5.16) in an obvious way to produce

$$\text{bias}\,(\hat{\beta}_{ij}) \simeq \beta_{ij} \sum_{r=1}^{K} \sum_{s=1}^{K} \alpha_{rs} - \alpha_{ij}, \qquad (13.13)$$

$$\text{var}\,(\hat{\beta}_{ij}) \simeq \beta_{ij}^2 \sum_{r=1}^{K} \sum_{s=1}^{K} \alpha_{rs} + (1 - 2\beta_{ij})\alpha_{ij}, \qquad (13.14)$$

and

$$\text{cov}\,(\hat{\beta}_{ij}, \hat{\beta}_{uv}) \simeq \beta_{ij}\beta_{uv} \sum_{r=1}^{K} \sum_{s=1}^{K} \alpha_{rs} - \beta_{ij}\alpha_{uv} - \beta_{uv}\alpha_{ij} \qquad (13.15)$$

for $(i, j) \neq (u, v)$, where $\alpha_{rs} = \beta_{rs}^2/E(d_{rs})$ and $E(d_{ij})$ is the expected value of d_{ij}.

In practice, most multiple choice experiments have involved only two types of male and female. The coefficient

$$\hat{Z}_I = \{(\hat{\beta}_{11}\hat{\beta}_{22})/(\hat{\beta}_{12}\hat{\beta}_{21})\}^{1/2} = \{(d_{11}d_{22})/(d_{12}d_{21})\}^{1/2} \qquad (13.16)$$

is then Levene's measure of sexual isolation, which was first used by Ehrman and Petit (1968). This will be high when most matings are type 1 × type 1 or type 2 × type 2, and low when most matings are type 1 × type 2. The expected value of $\hat{Z}_I$ should be close to 1 for random mating. Levene's measure of male selection is

$$\hat{Z}_m = \{(\hat{\beta}_{11}\hat{\beta}_{12})/(\hat{\beta}_{21}\hat{\beta}_{22})\}^{1/2}. \qquad (13.17)$$

This will be high or low according to whether type 1 or type 2 are most likely to mate, with an expected value of about 1 with random mating. Similarly,

$$\hat{Z}_f = \{(\hat{\beta}_{11}\hat{\beta}_{21})/(\hat{\beta}_{12}\hat{\beta}_{22})\}^{1/2} \qquad (13.18)$$

is Levene's measure of female selection.

Assuming that the d_{ij} values are independent Poisson variates, the Taylor series approximation method (Appendix, Section A.5) yields

$$\text{var}(\hat{Z}) \simeq Z^2\{1/E(d_{11}) + 1/E(d_{12}) + 1/E(d_{21}) + 1/E(d_{22})\}/4 \qquad (13.19)$$

for all three of the indices $\hat{Z}_\mathrm{I}$, $\hat{Z}_\mathrm{m}$ and $\hat{Z}_\mathrm{f}$, where Z indicates the expected value of $\hat{Z}$.

These $\hat{Z}$ coefficients are similar to the cross-product ratio that was discussed in Section 12.1 for the relationship between blood groups and diseases. This suggests that Haldane's (1956) modified coefficients may have better statistical properties. The modified coefficient of sexual isolation is

$$\hat{Y}_\mathrm{I} = \log_e\left\{\frac{(d_{11}+\frac{1}{2})(d_{22}+\frac{1}{2})}{(d_{12}+\frac{1}{2})(d_{21}+\frac{1}{2})}\right\}, \tag{13.20}$$

with variance

$$\mathrm{var}(\hat{Y}_\mathrm{I}) \simeq \frac{1}{E(d_{11})+1} + \frac{1}{E(d_{22})+1} + \frac{1}{E(d_{12})+1} + \frac{1}{E(d_{21})+1}. \tag{13.21}$$

This modified coefficient ranges from minus to plus infinity, with zero corresponding to no isolation. Large values imply a high degree of isolation. Similarly, a modified coefficient of male selection is

$$\hat{Y}_\mathrm{m} = \log_e\left[\left\{\left(\frac{d_{11}+\frac{1}{2}}{A_1 B_1}\right)\left(\frac{d_{12}+\frac{1}{2}}{A_1 B_2}\right)\right\}\Big/\left\{\left(\frac{d_{21}+\frac{1}{2}}{A_2 B_1}\right)\left(\frac{d_{22}+\frac{1}{2}}{A_2 B_2}\right)\right\}\right], \tag{13.22}$$

while a modified coefficient of female selection is

$$\hat{Y}_\mathrm{f} = \log_e\left[\left\{\left(\frac{d_{11}+\frac{1}{2}}{A_1 B_1}\right)\left(\frac{d_{21}+\frac{1}{2}}{A_2 B_1}\right)\right\}\Big/\left\{\left(\frac{d_{12}+\frac{1}{2}}{A_1 B_2}\right)\left(\frac{d_{22}+\frac{1}{2}}{A_2 B_2}\right)\right\}\right] \tag{13.23}$$

These have the same variance as $\hat{Y}_\mathrm{I}$, as given by equation (13.21).

The principal advantages of the modified coefficients are their symmetry and the fact that they are always finite. Thus, reversing the labels 1 and 2 on the two types of males and females changes $\hat{Z}_\mathrm{I}$ to $1/\hat{Z}_\mathrm{I}$, but only changes the sign of $\hat{Y}_\mathrm{I}$. If $d_{12} = 0$ then $\hat{Z}_\mathrm{I}$ becomes infinite but $\hat{Y}_\mathrm{I}$ can still be calculated. Goux and Anxolabehere (1980) have studied the indices $\hat{W}_\mathrm{I} = \hat{Z}_\mathrm{I}^2$, $\hat{W}_\mathrm{m} = \hat{Z}_\mathrm{m}^2$ and $\hat{W}_\mathrm{f} = \hat{Z}_\mathrm{f}^2$ and concluded that they have severe limitations in measuring what they purport to measure. Their criticisms will probably apply equally well to $\hat{Z}_\mathrm{I}$, $\hat{Z}_\mathrm{m}$ and $\hat{Z}_\mathrm{f}$. It seems possible that the indices $\hat{Y}_\mathrm{I}$, $\hat{Y}_\mathrm{m}$ and $\hat{Y}_\mathrm{f}$ may overcome some of the problems but this is something that still requires investigation.

In obtaining the estimates $\hat{\beta}_{ij}$ using equation (13.12) it has been assumed that the numbers of available males and females $A_1, A_2, .., A_K$, $B_1, B_2, \ldots, B_K$ remain constant. In fact it is sufficient that the relative frequencies remain constant. However, in some multiple choice experiments it has been the practice to remove mating pairs, in which case the relative frequencies may change substantially when there are a large number of matings. Also, females in particular may not be ready to mate again for some time after one mating and this m::y change the relative frequencies of different types of available females.

What is needed here is a modification of the estimator (13.12) to take into account reducing numbers of available males and females. Unfortunately, this modification is not straightforward and does not seem to have been attempted by anyone. For the present time, the best approach for estimation with experiments involving changing numbers of available males and females seems to be to apply the equations given so far in this section but to use them with 'average' values of $A_1, A_2, \ldots, A_K$ and $B_1, B_2, \ldots, B_K$ rather than the values at the start of the experiment. That is to say, if at the start of an experiment there are A_{i0} of type i males available for mating, and this reduces to A_{i1} by the end of the experiment, then take $A_i = \frac{1}{2}(A_{i0} + A_{i1})$. The same can be done for females. If males can mate several times and females only once, then this adjustment will only need to be done for females. Since the coefficients of sexual isolation $\hat{Z}_l$ and $\hat{Y}_l$ are not functions of the available numbers of males and females, they are not affected by adjustments of this type.

Example 13.2   Multiple choice experiments
with *Drosophila equinoxialis*

This example is concerned with some of the multiple choice experiments carried out by Ehrman and Petit (1968) on *Drosophila* from the *willistoni* group. Samples of *D. equinoxialis* were collected from three locations, Cordoba, Piojo and Turbo, in Colombia, South America. After a few generations in the laboratory, mating experiments were carried out in an observation chamber of the type devised by Elens and Wattiaux (1964). Each experiment involved males and females from two different locations, as shown in Table 13.3. Each experiment lasted three hours, during which time a single male could mate several times but most females only once.

Consider the experiment involving flies from Cardoba and Turbo with equal initial frequencies. This started with 108 males and females from Cordoba and also 108 males and females from Turbo. By the end of the experiment there were 111 matings. Presumably this did not have much effect on the availability of males for mating because each male could mate more than once. However, if the females become unavailable after mating then by the end of the experiment there were $108 - 28 - 29 = 51$ females from Cordoba and $108 - 23 - 31 = 54$ females from Turbo left. To allow for this, the values of B_1 and B_2 used for estimating selection and isolation parameters can be taken as $\frac{1}{2}(108 + 51) = 79.5$ for Cordoba and $\frac{1}{2}(108 + 54) = 81.0$ for Turbo. A similar correction can be made when analysing the other experimental results.

Table 13.4 shows estimates of β values and also the three indices $\hat{Y}_I$, $\hat{Y}_m$ and $\hat{Y}_f$. With random mating $\beta_{11} = \beta_{12} = \beta_{21} = \beta_{22} = 0.25$. The main point that emerges in looking at the $\hat{\beta}$ values is that when flies from one source are rare then these flies tend to mate among themselves rather than with flies from another source. For example, in the three experiments with flies from Cordoba and Turbo, $\hat{\beta}_{11}$ increases from 0.07 to 0.55 as the percentage of Cordoba flies

Table 13.3 Results from mating experiments with *Drosophila equinoxialis*.

Sources of Drosophila	Initial numbers*		Matings (male × female)				
	a	b	a × a	a × b	b × a	b × b	Total
a: Cordoba	50	200	12	8	19	64	103
b: Turbo	108	108	28	23	29	31	111
	160	40	41	9	30	20	100
a: Cordoba	30	120	10	26	14	53	103
b: Piojo	84	84	40	17	16	27	100
	160	40	63	18	22	10	113
a: Turbo	35	140	17	31	17	36	101
b: Piojo	84	84	33	24	25	18	100
	160	40	65	13	17	6	101

* The initial number were the same for males and females. In fact these numbers, and the mating numbers, are totals of from seven to ten smaller experiments. However, this is ignored in the present discussion.

Table 13.4 Estimates of β coefficients calculated from equation (13.12) and the indices $\hat{Y}_I, \hat{Y}_m$ and $\hat{Y}_f$ calculated from equations (13.20) to (13.23) for the data of Table 13.3. The numbers of females available for mating have been determined as described in the text.

Sources of Drosophila	Initial percentage of a	$\hat{\beta}_{11}$	$\hat{\beta}_{12}$	$\hat{\beta}_{21}$	$\hat{\beta}_{22}$	$\hat{Y}_I$	$\hat{Y}_m$	$\hat{Y}_f$	SE*
a: Cordoba	20	0.55	0.08	0.22	0.15	1.58	0.30	2.30	0.50
b: Turbo	50	0.25	0.21	0.26	0.28	0.26	−0.32	0.16	0.38
	80	0.07	0.07	0.20	0.66	1.07	−3.23	−1.30	0.45
a: Cordoba	20	0.45	0.26	0.26	0.13	0.38	1.75	0.76	0.46
b: Piojo	50	0.42	0.16	0.17	0.25	1.35	0.45	0.53	0.42
	80	0.15	0.20	0.21	0.44	0.47	−1.17	−1.02	0.45
a: Turbo	20	0.61	0.19	0.15	0.05	0.15	2.63	2.23	0.41
b: Piojo	50	0.35	0.22	0.26	0.17	−0.01	0.55	0.91	0.40
	80	0.23	0.18	0.25	0.34	0.59	−0.72	−0.15	0.53

* The standard errors are the same for $\hat{Y}_I, \hat{Y}_m$ and $\hat{Y}_f$.

decreases from 80% down to 20%. Here $\hat{\beta}_{11}$ reflects the probability of a Cordoba × Cordoba mating. For the same experiments $\hat{\beta}_{22}$ increases from 0.15 to 0.66 as the percentage of Turbo flies decreases from 80% down to 20%. However $\hat{\beta}_{12}$ and $\hat{\beta}_{21}$ show no clear frequency-dependent changes so that rare individuals seem to only have an advantage with individuals of the opposite sex from the same source.

The coefficient of sexual isolation $\hat{Y}_I$ shows no consistent pattern for these data. Cordoba and Turbo individuals are isolated at extreme frequencies because of the preference of each for individuals of their own type when they are rare. However, Cordoba and Piojo are only isolated when they are equally frequent. Turbo and Piojo are not particularly isolated at all. The coefficients of male selection and female selection are strongly frequency dependent, reflecting the relatively large numbers of matings between individuals from the same source when they are rare.

13.4 DATA FROM NATURAL POPULATIONS

It is sometimes possible to sample mating pairs in wild populations. A comparison of morph frequencies of mating individuals with morph frequencies of non-mating individuals will then give an indication of whether or not mating is at random.

Suppose that a random sample of non-mating individuals contains a_i males and b_i females of morph type i, for $i = 1, 2, \ldots, K$. Suppose also that a random sample of mating individuals contains d_{ij} pairs of morph i males with morph j females. Then it may be reasonable to regard the a_i and b_i values as estimates of the relative frequencies of available males and females, and modify equation (13.12) to

$$\hat{\beta}_{ij} = \{d_{ij}/(a_i b_j)\} \bigg/ \left\{ \sum_{r=1}^{K} \sum_{s=1}^{K} d_{rs}/(a_r b_s) \right\}. \tag{13.24}$$

In this case, sampling variation in d_{ij}, a_i and b_i values will cause sampling errors in the estimated β values. Assuming that these sample counts all have independent Poisson distributions, the Taylor series method (Appendix, Section A.5) yields

$$\text{bias} (\hat{\beta}_{ij}) \simeq \beta_{ij} \sum_{r=1}^{K} \sum_{s=1}^{K} \theta_{rs} - \theta_{ij}, \tag{13.25}$$

$$\text{var} (\hat{\beta}_{ij}) \simeq \beta_{ij}^2 \sum_{r=1}^{K} \sum_{s=1}^{K} \theta_{rs} + (1 - 2\beta_{ij})\theta_{ij}, \tag{13.26}$$

and

$$\text{cov} (\hat{\beta}_{ij}, \hat{\beta}_{uv}) \simeq \beta_{ij}\beta_{uv} \sum_{r=1}^{K} \sum_{s=1}^{K} \theta_{rs} - \beta_{ij}\theta_{uv} - \beta_{uv}\theta_{ij}, \tag{13.27}$$

where

$$\theta_{rs} = \beta_{rs}^2 \{1/E(d_{rs}) + 1/E(a_r) + 1/E(b_s)\},$$

with E denoting expected (mean) values.

When there are only $K = 2$ morphs, the indices of sexual isolation and selection that have been defined in Section 13.3 can still be calculated. The index of sexual isolation $\hat{Y}_I$ of equation (13.20) does not depend upon the a_i and b_i values, so its variance is still given by equation (13.21). However, the indices of male and female selection now become

$$\hat{Y}_m = \log_e\left[\left\{\left(\frac{d_{11}+\frac{1}{2}}{a_1 b_1}\right)\left(\frac{d_{12}+\frac{1}{2}}{a_1 b_2}\right)\right\}\middle/\left\{\left(\frac{d_{21}+\frac{1}{2}}{a_2 b_1}\right)\left(\frac{d_{22}+\frac{1}{2}}{a_2 b_2}\right)\right\}\right], \qquad (13.28)$$

and

$$Y_f = \log_e\left[\left\{\left(\frac{d_{11}+\frac{1}{2}}{a_1 b_1}\right)\left(\frac{d_{21}+\frac{1}{2}}{a_2 b_1}\right)\right\}\middle/\left\{\left(\frac{d_{12}+\frac{1}{2}}{a_1 b_2}\right)\left(\frac{d_{22}+\frac{1}{2}}{a_2 b_2}\right)\right\}\right], \qquad (13.29)$$

with variances
$$\text{var}(\hat{Y}_m) \simeq \text{var}(\hat{Y}_I) + 4\{1/E(a_1) + 1/E(a_2)\}, \qquad (13.30)$$

and
$$\text{var}(\hat{Y}_f) \simeq \text{var}(\hat{Y}_I) + 4\{1/E(b_1) + 1/E(b_2)\}. \qquad (13.31)$$

The second components on the right-hand sides of these equations are the contribution to the variance caused by sampling errors in the proportions of different types of males and females available for mating.

Example 13.3   Non-random mating of wild ladybirds

A visual polymorphism exists in populations of the two-spot ladybird *Adalia bipunctata*. There are a number of morphs for the pattern and colour of the body under the control of at least 11 alleles at a single locus. These range from black with two red spots, to all red, with black forms (melanics) being dominant to red forms (non-melanics). In England the commonest melanic forms are *quadrimaculata* and *sexpustulata* (black with four and six spots, respectively), and the commonest non-melanics are *typica* (red with two round black spots) and *annulata* (red with two irregular black patches or with two large black spots, each having one or more small satellite spots). There is a simple dominance hierarchy in the genetic system, with *quadrimaculata* being dominant to *sexpustulata*, which is dominant to *typica*, which is dominant to *annulata*.

Nearly all populations are polymorphic. In the British Isles the percentage of melanics varies geographically and exceeds 90% in Manchester and Glasgow (Bishop *et al.*, 1978). A number of factors have been suggested as being responsible for maintaining the polymorphism (Muggleton, 1978), including a more efficient absorption of solar radiation by melanics (Muggleton *et al.*, 1975) and greater tolerance of air pollution by melanics (Creed, 1975).

Mating pairs are often seen in the field. Lusis (1961) published data showing that the frequency of melanics among mating pairs from Riga and Moscow was greater than in the populations from which they came. He suggested that this was due to melanics being more active in sunlight, and related it to Timofeeff-Ressovsky's (1940) data which show that the frequency of melanics increased in Berlin during the summer months. Creed (1975) found no excess of melanics in Britain or Potsdam, but his method of analysis has been criticized by Muggleton (1979) because it confounds different effects that could lead to non-random mating and also because sampling errors are not properly taken into account.

Muggleton (1979) reanalysed previously published data and new data from England and concluded that non-random mating in *A. bipunctata* is frequency-dependent: melanics are in excess when at a frequency of 40% or less in a population and non-melanics are in excess when melanics comprise over 50%. O'Donald and Muggleton (1979) fitted a model of constant and variable mating preferences to the data. They found that the frequencies in Potsdam matings gave rise to strongly frequency-dependent sexual selection, while the frequency dependence of the English data was less pronounced. More recently, Majerus *et al.* (1982b) collected data on the mating frequencies

Table 13.5   Data on matings in a wild population of *Adalia bipunctata*.

(a) Non-mating individuals

| | Phenotype | | | | |
	Q	S	T	A	Total
Male	57	7	146	67	277
Female	58	8	123	48	237

(b) Mating pairs

| *Male* phenotype | Female phenotype | | | | |
	Q	S	T	A	Total
Q	19	2	31	14	66
S	3	1	6	2	12
T	20	2	28	8	58
A	11	1	9	12	33
Total	53	6	74	36	169

Q = *quadrimaculata*   S = *sexpustulata*
T = *typica*   A = *annulata*

of the phenotypes *quadrimaculata, sexpustulata, typica* and *annulata*. It is part of their data that are the subject of the present example.

Table 13.5 shows the frequencies of the four phenotypes in samples taken from a wild population of *A. bipunctata* in the grounds of Keele University, Staffordshire, England, over the period 3–5 August 1981. Applying equations (13.24) and (13.26) to these data results in the estimates shown in Table 13.6. Because of the small numbers of *sexpustulata*, these have been combined with *quadrimaculata*, to form a single class of melanics. Biases calculated from equation (13.25) are negligible.

In the absence of selection, all the nine β values should be equal to $1/9 = 0.111$. There are seven estimated β values that are more than two standard errors from 0.111, corresponding roughly to a 5% level of significance. It seems quite clear that melanic (*quadrimaculata + sexpustulata*) males were over-represented in mating pairs, largely at the expense of *typica* males. A similar pattern, but to a much smaller extent, is apparent for females. A large effect for males and small effect for females is, of course, consistent with the idea that it is female choice rather than male choice that determines the frequencies of different matings.

Table 13.6 Estimates of the preference coefficients β using equations (13.24) and (13.26).

| Male phenotype | Female phenotype | | | | | | Total |
| | Q + S | | T | | A | | |
	$\hat{\beta}$	SE	$\hat{\beta}$	SE	$\hat{\beta}$	SE	$\hat{\beta}$
Q + S	0.209*	0.025	0.166*	0.019	0.184*	0.022	0.559
T	0.081*	0.009	0.055*	0.006	0.040*	0.004	0.176
A	0.096	0.011	0.039*	0.004	0.132	0.015	0.267
	0.386		0.260		0.356		

Q + S = *quadrimaculata + sexpustulata*
T = *typica*   A = *annulata*

* Significantly different from the no-selection value of $1/9 = 0.111$ at approximately the 5% level of significance.

13.5 O'DONALD'S MODELS FOR SEXUAL SELECTION

To conclude this chapter it is appropriate to make mention of O'Donald's (1980) models for sexual selection. These models involve the idea that certain proportions of females have preferences for different male morphs and only mate with males that they prefer. The remaining females mate at random. O'Donald has shown that a constant female preference for certain types of male must necessarily give rise to frequency dependence in a male's mating advantage since males of a preferred type must mate more easily when they are rare than when they are common. Thus his models explain the phenomenon of rare-male advantage, although this explanation is doubted in some quarters (Searcy, 1982; Spiess, 1982).

A major stimulus for O'Donald's development of models for sexual selection has been his long-term study of the Arctic skua on Fair Isle (see Example 12.1). However, his book (O'Donald, 1980) describes applications of the models to many other species.

14 Concluding remarks

14.1 DIFFICULTIES IN INTERPRETING EVIDENCE FOR SELECTION

The variation in time and space of a natural animal population is determined by many factors. There are systematic effects caused by historical events, migration and selection. There are random effects of mutation and genetic drift. One of the problems in interpreting evidence for selection is therefore making allowance for other factors. Little or nothing will usually be known about historical events. Migration patterns and effective population sizes will be known only roughly. Making an allowance may therefore be very difficult, if not impossible. It seems inevitable that evidence for selection will tend to be somewhat ambiguous, with alternative non-selective explanations being readily available. This is particularly true with electrophoretic data, for which a usable generally accepted model of random drift is not yet available.

Even in cases where selection seems to be definitely occurring, there may be considerable doubt about what precisely is being selected. This is because much evidence for selection is an observed significant deviation from randomness, with no indication of the cause. Selection may not even have anything to do with the character being studied, with this just happening to be correlated with another character that is being selected.

An important point here is that just because a particular model fits some data, this does not mean that it corresponds to reality. An obvious case in point occurs when there are a series of gene frequencies in successive generations of a population. There may then be many different models (constant selective values, frequency-dependent selection, etc.) that fit about equally well. Without further information on different components of fitness there will be no way of deciding which model is correct.

A further complication occurs with sample data. Populations are not uniformly distributed in space. As discussed in Section 1.7, apparent changes in a population that are attributed to selection may then be nothing more than differences between non-random samples taken from two or more constant populations.

14.2 DIFFICULTIES IN TAKING LARGE ENOUGH SAMPLES

Even when random samples can be taken, it may not be possible to make them large enough to detect a moderate amount of selection. The amount of work done on the design of studies has not been great, as is reflected in a paucity of references in the previous chapters. However, in cases where sample size requirements have been studied, it appears that these sizes may have to be rather large (thousands, rather than hundreds) in order to detect moderate selection. For example, Reed (1975) has discussed this question in the context of selection on blood group polymorphisms. He notes that the studies carried out in earlier years were simply not large enough to detect selective effects of the order of 10–20%. There seems little doubt that the need for very large samples to study small selective effects is the main reason why little is known at the present time about the general level of selective effects on populations. By their very nature, studies only tend to pick up the most clear-cut cases of extreme selection.

14.3 SIMULATION MODELS

As mentioned above, the distribution of an animal over time and space is a result of historical events, migration, selection, mutation and random drift. In the past this has meant that in order to gain any understanding of population dynamics it has been necessary to make simplifying assumptions about the effects of these factors. However, with large computers becoming readily available it has now become possible to adopt an alternative approach. Assumptions can be made as realistic as possible and a population can be simulated on a computer. By adjusting parameters and modifying assumptions, an attempt can then be made to find a model that reproduces the behaviour of the real population. In this way simulation is a powerful tool for gaining an understanding of the dynamics of particular organisms.

This approach has been used to study the distribution of melanic forms of the moth *Biston betularia* in England and Wales. Bishop (1972) estimated selective values experimentally and then simulated the distribution of the cline in the frequency of melanics that runs from Liverpool to rural North Wales. He was unable to reproduce the actual distribution that exists. However, the model came closer to reality when heterosis was introduced. More recently, Cook and Mani (1980) developed a simulation model for *B. betularia* over the whole of England and Wales, taking into account visual selection and migration. They also found serious discrepancies between the predictions from their model and observations, and concluded that some other factor or factors must be influencing the population. Later Mani (1980) added non-visual selection and frequency-dependent predation to the model and was able to account satisfactorily for the clines and melanic frequencies from Liverpool to North Wales for the moths *B. betularia* and *Phigalia pilosaria*.

Statistical appendix

A.1 THE STATISTICS OF QUANTITATIVE VARIATION

Many studies of selection involve comparing the distribution of one or more quantitative variables $X_1, X_2, \ldots, X_p$ at different places or at different times. It is therefore useful to begin this appendix with a brief summary of methods that are used to describe these distributions. To begin with, a single variable X will be considered.

Given a random sample of values $X_1, X_2, \ldots, X_n$ from a distribution, the sample mean and variance and the third and fourth moments about the mean are

$$\left. \begin{array}{ll} \hat{\mu} = \sum_{i=1}^{n} X_i/n, & \hat{\sigma}^2 = \sum_{i=1}^{n} (X_i - \hat{\mu})^2/n, \\[2em] \hat{\mu}_3 = \sum_{i=1}^{n} (X_i - \hat{\mu})^3/n, \quad \text{and} \quad & \hat{\mu}_4 = \sum_{i=1}^{n} (X_i - \hat{\mu})^4/n, \end{array} \right\} \quad \text{(A1)}$$

respectively. The sample skewness can then be defined as $\hat{\gamma}_1 = \hat{\mu}_3/\hat{\sigma}^3$ and the sample kurtosis as $\hat{\gamma}_2 = \hat{\mu}_4/\hat{\sigma}^4 - 3$. The skewness is a measure of the extent to which the distribution of X is not symmetric about the mean, while the kurtosis is a measure of the 'flatness' of the distribution. It is a standard practice to use a caret to distinguish a sample estimate from a true population value. Thus $\hat{\mu}$ is an estimate of the true population mean μ, $\hat{\sigma}^2$ is an estimate of the population variance σ^2, etc.

The mean value of a random variable R is also called its expected value and denoted by $E(R)$. It can be shown that $E(\hat{\mu}) = \mu$, so that the sample mean $\hat{\mu}$ is an unbiased estimator of the population mean μ. However, $E(\hat{\sigma}^2) = (n-1)\sigma^2/n$ so that $\hat{\sigma}^2$ is a biased estimator of σ^2. Hence for some purposes the sample variance is taken as

$$s^2 = \sum_{i=1}^{n} (X_i - \hat{\mu})^2/(n-1) \quad \text{(A2)}$$

instead of $\hat{\sigma}^2$. Then $E(s^2) = \sigma^2$ so that s^2 is an unbiased estimator of σ^2.

The most commonly used distribution is the *normal* distribution, which is completely determined by its mean μ and standard deviation σ. This distribution is bell-shaped and symmetrical about μ. The skewness γ_1 and kurtosis γ_2 are both zero. Mathematically, the distribution is described by the probability density function

$$f(x) = \frac{1}{\sigma\sqrt{(2\pi)}} \exp\left\{-\frac{(x-\mu)^2}{2\sigma^2}\right\}, \qquad -\infty < x < +\infty. \qquad \text{(A3)}$$

Then the probability of a value of X between two limits A and B is given by the integral

$$P(A < X < B) = \int_A^B f(x)\,dx, \qquad A < B.$$

Roughly $f(x)\,\delta x$ can be thought of as the probability of a value between x and $x + \delta x$, where δx is small.

Suppose next that there are several variables $X_1, X_2, \ldots, X_p$ to be considered simultaneously, and that a random sample of size n is available. Then the X-values for the jth individual in the sample can be denoted as X_{1j}, $X_{2j}, \ldots, X_{nj}$ and the sample mean and variance for the ith variable can be defined as

$$\hat{\mu}_i = \sum_{j=1}^n X_{ij}/n \quad \text{and} \quad \hat{V}_{ii} = \sum_{j=1}^n (X_{ij} - \hat{\mu}_i)^2/n, \qquad \text{(A4)}$$

while the sample covariance between the two variables X_i and X_k is

$$\hat{V}_{ik} = \sum_{j=1}^n (X_{ij} - \hat{\mu}_i)(X_{kj} - \hat{\mu}_k)/n. \qquad \text{(A5)}$$

The sample mean vector and covariance matrix are then

$$\mu = \begin{bmatrix} \hat{\mu}_1 \\ \hat{\mu}_2 \\ \vdots \\ \hat{\mu}_p \end{bmatrix} \quad \text{and} \quad \hat{V} = \begin{bmatrix} \hat{V}_{11} & \hat{V}_{12} & \cdots & \hat{V}_{1p} \\ \hat{V}_{21} & \hat{V}_{22} & \cdots & \hat{V}_{2p} \\ \vdots & \vdots & & \vdots \\ \hat{V}_{p1} & \hat{V}_{p2} & \cdots & \hat{V}_{pp} \end{bmatrix} \qquad \text{(A6)}$$

The corresponding population vector of means and the covariance matrix can then be denoted by μ and V. The sample correlation between X_i and X_k is $r_{ik} = \hat{V}_{ik}/\sqrt{(\hat{V}_{ii}\hat{V}_{kk})}$, with the corresponding population value being $V_{ik}/\sqrt{(V_{ii}V_{kk})}$. To avoid biases, the sample values $\hat{V}_{ii}$ and $\hat{V}_{ik}$ of equations (A4) and (A5) are sometimes defined with a division by $n-1$ instead of n. This makes the expected value of $\hat{V}$ equal to V.

The multivariate normal distribution is undoubtedly the most important for applications. Here it is merely necessary to note that this distribution is completely determined by its mean vector and its covariance matrix.

A.2 THE STATISTICS OF POLYMORPHIC VARIATION

With polymorphic variation there are K morphs and sample data consist of frequency counts of these. Two models are then particularly important.

Firstly, there is the *multinomial* model. This arises when a random sample of n individuals is taken from a large population in which a proportion p_i of individuals are of morph i, with $\Sigma p_i = 1$. Then the probability of the sample containing r_i individuals of morph i, for $i = 1, 2, \ldots, K$, is

$$P(r_1, r_2, \ldots, r_K) = \frac{n!}{r_1! \, r_2! \ldots r_K!} \, p_1^{r_1} p_2^{r_2} \ldots p_K^{r_K}. \tag{A7}$$

With $K = 2$ morphs it is possible to write $p_1 = p$, $p_2 = 1 - p$, $r_1 = r$ and $r_2 = n - r$, so that

$$P(r, n - r) = \frac{n!}{r! \, (n - r)!} \, p^r (1 - p)^{n - r}, \tag{A8}$$

this being the probability function for the well known *binomial* distribution.

It can be proved that the mean value of r_i is np_i, the variance of r_i is $np_i(1 - p_i)$, and the covariance of r_i and r_k is $-np_i p_k$. It is also well known that the multinomial distribution can be well approximated by a multivariate normal distribution provided that all of the mean values for $r_1, r_2, \ldots, r_K$ are larger than about five. That is, the vector $\mathbf{R}' = (r_1, r_2, \ldots, r_K)$ can be treated as coming from a multivariate normal distribution with mean vector and covariance matrix

$$\mu = \begin{bmatrix} np_1 \\ np_2 \\ \vdots \\ np_K \end{bmatrix}, \text{ and } \mathbf{V} = \begin{bmatrix} np_1(1 - p_1) & -np_1 p_2 & \cdots & -np_1 p_K \\ -np_2 p_1 & np_2(1 - p_2) & \cdots & -np_2 p_K \\ \vdots & \vdots & & \vdots \\ -np_K p_1 & -np_K p_2 & \cdots & np_K(1 - p_K) \end{bmatrix}. \tag{A9}$$

The second important model is the *Poisson* model. In this case it is assumed that the morph frequencies $r_1, r_2, \ldots, r_K$ are independent Poisson variates with mean values $\mu_1, \mu_2, \ldots, \mu_K$. The probability of observing the frequencies is then the product of the K Poisson probabilities,

$$P(r_1, r_2, \ldots, r_K) = \prod_{i=1}^{K} \exp(-\mu_i) \, \mu_i^{r_i} / r_i!. \tag{A10}$$

The variance of r_i is μ_i and the covariance of r_i and r_k is zero. If μ_i is larger than about five, then r_i can be regarded for many purposes as having a normal distribution.

There is a relationship between the multinomial and Poisson models which means that they are in a way equivalent. Suppose that the Poisson model is correct. Then it is possible to rewrite the probability of equation (A10) in the form

$$P(r_1, r_2, \ldots, r_K) = P(r_1, r_2, \ldots, r_K | n) \, P(n), \tag{A11}$$

where

$$P(r_1, r_2, \ldots, r_K | n) = \frac{n!}{r_1! \, r_2! \ldots r_K!} p_1^{r_1} p_2^{r_2} \ldots p_K^{r_K},$$

and

$$P(n) = \exp\left(-\sum_{i=1}^{K} \mu_i\right)\left(\sum_{i=1}^{K} \mu_i\right)^n / n!,$$

where

$$n = \sum_{i=1}^{K} r_i \quad \text{and} \quad p_j = \mu_j / \sum_{i=1}^{K} \mu_i.$$

Here $P(r_1, r_2, \ldots r_K | n)$ is the probability of observing the counts r_1, $r_2, \ldots, r_K$, conditional upon the total count being equal to n, while $P(n)$ is the probability of the total count being n. Since $P(r_1, r_2, \ldots, r_K | n)$ is a multinomial probability, it follows that sample values $r_1, r_2, \ldots, r_K$ can be analysed either by regarding them as independent Poisson variates, or by regarding them as a sample of size n from a multinomial distribution. It frequently happens that one of these two approaches is rather easier to use than the other. Therefore the ability to use either is of real value.

A.3   THE METHOD OF MAXIMUM LIKELIHOOD

The method of maximum likelihood for estimation involves finding the probability of some data as a function of any parameters that need to be estimated, and then finding the parameter values that maximize this probability. These values are the maximum likelihood estimates. The probability of the data, regarded as a function of the unknown parameters, is called the *likelihood function*.

The important properties that make this method so useful are that under fairly general conditions, with large samples, maximum likelihood estimators are unbiased, have the smallest possible variances, are normally distributed, and have variances and covariances that can easily be approximated. The meaning of 'large' in this connection depends upon the particular case being considered. Generally, the user of maximum likelihood estimators will hope that all of these properties will approximately hold.

The way that a likelihood function is constructed is perhaps best understood by considering two simple examples. Suppose that a random sample of n counts $r_1, r_2, \ldots, r_n$ is taken from a Poisson distribution with an unknown mean μ. The probability of obtaining such a sample is then the product of the probabilities of the individual sample values,

$$P(r_i) = \exp(-\mu)\mu^{r_i}/r_i!.$$

That is to say, the full likelihood function is in this case

$$L(\mu) = \prod_{i=1}^{n} P(r_i) = \prod_{i=1}^{n} \exp(-\mu)\mu^{r_i}/r_i!.$$

It is viewed as being a function of the unknown mean μ since everything else is known. The maximum likelihood estimate is the value that maximizes this function, which can be shown to be $\hat{\mu} = \Sigma r_i/n$.

For a second example, suppose that a random sample of size n is taken from a normal distribution with mean μ and variance V. If the ith sample value is X_i then the probability of this is proportional to

$$(2\pi V)^{-1/2} \exp\{-(X_i - \mu)^2/2V\}.$$

The full likelihood function is the product of n expressions like this, regarded as a function of μ and V:

$$L(\mu, V) = \prod_{i=1}^{n} (2\pi V)^{-1/2} \exp\{-(x_i - \mu)^2/2V\}.$$

The maximum likelihood estimates of μ and V are the values that maximize this function, these being $\hat{\mu} = \Sigma X_i/n$ and $\hat{V} = \Sigma(X_i - \hat{\mu})^2/n$.

In many cases a likelihood function $L(\theta)$ can be maximized using calculus methods. In doing this it can be recognized that the log-likelihood function $l(\theta) = \log_e\{L(\theta)\}$ is maximized for the same value $\hat{\theta}$, of θ, so that $\hat{\theta}$ can be found by solving the equation

$$\frac{dl}{d\theta}(\hat{\theta}) = 0. \tag{A12}$$

An approximation for the variance of $\hat{\theta}$ is then given by

$$\text{var}(\hat{\theta}) = -\left[E\left\{\frac{d^2l}{d\theta^2}(\theta)\right\}\right]^{-1} \simeq -\left\{\frac{d^2l}{d\theta^2}(\theta)\right\}^{-1}, \tag{A13}$$

where E indicates the expected (mean) value of the second derivative.

Very often, equation (A12) has no explicit solution. A numerical solution can then be found using a technique such as root-bisection or by plotting the graph of $dl/d\theta$ against θ. Alternatively, the Newton–Raphson iterative method can be employed. This involves starting with an approximation $\hat{\theta}_0$ for $\hat{\theta}$ and using the first two terms in a Taylor series expansion for $dl/d\theta$, to write

$$\frac{dl}{d\theta}(\hat{\theta}) \simeq \frac{dl}{d\theta}(\hat{\theta}_0) + \frac{d^2l}{d\theta^2}(\hat{\theta}_0)\,\delta\theta$$

where $\hat{\theta} = \hat{\theta}_0 + \delta\theta$. From equation (A12) it then follows that

$$\delta\theta \simeq \frac{dl}{d\theta}(\hat{\theta}_0) \bigg/ \left\{-\frac{d^2l}{d\theta^2}(\hat{\theta}_0)\right\}. \tag{A14}$$

The Newton–Raphson method entails repeatedly using this equation to correct successive approximations to $\hat{\theta}$. When $\delta\theta$ become infinitesimally small, the maximum likelihood estimate has been found. Convergence is not guaranteed but the Newton–Raphson method usually converges providing that $\hat{\theta}_0$ is not too far from $\hat{\theta}$.

If there are several unknown parameters $\theta_1, \theta_2, \ldots, \theta_m$, then their maximum likelihood estimators $\hat{\theta}_1, \hat{\theta}_2, \ldots, \hat{\theta}_m$ are found by maximizing the log-likelihood function $l(\theta_1, \theta_2, \ldots, \theta_m) = l(\boldsymbol{\theta})$. Generally, this means solving the set of m equations

$$\frac{\partial l}{\partial \theta_i}(\hat{\boldsymbol{\theta}}) = 0, \qquad i = 1, 2, \ldots, m. \tag{A15}$$

For large samples the estimators generally have a multivariate normal distribution with a covariance matrix given by

$$\mathbf{V} \simeq E(\mathbf{D})^{-1} \simeq \mathbf{D}^{-1}. \tag{A16}$$

where $\mathbf{D}$ is the symmetric matrix with

$$d_{ij} = -\frac{\partial^2 l}{\partial \theta_i \partial \theta_j}(\boldsymbol{\theta})$$

in the ith row and jth column.

If equations (A15) have no explicit solution then the Newton–Raphson method can be used to improve initial approximations $\hat{\theta}_{10}, \hat{\theta}_{20}, \ldots, \hat{\theta}_{m0}$ for the maximum likelihood estimates $\hat{\theta}_1, \hat{\theta}_2, \ldots, \hat{\theta}_m$. The generalization of equation (A14) is

$$\begin{bmatrix} \delta\theta_1 \\ \delta\theta_2 \\ \vdots \\ \delta\theta_m \end{bmatrix} \simeq \mathbf{D}^{-1} \begin{bmatrix} \partial l/\partial\theta_1 \\ \partial l/\partial\theta_2 \\ \vdots \\ \partial l/\partial\theta_m \end{bmatrix}, \tag{A17}$$

where the first and second derivatives on the right-hand side are evaluated using the initial approximation $\hat{\boldsymbol{\theta}}_0$ for $\hat{\boldsymbol{\theta}}$. Equation (A17) is used to improve the approximations for the estimates by taking the new values $\hat{\theta}_{i1} = \hat{\theta}_{i0} + \delta\theta_i$. These are then used as initial approximations and are themselves improved using equation (A17). The cycle is continued until the corrections are negligible, at which point the maximum likelihood estimates have been obtained. Convergence is usually obtained provided that the initial approximations are reasonable.

For many of the applications of maximum likelihood that are discussed in this book, the data are assumed to come from a multinomial distribution so that the likelihood function takes the form of equation (A7). That is, the data consists of observed counts $r_1, r_2, \ldots, r_K$ in K classes, where the class probabilities $p_1, p_2, \ldots, p_K$ depend upon some unknown parameters

$\theta_1, \theta_2, \ldots, \theta_m$. Then the likelihood function is

$$L(\theta_1, \theta_2, \ldots, \theta_m) = C \prod_{j=1}^{K} p_i(\theta_1, \theta_2, \ldots, \theta_m)^{r_j},$$

where C is a constant that depends upon the class frequencies but not on the unknown parameters. Hence the log-likelihood function is

$$l(\theta_1, \theta_2, \ldots, \theta_m) = \log_e(C) + \sum_{j=1}^{K} r_j \log_e\{p_j(\theta_1, \theta_2, \ldots, \theta_m)$$

and equations (A15) for the maximum likelihood estimates of the parameters θ take the form

$$\sum_{j=1}^{K} \frac{r_j}{p_j} \frac{\partial p_j}{\partial \theta_i} = 0, \qquad i = 1, 2, \ldots, m. \qquad (A18)$$

A computer program (MAXLIK) for calculating maximum likelihood estimates for multinomial situations has been developed by Reed and Schull (1968). The code is provided in Section A.16, together with details about how to use it. Another useful computer program is GLIM, which carries out maximum likelihood estimation for a wide range of regression-type models. More information about this program is provided in Section A.10. The program MLP (Ross, 1980) is also worth mentioning.

A.4 CHI-SQUARE LIKELIHOOD RATIO TESTS

One of the advantages of fitting models to data using the principle of maximum likelihood is that the fit of different models may be compared using likelihood ratio tests. The idea is as follows. Suppose that there are two models, I and II, that are being considered for a set of data, where II incorporates I as a special case. Then a likelihood ratio test can be used to determine whether the extra flexibility in model II allows a significantly better fit to the data. If the parameters of the models are estimated by maximum likelihood then $l_I \leqslant l_{II}$, where l_I is the maximized log-likelihood for model I and l_{II} is the maximized log-likelihood for model II. The statistic $2(l_{II} - l_I)$ will approximately follow a chi-square distribution with $v_{II} - v_I$ degrees of freedom if model I is appropriate for the data. Here v_I and v_{II} are the numbers of estimated parameters for models I and II, respectively (Kendall, 1975, p. 129). A significantly large value for $2(l_{II} - l_I)$ shows that model II gives a significantly better fit to the data than model I.

A.5 TAYLOR SERIES APPROXIMATIONS FOR BIASES, VARIANCES AND COVARIANCES

Let $X_1, X_2, \ldots, X_n$ be random variables with means $\mu_1, \mu_2, \ldots, \mu_n$, respectively. Suppose that it is required to know the mean value of some function $f(X_1, X_2, \ldots, X_n) = f(\mathbf{X})$. Then, assuming that $f(\mathbf{X})$ can be differentiated twice, there is a Taylor series approximation

$$f(\mathbf{X}) \simeq f(\mu) + \sum_{i=1}^{n} (X_i - \mu_i) \frac{\partial f}{\partial X_i} + \frac{1}{2} \sum_{i=1}^{n} \sum_{j=1}^{n} (X_i - \mu_i)(X_j - \mu_j) \frac{\partial^2 f}{\partial X_i \partial X_j},$$

where all the partial derivatives are evaluated using the means of the X variables. Taking expected (mean) values shows that

$$\text{bias}\{f(\mathbf{X})\} = E\{f(\mathbf{X})\} - f(\mu) \simeq \frac{1}{2} \sum_{i=1}^{n} \sum_{j=1}^{n} \frac{\partial^2 f}{\partial X_i \partial X_j} \text{cov}(X_i, X_j). \quad \text{(A19)}$$

Using only the first two terms of the Taylor series approximation gives

$$\{f(\mathbf{X}) - f(\mu)\}^2 \simeq \sum_{i=1}^{n} \sum_{j=1}^{n} (X_i - \mu_i)(X_j - \mu_j) \frac{\partial f}{\partial X_i} \frac{\partial f}{\partial X_j},$$

so that taking expected values produces

$$\text{var}\{f(\mathbf{X})\} \simeq \sum_{i=1}^{n} \sum_{j=1}^{n} \frac{\partial f}{\partial X_i} \frac{\partial f}{\partial X_j} \text{cov}(X_i, X_j), \quad \text{(A20)}$$

assuming that the bias in $f(\mathbf{X})$ is negligible.

If there is a second function $g(\mathbf{X})$ to be considered, then

$$\{f(\mathbf{X}) - f(\mu)\}\{g(\mathbf{X}) - g(\mu)\} \simeq \sum_{i=1}^{n} \sum_{j=1}^{n} (X_i - \mu_i) \frac{\partial f}{\partial X_i} (X_j - \mu_j) \frac{\partial g}{\partial X_j}.$$

Taking expected values, assuming that both functions have negligible bias, then gives

$$\text{cov}\{f(\mathbf{X}), g(\mathbf{X})\} \simeq \sum_{i=1}^{n} \sum_{j=1}^{n} \frac{\partial f}{\partial X_i} \frac{\partial g}{\partial X_j} \text{cov}(X_i, X_j). \quad \text{(A21)}$$

Equations (A19) to (A21) have been used to obtain many of the formulae for biases, variances and covariances quoted in this book. They are equivalent to using what is sometimes called the δ-method.

In many cases the variables $X_1, X_2, \ldots, X_n$ will be uncorrelated, so that the equations reduce to

$$\text{bias}\{f(\mathbf{X})\} \simeq \frac{1}{2} \sum_{i=1}^{n} \frac{\partial^2 f}{\partial X_i^2} \text{var}(X_i), \quad \text{(A22)}$$

$$\text{var}\{f(\mathbf{X})\} \simeq \sum_{i=1}^{n} \left(\frac{\partial f}{\partial X_i} \right)^2 \text{var}(X_i), \quad \text{(A23)}$$

and

$$\operatorname{cov}\{f(\mathbf{X}), g(\mathbf{X})\} \simeq \sum_{i=1}^{n} \frac{\partial f}{\partial X_i} \frac{\partial g}{\partial X_j} \operatorname{var}(X_i) \tag{A24}$$

These equations can then be simplified even further as a basis for purely numerical approximations. The idea is to take the numerical derivatives

$$\frac{\partial f}{\partial X_i} \simeq \{f(\mu_1, \ldots, \mu_{i-1}, \mu_i + \delta_i, \mu_{i+1}, \ldots, \mu_n) - f(\boldsymbol{\mu})\}/\delta_i,$$

and

$$\frac{\partial g}{\partial X_i} \simeq \{g(\mu_1, \ldots, \mu_{i-1}, \mu_i + \delta_i, \mu_{i+1}, \ldots, \mu_n) - g(\boldsymbol{\mu})\}/\delta_i,$$

where δ_i is the standard deviation of X_i. Then equations (A23) and (A24) become

$$\operatorname{var}\{f(\mathbf{X})\} \simeq \sum_{i=1}^{n} (\Delta f_i)^2, \tag{A25}$$

and

$$\operatorname{cov}\{f(\mathbf{X}), g(\mathbf{X})\} \simeq \sum_{i=1}^{n} \Delta f_i \Delta g_i, \tag{A26}$$

where Δf_i and Δg_i are the changes in $f(\boldsymbol{\mu})$ and $g(\boldsymbol{\mu})$, respectively, resulting from a one standard deviation change in X_i.

For the bias equation, a numerical approximation for $\partial^2 f / \partial X_i^2$ is

$$\frac{\partial^2 f}{\partial X_i^2} \simeq [\{f(\mu_1, \ldots, \mu_{i-1}, \mu_i + \delta_i, \mu_{i+1}, \ldots, \mu_n) - f(\boldsymbol{\mu})\}/\delta_i$$
$$- \{f(\boldsymbol{\mu}) - f(\mu_1, \ldots, \mu_{i-1}, \mu_i - \delta_i, \mu_{i+1}, \ldots, \mu_n)\}/\delta_i]/\delta_i,$$

this being the difference between two numerical approximations to $\partial f / \partial X_i$. Substitution into equation (A22) then gives

$$\operatorname{bias}\{f(\mathbf{X})\} \simeq \tfrac{1}{2} \sum_{i=1}^{n} (\Delta f_i^+ + \Delta f_i^-), \tag{A27}$$

where Δf_i^+ is the change in $f(\boldsymbol{\mu})$ that results from a one standard deviation increase in X_i, while Δf_i^- is the change that results from a one standard deviation decrease in X_i.

A.6 WEIGHTED MEANS WITH MINIMUM VARIANCE

Suppose that $Y_1, Y_2, \ldots, Y_n$ are uncorrelated, unbiased estimators of the same parameter μ. Then it is a standard result that the weighted mean with minimum variance is

$$\hat{\mu} = \sum_{i=1}^{n} h_i Y_i \bigg/ \sum_{i=1}^{n} h_i, \tag{A28}$$

where $h_i = 1/\text{var}(Y_i)$, and the variance of $\hat{\mu}$ is given by

$$\text{var}(\hat{\mu}) = 1 \bigg/ \sum_{i=1}^{n} h_i. \tag{A29}$$

The weighted sum of squares of the Y_i values about the origin can be partitioned as

$$\sum_{i=1}^{n} h_i Y_i^2 = \sum_{i=1}^{n} h_i (Y_i - \hat{\mu})^2 + \hat{\mu}^2 \sum_{i=1}^{n} h_i, \tag{A30}$$

where the first term on the right-hand side is the weighted residual sum of squares, with $n-1$ degrees of freedom, and the second term is the sum of squares for the mean, with 1 degree of freedom. If the Y values are normally distributed then two potentially useful tests are available from this result:

(a) The residual sum of squares $\Sigma h_i (Y_i - \hat{\mu})^2$ can be compared with the chi-square distribution with $n-1$ degrees of freedom. If the sum of squares is significantly large then this indicates that the differences between the Y values are too large to be attributed to sampling errors. This casts doubts on the assumption that the Y values are all estimating the same parameter.

(b) The mean sum of squares $\hat{\mu}^2 \Sigma h_i$ can be compared with the chi-square distribution with 1 degree of freedom. If the sum of squares is significantly large then this is evidence that the parameter μ is non-zero.

If the estimators $Y_1, Y_2, \ldots, Y_n$ are correlated then equations (A28) to (A30) no longer apply. However, in this case it can be shown that the unbiased weighted mean with minimum variance is

$$\hat{\mu} = \sum_{i=1}^{n} \sum_{j=1}^{n} h_{ij} Y_i \bigg/ \sum_{i=1}^{n} \sum_{j=1}^{n} h_{ij} \tag{A31}$$

where h_{ij} is the element in the ith row and jth column of the matrix $\mathbf{H} = \mathbf{V}^{-1}$, with $\mathbf{V}$ being the covariance matrix for the Y values; that is, $\text{cov}(Y_i, Y_j)$ is the element in the ith row and jth column of $\mathbf{V}$. The weighted total sum of squares about the origin for the Y values can then be written as

$$\sum_{i=1}^{n} \sum_{j=1}^{n} h_{ij} Y_i Y_j = \sum_{i=1}^{n} \sum_{j=1}^{n} h_{ij} (Y_i - \hat{\mu})(Y_j - \hat{\mu}) + \hat{\mu}^2 \sum_{i=1}^{n} \sum_{j=1}^{n} h_{ij}, \tag{A32}$$

where the first term on the right-hand side is the 'residual' sum of squares with $n-1$ degrees of freedom, while the second term is the sum of squares for the mean with 1 degree of freedom. If the Y values can be assumed to be from a multivariate normal distribution then the residual sum of squares can be tested against the chi-square distribution; a significantly large value indicating that Y values do not all have the same mean. Similarly, the mean sum of squares can be tested against the chi-square distribution with 1 degree of freedom; a significantly large value indicating that μ is non-zero.

A special case of correlation occurs with estimators of survival probabilities (or logarithms of survival probabilities) from mark–recapture data. If $Y_1, Y_2, \ldots, Y_n$ represent a sequence of estimators for consecutive time periods, then their covariance matrix has the symmetric tri-diagonal form

$$
\mathbf{V} = \begin{bmatrix}
\delta_1 & \varepsilon_2 & & & & & \\
\varepsilon_2 & \delta_2 & \varepsilon_3 & & & 0 & \\
& \varepsilon_3 & \delta_3 & \cdot & & & \\
& & & \cdot & \cdot & \cdot & \\
& & & & \cdot & \cdot & \\
& & & & & \cdot & \\
& & & \varepsilon_{n-1} & \delta_{n-1} & \varepsilon_n & \\
& 0 & & & \varepsilon_n & \delta_n &
\end{bmatrix}.
$$

Jolly (1982) provides the inverse matrix for this special case. Having an explicit form for $\mathbf{H}$ may make the determination of the minimum variance weighted mean somewhat easier than it would otherwise be.

A.7 MULTIPLE LINEAR REGRESSION

Multiple linear regression involves the assumption that a random variable Y is related by the equation

$$
Y = \beta_0 + \beta_1 X_1 + \beta_2 X_2 + \ldots + \beta_p X_p + e \tag{A33}
$$

to the variables $X_1, X_2, \ldots, X_p$, where e represents an error term with mean value zero, and $\beta_0, \beta_1, \ldots, \beta_p$ are unknown constants. The usual situation is that data are available for $n > p$ values of Y, so that

$$
y_i = \beta_0 + \beta_1 x_{i1} + \beta_2 x_{i2} + \ldots + \beta_p x_{ip} + e_i,
$$

for $i = 1, 2, \ldots, n$. The model can then be expressed in matrix terms as

$$
\mathbf{Y} = \mathbf{X}\mathbf{B} + \mathbf{E}, \tag{A34}
$$

where

$$
\mathbf{Y} = \begin{bmatrix} y_1 \\ y_2 \\ \vdots \\ y_n \end{bmatrix}, \qquad
\mathbf{X} = \begin{bmatrix}
1 & x_{11} & x_{12} & \cdots & x_{1p} \\
1 & x_{21} & x_{22} & \cdots & x_{2p} \\
\vdots & \vdots & \vdots & & \vdots \\
1 & x_{n1} & x_{n2} & \cdots & x_{np}
\end{bmatrix},
$$

$$
\mathbf{B} = \begin{bmatrix} \beta_0 \\ \beta_1 \\ \vdots \\ \beta_p \end{bmatrix}, \qquad \text{and} \qquad
\mathbf{E} = \begin{bmatrix} e_1 \\ e_2 \\ \vdots \\ e_n \end{bmatrix}.
$$

If it is assumed that the errors $e_1, e_2, \ldots, e_n$ are independently distributed, with a constant variance σ^2, then it is a standard result that the least squares estimator of $\mathbf{B}$ is

$$\hat{\mathbf{B}} = (\mathbf{X}'\mathbf{X})^{-1}\mathbf{X}'\mathbf{Y}, \tag{A35}$$

providing that the matrix $\mathbf{X}'\mathbf{X}$ can be inverted. The covariance matrix for the estimators $\hat{\beta}_0, \hat{\beta}_1, \ldots, \hat{\beta}_p$ is then

$$\mathbf{V}_\beta = (\mathbf{X}'\mathbf{X})^{-1}\sigma^2, \tag{A36}$$

where σ^2 can be estimated as

$$\hat{\sigma}^2 = (\mathbf{Y}'\mathbf{Y} - \mathbf{Y}'\mathbf{X}\hat{\mathbf{B}})/(n-p-1), \tag{A37}$$

with $n-p-1$ degrees of freedom.

To test whether the individual regression coefficients $\beta_0, \beta_1, \ldots, \beta_p$ are significantly different from zero, it is necessary to assume that the errors $e_1, e_2, \ldots, e_n$ are normally distributed in addition to being independent. The statistic

$$(\hat{\beta}_i - \beta_i)/\sqrt{\{\text{vâr}(\hat{\beta}_i)\}} \tag{A38}$$

will then follow a t-distribution with $n-p-1$ degrees of freedom where the estimated variance $\text{vâr}(\hat{\beta}_i)$ is determined from equations (A36) and (A37). Hence a test for $\beta_i = 0$ involves calculating $\hat{\beta}_i/\sqrt{\{\text{vâr}(\hat{\beta}_i)\}}$ and seeing whether this is significantly different from zero when compared to the t-distribution. An overall test for whether the variables $X_1, X_2, \ldots, X_p$ between them explain a significant part of the variation in Y involves carrying out an analysis of variance (Table A1).

Table A1 Analysis of variance for a multiple regression.

Source of variation	Sum of squares	Degrees of freedom	Mean square	F
Regression variables $X_1, X_2, \ldots, X_p$	$\mathbf{Y}'\mathbf{X}\mathbf{B} - n\bar{y}^2$	p	M_1	M_1/M_0
Residual	$\mathbf{Y}'\mathbf{Y} - \mathbf{Y}'\mathbf{X}\mathbf{B}$	$n-p-1$	$M_0 = \hat{\sigma}^2$	
Total	$\sum y_i^2 - ny^2 = \mathbf{Y}'\mathbf{Y} - n\bar{y}^2$	$n-1$		

$$\bar{y} = \sum y_i/n$$

A test for a relationship between Y and $X_1, X_2, \ldots, X_p$ involves seeing whether M_1/M_0 is significantly large when compared with the F-distribution with p and $n-p-1$ degrees of freedom.

A.8   WEIGHTED REGRESSION

It sometimes happens that it is known that the errors $e_1, e_2, \ldots, e_n$ do not have the same variance. Rather, $\text{var}(e_i) = g_i \sigma^2$, where $g_1, g_2, \ldots, g_n$ are known values. In that case a weighted regression is appropriate, with y_i being given a weight $1/g_i$. The weighted least squares estimator of $\mathbf{B}$ is then

$$\hat{\mathbf{B}} = (\mathbf{X'HX})^{-1}\mathbf{X'HY}, \tag{A39}$$

with the covariance matrix for the terms in $\hat{\mathbf{B}}$ being

$$\hat{\mathbf{V}}_\beta = (\mathbf{X'HX})^{-1}\sigma^2, \tag{A40}$$

where σ^2 can be estimated as

$$\hat{\sigma}^2 = (\mathbf{Y'HY} - \mathbf{Y'HX\hat{B}})/(n - p - 1). \tag{A41}$$

In these equations $\mathbf{H}$ is the matrix of weights

$$\mathbf{H} = \begin{bmatrix} h_1 & 0 & \cdots & 0 \\ 0 & h_2 & \cdots & 0 \\ \vdots & \vdots & & \\ 0 & 0 & \cdots & h_n \end{bmatrix} = \begin{bmatrix} 1/g_1^2 & 0 & \cdots & 0 \\ 0 & 1/g_2^2 & \cdots & 0 \\ \vdots & \vdots & & \\ 0 & 0 & \cdots & 1/g_n^2 \end{bmatrix}$$

A test for whether $\hat{\beta}_i$ is significantly different from zero still involves comparing $\hat{\beta}_i/\sqrt{\{\text{vâr}(\hat{\beta}_i)\}^{1/2}}$ with the t-distribution with $n - p - 1$ degrees of freedom. The estimated variance $\text{vâr}(\hat{\beta}_i)$ is determined from equations (A40) and (A41). An analysis of variance is shown in Table A2.

Sometimes the variances of $e_1, e_2, \ldots, e_n$ will be known exactly. In that case it is possible to put $\sigma^2 = 1$ and $g_i = \text{var}(e_i)$ in equations (A39) and (A40).

Table A2   Analysis of variance for a weighted multiple regression.

Source of variation	Sum of squares	Degrees of freedom	Mean square	F
Regression variables $X_1, X_2, \ldots, X_p$	$\mathbf{Y'HX\hat{B}} - \hat{\mu}^2 \sum_{i=1}^{n} h_i$	p	M_1	M_1/M_0
Residual	$\mathbf{Y'HY} - \mathbf{Y'HX\hat{B}}$	$n - p - 1$	$M_0 = \hat{\sigma}^2$	
Total	$\mathbf{Y'HY} - \hat{\mu}^2 \sum_{i=1}^{n} h_i$	$n - 1$		

Here $\hat{\mu} = \sum h_i y_i / \sum h_i$ is the weighted least-squares estimate of the mean of Y. A test for a relationship between Y and $X_1, X_2, \ldots, X_p$ involves seeing whether M_1/M_0 is significantly large when compared with the F-distribution with p and $n - p - 1$ degrees of freedom.

Then the significance of β_i can be tested by comparing $\hat{\beta}_i/\sqrt{\{\operatorname{var}(\hat{\beta}_i)\}}$ against the standard normal distribution since σ^2 does not have to be estimated. Alternatively, $\hat{\beta}_i^2/\operatorname{var}(\hat{\beta}_i^2)$ can be treated as a chi-squared variate with one degree of freedom. A test for whether the regression model fits the data involves seeing whether $\hat{\sigma}^2$ is significantly larger than one, as indicated in Table A3. (See also the following section.)

Table A3   Analysis of variance when the covariance matrix of the regression errors $e_1, e_2, \ldots, e_n$ is known to be $\mathbf{V}_e$, which is non-singular with $\mathbf{H} = \mathbf{V}_e^{-1}$.

Source of variation	Sum of squares	Degrees of freedom	Mean square	F
Regression variables $X_1, X_2, \ldots, X_p$	$\mathbf{Y'HX\hat{B}} - \mu^2 \sum\limits_{i=1}^{n} \sum\limits_{j=1}^{n} h_{ij}$	p	M_1	M_1
Residual	$\mathbf{Y'HY} - \mathbf{Y'HX\hat{B}}$	$n-p-1$	M_0	M_0
Theoretical error	—	∞	1.00	
Total	$\mathbf{Y'HY} - \hat{\mu}^2 \sum\limits_{i=1}^{n} \sum\limits_{j=1}^{n} h_{ij}$			

Here h_{ij} is the element in the ith row and jth column of $\mathbf{H}$ and

$$\hat{\mu} = \sum_{i=1}^{n} \sum_{j=1}^{n} h_{ij} y_i \bigg/ \sum_{i=1}^{n} \sum_{j=1}^{n} h_{ij}$$

is the generalized least squares estimate of the mean of Y. A test for the goodness of fit of the regression model involves comparing M_0 with the F-distribution with $n-p-1$ and infinite degrees of freedom or, what amounts to the same thing, seeing whether the residual sum of squares $(n-p-1)M_0$ is significantly large compared with the chi-square distribution with $n-p-1$ degrees of freedom. If the regression model is a good fit, then a test for a relationship between Y and $X_1, X_2, \ldots, X_p$ involves either comparing M_1 with the F-distribution with p and infinite degrees of freedom or comparing pM_1 with the chi-square distribution with p degrees of freedom.

A.9   GENERALIZED LEAST SQUARES

A further generalization of the regression principle is needed in cases where the regression errors $e_1, e_2, \ldots, e_n$ are not independent. Thus, suppose now that $\mathbf{V}_e$ is the $n \times n$ covariance matrix of these errors with $\operatorname{cov}(e_i, e_j)$ in the ith row and jth column, this matrix being assumed to be known. Then equation (A39) with $\mathbf{H} = \mathbf{V}_e^{-1}$ gives the generalized least squares estimator of $\mathbf{B}$. The covariance matrix for the estimators $\hat{\beta}_0, \hat{\beta}_1, \ldots, \hat{\beta}_p$ is given by equation (A40) with $\sigma^2 = 1$. Table A3 shows the analysis of variance for this case.

A.10   THE COMPUTER PROGRAM GLIM

The computer program GLIM (general linear interactive modelling) has been used for many of the examples in this book. It is particularly useful for fitting log-linear models (first used in Chapter 3) and proportional hazards models (first used in Chapter 5). However, it can also be used for ordinary multiple regression, with or without weighting, and analysis of variance. Here only a brief summary of the main facilities of the program will be given. For further details see the *GLIM Manual* (Nelder, 1975) and the papers by Nelder and Wedderburn (1972), Nelder (1974), and Manly (1977d).

The standard multiple regression model has been reviewed above. It is of the form

$$Y = \beta_0 + \beta_1 X_1 + \beta_2 X_2 + \ldots + \beta_p X_p + e$$

where Y is an observation, the variables $X_1, X_2, \ldots, X_p$ are intended to account for the variation in Y, e is a random error with a normal distribution with zero mean and constant variance, and the coefficients $\beta_0, \beta_1, \ldots, \beta_p$ are to be estimated. In GLIM this model is indicated by '$X_1 + X_2 + \ldots + X_p$'. (It is assumed that there is a constant term β_0 in all models unless it is specified that there should not be.)

Analysis of variance models can be defined in the multiple regression form by introduction variables having the values 0 or 1 depending upon the levels of factors. Thus a one-factor model with three levels for the factor, A say, is set up as

$$Y = \beta_0 + \beta_1 A_1 + \beta_2 A_2 + e$$

where A_i is defined to be 1 at the ith level of the factor and zero otherwise. The third and last level of the factor is taken as a standard for which $Y = \beta_0 + e$. The last level of a factor is always taken to be the standard level in GLIM.

If two factors A and B are defined, with three and two levels, respectively, then the model $A + B$ is set up as

$$Y = \beta_0 + \beta_1 A_1 + \beta_2 A_2 + \beta_3 B_1 + e$$

where B_1 is 1 for the first level of B and zero otherwise. Thus the standard situation, for which $Y = b_0 + e$, is that for which A is at its third level and B at its second level. An interaction between factors A and B is indicated as $A.B$ so that the model $A + B + A.B$ is set up as

$$Y = \beta_0 + \beta_1 A_1 + \beta_2 A_2 + \beta_3 B_1 + \beta_4 (AB)_{11} + \beta_5 (AB)_{21} + e$$

where $(AB)_{i1}$ is zero unless A is at its ith level and B at its first level, in which case it is one. Higher-order interactions are handled in a similar way, so that $A.B.C$ indicates a second-order interaction between three factors A, B and C.

If X is defined to be a variable, then $X.A$ indicates that the regression coefficient for X varies according to the level of factor A. Thus the model $A + X + X.A$ is

$$Y = \beta_0 + \beta_1 A_1 + \beta_2 A_2 + \beta_3 X + \beta_4 A_1 X + \beta_5 A_2 X + e.$$

This means that $Y = \beta_0 + \beta_1 + \beta_3 X + \beta_4 X + e$ for the first level of A, $Y = \beta_0 + \beta_2 + \beta_3 X + \beta_5 X + e$ for the second level of A, and $Y = \beta_0 + \beta_3 X + e$ for the third level of A.

A generalized linear model is one of the form

$$Y = f\left(\sum \beta_i X_i\right) + e$$

where f is a specified function and e is an error term. GLIM allows various combinations of seven f functions and four error distributions, as shown in Table A4, and fits these models to data using the principle of maximum likelihood.

The specification of f is made by means of a LINK directive where this actually defines the inverse of f. If the expected value of Y is

$$\mu = f\left(\sum \beta_i X_i\right),$$

with the inverse relationship

$$\sum \beta_i X_i = g(\mu),$$

then g is called the link function. For example, with a Poisson log-linear model the data consists of counts and

$$Y = \exp\left(\sum \beta_i X_i\right) + e,$$

where Y is assumed to have a Poisson distribution. In this case $\mu = \exp\left(\sum \beta_i X_i\right)$ so that $\sum \beta_i X_i = \log_e(\mu)$. This model has a logarithmic link.

Binomial data of the form 'Y respond out of n tested' are allowed three specialized links. If Y/n is a proportion dying, then the double logarithmic link is equivalent to the proportional hazards model for survival data.

To use GLIM it is necessary to define: (1) the variables and factors that Y is to be related to, (2) the nature of errors, and (3) the link function. If the model so defined is allowed, as indicated in Table A4, then it is fitted to the data. The goodness of fit is indicated by the 'deviance'. For models with a normal distribution for errors, this is simply the usual residual sum of squares. For models with Poisson or binomial distributions, the deviance is approximately a chi-square variate if the model is correct. Hence in these cases a direct test of goodness of fit is possible by seeing whether the deviance is significantly large. Actually, with Poisson or binomial distributions it is possible to do more than this. The deviance is equal to minus twice the log-likelihood of the data so that if two models, I and II, are being considered for the data, where II incorporates

Table A4 Generalized linear models fitted by GLIM.

Link function	Model	Error distribution			
		Normal	Poisson	Binomial	Gamma
Identity: $g(\mu) = \mu$	$Y = \Sigma\beta_i X_i + e$	allowed[1]	allowed	not allowed	allowed
Log: $g(\mu) = \log_e(\mu)$	$Y = \exp(\Sigma\beta_i X_i) + e$	allowed	allowed[2]	not allowed	allowed
Reciprocal: $g(\mu) = 1/\mu$	$Y = 1/\Sigma\beta_i X_i + e$	allowed	allowed	not allowed	allowed
Square root: $g(\mu) = \mu^{1/2}$	$Y = (\Sigma\beta_i X_i)^2 + e$	allowed	allowed	not allowed	allowed
Logit: $g(p) = \log_e\left(\dfrac{p}{1-p}\right)$	$Y/n = \dfrac{\exp(\Sigma\beta_i X_i)}{1 + \exp(\Sigma\beta_i X_i)} + e$	not allowed	not allowed	allowed	not allowed
Probit: $g(p) = \Phi^{-1}(p)$ (where Φ denotes the standard normal integral)	$Y/n = \Phi(\Sigma\beta_i X_i) + e$	allowed	allowed	allowed	allowed
Double log: $g(p) = \log_e\{-\log_e(1-p)\}$	$Y/n = 1 - \exp\{-\exp(\Sigma\beta_i X_i)\} + e$	not allowed	not allowed	allowed[3]	not allowed

[1] Standard multiple regression and analysis of variance
[2] Poisson log-linear models
[3] For fitting the proportional hazards model

I as a special case, then the difference between the deviances can be tested for significance against the chi-square distribution, as discussed in Section 4 above.

For example, suppose that model I (the special case) has a deviance of 25.0, with 18 degrees of freedom. Regarded as a chi-square value this is not significantly large at the 5% level; the model fits the data. On the other hand, suppose that model II has a deviance of 15.0 with 12 degrees of freedom. This is also not significantly large. The improvement in fit obtained by fitting the more general model is $25.0 - 15.0 = 10.0$, with $18 - 12 = 6$ degrees of freedom. Regarded as a chi-square variate this improvement is not significantly large at the 5% level. Consequently it can be argued that the special case model I fits the data about as well as the more general model II. It must be stressed that these chi-square tests are only appropriate with Poisson and binomial distributions. With a normal distribution for errors, analysis of variance can be used for testing the significance of different effects.

A.11 FITTING THE PROPORTIONAL HAZARDS MODEL TO DATA

Two methods for fitting the proportional hazards model to data are suggested. One involves using a computer program such as GLIM or BMDP2L (Dixon, 1981), while the other is based upon a weighted multiple regression.

Suppose that the data consist of survival numbers for individuals in K classes, where the individuals within one class have the same values for certain quantitative variables $X_1, X_2, \ldots, X_p$. Suppose also that at time t_0 there are a_{i0} individuals in the ith class and that the numbers of survivors are observed at times $t_1, t_2, \ldots, t_s$ to be $a_{i1}, a_{i2}, \ldots, a_{is}$, respectively. In addition, let $b_{ij} = a_{ij-1} - a_{ij}$ denote the number of class i individuals dying in the interval (t_{j-1}, t_j).

Provided that $a_{ij-1} \neq 0$, $\hat{\phi}_{ij} = a_{ij}/a_{ij-1}$ is an obvious estimator of ϕ_{ij}, the probability of a class i individual surviving until time t_j, given that it is alive at time t_{j-1}. It seems quite clear that $\hat{\phi}_{ij}$ and $\hat{\phi}_{kl}$ are independent if $i \neq k$. Also, Chiang (1960) has shown that $\hat{\phi}_{ij}$ and $\hat{\phi}_{il}$ are uncorrelated for $j \neq l$. Furthermore, the likelihood function for the data is essentially the same as a likelihood function for data consisting of completely independent proportions surviving. To see this, note that for individuals in the ith class the probability of dying in the interval (t_{j-1}, t_j) is

$$\beta_{ij} = \begin{cases} 1 - \phi_{i1}, & j = 1 \\ \phi_{i1}\phi_{i2} \ldots \phi_{ij-1}(1 - \phi_{ij}), & j = 2, 3, \ldots, s \end{cases}$$

while the probability of surviving through to time t_s is

$$\pi_i = \phi_{i1}\phi_{i2} \ldots \phi_{is} = 1 - \beta_{i1} - \beta_{i2} - \ldots - \beta_{is}.$$

The numbers observed dying in the different time intervals, $b_{i1}, b_{i2}, \ldots, b_{is}$, and the number of survivors at time t_s, a_{is}, will therefore follow a multinomial distribution with probability function

$$P(b_{i1}, b_{i2}, \ldots, b_{is}, a_{is}) = \frac{a_{i0}!}{a_{is}! \prod\limits_{j=1}^{s} b_{ij}!} \prod_{j=1}^{s} \beta_{ij}^{b_{ij}} \pi_i^{a_{is}}.$$

This probability function can be rewritten as

$$P(b_{i1}, b_{i2}, \ldots, b_{is}, a_{is}) = C \phi_{i1}^{a_{i1}} (1 - \phi_{i1})^{a_{i0} - a_{i1}} \phi_{i2}^{a_{i2}} (1 - \phi_{i2})^{a_{i1} - a_{i2}}$$
$$\ldots \phi_{is}^{a_{is}} (1 - \phi_{is})^{a_{is-1} - a_{is}},$$

where C is a constant. This is proportional to the probability of observing s independent binomial proportions a_{i1}/a_{i0}, $a_{i2}/a_{i1}, \ldots, a_{is}/a_{is-1}$. Thus the likelihood function for the observed survival data is, apart from a constant, the same as the likelihood function for s independent binomial proportions. The constant does not affect maximum likelihood estimation and therefore computer programs such as GLIM or BMDP2L can be used to fit models to data, with the proportions surviving different time intervals treated as being independent.

An alternative approach involves using standard weighted multiple regression. The idea in this case is that if a probability of survival is of the form

$$\phi = \exp\{-\exp(\beta_0 + \beta_1 X_1 + \ldots + \beta_p X_p)\},$$

then $\log\{-\log(\hat{\phi})\}$ is an estimate of $\beta_0 + \beta_1 X_1 + \ldots + \beta_p X_p$, where $\hat{\phi}$ is the observed proportion surviving selection. Therefore the coefficients $\beta_0, \beta_1, \ldots, \beta_p$ can be estimated by carrying out a multiple regression analysis using $\log\{-\log(\hat{\phi})\}$ as the dependent variable. One problem with this method is that if $\hat{\phi}$ is 0 or 1 then the resulting value of $\log\{-\log(\hat{\phi})\}$ is infinite. However, this problem can be overcome by using the modified dependent variable defined by

$$\hat{y} = \begin{cases} \log[-\log\{1/2n)\}], & \hat{\phi} = 0 \\ \log\left[-\log\left\{\hat{\phi} + \dfrac{(1-\hat{\phi})\{\log(\hat{\phi})+1\}}{2n\log(\hat{\phi})}\right\}\right], & 0 < \hat{\phi} < 1, \\ \log[-\log\{1 - 1/2n\}], & \hat{\phi} = 1, \end{cases}$$

where n is the initial number of individuals, of which a proportion $\hat{\phi}$ survive. The estimator $\hat{y}$ is virtually unbiased and has variance

$$\mathrm{var}(\hat{y}) \simeq (1 - \phi)/[n\phi\{\log(\phi)\}^2]$$

(Manly, 1978b). The reciprocal of var $(\hat{y})$ is the appropriate weight to give $\hat{y}$ for multiple regression.

A.12   A COMPUTER PROGRAM FOR A RANDOMIZATION TEST ON POLYMORPHIC DATA

A listing for the FORTRAN computer program HABITATS is provided here. This program carries out the randomization test described in Section 6.3. The purpose of the test is to see whether differences between habitats are too large to be accounted for by variations within habitats, for a polymorphic population.

The input to the program is explained on comment cards at the start of the listing. There can be up to 20 types of habitat (NH), up to 20 morph or gene frequencies (NT), and up to 100 sampling locations. The number of random allocations of locations to habitats can be 50, 100, 500 or 1000. The format of data frequencies is specified by a FORTRAN format which is input to the program on the type 3 data card.

Output from the program consists of: (a) a summary of the data; (b) chi-square values summed for each habitat separately; (c) the overall total chi-square value; (d) chi-square values summed for each morph or gene separately; (e) the randomized order of locations of the first ten randomizations, with the resulting total chi-square values; (f) selected rank order values for the total chi-square values obtained by randomization; and (g) selected rank order values for the chi-square values obtained by randomization for each morph or gene separately.

The total chi-square value from the data is assessed for significance by comparison with the output (f). Individual morphs or genes are assessed by comparing their observed chi-square values with the output (g). As noted in Example 6.1, it may be appropriate to use a higher level of significance when testing individual morphs or genes than is used for testing the overall total chi-square value.

Program listing

```
C
C        ********************
C        * PROGRAM HABITATS *
C        ********************
C
C        THIS PROGRAM IS FOR THE ANALYSIS OF POISSON DATA WITH
C        EXTRANEOUS VARIANCE BY MEANS OF RANDOMIZATION TESTS.  THE DATA
C        CONSISTS OF COUNTS OF, FOR EXAMPLE, NT DIFFERENT TYPES OF
C        ANIMAL IN SAMPLES FROM NH HABITATS.  THE QUESTION ASKED IS
C        WHETHER THE DISTRIBUTION VARIES WITH THE HABITAT.  THIS IS
C        TESTED BY ALLOCATING THE SAMPLES AT RANDOM TO THE HABITATS,
C        KEEPING THE NUMBER OF SAMPLES FROM EACH HABITAT FIXED.  THE
C        RANDOM DISTRIBUTION OF CHI-SQUARE CAN THEN BE  USED TO DETERMINE
C        HOW SIGNIFICANT THE OBSERVED VALUE IS.  THE SIGNIFICANCE FOR
C        INDIVIDUAL ANIMALS CAN BE EXAMINED BY THEIR CONTRIBUTION TO
C        CHI-SQUARE.  THE FUNCTION RANDOM(ISEED) IS ONLY AVAILABLE ON
```

```
C       SOME BURROUGHS COMPUTERS AND CODE FOR THIS WILL HAVE TO BE
C       PROVIDED ON OTHER MACHINES.  THIS FUNCTION PRODUCES A UNIFORM
C       RANDOM NUMBER IN THE INTERVAL (0,1).  THE PROGRAM HABITATS
C       WAS WRITTEN BY B.F.J. MANLY, UNIVERSITY OF OTAGO, DUNEDIN
C       NEW ZEALAND.  THIS VERSION IS DATED APRIL, 1985.
C
C       **************
C       * DATA CARDS *
C       **************
C
C       CARD TYPE 1
C          COLS  1- 5  NH, NUMBER OF HABITATS(MAX=20)
C                6-10  NT, NUMBER OF TYPES OF INDIVIDUAL (MAX=20)
C                11-15 MSIM, DETERMINES THE NUMBER OF RANDOM ALLOCATIONS
C                      IT MUST BE 1, 2, 3, OR 4 FOR 50, 100, 500 OR 1000
C                      ALLOCATIONS, RESPECTIVELY
C                16-20 ISEED, RANDOM NUMBER STARTER, INTEGER IN RANGE 1 TO
C                      99999
C
C       CARD(S) TYPE 2
C                1- 5  NL(1), NUMBER OF LOCATIONS FOR HABITAT 1
C                6-10  NL(2), NUMBER OF LOCATIONS FOR HABITAT 2
C                  :       :
C                66-70 (UP TO 14 NUMBERS PER CARD)
C
C       CARD TYPE 3
C         A FORMAT CARD FOR THE DATA FROM ONE LOCATION, E.G. (16I5)
C
C       CARD TYPE 4
C         A TITLE FOR THIS SET OF DATA IN UP TO 72 COLS
C
C       CARDS TYPE 5
C         THE DATA.  THE DATA FREQUENCIES FOR EACH LOCATION ARE READ ONE
C         BY ONE USING THE FORMAT ON THE TYPE 3 CARD.  THEY ARE STORED IN
C         THE ARRAY FREQ.
C
        REAL S(100,20),STAT1(1000),CONT(1000,20),CHIM(20),CHIH(20)
        INTEGER IPF(18),FREQ(100,20),TITLE(18),NL(20),LOC(100),IPR(4,15),
       1 IPRA(15),IA(100)
        DATA ((IPR(I,J), J=1,15), I=1,4)/ 1,2,3,4,5,10,15,25,36,41,46,47,
       1 48,49,50, 1,2,3,4,5,10,30,50,71,91,96,97,98,99,100, 1,2,5,10,25,
       2 50,100,250,401,451,476,491,496,499,500, 1,2,5,10,50,100,250,500,
       3 751,901,951,991,996,998,1000/
C
C****   READ AND PRINT DATA
        READ(5,1000) NH,NT,MSIM,ISEED
   1000 FORMAT(14I5)
        IF (MSIM.LT.1) MSIM=1
        IF (MSIM.GT.4) MSIM=4
        NSIM=50
        IF (MSIM.EQ.2) NSIM=100
        IF (MSIM.EQ.3) NSIM=500
        IF (MSIM.EQ.4) NSIM=1000
        READ(5,1000)(NL(I), I=1,NH)
        READ(5,1001) (IPF(I), I=1,18)
   1001 FORMAT(18A4)
        READ(5,1001) (TITLE(I), I=1,18)
        DO 40 I=1,15
     40 IPRA(I)=IPR(MSIM,I)
```

```
           IL=0
           DO 1 IH=1,NH
           M=NL(IH)
           DO 1 I=1,M
           IL=IL+1
         1 READ(5,IPF) (FREQ(IL,J), J=1,NT)
           WRITE(6,9000) (TITLE(I), I=1,18),ISEED
      9000 FORMAT(' RANDOMIZATION TEST FOR POISSON DATA WITH EXTRANEOUS VARIA
          1NCE'/' DATA TITLE = ',18A4/' RANDOM NUMBER STARTER = ',I5)
           IL=0
           DO 2 IH=1,NH
           WRITE(6,9001) IH
      9001 FORMAT(/' FREQUENCIES FOR HABITAT ',I3)
           M=NL(IH)
           DO 2 I=1,M
           IL=IL+1
         2 WRITE(6,9002) (FREQ(IL,J), J=1,NT)
      9002 FORMAT(' ',20I6)
           NLT=IL
   C
   C**** CHI-SQUARED VALUES FOR ACTUAL DISTRIBUTION OF LOCATIONS
           DO 4 I=1,NLT
           DO 4 J=1,NT
         4 S(I,J)=FREQ(I,J)
           CALL CHILL(STAT01,CHIM,CHIH,S,NH,NL,NT)
           WRITE(6,9050) (CHIH(I), I=1,NH)
      9050 FORMAT(//' CHI-SQUARED VALUES FOR THE HABITATS, IN ORDER'/
          1 (15F8.2))
           WRITE(6,9003) STAT01,(CHIM(I), I=1,NT)
      9003 FORMAT(/' TOTAL CHI-SQUARED VALUE FOR ACTUAL LOCATION DISTRIBUTION
          1 = ',F8.2/' VALUES FOR THE DIFFERENT MORPHS, IN ORDER'/(15F8.2))
   C
   C**** GENERATE RANDOM DISTRIBUTION OF LOCATIONS WITH CORRESPONDING
   C**** CHI-SQUARED VALUES
           WRITE(6,9040)
      9040 FORMAT(//' RANDOMIZATION TEST RESULTS FOLLOW'/
          1' DISTRIBUTIONS HAVE BEEN RANKED AND ONLY SPECIFIC RANK VALUES ARE
          2 GIVEN'/
          2' E.G., RANK 5 MEANS THE 5TH LARGEST VALUE FROM RANDOMIZATION'/
          4' THE FIRST TEN RANDOM ALLOCATIONS ARE'/
          5' ALLOC    CHI**2      RANDOM ALLOCATION')
           DO 5 ISIM=1,NSIM
           DO 6 I=1,NLT
         6 LOC(I)=0
           DO 8 I=1,NLT
         7 IRAN=IFIX(RANDOM(ISEED)*NLT)+1
           IF (LOC(IRAN).GT.0) GO TO 7
           IA(I)=IRAN
           LOC(IRAN)=1
           DO 8 J=1,NT
         8 S(IRAN,J)=FREQ(I,J)
           CALL CHILL(STAT1(ISIM),CHIM,CHIH,S,NH,NL,NT)
           DO 15 I=1,NT
        15 CONT(ISIM,I)=CHIM(I)
         5 IF (ISIM.LT.11) WRITE(6,9041) ISIM,STAT1(ISIM),(IA(I), I=1,NLT)
      9041 FORMAT(I6,F10.2,5X,30I3/(21X,30I3))
   C
   C**** SORT CHI-SQUARE VALUES AND PRINT
           CALL SORT(STAT1,1000,1,NSIM,1)
```

```
      SUM=0.0
      SUMA=0.0
      DO 10 I=1,NSIM
      SUM=SUM+STAT1(I)
   10 SUMA=SUMA+STAT1(I)**2
      RMEAN=SUM/NSIM
      SE=SQRT((SUMA-NSIM*RMEAN**2)/((NSIM-1.0)*NSIM))
      WRITE(6,9004) NSIM,(IPRA(I),STAT1(IPRA(I)), I=1,15)
 9004 FORMAT(/' TOTAL CHI-SQUARE VALUES OBTAINED BY ',I4,' RANDOM ALLOCA
     1TIONS OF LOCATIONS TO HABITATS'/
     2'    RANK    VALUE     RANK    VALUE     RANK    VALUE     RANK    VALUE
     3    RANK    VALUE     RANK    VALUE     RANK    VALUE     RANK    VALUE'/
     4(' ',8(I8,F8.1)))
      WRITE(6,9005) RMEAN,SE
 9005 FORMAT(' MEAN TOTAL CHI-SQUARE = ',F8.2,' WITH STANDARD ERROR = '
     1,F8.2)
C
C**** SORT AND PRINT MORPH CONTRIBUTIONS
      CALL SORT(CONT,1000,20,NSIM,NT)
      DO 30 J=1,NT
   30 WRITE(6,9030) J,(IPRA(I),CONT(IPRA(I),J), I=1,15)
 9030 FORMAT(/' CONTRIBUTIONS OBTAINED BY RANDOM ALLOCATIONS FOR THE TYP
     1E',I3,' INDIVIDUALS'/
     1'    RANK    VALUE     RANK    VALUE     RANK    VALUE     RANK    VALUE
     2    RANK    VALUE     RANK    VALUE     RANK    VALUE     RANK    VALUE'/
     3(' ',8(I8,F8.1)))
      STOP
      END
C
C
C     ****************************************************************
C     * SUBROUTINE CHILL TO CALCULATE CHI-SQUARE FOR ONE SET OF DATA *
C     ****************************************************************
C
      SUBROUTINE CHILL(TCHI,CHIM,CHIH,S,NH,NL,NT)
C
      REAL S(100,20),R(100),C(20,20),CHIM(20),SS(20),CHIH(20)
      INTEGER NL(20)
C
C**** INITIAL SETTING UP
      L=0
      LA=0
      DO 6 J=1,NT
    6 CHIM(J)=0.0
C
C**** CALCULATE CHI-SQUARE
      DO 1 IH=1,NH
      CHIH(IH)=0.0
      TFREQ=0.0
      M=NL(IH)
      DO 2 I=1,M
    2 R(L+I)=0.0
      DO 3 J=1,NT
    3 C(IH,J)=0.0
      DO 4 I=1,M
      L=L+1
      DO 4 J=1,NT
      R(L)=R(L)+S(L,J)
      C(IH,J)=C(IH,J)+S(L,J)
    4 TFREQ=TFREQ+S(L,J)
```

```
        DO 5 I=1,M
        LA=LA+1
        DO 5 J=1,NT
        EX=R(LA)*C(IH,J)/TFREQ
        CONT=0.0
        IF (EX.GT.0.0) CONT=(S(LA,J)-EX)**2/EX
        CHIM(J)=CHIM(J)+CONT
    5   CHIH(IH)=CHIH(IH)+CONT
    1   CONTINUE
        TCHI=0.0
        DO 7 I=1,NT
    7   TCHI=TCHI+CHIM(I)
        RETURN
        END
C
C       ****************
C       *SUBROUTINE SORT*
C       ****************
C
        SUBROUTINE SORT(A,IR,IC,NI,NJ)
C
        REAL A(IR,IC)
C
C****   SHELLSORT
        DO 1 J=1,NJ
        IG=NI
    4   IG=IG/2
        IF (IG.EQ.0) GO TO 1
        DO 2 I=1,NI-IG
        DO 5 L=1,I,IG
        K=I-L+1
        K1=K+IG
        IF (A(K,J).GT.A(K1,J)) GOTO 3
        GO TO 2
    3   W=A(K,J)
        A(K,J)=A(K1,J)
        A(K1,J)=W
    5   CONTINUE
    2   CONTINUE
        GO TO 4
    1   CONTINUE
        RETURN
        END
```

A.13 A COMPUTER PROGRAM FOR MANTEL'S TEST

A listing of the FORTRAN program MANTEL is provided below. This carries out Mantel's (1967) test for association between two symmetric distance matrices, as described in Section 6.8.

The input to the program is detailed on comment cards that are given at the start of the listing. The two matrices compared are of size $N \times N$, where N can be up to 100. The number of random permutations of the rows and columns of the second matrix relative to the first (NSIM) can be up to 1000. The two matrices are input together on the type 4 cards, using the FORTRAN format given on the type 3 card. For example, if $N = 5$ then the distances are input on five cards in the order

$$
\begin{array}{ccccc}
0 & a_{12} & a_{13} & a_{14} & a_{15} \\
b_{21} & 0 & a_{23} & a_{24} & a_{25} \\
b_{31} & b_{32} & 0 & a_{34} & a_{35} \\
b_{41} & b_{42} & b_{43} & 0 & b_{45} \\
b_{51} & b_{52} & b_{53} & b_{54} & 0
\end{array}
$$

where a_{ij} is the distance from colony i to colony j for the first distance matrix and b_{ij} is the distance between the same two colonies for the second distance matrix. A typical FORTRAN format would be (5F6.0).

Output from the program consists of: (a) a print of the data matrices; (b) the test statistic Z, its expected value and standard error (assuming independent matrices), an approximately standard normal deviate $g = (Z - \text{expected value})/(\text{standard error})$, and the correlation coefficient r between the distances in the two matrices; (c) details of the first ten randomizations, with their values for g and r; and (d) the full list of g values obtained by randomization, ranked from 1 to NSIM.

The g value calculated from the original data can be assessed for significance by comparing it with the standard normal distribution. Thus a value greater than 1.64 is significantly large at the 5% level. However, this only gives an approximate test. It is better to compare the observed g with the ranked randomized distribution. For example, if NSIM = 1000 randomizations are made then the observed g is significantly large at about the 5% level if it exceeds 950 of the randomized values.

Program listing

```
C
C       ******************
C       * PROGRAM MANTEL *
C       ******************
C
C       THIS PROGRAM CARRIES OUT MANTEL'S TEST ON TWO SYMMETRIC
C       MATRICES.  ONE OF THE MATRICES MIGHT BE, FOR EXAMPLE, DISTANCES
C       BETWEEN N COLONIES BASED UPON MORPH DISTRIBUTIONS SO THAT "CLOSE"
C       COLONIES HAVE SIMILAR MORPH PROPORTIONS WHILE "DISTANT" COLONIES
C       HAVE VERY DIFFERENT MORPH PROPORTIONS.  THE OTHER MATRIX MIGHT
C       BE ENVIRONMENTAL DISTANCES.  MANTEL'S TEST WOULD THEN SEE WHETHER
C       THE TWO TYPES OF DISTANCE ARE CORRELATED.  THE TEST IS BASED UPON
C       SEEING WHETHER A STATISTIC Z IS SIGNIFICANT WHEN COMPARED TO THE
C       DISTRIBUTION FOR Z OBTAINED BY RANDOM PERMUTATIONS OF THE ROWS
C       AND COLUMNS OF THE SECOND MATRIX COMPARED TO THE FIRST MATRIX.
C       THE FUNCTION RANDOM(ISEED) IS ONLY AVAILABLE ON SOME BURROUGHS
C       COMPUTERS AND CODE FOR THIS WILL HAVE TO BE PROVIDED ON OTHER
C       MACHINES.  THIS FUNCTION PRODUCES A UNIFORM RANDOM NUMBER IN THE
C       RANGE (0,1).  THE PROGRAM MANTEL WAS WRITTEN BY B.F.J. MANLY,
C       UNIVERSITY OF OTAGO, DUNEDIN, NEW ZEALAND.  THIS VERSION IS
C       DATED OCTOBER, 1984.
```

```
C
C       **************
C       * DATA CARDS *
C       **************
C
C       CARD TYPE 1
C         A TITLE FOR THE DATA (UP TO 72 CHARACTERS)
C
C       CARD TYPE 2
C         COLS   1- 5  N, THE SIZE OF THE MATRICES (MAX=100)
C                6-10  NSIM, NUMBER OF SIMULATIONS (MAX=1000)
C               11-15  ISEED. RANDOM NUMBER STARTER IN RANGE 1-99999
C
C       CARD TYPE 3
C         A FORMAT CARD FOR THE MATRIX ELEMENTS ON CARDS TYPE 4
C
C       CARDS TYPE 4 (N OF THEM)
C         THE MATRIX OF DISTANCES.  THE UPPER TRIANGULAR MATRIX GIVES
C         ONE SET OF DISTANCES (A).  THE LOWER TRIANGULAR MATRIX GIVES
C         THE OTHER SET (B).  THE DIAGONALS ARE ZERO.
C
        REAL A(100,100),B(100,100),KA,KB,DATA(100,100),GRAN(1000),
       1 BRAN(100,100)
        INTEGER TITLE(18),IPF(18),IA(100)
C
C**** READ AND PRINT DATA
        READ(5,1000) (TITLE(I), I=1,18)
   1000 FORMAT(18A4)
        READ(5,1001) N,NSIM,ISEED
   1001 FORMAT(3I5)
        READ(5,1000) (IPF(I), I=1,18)
        DO 1 I=1,N
        READ(5,IPF) (DATA(I,J), J=1,N)
      1 DATA(I,I)=0.0
        DO 2 I=1,N
        DO 2 J=1,I
        A(I,J)=DATA(I,J)
        A(J,I)=A(I,J)
        B(I,J)=DATA(J,I)
      2 B(J,I)=B(I,J)
        WRITE(6,9000) TITLE,(I, I=1,N)
   9000 FORMAT(' MANTEL TEST FOR MATRIX ASSOCIATIONS'/
       1' DATA TITLE: ',18A4//
       2' DATA MATRICES (A ABOVE, B BELOW DIAGONAL)'/(11X,15I8))
        DO 3 I=1,N
      3 WRITE(6,9001) I,(DATA(I,J), J=1,N)
   9001 FORMAT(' ',I5,5X,15F8.2/(11X,15F8.2))
C
C**** CALCULATE AND PRINT STATISTICS FOR THESE DATA
        CALL STATIS(A,B,N,Z,EXPZ,SEZ,G,R)
        WRITE(6,9004) Z,EXPZ,SEZ,G,R
   9004 FORMAT(//' TEST RESULTS:  Z = ',F10.1,'    EXPECTED VALUE = ',
       1F10.1,'    STANDARD ERROR = ',F10.1/
       2' G = (Z - EXPECTED VALUE)/(STANDARD ERROR) = ',F10.3,'   R = ',
       3F10.3)
C
C**** GENERATE RANDOM PERMUTATIONS
        WRITE(6,9010) ISEED
   9010 FORMAT(//' RANDOM NUMBER STARTER = ',I5/
```

```
    1' THE FIRST TEN RANDOM ALLOCATIONS ARE AS FOLLOWS'/
    2 '   I        G        R           ORDER OF LOCATIONS')
      DO 4 ISIM=1,NSIM
      DO 5 I=1,N
    5 IA(I)=I
      DO 6 I=1,N
      J=N-I+1
      K=INT(J*RANDOM(ISEED)+0.999999)
      L=IA(J)
      IA(J)=IA(K)
    6 IA(K)=L
      DO 8 I=1,N
      DO 8 J=1,N
    8 BRAN(I,J)=B(IA(I),IA(J))
      CALL STATIS(A,BRAN,N,Z,EXPZ,SEZ,GRAN(ISIM),R)
    4 IF (ISIM.LE.10) WRITE(6,9011) ISIM,GRAN(ISIM),R,
    1 (IA(I), I=1,N)
 9011 FORMAT(I6,2F10.3,5X,30I3/(31X,30I3))
C
C**** SORT AND PRINT
      CALL SORT(GRAN,1000,1,NSIM,1)
      WRITE(6,9005) NSIM,(I,GRAN(I), I=1,NSIM)
 9005 FORMAT(//' G VALUES OBTAINED FROM ',I4,' RANDOM PERMUTATIONS'/
    2'  RANK    G    RANK    G    RANK    G    RANK    G    RANK
    3G    RANK    G    RANK    G    RANK    G    RANK    G    RANK    G
    4'/(' ',10(I7,F6.2)))
      STOP
      END
C
C     ****************************************
C     * SUBROUTINE TO CALCULATE MATRIX SUMS *
C     ****************************************
C
      SUBROUTINE SUMS(X,N,A,B,D,G,H,K)
C
      REAL X(100,100),K
C
      A=0.0
      B=0.0
      D=0.0
      DO 2 I=1,N
      ROW=0.0
      DO 3 J=1,N
      ROW=ROW+X(I,J)
    3 B=B+X(I,J)**2
      A=A+ROW
    2 D=D+ROW*ROW
      G=A*A
      H=D-B
      K=G+2.0*B-4.0*D
      RETURN
      END
C
C     *************************************
C     * SUBROUTINE TO CALCULATE STATISTICS *
C     *************************************
C
      SUBROUTINE STATIS(A,B,N,Z,EXPZ,SEZ,T,R)
C
      REAL A(100,100),B(100,100),KA,KB
```

```
C
C****  CALCULATE MATRIX CORRELATION
       CALL SUMS(A,N,AA,BA,DA,GA,HA,KA)
       CALL SUMS(B,N,AB,BB,DB,GB,HB,KB)
       Z=0.0
       TN=N*(N-1.0)
       DO 5 I=1,N
       DO 5 J=1,N
     5 Z=Z+A(I,J)*B(I,J)
       R=(Z-AA*AB/TN)/(SQRT((BA-AA**2/TN)*(BB-AB**2/TN)))
C
C****  CALCULATE EXPECTED VALUE AND STANDARD ERROR OF Z
       VARZ=(2.0*BA*BB+4.0*HA*HB/(N-2.0)+KA*KB/((N-2.0)*(N-3.0))
     1      -GA*GB/(N*(N-1.0)))/(N*(N-1.0))
       SEZ=SQRT(VARZ)
       EXPZ=AA*AB/(N*(N-1.0))
       T=(Z-EXPZ)/SEZ
       RETURN
       END
C
C      *****************
C      *SUBROUTINE SORT*
C      *****************
C
```

(This program should include the subroutine SORT that is included with the program HABITATS above.)

A.14 A COMPUTER PROGRAM FOR BOROWSKY'S TEST

A listing of the FORTRAN computer program BOROWSKY is provided below. This carries out Borowsky's randomization test for parallel variation, as described in Section 7.2.

The input to the program is summarized on comment cards that are given at the start of the listing. There can be up to 50 sympatric colonies (NL), up to 20 distribution variables (NA), and up to 1000 random pairings of species 2 data to species 1 data can be carried out. The format of the data for one species in one colony is specified by a FORTRAN format card.

The distribution variables are stored in an array DATA. These variables can be (a) gene or morph frequencies, or (b) colony mean values for quantitative variables. In case (a) the program converts the frequencies to proportions if that has not already been done, and uses Borowsky's distance function given by equation (7.1). In case (b) it is assumed that the quantitative variables have already been suitably coded (for example, to have unit standard deviations) and the Euclidean distance function of equation (7.4) is used.

Output from the program consists of: (a) a print of the data; (b) the matrix of distances D_{ik} from colony i to colony k; (c) the matrix of measures of parallel variation S_{ik} of equation (7.2), with their mean $\overline{S}$; (d) the randomized order of species 2 colonies for the first ten randomizations; and (e) the NSIM randomized values of $\overline{S}$, ranked from smallest to largest.

The observed $\bar{S}$ value can be assessed for significance relative to the randomized distribution. For example, it is significantly different from zero at the 5% level on a two-sided test if it is further from zero than 95% of randomized values.

Program listing

```
C
C      ********************
C      * PROGRAM BOROWSKY *
C      ********************
C
C         THIS PROGRAM CARRIES OUT BOROWSKY'S TEST FOR AN ASSOCIATED
C      DISTRIBUTION FOR TWO CLOSELY RELATED SPECIES.  DISTANCES
C      BETWEEN SPECIES CAN BE MEASURED BY GENE FREQUENCIES, MORPH
C      FREQUENCIES, OR MEAN VALUES FOR QUANTITATIVE VARIABLES IN
C      DIFFERENT COLONIES.
C         THE OPERATION OF THE PROGRAM DIFFERS SLIGHTLY ACCORDING TO
C      WHETHER DISTANCES ARE MEASURED BY (A) GENE OR MORPH PROPORTIONS,
C      OR (B) MEAN VALUES FOR QUANTITATIVE VARIABLES.  IN CASE (A) THE
C      PROGRAM BEGINS BY CONVERTING COUNTS TO PROPORTIONS.  BOROWSKY'S
C      MEASURE
C                D(I,K) = 0.5*SUM(ABS(X-Y))
C      IS THEN USED FOR THE DISTANCE FROM SPECIES 1 IN COLONY I TO
C      SPECIES 2 IN COLONY K.  IN CASE (B) IT IS ASSUMED THAT THE
C      QUANTITATIVE VARIABLE VALUES THAT ARE USED FOR DATA HAVE
C      ALREADY BEEN SUITABLY CODED.  THE EUCLIDEAN DISTANCE FUNCTION
C                D(I,K) = (SUM((X-Y)**2))**0.5
C      IS THEN USED.  BOROWSKY'S TEST INVOLVES SEEING WHETHER A TEST
C      STATISTIC SBAR IS SIGNIFICANT WHEN COMPARED TO THE DISTRIBUTION
C      OF SBAR THAT IS FOUND WHEN RANDOM PAIRINGS ARE MADE BETWEEN
C      THE DATA FOR SPECIES 1 AND 2.  THE FUNCTION RANDOM(ISEED) IS ONLY
C      AVAILABLE ON SOME BURROUGHS COMPUTERS AND CODE FOR IT WILL HAVE
C      TO BE PROVIDED ON OTHER MACHINES.  IT PRODUCES A RANDOM NUMBER
C      IN THE RANGE (0,1).  THE PROGRAM BOROWSKY WAS WRITTEN BY
C      B.F.J. MANLY, UNIVERSITY OF OTAGO, DUNEDIN, NEW ZEALAND.  THIS
C      VERSION IS DATED APRIL, 1985.
C
C      **************
C      * DATA CARDS *
C      **************
C
C      CARD TYPE 1
C         TITLE FOR THE DATA IN UP TO 72 COLS
C
C      CARD TYPE 2
C         COLS  1- 5  NL, NUMBER OF SYMPATRIC COLONIES (MAX=50)
C               6-10  NA, NUMBER OF DISTRIBUTION VARIABLES (MAX=20)
C               11-15 NSIM, NUMBER OF RANDOM ALLOCATIONS (MAX=1000)
C               16-20 ISEED, RANDOM NUMBER STARTER, INTEGER IN RANGE 1 TO
C                     99999
C                  25 1 FOR QUANTITATIVE DATA AND EUCLIDEAN DISTANCES
C                     (ANYTHING ELSE MEANS GENE OR MORPH PROPORTIONS)
C
C      CARD TYPE 3
C         FORMAT FOR DATA ON ONE SPECIES IN ONE COLONY, E.G. (14F5.0)
```

```
C
C       CARDS TYPE 4 (THERE ARE 2*NL OF THESE)
C         THE DATA FREQUENCIES ARE READ USING THE FORMAT ON CARD TYPE 3
C         IN THE ORDER SPECIES 1, COLONY 1; SPECIES 2, COLONY 1;
C         SPECIES 1, COLONY 2; ... SPECIES 2, COLONY NL.
C
        REAL DATA(50,20,2),D(50,50),S(50,50),DRAN(50,50),SRAN(1000)
        INTEGER IPF(18),TITLE(18),LOC(50),IA(50)
        DATA S/2500*0.0/
C
C**** READ DATA, CONVERT TO PROPORTIONS IF NECESSARY, AND PRINT
        READ(5,1001) (TITLE(I), I=1,18)
 1001 FORMAT(18A4)
        READ(5,1000) NL,NA,NSIM,ISEED,IQUANT
 1000 FORMAT(5I5)
        READ(5,1001) (IPF(I), I=1,18)
        DO 1 IL=1,NL
        DO 1 I=1,2
        READ(5,IPF) (DATA(IL,J,I), J=1,NA)
        IF (IQUANT.EQ.1) GO TO 1
        SUM=0.0
        DO 20 J=1,NA
   20 SUM=SUM+DATA(IL,J,I)
        DO 21 J=1,NA
   21 DATA(IL,J,I)=DATA(IL,J,I)/SUM
    1 CONTINUE
        WRITE(6,9000) (TITLE(I), I=1,18),ISEED,(J, J=1,NA)
 9000 FORMAT(' BOROWSKY (1977) TEST FOR PARALLEL VARIATION: ',18A4/
     1' RANDOM NUMBER STARTER = ',I5//
     2' DATA'/'  CLNY   SP  ',20I6)
        DO 2 IL=1,NL
        WRITE(6,9001)
 9001 FORMAT(' ')
        DO 2 I=1,2
    2 WRITE(6,9002) IL,I,(DATA(IL,J,I), J=1,NA)
 9002 FORMAT(' ',2I5,'  ',20F6.2)
        IF (IQUANT.NE.1) WRITE(6,9010)
 9010 FORMAT(/' THE ORIGINAL DATA ARE ASSUMED TO BE GENE OR MORPH FREQUE
     1NCIES.  THEY HAVE BEEN CONVERTED TO PROPORTIONS')
        IF (IQUANT.EQ.1) WRITE(6,9011)
 9011 FORMAT(/' THESE DATA ARE ASSUMED TO BE MEANS FOR QUANTITATIVE VARI
     1ABLES')
C
C**** CALCULATE AND PRINT DISTANCE AND S MATRICES
        DO 3 I=1,NL
        DO 3 J=1,NL
        D(I,J)=0.0
        DO 12 K=1,NA
        IF (IQUANT.NE.1) D(I,J)=D(I,J)+0.5*ABS(DATA(I,K,1)-DATA(J,K,2))
   12 IF (IQUANT.EQ.1) D(I,J)=D(I,J)+(DATA(I,K,1)-DATA(J,K,2))**2
    3 IF (IQUANT.EQ.1) D(I,J)=SQRT(D(I,J))
        WRITE(6,9003) (I, I=1,NL)
 9003 FORMAT(//' DISTANCE MATRIX'/(11X,20I6))
        DO 4 I=1,NL
    4 WRITE(6,9004) I,(D(I,J), J=1,NL)
 9004 FORMAT(I6,5X,20F6.2/(11X,20F6.2))
        CALL SCALC(NL,D,S,SBAR)
        WRITE(6,9005)
 9005 FORMAT(/' S MATRIX')
```

```
         DO 5 I=1,NL
      5  WRITE(6,9004) I,(S(I,J), J=1,I)
         WRITE(6,9006) SBAR
 9006 FORMAT(/' OBSERVED MEAN OF S (SBAR) = ',F10.4)
C
C**** GENERATES RANDOM ALLOCATIONS OF LOCATIONS WITH S VALUES
         WRITE(6,9008)
 9008 FORMAT(//' RESULTS FOR THE FIRST TEN RANDOM ALLOCATIONS'/
      1 ' ALLOC    SBAR     RANDOM ORDERS FOR SPECIES 2')
         DO 6 ISIM=1,NSIM
         DO 7 I=1,NL
      7  LOC(I)=0
         DO 8 I=1,NL
     10  IRAN=IFIX(RANDOM(ISEED)*NL)+1
         IF (LOC(IRAN).GT.0) GO TO 10
         LOC(IRAN)=1
      8  IA(I)=IRAN
         DO 9 I=1,NL
         DO 9 K=1,NL
      9  DRAN(I,K)=D(I,IA(K))
         CALL SCALC(NL,DRAN,S,SRAN(ISIM))
      6  IF (ISIM.LE.10) WRITE(6,9009) ISIM,SRAN(ISIM),(IA(K), K=1,NL)
 9009 FORMAT(I6,F8.4,5X,30I3/19X,20I3)
C
C**** SORT SBAR VALUES AND PRINT THEM
         CALL SORT(SRAN,1000,1,NSIM,1)
         WRITE(6,9007) NSIM,(I,SRAN(I), I=1,NSIM)
 9007 FORMAT(//' SBAR VALUES OBTAINED BY ',I4,' RANDOM PAIRINGS OF SPECI
      1ES 2 DATA TO SPECIES 1 DATA'/
      2'     RANK   VALUE     RANK   VALUE     RANK   VALUE     RANK   VALUE
      3     RANK   VALUE     RANK   VALUE     RANK   VALUE     RANK   VALUE'/
      4(' ',8(I8,F8.4)))
         STOP
         END
C
C     ********************************************************
C     * SUBROUTINE SCALC TO CALCULATE TEST STATISTIC SBAR *
C     ********************************************************
C
      SUBROUTINE SCALC(NL,D,S,SBAR)
C
      REAL D(50,50),S(50,50)
C
      SUM=0.0
      DO 1 I=2,NL
      DO 1 K=1,I-1
      S(I,K)=0.5*(D(I,K)+D(K,I)-D(I,I)-D(K,K))
    1 SUM=SUM+S(I,K)
      SBAR=SUM/(0.5*NL*(NL-1))
      RETURN
      END
C
C     *****************
C     *SUBROUTINE SORT*
C     *****************
C

         (This program should include the subroutine SORT that is included
         with the program HABITATS above.)
```

A.15 FISHER'S METHOD FOR COMBINING INDEPENDENT TEST RESULTS

Suppose that n independent significance tests result in probabilities p_1, $p_2, \ldots, p_n$, where p_i is the probability of obtaining a more extreme test statistic than the one actually observed. Then the ith test is said to be significant at the $\alpha\%$ level if $100\, p_i < \alpha$. If n is large then one or more of the series of tests may quite likely give a significant result even although the null hypothesis is true in all cases. On the other hand, it may happen that none of the tests are individually significant but together they are. It is therefore useful to have a general method for combining test results.

Fisher's (1970) method is one approach. If the null hypothesis is true for the ith test then p_i will be a random value from a uniform distribution over the interval $(0, 1)$. It can be shown that in that case $-2\log_e(p_i)$ is a chi-square variate with two degrees of freedom. Consequently, if the null hypothesis is true for all n tests, then

$$X^2 = -2 \sum_{i=1}^{n} \log_e(p_i)$$

will be a chi-square variate with $2n$ degrees of freedom. Large values of X^2 are associated with small values of p_i. Hence if X^2 is significantly large when compared with the chi-square distribution then this is evidence that the null hypothesis is wrong for one or more of the tests. In some cases, small values of X^2 are also evidence against the null hypothesis.

Some care is needed in determining the probabilities p_i. They depend upon whether the original n tests are one- or two-sided, and whether or not a deviation in the same direction is expected for all n tests. There are three possibilities, as follows.

Firstly, suppose that all of the n tests are two-sided and that there is no reason to suppose that significant test results should all be in the same direction. For example, the first test may give a test statistic which is nearly significantly high while the second test may give one nearly significantly low. These are both regarded as nearly giving evidence against the null hypothesis. In this case it is appropriate to take p_i as the probability of a more extreme result than that observed for the ith test, and test to see whether X^2 is significantly large.

Secondly, suppose that the situation is the same as for the first case except that if the null hypothesis is false then it will tend to give results in the same direction for all n tests: either all the test statistics will tend to be large or they will all tend to be small. However, the tests are still two-sided because it cannot be predicted in advance which way they will go. In that case it is appropriate to take p_i as the probability of a larger test result than the one actually observed. Then the values $p_1, p_2, \ldots, p_n$ may either tend to be all large or all small when there is evidence against the null hypothesis. It follows that a two-sided test is needed for X^2. That is to say, X^2 will be significant at the 5% level if it is less than the lower 2.5% point or greater than the upper 2.5% point of the chi-square distribution.

Finally, suppose that the n tests are all one-sided. In this case the situation is more straightforward. For each test, p_i is the probability of getting a more extreme result than the one observed, in the direction of evidence against the null hypothesis. Only large values of X^2 are evidence against the null hypothesis.

Borowsky's (1977) test results shown in Table 7.6 are an example of the second situation. There are 26 $\overline{S}$ values, together with estimates of the probabilities of getting larger values by chance alone. It can be argued that when natural selection occurs, an $\overline{S}$ value will tend to be different from zero but the direction of difference might be positive or negative. However, if selection affects all the tests in the same way, then either all the $\overline{S}$ values will tend to be positive or they will all tend to be negative. Thus for the ith test, p_i should be taken as the probability of exceeding the observed value by chance alone and X^2 should be tested against both tails of the chi-square distribution.

A.16   THE COMPUTER PROGRAM
MAXLIK: GENERAL

The FORTRAN computer program MAXLIK was designed to estimate the parameters of a multinomial distribution, using the Newton–Raphson iterative process that is described in Section 3 of this Appendix. It assumes that a total of TN individuals are independently classified into JMAX classes, in such a way that the probability of being in the Jth class is F(J), with F(1) + F(2) + ... + F(JMAX) = 1. The class probabilities are functions of parameters X(1), X(2), . . . , X(IMAX) that are to be estimated.

The original development of MAXLIK was due to Reed and Schull (1968). (See also Reed, 1969a.) The version provided here started off as the IBM 7094 code in double precision written by Reed in 1968. Reed has kindly agreed to its publication. However, it must be stressed that the program has evolved considerably over the years and Reed can no longer be considered responsible for it. Particular changes that have been made to Reed's original code have been a move to single precision, elimination of weights from the FREQ subroutine, and modifications to input and output. The matrix inversion routine included in MAXLIK is a modification of one given by McCormick and Salvadori (1964, p. 306).

In order to use MAXLIK for a particular problem it is necessary to write a subroutine FREQ(X, F, K). Two versions of FREQ are provided below, these being for estimation with the models discussed in Section 8.14. The purpose of FREQ is to determine the probabilities F(J), given the parameters X(I) and certain constants K(1), K(2), . . . , K(LMAX). The constants may or may not be needed for particular applications.

The input for MAXLIK is described on comment cards at the start of the listing. A number of similar sets of data may be analysed in one run. There can be up to 50 parameters estimated and up to 500 data classes. In some cases the parameters X(I) will be proportions that add to 1 so that

$$X(IMAX + 1) = 1 - X(1) - X(2) - \ldots - X(IMAX).$$

The parameter X(IMAX + 1) may then be of some interest, although it is not included in the estimation process since it can be expressed in terms of the other parameters. In such cases 'YES' can be placed in columns 18 to 20 of the first data card and standard errors and other information will then be output for X(IMAX + 1). For each separate set of data the following information is needed: a title, up to 72 characters long; observed class frequencies N(1), N(2), . . . , N(JMAX); initial approximations for X(1), X(2), . . . , X(IMAX); and, if LMAX $\neq$ 0, the constants K(1), K(2), . . . , K(LMAX). The maximum value of LMAX is 500.

Output from the program consists of: (a) a print of log-likelihood values at the start and end of each iteration; (b) a list of observed and expected class frequencies, with chi-square values; (c) parameters estimates with variances, standard errors and the last corrections made; (d) the parameter covariance matrix; and (e) the parameter correlation matrix. The program stops iterating if the log-likelihood does not increase. This may be because a maximum has been found or it may be because the iterative process fails to converge. The log-likelihood values will indicate which of these two possibilities has occurred. The program also stops if the corrections to the parameters are all less than 0.000 01 in absolute value. The maximum number of iterations allowed has been set at MAXIT = 20 in the statement following label 400. This can be changed, but 20 iterations should be more than enough for convergence if this is going to occur.

Differentiation is done numerically at statement 4 in the program. The partial derivative with respect to X(I) is found by increasing and decreasing X(I) by 0.000 01. This should be quite satisfactory providing that X(I) is not very close to zero nor very large. It is best to define parameters so that they are all about equal to one. Parameters equal to proportions should be quite satisfactory providing that they fall within the range 0.000 02 to 0.999 98.

There is one aspect of the practical use of MAXLIK that warrants a special mention. A type of situation that frequently occurs is that s random samples with sizes $n_1, n_2, \ldots, n_s$ are taken either from one population at s different times, or from s different populations. The sample members are then classified into K classes so that the data are of the form of Table A5. A certain model then says that the probability of an individual being of morph j in the ith sample is $p_{ij}(\theta_1, \theta_2, \ldots, \theta_r)$, a function of unknown parameters θ_1, $\theta_2, \ldots, \theta_r$. These parameters might, for example, be gene frequencies.

There are two special cases that are of particular interest. The first of these occurs when all the sample morph frequencies n_{ij} are independent Poisson variates. In this case it has been shown in Section 2 of this Appendix that the distribution of the n_{ij} values conditional upon the observed total frequency $n = \Sigma n_i$ is multinomial. The probability of an individual being in the ith sample and also being classified as morph j is then

$$f_{ij} = p_{ij}(\theta_1, \theta_2, \ldots, \theta_r)\gamma_i, \tag{A42}$$

Table A5   Morph frequencies obtained from s samples.

	Morph				
Sample	1	2	...	K	Total
1	n_{11}	n_{12}	...	n_{1K}	n_1
2	n_{21}	n_{22}	...	n_{2K}	n_2
$\vdots$	$\vdots$	$\vdots$			$\vdots$
s	n_{s1}	n_{s2}	...	n_{sK}	n_s
					n

where γ_i is the expected value of n_i/n. The likelihood function for the data is

$$P(n_{11}, n_{12}, \ldots, n_{sK}) = \frac{n!}{\prod\limits_{i=1}^{s} \prod\limits_{j=1}^{K} n_{ij}!} \prod\limits_{i=1}^{s} \prod\limits_{j=1}^{K} f_{ij}^{n_{ij}}$$

$$= C_1 \prod\limits_{i=1}^{s} \prod\limits_{j=1}^{K} p_{ij}(\theta_1, \theta_2, \ldots, \theta_r)^{n_{ij}} \prod\limits_{i=1}^{n} \gamma_i^{n_i}, \tag{A43}$$

where C_1 is a constant that does not involve any of the parameters $\theta_1, \theta_2, \ldots, \theta_r$ or $\gamma_1, \gamma_2, \ldots, \gamma_s$.

Because of its multinomial nature, this model can be estimated using MAXLIK. In doing this the numerical process can be simplified somewhat by recognizing that simple estimators of $\gamma_1, \gamma_2, \ldots, \gamma_{s-1}$ and $\gamma_s = 1 - \gamma_1 - \gamma_2 - \ldots - \gamma_{s-1}$ are available, irrespective of how p_{ij} depends upon the other parameters. Application of the standard maximum likelihood methods described in Section 3 of this Appendix shows that the estimator of γ_i is $\hat{\gamma}_i = n_i/n$, and that there is no correlation between the γ and θ parameters. In practice, this means that when using MAXLIK it is valid to set $\gamma_i = n_i/n$ in equation (A42) and estimate the θ parameters as if the γ parameters are known exactly. The estimates $\hat{\theta}_1, \hat{\theta}_2, \ldots, \hat{\theta}_r$ and their estimated variances and covariances will be unaffected but computation times can be reduced considerably by this 'trick'.

The second special case occurs when the sample sizes $n_1, n_2, \ldots, n_s$ are fixed in advance. This occurs particularly with laboratory experiments, where there is more control than there is in the field. If n_i is fixed then the morph frequencies $n_{i1}, n_{i2}, \ldots, n_{iK}$ will be multinomially distributed. Hence the probability function for all s samples will be the product of s multinomials,

$$P(n_{11}, n_{12}, \ldots, n_{sK}) = \prod_{i=1}^{s} \frac{n_i!}{\prod\limits_{j=1}^{K} n_{ij}!} \prod_{j=1}^{K} P_{ij}(\theta_1, \theta_2, \ldots, \theta_r)^{n_{ij}}$$

$$= C_2 \prod_{i=1}^{s} \prod_{j=1}^{K} P_{ij}(\theta_1, \theta_2, \ldots, \theta_r)^{n_{ij}} \qquad \text{(A44)}$$

where C_2 is a constant that does not depend upon the parameters θ_1, $\theta_2, \ldots, \theta_r$. Comparison of equations (A43) and (A44) shows that they are proportional to each other when each is regarded as a function of the θ parameters. Consequently, both functions must be maximized for the same values of these parameters. It also follows from the general theory of maximum likelihood estimation that estimated variances and covariances will be the same. In other words, in terms of maximum likelihood estimation of the θ parameters it is irrelevant whether the sample sizes $n_1, n_2, \ldots, n_s$ are fixed in advance or are random Poisson variables. The same estimation procedure – take $\gamma_i = n_i/n$ and calculate class probabilities using equation (A42) – can be used in either case.

Program listing

```
C
C      ******************
C      * PROGRAM MAXLIK *
C      ******************
C
C      A PROGRAM FOR GENERAL MAXIMUM LIKELIHOOD ESTIMATION, WRITTEN BY
C      T.E. REED, DEPARTMENT OF ZOOLOGY, UNIVERSITY OF TORONTO, JUNE
C      JUNE 1968. MODIFIED BY B.F.J. MANLY, UNIVERSITY OF OTAGO.  THIS
C      VERSION IS DATED MAY, 1983.
C
C      THIS PROGRAM ESTIMATES THE VALUES OF PARAMETERS X(I) WHICH
C      DETERMINE THE FREQUENCIES F(J) OF JMAX CLASSES.  THE PROGRAM IS
C      GENERAL IN THE SENSE THAT THE MAXIMUM NUMBER OF PARAMETERS (IMAX)
C      AND CLASSES (JMAX) AS WELL AS THE ALGEBRAIC RELATIONSHIPS BETWEEN
C      THE X(I) AND F(J) ARE ALL UNSPECIFIED.  SUBROUTINE FREQ(X,F,W,K),
C      SUPPLIED BY THE USER, EXPRESSES THE RELATIONSHIPS BETWEEN F(J) AND
C      X(I).  ESTIMATION IS BY MAXIMUM LIKELIHOOD, USING EFFICIENT
C      SCORES (RAO, 1952, P. 168).  DIFFERENTIATION IS DONE NUMERICALLY.
C
C      *************
C      * DATA CARDS *
C      *************
C
C      CARD TYPE 1
```

```
C        COLS   1- 5    NO. OF PROBLEMS (NPTODO)
C               9-10    NO. PARAMETERS (IMAX), MAXIMUM 50.
C               13-15   NO. CLASSES (JMAX), MAXIMUM 500.
C               18-20   YES IF (1 - SUM OF ESTIMATED PARAMETERS) IS TO
C                       BE CALCULATED - BLANK OTHERWISE.
C               23-25   NO. OF CONSTANTS K(L) FOR FREQ (LMAX), MAXIMUM
C                       500.  MAY BE 0.   CONSTANTS CAN VARY WITH DATA.
C
C     CARD(S) TYPE 2 (AS MANY AS NEEDED)
C        COLS   2- 5    PARAMETER NAMES (KPAR).  THE LAST NAME
C               7-10    IS FOR '1 - SUM OF OTHER PARAMETERS' IF
C                :      THIS IS WANTED.  THERE ARE UP TO 14 NAMES
C               67-70   PER CARD, OF FOUR CHARACTERS EACH.
C
C     FOR EACH PROBLEM THERE ARE THE FOLLOWING CARDS
C        1 - COMMENTS IN COLS. 1-72
C        2 - OBSERVED FREQUENCIES N(J) IN COLS. 1-5, ..., 66-70, WITH AS
C            MANY CARDS AS NEEDED.
C        3 - TRIAL ESTIMATES FOR X(I) IN COLS. 1-5, ..., 66-70, WITH AS
C            MANY CARDS AS NEEDED.
C        4 - CONSTANTS K(L) FOR SUBROUTINE FREQ.  THERE ARE LMAX OF THESE
C            WITH K(1) IN COLS. 1-5, K(2) IN COLS. 6-10, ..., K(14) IN
C            COLS. 66-70, WITH AS MANY CARDS AS NEEDED.  IN THIS VERSION
C            OF MAXLIK THESE CONSTANTS ARE REAL VARIABLES.  THERE ARE
C            NONE OF THESE CARDS IF LMAX=0.
C
      DIMENSION EST(50),X(50),F(500),DF(50,500),U(50),RM(50,50),XN(500),
     1C(50,50),DX(50),VX(50),SEX(50),EN(500),PR(500),CHI(500),
     2TEM(500),N(500),COM(18),KPAR(50)
      REAL K(500)
      INTEGER TN
      DATA SUM/3HYES/
C
C**** READ DATA
  203 READ(5,200) NPTODO,IMAX,JMAX,SUMM,LMAX
  200 FORMAT(3I5,2X,A3,I5)
      IMAX1=IMAX+1
      IF (SUMM.EQ.SUM) READ(5,1) (KPAR(I), I=1,IMAX1)
      IF (SUMM.NE.SUM) READ(5,1) (KPAR(I), I=1,IMAX)
    1 FORMAT(14(1X,A4))
      NPDONE=0
  201 IF (NPDONE.EQ.NPTODO) STOP
  202 READ(5,14) (COM(I), I=1,18)
   14 FORMAT(18A4)
      READ(5,102) (N(J), J=1,JMAX)
  102 FORMAT(14I5)
      NPDONE=NPDONE+1
      READ(5,101) (EST(I), I=1,IMAX)
  101 FORMAT(14F5.0)
      IF (LMAX.GT.1) READ(5,101) (K(L), L=1,LMAX)
      DO 2 I=1,IMAX
    2 X(I)=EST(I)
      DO 21 J=1,JMAX
   21 XN(J)=N(J)
      TXN=0.0
      DO 9 J=1,JMAX
    9 TXN=TXN+XN(J)
      WRITE(6,36) NPDONE,(COM(I), I=1,18)
   36 FORMAT('1PROBLEM NO. ',I2,5X,18A4//'                 INITIAL    FINA
```

```
     1L'//  ITERATION  LOG(LIK)  LOG(LIK)')
C
C**** FIND INITIAL LOG LIKELIHOOD
      CALL FREQ(X,F,K)
      ALIK1=0.0
      DO 400 J=1,JMAX
  400 IF (XN(J).GT.0.0) ALIK1=ALIK1+XN(J)*ALOG(TXN*F(J)/XN(J))
C
C**** START OF ITERATION
      MAXIT=20
      DO 20 ITER=1,MAXIT
      NITER=ITER
C
C**** BEGIN NUMERICAL DIFFERENTIATION
      DO 4 I=1,IMAX
      X(I)=X(I)+0.00001
      CALL FREQ(X,TEM,K)
      X(I)=X(I)-0.00002
      CALL FREQ(X,F,K)
      DO 4 J=1,JMAX
    4 DF(I,J)=(TEM(J)-F(J))/0.00002
      DO 5 I=1,IMAX
      U(I)=0.0
      DO 5 II=1,IMAX
    5 RM(I,II)=0.0
      DO 6 I=1,IMAX
      DO 6 J=1,JMAX
    6 U(I)=U(I)+XN(J)*DF(I,J)/F(J)
      DO 7 I=1,IMAX
      DO 7 II=1,IMAX
      DO 7 J=1,JMAX
    7 RM(I,II)=RM(I,II)+DF(I,J)*DF(II,J)/F(J)
      DO 8 I=1,IMAX
      DO 8 II=1,IMAX
    8 C(I,II)=RM(I,II)
C
C**** INVERT INFORMATION MATRIX RM (USING MATIV) TO OBTAIN COVARIANCE
C**** MATRIX C
      CALL MATIV(C,IMAX)
      DO 10 I=1,IMAX
   10 DX(I)=0.0
C
C**** CALCULATE CORRECTIONS DX(I) FOR X(I)
      DO 11 I=1,IMAX
      DO 11 II=1,IMAX
   11 DX(I)=DX(I)+C(I,II)*U(II)/TXN
      DO 12 I=1,IMAX
      IF (ABS(DX(I))-0.00001) 12,13,13
   12 CONTINUE
      GO TO 30
   13 DO 420 I=1,IMAX
  420 X(I)=X(I)+DX(I)
      CALL FREQ(X,TEM,K)
      ALIK2=0.0
      DO 421 J=1,JMAX
  421 IF (XN(J).GT.0.0) ALIK2=ALIK2+XN(J)*ALOG(TXN*TEM(J)/XN(J))
      WRITE(6,422) ITER,ALIK1,ALIK2
  422 FORMAT(I11,2F10.2)
      IF (ALIK2.LT.ALIK1) GO TO 423
```

```
   20 ALIK1=ALIK2
      GO TO 30
C
C**** NO CONVERGENCE
  423 WRITE(6,424)
  424 FORMAT(//' THE ITERATIVE PROCESS IS STOPPED BECAUSE THE LIKELIHOOD
     1 IS NOT INCREASING.  LAST ITERATION IS OMITTED')
      DO 425 I=1,IMAX
  425 X(I)=X(I)-DX(I)
C
C**** END OF ITERATION
   30 DO 32 I=1,IMAX
      VX(I)=C(I,I)/TXN
      SEX(I)=0.0
   32 IF (VX(I).GT.0.0) SEX(I)=SQRT(VX(I))
      IF (SUMM.NE.SUM) GO TO 18
C
C**** CALCULATE VALUES FOR (1 - SUM OF PARAMETERS)
      TEST=0.0
      TX=0.0
      VART=0.0
      DO 321 I=1,IMAX
      TEST=TEST+EST(I)
      TX=TX+X(I)
      DO 321 II=1,IMAX
  321 VART=VART+(1.0/TXN)*C(I,II)
      ZTREST=1.0-TEST
      ZFIEST=1.0-TX
      SET=SQRT(VART)
C
C**** OUTPUT RESULTS
   18 TNE=0.0
      TPR=0.0
      TF=0.0
      TCHI=0.0
      TN=0.0
      DO 33 J=1,JMAX
      EN(J)=F(J)*TXN
      TNE=TNE+EN(J)
      PR(J)=XN(J)/TXN
      TPR=TPR+PR(J)
      TF=TF+F(J)
   33 TN=TN+N(J)
      DO 34 J=1,JMAX
      CHI(J)=9999.0
      IF (ABS(EN(J)).LT.1.0E-10) GO TO 35
      CHI(J)=(XN(J)-EN(J))**2/EN(J)
   34 TCHI=TCHI+CHI(J)
   35 WRITE(6,37)
   37 FORMAT(//55X,29HDISTRIBUTION IN THE J CLASSES/)
      SCHI=TCHI
      WRITE(6,41) (J,N(J),EN(J),PR(J),F(J),CHI(J), J=1,JMAX)
   41 FORMAT(4X,5HCLASS,3X,8HOBS. NO.,10X,8HEXP. NO.,10X,10HOBS. PROB.,
     110X,10HEXP. PROB.,6X,34HCHI-SQUARE (CELL VALUES AND TOTAL)/(1X,
     218,I10,3E20.6,4X,E20.6))
      WRITE(6,42) TN,TNE,TPR,TF,TCHI
   42 FORMAT(3X,6H TOTAL,I10,3E20.6,4X,E20.6)
      WRITE(6,82) NITER
   82 FORMAT(/5X,23HNUMBER OF ITERATIONS = ,I2)
```

```
  180 WRITE(6,80) (KPAR(I),EST(I),X(I),VX(I),SEX(I),DX(I), I=1,IMAX)
   80 FORMAT(//40X,21HPARAMETER   ESTIMATES//1X,9HPARAMETER,6X,
      110HTRIAL EST.,10X,10HFINAL EST.,12X,8HVARIANCE,11X,9HSTD. ERR.,
      27X,24HDELTA X(LAST CORRECTION)/(A7,5E20.6))
      IF (SUMM.NE.SUM) GO TO 19
      WRITE(6,81) KPAR(IMAX1),ZTREST,ZFIEST,VART,SET
   81 FORMAT(1X,A6,4E20.6)
   19 WRITE(6,500)
  500 FORMAT(//20X,32HCOVARIANCE MATRIX FOR PARAMETERS/)
      DO 501 I=1,IMAX
      DO 501 II=1,IMAX
  501 C(I,II)=C(I,II)/TXN
      DO 502 I=1,IMAX
  502 WRITE(6,503) KPAR(I),(C(I,II), II=1,IMAX)
  503 FORMAT(1X,A6,10E12.4/(7X,10E12.4))
      WRITE(6,84)
   84 FORMAT(//20X,33HCORRELATION MATRIX FOR PARAMETERS/)
      DO 50 I=1,IMAX
      DO 50 II=1,IMAX
      PROD=C(I,I)*C(II,II)
      RM(I,II)=0.0
   50 IF (PROD.GT.0.0) RM(I,II)=C(I,II)/SQRT(PROD)
      DO 60 I=1,IMAX
   60 WRITE(6,46) KPAR(I),(RM(I,II), II=1,IMAX)
   46 FORMAT(1X,A6,15F8.3/(7X,15F8.3))
      IF (LMAX.GT.0) WRITE(6,118) (K(L), L=1,LMAX)
  118 FORMAT(//15H FREQ CONSTANTS,10F10.3/(15X,10F10.3))
      GO TO 201
      END
C
C
C     *****************************************
C     * SUBROUTINE MATIV FOR MATRIX INVERSION *
C     *****************************************
C
      SUBROUTINE MATIV(A,N1)
C
C     THE MATIV INVERSION PROGRAM IS GIVEN IN MC CORMICK AND
C     SALVADORI (1964, P. 306). IT HAS BEEN ALTERED TO DO ONLY MATRIX
C     INVERSION.
C
      DIMENSION A(50,50),INDEX(50,3)
      EQUIVALENCE (IROW,JROW),(ICOLUM,JCOLUM), (AMAX,T,SWAP)
C
C**** INITIALIZATION
      N=N1
      DO 20 J=1,N
   20 INDEX(J,3)=0
      DO 550 I=1,N
C
C**** SEARCH FOR PIVOT ELEMENT
      AMAX=0.0
      DO 105 J=1,N
      IF (INDEX(J,3)-1) 60,105,60
   60 DO 100 K=1,N
      IF (INDEX(K,3)-1) 80,100,740
   80 IF (AMAX-ABS(A(J,K))) 85,100,100
   85 IROW=J
      ICOLM=K
      AMAX=ABS(A(J,K))
```

```
100 CONTINUE
105 CONTINUE
    INDEX(ICOLM,3)=INDEX(ICOLM,3)+1
    INDEX(I,1)=IROW
    INDEX(I,2)=ICOLM
C
C**** INTERCHANGE ROWS TO PUT PIVOT ELEMENT ON DIAGONAL
    IF (IROW-ICOLM) 140,310,140
140 DO 200 L=1,N
    SWAP=A(IROW,L)
    A(IROW,L)=A(ICOLM,L)
200 A(ICOLM,L)=SWAP
310 PIVOT=A(ICOLM,ICOLM)
C
C**** DIVIDE PIVOT ROW BY PIVOT ELEMENT
    A(ICOLM,ICOLM)=1.0
    DO 350 L=1,N
350 A(ICOLM,L)=A(ICOLM,L)/PIVOT
C
C**** REDUCE NON-PIVOT ROWS
380 DO 550 L1=1,N
    IF (L1-ICOLM) 400,550,400
400 T=A(L1,ICOLM)
    A(L1,ICOLM)=0.0
    DO 450 L=1,N
450 A(L1,L)=A(L1,L)-A(ICOLM,L)*T
550 CONTINUE
C
C**** INTERCHANGE COLUMNS
    DO 710 I=1,N
    L=N+1-I
    IF(INDEX(L,1)-INDEX(L,2)) 630,710,630
630 JROW=INDEX(L,1)
    JCOLM=INDEX(L,2)
    DO 705 K=1,N
    SWAP=A(K,JROW)
    A(K,JROW)=A(K,JCOLM)
    A(K,JCOLM)=SWAP
705 CONTINUE
710 CONTINUE
    DO 730 K=1,N
    IF (INDEX(K,3)-1) 715,720,715
715 GO TO 740
720 CONTINUE
730 CONTINUE
740 RETURN
    END
```

A.17   THE COMPUTER PROGRAM
MAXLIK: FREQ SUBROUTINE (a)

The computer program MAXLIK requires a special version of the subroutine FREQ for each application. The version provided in this section is for the models expressed by equations (8.47) and (8.49).

The purpose of the FREQ subroutine is to calculate the probabilities associated with different classes as functions of the parameters to be estimated.

As explained in the previous section, if the data being considered consists of a number of independent random samples with sizes $n_1, n_2, \ldots, n_s$, with $\sum n_i = n$, then n_i/n can be taken as the 'probability' of an individual being in the ith sample for the calculation of the class probabilities. This is true irrespective of whether the sample sizes are considered to be random variables or fixed.

In the application being considered the observed class frequencies are n_{kil}, for $k = 1, 2, i = 0, t$, and $l = 1, 2, \ldots, L$. That is, n_{1il} is the frequency of $A_1 A_-$ individuals and n_{2il} is the frequency of $A_2 A_2$ individuals in a sample taken at time i from population l. Assuming that samples are taken before selection in a generation, equations (8.47) show that the class probability associated with observed frequency n_{kil} is given by

$$f_{kil} = \begin{cases} (1 - q_{il}^2)n_{il}/n, & k = 1, \\ q_{il}^2 n_{il}/n, & k = 2, \end{cases} \tag{A45}$$

where q_{il} is the proportion of the A_2 allele in population l at time i. It is the probabilities f_{kil} that FREQ evaluates. They are functions of the proportions of A_2 at time zero, $q_{01}, q_{02}, \ldots, q_{0L}$, and the selection parameter u. The allele proportions at time t, $q_{t1}, q_{t2}, \ldots, q_{tL}$, are determined from the repeated application of equation (8.46), taking $s = 1 - \exp(u)$.

If samples are taken after selection in generations 0 and t, then equations (8.47) have to be replaced with equations (8.49). The class frequencies then become

$$f_{kil} = \begin{cases} \{(1 - q_{il}^2)n_{il}/n\}(1 - s)/\{1 - s(1 - q_{il}^2)\}, & k = 1, \\ \{q_{il}^2 n_{il}/n\}/\{1 - s(1 - q_{il}^2)\}, & k = 2, \end{cases}$$

still with $s = 1 - \exp(u)$. No other changes are required. In the FREQ subroutine the input constant $K(3)$ has to be set equal to 1 for samples after selection.

In testing for goodness of fit by a chi-square test, the number of degrees of freedom is the number of classes, minus one for every total that is the same for observed and expected frequencies, minus one for every estimated parameter. In the present case there are $4L$ phenotype frequencies. Total observed and expected frequencies agree for each sample from each population, which results in the loss of $2L$ degrees of freedom. There are $L + 1$ estimated parameters. Hence the degrees of freedom for chi-square are $4L - 2L - (L + 1) = L - 1$.

Some example input data for MAXLIK are shown in Table A6. This is for *Cepaea nemoralis* populations at ten locations on the Berrow sand dunes that were sampled in 1926 and again in 1959/60 (from Table 8.10). There are in this case 11 parameters $u, q_{01}, q_{02}, \ldots, q_{0\,10}$. The data are analysed first assuming samples before selection ($K(3) = 0$), and then assuming samples after selection ($K(3) = 1$). The sample phenotype frequencies are input in the order n_{101}, $n_{201}, n_{1t1}, n_{2t1}, \ldots, n_{10L}, n_{20L}, n_{1tL}$ and n_{2tL}. This is row by row following the

Table A6 Example input data for MAXLIK using version (a) of the FREQ subroutine. Original data come from Table 8.10.

2	11	40		23									
U	Q1	Q2	Q3	Q4	Q5	Q6	Q7	Q8	Q9	Q10			

LOCATIONS D19-D56 FOR 1926–1959/60 SAMPLES BEFORE SELECTION

5	137	0	132	1	47	0	342	2	228	0	220	5	175
3	103	7	161	1	58	63	406	9	80	22	111	20	125
95	332	41	316	18	231	11	569	14	101	18	286		
0	.982	.989	.996	.986	.979	.930	.913	.882	.963	.937			
10	11.3	0	142	132	48	342	230	220	180	106	168	59	469
89	133	145	427	357	249	580	115	304					

LOCATIONS D19-D56 FOR 1926–1959/60 SAMPLES AFTER SELECTION

5	137	0	132	1	47	0	342	2	228	0	220	5	175
3	103	7	161	1	58	63	406	9	80	22	111	20	125
95	332	41	316	18	231	11	569	14	101	18	286		
0	.982	.989	.996	.986	.979	.930	.913	.882	.963	.937			
10	11.3	1	142	132	48	342	230	220	180	106	168	59	469
89	133	145	427	357	249	580	115	304					

order shown in Table 8.10. Initial approximations for the parameters are taken as $\hat{u} = 0$ and $q_{0l} \simeq \sqrt{(n_{20l}/n_{0l})}$.

There are 23 constants needed for FREQ for this particular set of data; in general, data for L locations requires $2L + 3$ constants. These constants are, in the order $K(1)$ to $K(23)$: the number of locations (10), the time t of the second sample measured in generations (11.3 generations of three years from 1926 to

Table A7   Estimates obtained from MAXLIK using the input data shown in Table A6.

	Samples before selection		Samples after selection	
Parameter	Estimate	Standard error	Estimate	Standard error
U	−0.0693	0.0122	−0.0695	0.0122
Q1	0.9875	0.0055	0.9867	0.0059
Q2	0.9975	0.0025	0.9973	0.0027
Q3	0.9969	0.0022	0.9967	0.0023
Q4	0.9823	0.0062	0.9811	0.0066
Q5	0.9793	0.0072	0.9778	0.0077
Q6	0.9272	0.0083	0.9225	0.0088
Q7	0.8938	0.0157	0.8874	0.0169
Q8	0.8832	0.0101	0.8763	0.0112
Q9	0.9719	0.0054	0.9700	0.0059
Q10	0.9379	0.0112	0.9339	0.0122

1960), an indicator for samples before or after selection (0 for before, 1 for after), the sample size at location 1 for generation 0, the sample size for location 1 for generation $t, \ldots$, the sample size at location 10 for generation 0, the sample size at location 10 for generation t.

Table A7 shows the estimates produced by MAXLIK for these data. Almost the same results are obtained when samples are assumed to be before or after selection.

Listing for version (a) of the FREQ subroutine

```
      SUBROUTINE FREQ(X,F,K)
C
C     THIS VERSION IS FOR ESTIMATING A CONSTANT SELECTIVE VALUE FOR
C     SEVERAL LOCATION, USING CLARKE AND MURRAY'S DOMINANCE SELECTION
C     MODEL.
C
C     X(1)=LOG(1-S), X(2)=Q0(1), X(2)=Q0(2),..., X(NL+1)=Q0(NL).
C     K(1)=NUMBER OF LOCATIONS, K(2)=T, K(3)=1 FOR SAMPLES AFTER
C     SELECTION, K(4), K(5), ETC. ARE SAMPLE SIZES IN THE ORDER LOCATION
C     1, TIME 0; LOCATION 1, TIME T; LOCATION 2, TIME 0, AND SO ON.
C
      REAL X(50),F(500),K(500),Q0(20),GA(20,2),Q(50),QT(20)
      DATA GA/40*0.0/, NL,NLP,NG,T/0,0,0,0.0/
C
C**** INITIAL SET-UP (ONLY ON FIRST ENTRY)
      IF (GA(1,1).NE.0.0) GO TO 1
      NL=K(1)
      SUM=0.0
      DO 2 I=1,NL
      J=2*I+1
      GA(I,1)=K(J+1)
      GA(I,2)=K(J+2)
    2 SUM=SUM+GA(I,1)+GA(I,2)
      DO 3 I=1,NL
      DO 3 J=1,2
    3 GA(I,J)=GA(I,J)/SUM
      T=K(2)
      NLP=NL+1
      NG=INT(T)+1
C
C**** SORT OUT PARAMETERS
    1 S=1.0-EXP(X(1))
      DO 4 I=2,NLP
    4 Q0(I-1)=X(I)
C
C**** CALCULATE GENE FREQUENCIES
      DO 5 I=1,NL
      Q2=Q0(I)**2
      Q(1)=Q0(I)+S*Q2*(1.0-Q0(I))/(1.0-S*(1.0-Q2))
      DO 6 J=2,NG
      Q2=Q(J-1)**2
    6 Q(J)=Q(J-1)+S*Q2*(1.0-Q(J-1))/(1.0-S*(1.0-Q2))
      DT=T-NG+1
    5 QT(I)=(1.0-DT)*Q(NG-1)+DT*Q(NG)
      IF (K(3).EQ.1) GO TO 10
```

```
C
C**** CALCULATE PROBABILITIES ASSUMING SAMPLES BEFORE SELECTION
      DO 7 I=1,NL
      J=4*(I-1)
      Q2=Q0(I)**2
      F(J+1)=GA(I,1)*(1.0-Q2)
      F(J+2)=GA(I,1)*Q2
      Q2=QT(I)**2
      F(J+3)=GA(I,2)*(1.0-Q2)
    7 F(J+4)=GA(I,2)*Q2
      GO TO 12
C
C**** CALCULATE PROBABILITIES ASSUMING SAMPLES AFTER SELECTION
   10 DO 11 I=1,NL
      J=4*(I-1)
      Q2=Q0(I)**2
      A2A2=Q2/(1.0-S*(1.0-Q2))
      F(J+1)=GA(I,1)*(1.0-A2A2)
      F(J+2)=GA(I,1)*A2A2
      Q2=QT(I)**2
      A2A2=Q2/(1.0-S*(1.0-Q2))
      F(J+3)=GA(I,2)*(1.0-A2A2)
   11 F(J+4)=GA(I,2)*A2A2
   12 DO 13 I=1,4*NL
   13 IF (F(I).LT.0.000001) F(I)=0.000001
      RETURN
      END
```

A.18   THE COMPUTER PROGRAM MAXLIK: FREQ SUBROUTINE (b)

The second version of the FREQ subroutine that is provided is for the situation where either equations (8.50) or (8.51) apply for selection on a single population. That is, the phenotype $A_1 A_-$ has a constant selective value of $1 - s = \exp(u)$ relative to $A_2 A_2$. The selection parameter u is to be estimated on the basis of a series of samples taken from generations $t_1, t_2, \ldots, t_m$ of the population.

Let n_{1i} denote the number of $A_1 A_-$ individuals and n_{2i} denote the number of $A_2 A_2$ individuals in a sample of size n_i taken before selection in generation t_i. Then the class probability associated with the observed frequency n_{ki} is given by equations (8.50) to be

$$f_{ki} = \begin{cases} (1 - q_i^2)n_i/n, & k = 1 \\ q_i^2\, n_i/n, & k = 2 \end{cases} \tag{A46}$$

taking n_i/n as the 'probability' of an individual being in the ith sample, where $n = \sum n_i$ is the total number of individuals sampled. It is these class probabilities that are evaluated by FREQ. Here there are only two parameters to be estimated, the selection parameter, u, and q_0, the proportion of the allele A_2 at the start of generation 0. The gene frequencies $q_1, q_2, q_3, \ldots$ in other generations can be determined using equation (8.46), taking $s = 1 - \exp(u)$.

If samples are taken after selection in each generation, equations (8.50) have to be replaced by equations (8.51). Then equation (A46) changes to

$$f_{ki} = \begin{cases} \{(1-q_i^2)n_i/n\}(1-s)/\{1-s(1-q_i^2)\}, & k = 1 \\ \{q_i^2 n_i/n\}/\{1-s(1-q_i^2)\}, & k = 2 \end{cases}$$

Samples after selection are indicated by $K(2) = 1$ in FREQ.

As usual, the number of degrees of freedom for a chi-square goodness of fit test is the number of observed and expected frequencies being compared, minus the number of totals that agree, minus the number of estimated parameters. In the present situation there are $2m$ phenotype frequencies, observed and expected total frequencies agree for each of the m samples, and there are two estimated parameters. Hence the degrees of freedom are $2m - m - 2 = m - 2$.

Table A8 Example input for MAXLIK using version (b) of the FREQ subroutine. The original data come from Table 8.13.

2	2	10		12						
U	Q0									
LOCATION D40 1926–75 WITH SAMPLES BEFORE SELECTION										
10	165	13	90	5	50	10	88	11	81	
0.0	.971									
5	0	0	11.3	12.3	14.3	16.3	175	103	55	98	92
LOCATION D40 1926–75 WITH SAMPLES AFTER SELECTION										
10	165	13	90	5	50	10	88	11	81	
0.0	.971									
5	1	0	11.3	12.3	14.3	16.3	175	103	55	98	92

Table A8 shows some example input data for MAXLIK with this version of FREQ. This is for the estimation of u for location D40 of the data shown in Table 8.13. The constants $K(1), K(2), \ldots, K(12)$ are explained in the comment given at the start of the subroutine. Initial approximations for the parameters have been set at $\hat{u} \simeq 0$ and $\hat{q}_0 \simeq \sqrt{(n_{21}/n_1)} = \sqrt{(165/175)} = 0.971$. (The first sample is taken at time $t_1 = 0$.) Assuming samples before selection, MAXLIK gives $\hat{u} = 0.0496$, with standard error 0.0256, and $\hat{q}_0 = 0.9695$, with standard error 0.0090, after five iterations. Assuming samples after selection, $\hat{u}$ and its standard error are unchanged while $\hat{q}_0$ becomes 0.9706, with standard error 0.0092.

Listing for version (b) of the FREQ subroutine

```
      SUBROUTINE FREQ(X,F,K)
C
C     THIS VERSION IS FOR ESTIMATING A CONSTANT SELECTIVE VALUE FOR
C     ONE LOCATION, USING CLARKE AND MURRAY'S DOMINANCE SELECTION
C     MODEL.
C
C     X(1)=LOG(1-S), X(2)=Q0.
C     K(1)=NUMBER OF SAMPLES=NS, K(2)=1 FOR SAMPLES AFTER SELECTION,
C     K(3)=T(1), K(4)=T(2), ..., K(NS+2)=T(NS), K(NS+3)=1ST SAMPLE
C     SIZE, K(NS+4)=2ND SAMPLE SIZE, ...,K(2NS+2)=LAST SAMPLE SIZE.
C
      REAL X(50),F(500),K(500),GA(50),Q(50),QS(50),T(50)
      DATA GA/50*0.0/, T/50*0.0/, NS,NG,NSM,NSP/4*0/
C
C**** INITIAL SET-UP (ONLY ON FIRST ENTRY)
      IF (K(1).EQ.0.0) GO TO 1
      NS=K(1)
      K(1)=0.0
      SUM=0.0
      DO 2 I=1,NS
      T(I)=K(I+2)
      GA(I)=K(NS+2+I)
    2 SUM=SUM+GA(I)
      DO 3 I=1,NS
    3 GA(I)=GA(I)/SUM
      NG=INT(T(NS))+2
      NSM=NS-1
      NSP=NS+1
C
C**** SORT OUT PARAMETERS
    1 S=1.0-EXP(X(1))
      Q0=X(2)
C
C**** CALCULATE GENE FREQUENCIES
      Q(1)=Q0
      DO 4 J=2,NG
      Q2=Q(J-1)**2
    4 Q(J)=Q(J-1)+S*Q2*(1.0-Q(J-1))/(1.0-S*(1.0-Q2))
      DO 5 I=1,NS
      IT=INT(T(I))+1
      DT=T(I)-IT+1
    5 QS(I)=(1.0-DT)*Q(IT)+DT*Q(IT+1)
      IF (K(2).EQ.1) GO TO 10
C
C**** CALCULATE PROBABILITIES ASSUMING SAMPLES BEFORE SELECTION
      DO 6 I=1,NS
      J=2*(I-1)
      Q2=QS(I)**2
      F(J+1)=GA(I)*(1.0-Q2)
    6 F(J+2)=GA(I)*Q2
      GO TO 12
C
C**** CALCULATE PROBABILITIES ASSUMING SAMPLES AFTER SELECTION
   10 DO 11 I=1,NS
      J=2*(I-1)
      Q2=QS(I)**2
```

```
      A2A2=Q2/(1.0-S*(1.0-Q2))
      F(J+1)=GA(I)*(1.0-A2A2)
   11 F(J+2)=GA(I)*A2A2
   12 DO 13 I=1,2*NS
   13 IF (F(I).LT.0.000001) F(I)=0.000001
      RETURN
      END
```

A.19 A COMPUTER PROGRAM FOR THE EWENS–WATTERSON TEST

Stewart (1977) has developed a fast algorithm for simulating the distribution of the homozygosity statistic $\hat{F} = \sum n_i^2/n^2$ for samples from the Ewens sampling distribution of equation (9.8). This algorithm was used to obtain the percentage points shown in Table 9.3. It can be used generally to test the significance of observed values of the test statistic. The FORTRAN program EWENS given below is based upon this algorithm. Input to the program is described on initial comment cards.

Program listing

```
C
C     ****************
C     * PROGRAM EWENS *
C     ****************
C
C     PROGRAM TO GENERATE THE DISTRIBUTION OF THE EWENS-WATTERSON TEST
C     STATISTIC BY SIMULATION USING STEWART'S (1977) ALGORITHM.  NOTE
C     THAT A FUNCTION RANDOM MAY NEED TO BE ADDED FOR RUNNING ON
C     MACHINES OTHER THAN BURROUGHS.  THE PROGRAM EWENS WAS WRITTEN BY
C     B.F.J. MANLY, UNIVERSITY OF OTAGO, DUNEDIN, NEW ZEALAND.  THIS
C     VERSION IS DATED APRIL, 1985.
C
C     *************
C     * DATA CARD *
C     *************
C
C        COLS  1- 5  K, NUMBER OF ALLELES (MAX=20)
C              6-10  N, SAMPLE SIZE (MAX=2000)
C             11-15  NSIM, NUMBER OF SIMULATIONS WANTED (MAX=2000)
C             16-20  ISEED, THE RANDOM NUMBER STARTER IN THE RANGE 1 TO
C                    99999 FOR THE BURROUGHS RANDOM NUMBER GENERATOR
C
      REAL F(2000),X(20),SUMX(20)
      INTEGER MEANX(20)
C
C**** READ DATA
      READ(5,1000) K,N,NSIM,ISEED
 1000 FORMAT(16I5)
      WRITE(6,9000) K,N,NSIM,ISEED
 9000 FORMAT(' SIMULATION OF THE DISTRIBUTION OF THE EWENS-WATTERSON TES
     1T STATISTIC'//
     2' NUMBER OF ALLELES (K) = ',I2,'    SAMPLE SIZE (N) = ',I4,'
     3NUMBER OF VALUES GENERATED = ',I4,'    RANDOM NUMBER STARTER = ',
     4I5/)
```

```
C
C**** GENERATE AND SORT NSIM VALUES OF F
      SUMF=0.0
      SSF=0.0
      DO 1 I=1,K
    1 SUMX(I)=0.0
      DO 2 ISIM=1,NSIM
      CALL EWENS(K,N,ISEED,F(ISIM),X)
      SUMF=SUMF+F(ISIM)
      SSF=SSF+F(ISIM)**2
      CALL SORT(X,20,1,K,1)
      DO 2 I=1,K
    2 SUMX(I)=SUMX(I)+X(I)
      CALL SORT(F,2000,1,NSIM,1)
C
C**** CALCULATE MEANS FOR ALLELE FREQUENCIES AND THE MEAN AND SD OF F
      FMEAN=SUMF/NSIM
      FSD=SQRT((SSF-SUMF*SUMF/NSIM)/(NSIM-1.0))
      DO 3 I=1,K
    3 MEANX(I)=SUMX(I)/NSIM
C
C**** PRINT RESULTS
      WRITE(6,9001) (I,F(I), I=1,NSIM)
 9001 FORMAT(' HOMOZYGOSITY (F) VALUES OBTAINED, RANKED IN ORDER'/
     1'    RANK    VALUE    RANK    VALUE    RANK    VALUE    RANK    VALUE
     2    RANK    VALUE    RANK    VALUE    RANK    VALUE    RANK    VALUE'/
     3('  ',8(I8,F8.5)))
      WRITE(6,9002) FMEAN,FSD,(MEANX(I), I=1,K)
 9002 FORMAT(///' MEAN OF F = ',F8.5,'    STD DEV OF F = ',F8.5/
     1///' MEAN VALUES OF ALLELE FREQUENCIES WHEN THEY ARE RANKED IN ORD
     2ER IN EACH SAMPLE'//'  ',2016)
      STOP
      END


C
C
C
C
C     ******************************************************************
C     *SUBROUTINE EWENS TO GENERATE ONE VALUE OF F FOR THE GIVEN VALUES*
C     *OF K AND N.  RANDOM(ISEED) PRODUCES A RANDOM VALUE IN THE RANGE *
C     *(0,1) ON THE BURROUGHS COMPUTERS.  A FUNCTION FOR HAVE TO BE    *
C     * MADE AVAILABLE FOR THIS ON OTHER MACHINES.                     *
C     ******************************************************************
C
      SUBROUTINE EWENS(K,N,ISEED,F,X)
C
      REAL B(20,2000),A(19),X(20)
      DATA IND/0/, B/40000*0.0/
C
C**** DO INITIAL SETTING UP ON FIRST ENTRY ONLY
      IF (IND.EQ.1) GO TO 10
      IND=1
      DO 1 J=1,N
    1 B(1,J)=1.0/J
      DO 2 I=2,K
      B(I,I)=1.0
      NM=N-1
      DO 2 J=I,NM
    2 B(I,J+1)=(I*B(I-1,J)+J*B(I,J))/(J+1)
```

```
C
C**** GENERATE K-1 RANDOM NUMBERS
   10 KM=K-1
      DO 11 I=1,KM
   11 A(I)=RANDOM(ISEED)
C
C**** FIND X(1), X(2), ..., X(K), THE SAMPLE ALLELE FREQUENCIES
      NLEFT=N
      DO 14 L=1,KM
      CUM=0.0
      DO 12 I=1,NLEFT
      CUM=CUM+B(K-L,NLEFT-I)/(B(K-L+1,NLEFT)*I)
      IF (CUM.GE.A(L)) GO TO 13
   12 CONTINUE
   13 X(L)=I
   14 NLEFT=NLEFT-X(L)
      X(K)=NLEFT
C
C**** CALCULATE THE SAMPLE HOMOZYGOSITY F
      F=0.0
      DO 15 I=1,K
   15 F=F+(X(I)/N)**2
      RETURN
      END
C
C     ****************
C     *SUBROUTINE SORT*
C     ****************
C

      (This program should include the subroutine SORT that is included
      with the program HABITATS above.)
```

References

Adams, J. and Ward, R. H. (1973). Admixture studies and the detection of selection. *Science* **180**: 1137–43.

Aird, I., Bentall, H. H. and Roberts, J. A. F. (1953). A relationship between cancer of stomach and the ABO blood groups. *Brit. Med. J.* **i**: 799–801.

Aird, I., Bentall, H. H., Mehigan, J. A. and Roberts, J. A. F. (1954). The blood groups in relation to peptic ulceration and carcinoma of colon, rectum, breast and bronchus. *Brit. Med. J.* **ii**: 315–21.

Allendorf, F. W. (1983). Linkage disequilibrium generated by selection against null alleles at duplicate loci. *Amer. Nat.* **121**: 588–92.

Alvarez, G., Santos, M. and Zapata, C. (1983). Selection at sex-linked loci. 1. A method of estimating total fitnesses. *Heredity* **50**: 147–57.

Anderson, S., Auquier, A., Hauck, W. W., Oakes, D., Vandaele, W. and Weisberg, H. I. (1980). *Statistical Methods for Comparative Studies*. Wiley, New York.

Anderson, T. W. (1958). *An Introduction to Multivariate Statistical Analysis*. Wiley, New York.

Anxolabehere, D., Goux, J. M. and Periquet, G. (1982). A bias in estimation of viabilities from competition experiments. *Heredity* **48**: 271–82.

Arnold, S. J. and Wade, M. J. (1984a). On the measurement of natural and sexual selection: theory. *Evolution* **38**: 709–19.

Arnold, S. J. and Wade, M. J. (1984b). On the measurement of natural and sexual selection: applications. *Evolution* **38**: 720–34.

Arnason, A. N. and Baniuk, L. (1978). *POPAN-2, A Data Maintenance and Analysis System for Mark–Recapture Data*. Charles Babbage Research Centre, Manitoba.

Arthur, W. (1978). Morph frequency and coexistence in *Cepaea*. *Heredity* **41**: 335–46.

Arthur, W. (1980). Further associations between morph frequencies and coexistence in *Cepaea*. *Heredity* **44**: 417–21.

Arthur, W. (1982a). The evolutionary consequences of interspecific competition. *Adv. Ecol. Res.* **12**: 127–87.

Arthur, W. (1982b). A critical examination of the case for competitive selection in *Cepaea*. *Heredity* **48**: 407–19.

Avise, J. C. and Aquadro, C. F. (1982). A comparative summary of genetic distances in vertebrates: patterns and correlations. *Evol. Biol.* **15**: 151–85.

Ayala, F. J. (1982). Genetic variation in natural populations: problem of electrophoretically cryptic alleles. *Proc. Nat. Acad. Sci., USA* **79**: 550–4.

Ayala, F. J. and Gilpin, M. E. (1974). Gene frequency comparisons between taxa: support for the natural selection of protein polymorphisms. *Proc. Nat. Acad. Sci., USA* **71**: 4847–49.

Ayala, F. J. and Tracey, M. L. (1974). Genetic differentiation within and between species of the *Drosophila willistoni* group. *Proc. Nat. Acad. Sci., USA* **71**: 999–1003.

Baker, A. J. (1980). Morphometric differentiation in New Zealand populations of the house sparrow (*Passer domesticus*). *Evolution* **34**: 638–53.

Baker, M. C. and Fox, S. F. (1978). Differential survival in common grackles sprayed with turgitol. *Amer. Nat.* **112**: 675–82.

Baker, W. K. (1975). Linkage disequilibrium over space and time in natural populations of *Drosophila montana*. *Proc. Nat. Acad. Sci., USA* **72**: 4095–9.

Baker, W. K. (1983). Reply to Allendorf. *Amer. Nat.* **121**: 593–4.

Baker, W. K. and Kaeding, E. A. (1981). Linkage disequilibrium at the alpha-esterase loci in a population of *Drosophila montana* from Utah. *Amer. Nat.* **117**: 804–9.

Bantock, C. R. (1974). Experimental evidence for non-visual selection by *Cepaea nemoralis*. *Heredity* **33**: 409–12.

Bantock, C. R. and Bayley, J. A. (1973). Visual selection for shell size in *Cepaea* (Held.). *J. Anim. Ecol.* **42**: 247–61.

Bantock, C. R., Bayley, J. A. and Harvey, P. H. (1976). Simultaneous selective predation on two features of a mixed sibling species population. *Evolution* **29**: 636–49.

Bantock, C. R. and Ratsey, M. (1980). Natural selection in experimental populations of the landsnail *Cepaea nemoralis* (L.). *Heredity* **44**: 37–54.

Barker, J. S. F. (1981). Selection at allozyme loci in cactophilic *Drosophila*. In *Genetic Studies of Drosophila Populations* (eds J. B. Gibson and J. G. Oakeshott), pp. 161–84. Aust. Nat. Univ. Press, Canberra.

Barker, J. S. F. and East, P. D. (1980). Evidence for selection following perturbation of allozyme frequencies in a natural population of *Drosophila*. *Nature* **284**: 166–8.

Bateman, A. J. (1949). Analysis of data on sexual isolation. *Evolution* **3**: 174–7.

Beardmore, J. A. and Karimi-Booshehri, F. (1983). ABO genes are differentially distributed in socio-economic classes in England. *Nature* **303**: 522–4.

Bedall, F. K. and Zimmermann, H. (1976). On the generation of multivariate normal distributed random vectors by N(0, 1) distributed random numbers. *Biom. J.* **18**: 467–71.

Begon, F. J. (1979). *Investigating Animal Abundance*. Edward Arnold, London.

Bell, G. (1974). The reduction of morphological variation in natural populations of smooth newt larvae. *J. Anim. Ecol.* **43**: 115–28.

Bernstein, P. (1931). *Comitato Italiano per lo Studio dei Problemi della Populazione*. Istituto Poligrafico deli Stato, Rome.

Berry, R. J., Bonner, W. N. and Peters, J. (1979). Natural selection in house mice (*Mus musculus*) from South Georgia (south Atlantic Ocean). *J. Zool., Lond.* **189**: 385–98.

Berry, R. J. and Crothers, J. H. (1968). Stabilising selection in the dog whelk (*Nucella lapillus*). *J. Zool., Lond.* **155**: 5–17.

Berry, R. J. and Crothers, J. H. (1970). Genotypic stability and physiological tolerance in the dog whelk (*Nucella lapillus*). *J. Zool., Lond.* **162**: 293–302.

Bishop, J. A. (1972). An experimental study of the cline of industrial melanism in *Biston betularia* (L.) (Lepidoptera) between urban Liverpool and rural North Wales. *J. Anim. Ecol.* **41**: 209–43.

Bishop, J. A., Cook, L. M. and Muggleton, J. (1978). The response of two species of moth to industrialisation in Northwest England. *Phil. Trans. Roy. Soc. Lond.* **B281**: 489–542.

Blower, J. G., Cook, L. M. and Bishop, J. A. (1981). *Estimating the Size of Animal*

Populations. George Allen and Unwin, London.

Blumberg, B. S. and Hesser, J. E. (1971). Loci differentially affected by selection in two American black populations. *Proc. Nat. Acad. Sci., USA* **68**: 2554–8.

Boag, P. T. and Grant, P. R. (1981). Intense natural selection in a population of Darwin's finches (*Geospizinae*) in the Galapagos. *Science* **214**: 82–5.

Borowsky, R. (1977). Detection of the effects of selection on protein polymorphisms in natural populations by means of distance analysis. *Evolution* **31**: 341–6. (Errata in *Evolution* **31**: 648.)

Brakefield, P. (1979). Spot-number in *Maniola jurtina* – variation between generations and selection in marginal populations. *Heredity* **42**: 259–66.

Brower, L. P., Brower, J. V. Z., Stiles, F. G., Croze, H. J. and Hower, A. S. (1964). Mimicry: differential advantage of colour patterns in the natural environment. *Science* **144**: 183–5.

Brower, L. P., Cook, L. M. and Croze, H. J. (1967). Predator responses to artificial Batesian mimics released in a neotropical environment. *Evolution* **21**: 11–23.

Brown, A. H. D. (1975). Sample sizes required to detect linkage disequilibrium between two or three loci. *Theor. Pop. Biol.* **8**: 184–201.

Brown, A. H. D. and Feldman, M. W. (1981). Population structure of multilocus associations. *Proc. Nat. Acad. Sci., USA* **78**: 5913–6.

Brown, A. H. D., Feldman, M. W. and Nevo, E. (1980). Multilocus structure of natural populations of *Hordeum spontaneum*. *Genetics* **96**: 523–36.

Brown, A. H. D. and Marshall, D. R. (1981). Evolutionary changes accompanying colonisation in plants. In *Evolution Today, proceedings of the 2nd International Congress of Systematic and Evolutionary Biology*. (eds G. G. E. Scudder and J. L. Reveal), pp 351–63. Hunt Institute for Botanical Documentation, Carnegie-Mellon University.

Brown, A. H. D., Marshall, D. R. and Albrecht, L. (1975). Profiles of electrophoretic alleles in natural populations. *Genet. Res., Camb.* **25**: 137–43.

Brown, A. H. D., Marshall, D. R. and Weir, B. S. (1981). Current status of the charge state model for protein polymorphism. In *Genetic Studies of Drosophila Populations* (eds J. B. Gibson and J. G. Oakshott), pp. 15–43. Aust. Nat. Univ. Press, Canberra.

Brown, W. L. and Wilson, E. O. (1956). Character displacement. *Syst. Zool.* **5**: 49–64.

Brownie, C., Anderson, D. R., Burnham, K. P. and Robson, D. S. (1978). *Statistical Inference from Band Recovery Data – a Handbook*. Fish and Wildlife Service, Resource Publ. 131, US Dept. of the Interior, Washington, DC.

Bryant, E. H., Kence, A. and Kimball, K. T. (1980). A rare-male advantage induced in the housefly by wing clipping and some general considerations about *Drosophila*. *Genetics* **96**: 975–93.

Buchanan, J. A. and Higley, E. T. (1921). The relationship of blood groups to disease. *Brit. J. Exp. Path.* **2**: 247–55.

Bumpus, H. C. (1898). The elimination of the unfit as illustrated by the introduced sparrow *Passer domesticus*. *Biol. Lect. from the Marine Lab., Woods Hole*, 11th Lecture, pp. 209–26.

Bungaard, J. and Christiansen, F. B. (1972). Dynamics of polymorphisms: I. Selection components in an experimental population of *Drosophila melanogaster*. *Genetics* **71**: 439–460.

Cain, A. J. and Currey, J. D. (1968). Studies on *Cepaea*. III. Ecogenetics of a population of *Cepaea nemoralis* (L.) subject to strong area effects. *Phil. Trans. Roy. Soc. Lond.* **B253**: 447–82.

Cain, A. J. and Sheppard, P. M. (1950). Selection in the polymorphic land snail *Cepaea nemoralis*. *Heredity* **4**: 275–94.

Cain, A. J. and Sheppard, P. M. (1954). Natural selection in *Cepaea*. *Genetics* **39**: 89–116.

Cameron, R. A. D. and Williamson, P. (1977). Estimating migration and the effects of disturbance in mark-recapture studies on the snail *Cepaea nemoralis* (L.). *J. Anim. Ecol.* **46**: 173–9.

Carter, M. A. (1967). Selection in mixed colonies of *Cepaea nemoralis* and *Cepaea hortensis*. *Heredity* **22**: 117–39.

Carter, M. A. (1968). Thrush predation of an experimental population of the snail *Cepaea nemoralis* (L.). *Proc. Linn. Soc.* **179**: 241–9.

Causton, D. R. (1977). *A Biologist's Mathematics*. Edward Arnold, London.

Cavalli, L. L. (1950). The analysis of selection curves. *Biometrics* **6**: 208–20.

Cavalli-Sforza, L. L. and Bodmer, W. F. (1972). *The Genetics of Human Populations*. W. H. Freeman and Co., San Francisco.

Ceppellini, R., Siniscalco, M. and Smith, C. A. B. (1955). The estimation of gene frequencies in a randomly-mating population. *Ann. Hum. Genet.* **20**: 97–115.

Chakraborty, R., Fuerst, P. A. and Nei, M. (1978). Statistical studies on protein polymorphisms in natural populations. II. Gene differentiation between populations. *Genetics* **88**: 367–90.

Chakraborty, R., Fuerst, P. A. and Nei, M. (1980). Statistical studies on protein polymorphisms in natural populations. III. Distribution of allele frequencies and the number of alleles per locus. *Genetics* **94**: 1039–63.

Chakraborty, R. and Griffiths, R. C. (1982). Correlation of heterozygosity and the number of alleles in different frequency classes. *Theor. Pop. Biol.* **21**: 205–18.

Charlesworth, B. (1976). Natural selection in age-structured populations. *Lect. Math. Life Sci.* (Amer. Math. Soc.) **8**: 69–87.

Charlesworth, B. and Charlesworth, D. (1973). The measurement of fitness and mutation rate in human populations. *Ann. Hum. Genet.* **37**: 175–87.

Chen, K. (1979). Moment analysis of the probit model for directional selection. *Biom. J.* **21**: 773–9.

Chesson, J. (1976). A non-central multivariate hypergeometric distribution arising from biased sampling with application to selective predation. *J. Appl. Prob.* **13**: 795–7.

Chesson, J. (1978). Measuring preference in selective predation. *Ecology* **59**: 211–5.

Chesson, J. (1983). The estimation and analysis of preference and its relationship to foraging models. *Ecology* **64**: 1297–304.

Chiang, C. L. (1960). A stochastic study of the life table and its applications. I. Probability distributions of the biometric functions. *Biometrics* **16**: 618–635.

Chinnici, J. P. and Sansing, R. C. (1977). Mortality rates, optimal and discriminating birthweights between white and nonwhite single births in Virginia (1955–1973). *Hum. Biol.* **49**: 335–48.

Christiansen, F. B. (1977). Population genetics of *Zoarces viviparus* (L.): a review. In *Measuring Selection in Natural Populations* (eds F. B Christiansen and T. M. Fenchel), pp. 21–47. Springer-Verlag, Berlin.

Christiansen, F. B. (1980). Studies on selection components in natural populations using samples of mother–offspring combinations. *Hereditas* **92**: 199–203.

Christiansen, F. B., Bungaard, J. and Barker, J. S. F. (1977). On the structure of fitness estimates under post-observational selection. *Evolution* **31**: 843–53.

Christiansen, F. B. and Frydenberg, O. (1973). Selection component analysis of natural polymorphisms using population samples including mother–child combinations. *Theor. Pop. Biol.* **4**: 425–45.

Christiansen, F. B. and Frydenberg, O. (1974). Geographical patterns of four polymorphisms in *Zoarces viviparus* as evidence of selection. *Genetics* **77**: 765–70.

Christiansen, F. B. and Frydenberg, O. (1976). Selection component analysis of natural polymorphisms using mother–offspring samples of successive cohorts. In *Population Genetics and Evolution* (eds S. Karlin and E. Nevo), pp. 277–301. Academic Press, New York.

Christiansen, F. B., Frydenberg, O., Gyldenholm, A. O. and Simonsen, V. (1974). Genetics of *Zoarces* populations. VI. Further evidence, based upon age group samples, of a heterozygote deficit in the Est III polymorphism. *Hereditas* **77**: 225–36.

Christiansen, F. B., Frydenberg, O. and Simonsen, V. (1973). Genetics of *Zoarces* populations. IV. Selection component analysis of an esterase polymorphism using population samples including mother–offspring combinations. *Hereditas* **73**: 291–304.

Christiansen, F. B., Frydenberg, O. and Simonsen, V. (1977). Genetics of *Zoarces* populations. X. Selection component analysis of the Est III polymorphism using samples of successive cohorts. *Hereditas* **87**: 129–50.

Clarke, A. G., Feldman, M. W. and Christiansen, F. B. (1981). The estimation of epistasis in components of fitness in experimental populations of *Drosophila melanogaster*. 1. A two-stage maximum likelihood model. *Heredity* **46**: 321–46.

Clarke, A. G. and Feldman, M. W. (1981). The estimation of epistasis in components of fitness in populations of *Drosophila melanogaster*. 2. Assessment of meiotic drive, viability, fecundity and sexual selection. *Heredity* **46**: 347–77.

Clarke, B. (1960). Divergent effects of natural selection on two closely-related polymorphic snails. *Heredity* **14**: 423–43.

Clarke, B. (1962a). Natural selection in mixed population of polymorphic snails. *Heredity* **17**: 319–45.

Clarke, B. (1962b). Balanced polymorphism and the diversity of sympatric species. In *Taxonomy and Geography* (ed. D. Nichols), pp. 47–70. Syst. Assoc., Oxford, Publ. 4.

Clarke, B. (1969). The evidence for apostatic selection. *Heredity* **24**: 347–52.

Clarke, B. (1975). The contribution of ecological genetics to evolutionary theory: detecting the direct effects of natural selection on particular polymorphic loci. *Genetics* **79**: 101–13.

Clarke, B. (1979). The evolution of genetic diversity. *Proc. Roy. Soc. Lond.* **B205**: 453–74.

Clarke, B., Arthur, W., Horsley, D. T. and Parkin, D. T. (1978). Genetic variation and natural selection in pulmonate molluscs. In *The Pulmonates* (ed. J. Peake), pp. 219–70. Academic Press, London.

Clarke, B. and Murray, J. (1962a). Change in gene frequency of *Cepaea nemoralis* (L.). *Heredity* **17**: 445–465.

Clarke, B. and Murray, J. (1962b). Changes of gene frequency in *Cepaea nemoralis*

456 Statistics of natural selection

(L.): the estimation of selective values. *Heredity* **17**: 467–76.

Clarke, C. A. and Sheppard, P. M. (1966). A local survey of the distribution of the moth *Biston betularia* and estimates of the selective values of these in an industrial environment. *Proc. Roy. Soc. Lond.* **B165**: 424–39.

Clegg, M. T., Kahler, A. L. and Allard, R. W. (1978). Estimation of life cycle components of selection in an experimental plant population. *Genetics* **89**: 765–92.

Cleghorn, T. E. (1960). MNSs gene frequencies in English blood donors. *Nature* **187**: 701.

Cobbs, G. (1979). A model allowing continuous variation in electrophoretic mobility of neutral alleles. *Genetics* **92**: 669–78.

Cobbs, G. and Prakash, S. (1977). An experimental investigation of the unit charge model of protein polymorphism and its relationship to the esterase-5 locus of *Drosophila pseudoobscura, D. persimilis* and *D. miranda. Genetics* **87**: 717–42.

Cock, M. J. W. (1978). The assessment of preference. *J. Anim. Ecol.* **47**: 805–16.

Cockerham, C. C. and Weir, B. S. (1977). Digenic descent measures in finite populations. *Genet. Res., Camb.* **30**: 121–47.

Confer, J. L., Applegate, G. and Evanik, C. A. (1980). Selective predation by zooplankton and the response of Cladoceran eyes to light. In *Evolution and Ecology of Zooplankton Communities* (ed. W. C. Kerfoot), pp. 604–8. Univ. Press of New England, Hanover, New Hampshire.

Conroy, B. A. and Bishop, J. A. (1980). Maintenance of the polymorphism for melanism in the moth *Phigalia pilosaria* in rural north Wales. *Proc. Roy. Soc. Lond.* **B210**: 285–98.

Cook, D. G. and Pocock, S. J. (1983). Multiple regression in geographical mortality studies, with allowance for spatially correlated errors. *Biometrics* **39**: 361–71.

Cook, L. M. (1969). An experiment on selection for mimicry. *Entomologist* **102**: 107–13.

Cook, L. M. (1971). *Coefficients of Natural Selection.* Hutchinson, London.

Cook, L. M. (1979). Variation in the Madeiran lizard *Lacerta dugesii. J. Zool., Lond.* **187**: 327–40.

Cook, L. M., Brower, L. P. and Alcock, J. (1969). An attempt to verify mimetic advantage in a neotropical environment. *Evolution* **23**: 339–45.

Cook, L. M. and Mani, G. S. (1980). A migration-selection model for the morph frequency variation in the peppered moth over England and Wales. *Biol. J. Linn. Soc.* **13**: 179–98.

Cook, L. M. and Miller, P. (1977). Density dependent selection on polymorphic prey – some data. *Amer. Nat.* **111**: 594–8.

Cook, L. M. and O'Donald, P. (1971). Shell size and natural selection in *Cepaea nemoralis.* In *Ecological Genetics and Evolution* (ed. R. Creed), pp. 93–108. Blackwell, Oxford.

Cormack, R. M. (1964). Estimates of survival from the sighting of marked animals. *Biometrics* **51**: 429–38.

Cormack, R. M. (1970). Statistical appendix. *J. Anim. Ecol.* **39**: 24–7.

Coulson, J. C. (1960). A study of the mortality of the starling based on ringing recoveries. *J. Anim. Ecol.* **29**: 251–71.

Cox, D. R. (1972). Regression models and life tables. *J. Roy. Statist. Soc.* **B34**: 187–220.

Crawford, M. H., Gonzalez, N. L., Schanfield, M. S., Dykes, D. D., Skradski, K. and

Polesky, H. F. (1981). The Black Caribs (Garifuna) of Livingston, Guatemala: genetic markers and admixture estimates. *Hum. Biol.* **53**: 87–103.

Creed, E. R. (1975). Melanism in the two-spot ladybird: the nature and intensity of selection. *Proc. Roy. Soc. Lond.* **B190**: 135–48.

Crosbie, S. F. and Manly, B. F. J. (1982). Capture–recapture models with restrictions on parameters. *Proc. Amer. Statist. Assoc., Statist. Comp. Sect.*, Detroit, August 1981, pp. 33–40.

Crosbie, S. F. and Manly, B. F. J. (1985). Parsimonious modelling of capture–mark–recapture studies. *Biometrics.* (to appear)

Crow, J. F. (1958). Some possibilities for measuring selection intensities in man. *Hum. Biol.* **30**: 1–13.

Crow, J. F. (1961). Population genetics. *Amer. J. Hum. Genet.* **13**: 137–50.

Curtis, C. F., Cook, L. M. and Wood, R. J. (1978). Selection for and against insecticide resistance and possible methods for inhibiting the evolution of resistance in mosquitoes. *Ecol. Ent.* **3**: 273–87.

Czelusniak, J., Goodman, M., Hewett-Emmett, D., Weiss, M. L., Venta, P. J. and Tashian, R. E. (1982). Phylogenetic origins and adaptive evolution of avian and mammalian haemoglobin genes. *Nature* **298**: 297–300.

Darwin, C. R. (1859). *The Origin of Species by Natural Selection, or the Preservation of Favoured Races in the Struggle for Life*, 1st Edit. John Murray, London. Reprinted by Watts, London, 1950.

Darwin, E. (1794). *Zoonomia, or the Laws of Organic Life.* J. Johnson, London.

Dawkins, R. (1982). *The Extended Phenotype.* W. H. Freeman, Oxford.

De Benedictis, P. A. (1977). Studies in the dynamics of genetically variable populations. I. Frequency and density-dependent selection in experimental populations of *Drosophila melanogaster*. *Genetics* **87**: 343–56.

De Benedictis, P. A. (1978). Are populations characterised by their genes or by their genotypes? *Amer. Nat.* **112**: 155–75.

Di Cesnola, A. P. (1906). A first study of natural selection in *Helix arbustorum* (Helicogena). *Biometrika* **5**: 387–99.

Dietz, E. J. (1983). Permutation tests for association between two distance matrices. *Syst. Zool.* **32**: 21–6.

Dixon, W. J. (ed.) (1981). *BMDP Statistical Software, 1981*. Univ. of California Press, Los Angeles.

Dixon, W. J. and Massey, F. J. (1969). *Introduction to Statistical Analysis*. McGraw-Hill, New York.

Dobzhansky, Th. (1943). Genetics of natural populations. IX. Temporal changes in the composition of populations of *Drosophila pseudoobscura*. *Genetics* **28**: 162–86.

Dobzhansky, Th. (1970). *Genetics of the Evolutionary Process.* Columbia Univ. Press, New York.

Dobzhansky, Th. and Epling, C. (1944). Contributions to the genetics, taxonomy and ecology of *Drosophila pseudoobscura* and its relatives. Carnegie Inst., Washington, DC, Publ. 554, pp. 47–144.

Dobzhansky, Th. and Levene, H. (1951). Development of heterosis through natural selection in experimental population of *Drosophila pseudoobscura*. *Amer. Nat.* **85**: 247–64.

Dobzhansky, Th. and Pavlovsky, O. (1953). Indeterminate outcome of certain experiments on *Drosophila* populations. *Evolution* **7**: 198–210.

Douglas, M. E. and Endler, J. A. (1982). Quantitative matrix comparison in ecological and evolutionary investigations. *J. Theor. Biol.* **99**: 777–95.

Du Mouchel, W. H. and Anderson, W. W. (1968). The analysis of selection in experimental populations. *Genetics* **58**: 435–49.

Edwards, J. H. (1965). The meaning of the association between blood groups and disease. *Ann. Hum. Genet., Lond.* **29**: 77–83.

Ehrlich, P. R. and White, R. R. (1980). Colorado checkerspot butterflies: isolation, neutrality, and the biospecies. *Amer. Nat.* **115**: 328–41.

Ehrman, L. and Petit, C. (1968). Genotype frequency and mating success in the *willistoni* species group of *Drosophila*. *Evolution* **22**: 649–58.

Ehrman, L. and Probber, J. (1978). Rare *Drosophila* males: the mysterious matter of choice. *Amer. Scient.* **66**: 216–22.

Elandt-Johnson, R. C. and Johnson, N. L. (1980). *Survival Models and Data Analysis.* Wiley, New York.

Eldredge, N. and Gould, S. J. (1972). Punctuated equilibria: an alternative to phyletic gradualism. In *Models in Paleobiology* (ed. T. J. M. Schopf), pp. 82–115. Freeman and Cooper, San Francisco.

Elens, A. A. and Wattiaux, J. M. (1964). Direct observation of sexual isolation. *Drosophila Inform. Serv.* **39**: 118–9.

Elston, R. C. (1971). The estimation of admixture in racial hybrids. *Ann. Hum. Genet., Lond.* **35**: 9–17.

Elston, R. C. and Forthofer, R. (1977). Testing for Hardy–Weinberg equilibrium in small samples. *Biometrics* **33**: 536–42.

Ewens, W. J. (1972). The sampling theory of selectively neutral alleles. *Theor. Pop. Biol.* **3**: 87–112.

Ewens, W. J. (1977a). Population genetics theory in relation to the neutralist–selectionist controversy. *Adv. Hum. Genet.* **8**: 67–134.

Ewens, W. J. (1977b). Selection and neutrality. In *Measuring Selection in Natural Populations* (eds F. B. Christiansen and T. M. Fenchel), pp. 159–75. Springer-Verlag, Berlin.

Ewens, W. J. (1979a). Testing the generalised neutrality hypothesis. *Theor. Pop. Biol.* **15**: 205–16.

Ewens, W. J. (1979b). *Mathematical Population Genetics.* Springer-Verlag, Berlin.

Ewens, W. J. and Feldman, M. (1976). The theoretical assessment of selective neutrality. In *Population Genetics and Evolution* (eds S. Karlin and E. Nevo), pp. 303–37. Academic Press, New York.

Ewens, W. J. and Gillespie, J. H. (1974). Some simulation results for the neutral alleles model, with interpretations. *Theor. Pop. Biol.* **6**: 35–57.

Falconer, D. S. (1981). *Introduction to Quantitative Genetics*, 2nd Edit. Longman, London.

Felsenstein, J. (1977). Multivariate normal genetic models with a finite number of loci. In *Proc. Int. Conf. Quant. Genet.* (eds E. Pollak, O. Kempthorne and T. B. Bailey), pp. 227–46. Iowa State Univ. Press, Ames, Iowa.

Fenchel, T. (1975a). Factors determining the distribution patterns of mud snails (*hydrobiidae*). *Oecologia* **20**: 1–17.

Fenchel, T. (1975b). Character displacement and coexistence in mud snails (*hydrobiidae*). *Oecologia* **20**: 19–32.

Fisher, R. A. (1930). *The Genetical Theory of Natural Selection*. Clarendon Press, Oxford.

Fisher, R. A. (1939). Selective forces in wild populations of *Paratettix texanus*. *Ann. Eugen.* **9**: 109–22.

Fisher, R. A. (1970) *Statistical Methods for Research Workers*, 14th Edit. Oliver and Boyd, Edinburgh.

Fisher, R. A. and Ford, E. B. (1947). The spread of a gene in natural conditions in a colony of the moth *Panaxia dominula* L. *Heredity* **1**: 143–74.

Fisher, R. A. and Taylor, G. L. (1940). Scandinavian influence in Scottish ethnology. *Nature* **145**: 590.

Fitch, W. M. (1982). The challenges to Darwinism since the last centennial and the impact of molecular studies. *Evolution* **36**: 1133–43.

Fitch, W. M. and Langley, C. H. (1976a). Evolutionary rates in proteins: neutral mutations and the molecular clock. In *Molecular Anthropology* (eds M. Goodman and R. E. Tashian), pp. 197–218. Plenum Press, New York.

Fitch, W. M. and Langley, C. H. (1976b). Protein evolution and the molecular clock. *Federation Proc.* **35**: 2092–7.

Fleischer, R. C. and Johnston, R. F. (1982). Natural selection on body size and proportions in house sparrows. *Nature* **298**: 747–9.

Ford, E. B. (1945). Polymorphism. *Biol. Rev.* **20**: 73–88.

Ford, E. B. (1975). *Ecological Genetics*, 4th Edit. Chapman and Hall, London.

Ford, E. B. and Sheppard, P. M. (1969). The medionigra polymorphism of *Panaxia dominula*. *Heredity* **24**: 561–9.

Fordham, R. A. (1970). Mortality and population change of dominican gulls in Wellington, New Zealand. *J. Anim. Ecol.* **39**: 13–27.

Fox, S. F. (1975). Natural selection on morphological phenotypes of the lizard *Uta stansburiana*. *Evolution* **29**: 95–107.

Fuerst, P. A., Chakraborty, R. and Nei, M. (1977). Statistical studies on protein polymorphisms in natural populations. I. Distribution of single locus heterozygosity. *Genetics* **86**: 455–83.

Fuerst, P. A. and Ferrell, R. E. (1980). The stepwise mutation model. An experimental evaluation with haemoglobin variants. *Genetics* **94**: 185–201.

Fullick, T. G. and Greenwood, J. J. D. (1979). Frequency-dependent food selection in relation to two models. *Amer. Nat.* **113**: 762–5.

Galton, F. (1869). *Hereditary Genius*. Macmillan, London.

Galton, F. (1889). *Natural Inheritance*. Macmillan, London.

Gillespie, J. H. and Langley, C. H. (1979). Are evolutionary rates really variable? *J. Mol. Evol.* **13**: 27–34.

Glasgow, J. P. (1961). Selection for size in tsetse flies. *J. Anim. Ecol.* **30**: 87–94.

Gojobori, J., Ishii, K. and Nei, M. (1982). Estimation of average number of nucleotide substitutions when the rate of substitution varies with the nucleotide. *J. Mol. Evol.* **18**: 414–23.

Goodman, M., Weiss, M. L. and Czelusniak, J. (1982). Molecular evolution above the species level: branching pattern, rates and mechanisms. *Syst. Zool.* **31**: 376–99.

Gordon, A. D. (1981). *Classification*. Chapman and Hall, London.

Gordon, C. (1935). An experiment on a released population of *Drosophila melanogaster*. *Amer. Nat.* **69**: 381.

Gould, S. J. (1980). Is a new and general theory of evolution emerging? *Paleobiology* **6**: 119–30.

Gould, S. J. (1981). But not Wright enough: reply to Orzack. *Paleobiology* **7**: 131–4.

Gould, S. J. and Eldredge, N. (1977). Punctuated equilibria: the tempo and mode of evolution reconsidered. *Paleobiology* **3**: 115–51.

Gould, S. J. and Johnston, R. F. (1972). Geographical variation. *Ann. Rev. Syst. Ecol.* **3**: 457–98.

Goux, J. M. and Anxolabehere, D. (1980). The measurement of sexual isolation and selection: a critique. *Heredity* **45**: 255–62.

Grant, P. R. (1972). Centripetal selection and the house sparrow. *Syst. Zool.* **21**: 23–30.

Greenwood, J. J. D. and Elton, R. A. (1979). Analysing experiments on frequency-dependent selection by predators. *J. Anim. Ecol.* **48**: 721–37.

Greenwood, J. J. D., Wood, E. M. and Batchlor, S. (1981). Apostatic selection on distasteful prey. *Heredity* **47**: 27–34.

Hagen, D. W. and Gilbertson, L. G. (1973). Selective predation and the intensity of selection acting upon the lateral plates of threespine sticklebacks. *Heredity* **30**: 273–87.

Haldane, J. B. S. (1924). A mathematical theory of natural and artificial selection. *Trans. Camb. Phil. Soc.* **23**: 19–40.

Haldane, J. B. S. (1953). Some animal life tables. *J. Inst. Actuar.* **79**: 83–9.

Haldane, J. B. S. (1954a). The measurement of natural selection. *Caryologia* Vol. Suppl., 480–7.

Haldane, J. B. S. (1954b). An exact test for ramdomness of mating. *J. Genet.* **52**: 632–5.

Haldane, J. B. S. (1955). The calculation of mortality rates from ringing data. *Proc. 11th Int. Orn. Congr.*, pp. 454–8.

Haldane, J. B. S. (1956). The estimation and significance of the logarithm of the ratio of frequencies. *Ann. Hum. Genet.* **20**: 309–11.

Haldane, J. B. S. (1959). Natural selection. In *Darwin's Biological Work* (ed. P. R. Bell), pp. 101–49. Cambridge Univ. Press, London.

Harris, R. J. (1975). *A Primer on Multivariate Statistics.* Academic Press, New York.

Hartl, D. L., Burla, H. and Jungen, H. (1980). Can statistical tests of neutrality detect selection? *Genetica* **54**: 185–9.

Haviland, M. D. and Pitt, F. (1919). The selection of *Helix nemoralis* by the song thrush (*Turdus musicus*). *Ann. Mag. Nat. Hist.* **3**: 525–31.

Hayami, I. (1978). Notes on the rates and patterns of size change in evolution. *Paleobiology* **4**: 252–60.

Haymer, D. S. and Hartl, D. L. (1981). Using frequency distribution to detect selection: inversion polymorphisms in *Drosophila pseudoobscura*. *Evolution* **35**: 1243–6.

Hecht, M. K. (1952). Natural selection in the lizard genus *Aristelliger*. *Evolution* **6**: 112–24.

Hed, H. M. E. (1984). Opportunity for selection during the 17–19th centuries in the diocese of Linkoping as estimated with Crow's index in a population of clergymen's wives. *Hum. Hered.* **34**: 378–87.

Hed, H. M. E. and Rasmuson, M. (1981). Cohort study of opportunity for selection on two Swedish 19th century parishes with a survey of other estimates. *Hum. Hered.* **31**: 78–83.

Hedrick, P. W. (1982). Genetic hitchhiking: a new factor in evolution? *Bioscience* **32**: 845–53.

Hedrick, P. W., Jain, S. and Holden, L. (1978). Multilocus systems in evolution. *Evol. Biol.* **11**: 101–84.

Hedrick, P. W. and Murray, E. (1983). Selection and measurement of fitness. In The Genetics and Biology of Drosophila, Vol. 3b (eds M. Ashburrer, H. L. Carson and J. N. Thompson), pp. 61–104. Academic Press, London.

Henderson, N. R. and Lambert, D. M. (1982). No significant deviation from random mating of worldwide populations of *Drosophila melanogaster*. *Nature* **300**: 437–40.

Henneberg, M. (1976). Reproductive possibilities and estimations of the biological dynamics of earlier human populations. *J. Hum. Evol.* **5**: 41–8.

Hertzog, K. P. and Johnston, F. E. (1968). Selection and the Rh polymorphism. *Hum. Biol.* **40**: 86–97.

Hill, W. G. (1974). Estimation of linkage disequilibrium in randomly mating populations. *Heredity* **33**: 229–39.

Hill, W. G. (1975). Tests for association of gene frequencies at several loci in random mating diploid populations. *Biometrics* **31**: 881–8.

Hill, W. G. (1976). Non-random association of neutral linked genes in finite populations. In *Population Genetics and Evolution* (eds S. Karlin and E. Nevo), pp. 339–76. Academic Press, New York.

Hoeffding, W. (1948). A class of statistics with asymptotic normal distribution. *Ann. Math. Stat.* **19**: 293–325.

Horsley, D. T., Lynch, B. M., Greenwood, J. J. D., Hardman, B. and Mosely, S. (1979). Frequency-dependent selection by birds when the density of prey is high. *J. Anim. Ecol.* **48**: 483–90.

Hubby, J. L. and Lewontin, R. C. (1966). A molecular approach to the study of genic heterozygosity in natural populations. I. The number of alleles at different loci in *Drosophila pseudoobscura. Genetics* **54**: 577–94.

Hudson, R. R. (1983). Testing the constant-rate neutral allele model with protein sequence data. *Evolution* **37**: 203–17.

Huxley, J. S. (1942). *Evolution, the Modern Synthesis*. Allen and Unwin, London.

Inger, R. F. (1943). Further notes on differential selection of variant juvenile snakes. *Amer. Nat.* **77**: 87–90.

Jackson, C. H. N. (1939). The analysis of an animal population. *J. Anim. Ecol.* **8**: 238–46.

Jackson, C. H. N. (1946). An artificially isolated generation of tsetse flies (*Diptera*). *Bull. Ent. Res.* **37**: 291–9.

Jackson, C. H. N. (1948). Some further isolated generations of tsetse flies. *Bull. Ent. Res.* **39**: 441–51.

Jeffords, M. R., Sternburg, J. G. and Waldbauer, G. P. (1979). Batesian mimicry: field demonstration of the survival value of pipevine swallowtail and monarch colour patterns. *Evolution* **33**: 275–86.

Jeffords, M. R., Waldbauer, G. P. and Sternburg, J. G. (1980). Determination of the time of day at which diurnal moths painted to resemble butterflies are attacked by birds. *Evolution* **34**: 1205–11.

Jelnes, J. E. (1983). A method of obtaining indices of distance and similarity from observations on differences in electrophoretic mobility of population samples.

Hereditas **98**: 281–6.

Johnson, C. (1976). *Introduction to Natural Selection.* Univ. Park Press, Baltimore.

Johnson, G. (1972). Enzyme polymorphisms: evidence that they are not selectively neutral. *Nature, New Biol.* **237**: 170–1.

Johnson, G. (1974). On the estimation of the effective number of alleles from electrophoretic data. *Genetics* **78**: 771–6.

Johnson, G. (1977). Evolution of the stepwise mutation model of electrophoretic mobility: comparison of the gel sieving behaviour of alleles at the esterase-5 locus of *Drosophila pseudoobscura. Genetics* **87**: 139–57.

Johnson, G. and Feldman, M. W. (1973). On the hypothesis that polymorphic enzyme alleles are selectively neutral. 1. The evenness of allele frequency distributions. *Theor. Pop. Biol.* **4**: 209–21.

Johnson, M. S. (1980). Association of shell banding and habitant in a colony of the land snail *Theba pisana. Heredity* **45**: 7–14.

Johnston, R. F., Niles, D. M. and Rohwer, S. A. (1972). Hermon Bumpus and natural selection in the house-sparrow *Passer domesticus. Evolution* **26**: 20–31.

Jolly, G. M. (1965). Explicit estimates from capture–recapture data with both death and immigration – stochastic model. *Biometrika* **52**: 225–47.

Jolly, G. M. (1979). A unified approach to mark–recapture stochastic models exemplified by a constant survival rate model. In *Sampling Biological Populations* (eds R. M. Cormack, G. P. Patil, and D. S. Robson), pp. 277–82. Inter. Coop. Publ. House, Maryland.

Jolly, G. M. (1982). Mark–recapture models with parameters constant in time. *Biometrics* **38**: 301–21.

Jolly, G. M. and Dickson, J. M. (1980). Mark recapture suite of programs. *Compstat 1980, Proc. in Comp. Statist.* Physica-Verlag, Wien.

Jones, J. S. (1982). Genetic differences in individual behaviour associated with shell polymorphism in the snail *Cepaea nemoralis. Nature* **298**: 749–50.

Jones, J. S., Leith, B. H. and Rawlings, P. (1977). Polymorphism in *Cepaea*: a problem with too many solutions? *Ann. Rev. Ecol. Syst.* **8**: 109–43.

Jones, J. S. and Parkin, D. T. (1977). Experimental manipulation of some snail population subject to climatic selection. *Amer. Nat.* **111**: 1014–7.

Jones, J. S., Selander, R. S. and Schnell, G. D. (1980). Patterns of morphological and molecular polymorphism in the land snail *Cepaea nemoralis. Biol. J. Linn. Soc.* **14**: 359–87.

Jones, R. (1958). Lee's phenomena of "apparent change in growth-rate" with particular reference to haddock and plaice. *Int. Comm. Northwest. Atl. Fish., Spec. Publ.* **1**: 229–42.

Jorgensen, C. D., Smith, H. D. and Scott, D. T. (1975). Small mammal estimates using recapture methods, with variables partitioned. *Acta Theor.* **20**: 303–18.

Kalbfleisch, J. D. and Prentice, R. L. (1980). *The Statistical Analysis of Failure Time Data.* Wiley, New York.

Kaplan, N. and Risko, K. (1982). A method for estimating rates of nucleotide substitution using DNA sequence data. *Theor. Pop. Biol.* **21**: 318–28.

Karlin, S. and Piazza, A. (1981). Statistical methods for assessing linkage dis-equilibrium at the HLA-A, B, C loci. *Ann. Hum. Genet.* **45**: 79–94.

Karn, M. N. and Penrose, L. S. (1951). Birth weight and gestation time in relation to maternal age, parity and infant survival. *Ann. Eugen.* **16**: 145–64.

Kearsey, M. J. and Barnes, B. W. (1970). Variation for metrical characters in *Drosophila* populations. II. Natural selection. *Heredity* **25**: 11–21.

Kempthorne, O. and Pollack, E. (1970). Concepts of fitness in Mendelian populations. *Genetics* **64**: 125–45.

Kence, A. (1981). A rare-male advantage in *Drosophila*: a possible source of bias in experimental design. *Amer. Nat.* **117**: 1027–8.

Kendall, M. G. (1975). *Multivariate Analysis*. Griffin, London.

Kerfoot, W. C. (1980). Commentary: transparency, body size and prey conspicuousness. In *Evolution and Ecology of Zooplankton Communities* (ed. W. C. Kerfoot), pp. 609–17. Univ. Press of New England, Hanover, New Hampshire.

Kettlewell, H. B. D. (1955). Selection experiments on industrial melanism in the Lepidoptera. *Heredity* **9**: 323–42.

Kettlewell, H. B. D. (1956). Further selection experiments on industrial melanism in the Lepidoptera. *Heredity* **10**: 287–301.

Kettlewell, H. B. D. (1961). Selection experiments on melanism in *Amathes glareosa* Esp. (Lepidoptera). *Heredity* **16**: 415–34.

Kettlewell, H. B. D. and Berry, R. J. (1961). The study of a cline. *Heredity* **16**: 403–14.

Kettlewell, H. B. D. and Berry, R. J. (1969). Gene flow in a cline. *Heredity* **24**: 1–14.

Kettlewell, H. B. D., Berry, R. J., Cadbury, C. J. and Phillips, G. C. (1969). Differences in behaviour, dominance and survival within a cline. *Heredity* **24**: 15–25.

Khamis, H. J. and Hinkelmann, K. (1984). Log-linear model analysis of the association between disease and genotype. *Biometrics* **40**: 177–88.

Kimura, M. (1968). Evolutionary rates at the molecular level. *Nature* **217**: 624–6.

Kimura, M. (1979). The neutral theory of molecular evolution. *Sci. Amer.* **241**: 98–126.

Kimura, M. and Crow, J. F. (1964). The number of alleles that can be maintained in a finite population. *Genetics* **49**: 725–38.

King, J. L. and Jukes, T. H. (1969). Non-Darwinian evolution. *Science* **164**: 788–98.

King-Hele, D. G. (1977). *Doctor of Revolution*. Faber and Faber, London.

Kirby, G. C. (1974). The bias in the regression of delta q on q. *Heredity* **33**: 93–7.

Kirby, G. C. and Halliday, R. B. (1973). Another view of neutral alleles in natural populations. *Nature* **241**: 463.

Kluge, A. G. and Kerfoot, W. C. (1973). The predictability and regularity of character divergence. *Amer. Nat.* **107**: 426–42.

Knights, R. W. (1979). Experimental evidence for selection on shell size in *Cepaea hortensis* (Mull). *Genetica* **50**: 51–60.

Knoppien, P. (1984). The rare male advantage: an artifact caused by marking procedures? *Amer. Nat.* **123**: 862–6.

Kopec, A. C. (1973). Blood group distributions in Britain. In *Genetic Variation in Britain* (eds D. F. Roberts and E. Sunderland), pp. 129–39. *Symp. Soc. Study Hum. Biol.* 12.

Krimbas, C. B. and Tsakas, S. (1971). The genetics of *Dacus oleae*. V. Changes of esterase polymorphism in a natural population following insecticide control—selection or drift? *Evolution* **25**: 454–60.

Lakhani, K. H. and Newton, I. (1983). Estimating age-specific bird survival rates from ring recoveries—can it be done? *J. Anim. Ecol.* **52**: 83–91.

Lande, R. (1976). Natural selection and random genetic drift in phenotypic characters. *Evolution* **30**: 314–34.

Lande, R. (1977). Statistical test for natural selection on quantitative characters. *Evolution* **31**: 442–4.

Lande, R. (1979). Quantitative genetic analysis of multivariate evolution applied to brain : body size allometry. *Evolution* **33**: 402–16.

Lande, R. (1980a). Microevolution in relation to macroevolution. *Paleobiology* **6**: 233–8.

Lande, R. (1980b). Genetic variation and phenotypic evolution during allopatric speciation. *Amer. Nat.* **116**: 463–79.

Lande, R. and Arnold, S. J. (1983). The measurement of selection on correlated characters. *Evolution* **37**: 1210–26.

Langley, C. H. (1977). Nonrandom associations between allozymes in natural populations of *Drosophila melanogaster*. In *Measuring Natural Selection* (eds F. B. Christiansen and T. M. Fenchel), pp. 265–73. Springer-Verlag, Berlin.

Langley, C. H. and Fitch, W. M. (1973). The constancy of evolution: a statistical analysis of the a and b haemoglobins, cytochrome c, and fibrinopeptide A. In *Genetic Structure of Populations* (ed. N. E. Morton), pp. 246–62. Univ. of Hawaii Press, Honolulu.

Langley, C. H. and Fitch, W. M. (1974). An examination of the constancy of the rate of evolution. *J. Mol. Evol.* **3**: 161–77.

Langley, C. H., Smith, D. B. and Johnson, F. M. (1978). Analysis of linkage disequilibrium between allozyme loci in natural populations of *Drosophila melanogaster*. *Genet. Res., Camb.* **32**: 215–29.

Langley, C. H., Tobari, Y. N. and Kojima, K. (1974). Linkage disequilibrium in natural populations of *Drosophila melanogaster*. *Genetics* **78**: 921–36.

Larson, A. and Highton, R. (1978). Geographical protein variation and divergence in the salamanders of the *Plethodon welleri* group (Amphibia, Plethodontidae). *Syst. Zool.* **27**: 431–48.

Leonard, J. E. and Ehrman, L. (1983). Does the rare male advantage result from faulty experimental design? *Genetics* **104**: 713–6.

Levene, H. (1949). A new measure of sexual isolation. *Evolution* **3**: 315–21.

Levene, H., Pavlovsky, O. and Dobzhansky, T. (1954). Interaction of the adaptive values in polymorphic experimental population of *Drosophila pseudoobscura*. *Evolution* **8**: 335–49.

Lewontin, R. C. (1974). *The Genetic Basis for Evolutionary Change*. Columbia Univ. Press, New York.

Lewontin, R. C. and Cockerham, C. C. (1959). The goodness of fit test for detecting natural selection in random mating populations. *Evolution* **13**: 561–4.

Lewontin, R. C. and Hubby, J. L. (1966). A molecular approach to the study of genic heterozygosity in natural populations. II. Amount of variation and degree of heterozygosity in natural populations of *Drosophila pseudoobscura*. *Genetics* **54**: 595–609.

Lewontin, R. C., Kirk, D. and Crow, J. (1968). Selective mating, assortative mating, and inbreeding: definitions and implications. *Eugen. Quart.* **15**: 141–3.

Lewontin, R. C. and Krakauer, J. (1973). Distribution of gene frequency as a test of the theory of the selective neutrality of polymorphisms. *Genetics* **74**: 175–95.

Lewontin, R. C., Moore, J. A., Provine, W. B. and Wallace, B. (eds) (1981). *Dobzhansky's Genetics of Natural Populations, I–XLIII*. Columbia Univ. Press, New York.

Lewontin, R. C. and White, M. J. D. (1960). Interaction between inversion polymorphisms of two chromosome pairs in the grasshopper, *Moraba scurra*. *Evolution* **14**: 116–29.

Li, W. H. (1976). A mixed model of mutation for electrophoretic identity of proteins within and between populations. *Genetics* **83**: 423–32.

Lowther, P. E. (1977). Selection intensity in North American house sparrows (*Passer domesticus*). *Evolution* **31**: 649–56.

Lusis, J. J. (1961). On the biological meaning of colour polymorphism of ladybeetle *Adalia bipunctata* L. *Latvijas Entomologs* **4**: 3–29.

McCommas, S. A. (1983). A taxonomic approach to evaluation of the charge state model using twelve species of sea anemone. *Genetics* **103**: 741–52.

McCormick, J. M. and Salvadori, M. G. (1964). *Numerical Methods in Fortran*. Prentice Hall, New Jersey.

McDonald, J. F. (1983). The molecular basis of adaption: a critical review of relevant ideas and observations. *Ann. Rev. Ecol. Syst.* **14**: 77–102.

McGilchrist, C. A. and Simpson, J. M. (1979). Effects of differential mortality rate on risk-variable distribution. *Math. Biosci.* **43**: 173–80.

McKechnie, S. W., Ehrlich, P. R. and White, R. R. (1975). Population genetics of *Euphydryas* butterflies. I. Genetic variation and the neutrality hypothesis. *Genetics* **81**: 571–94.

Majerus, M., O'Donald, P. and Weir, J. (1982a). Female mating preference is genetic. *Nature* **300**: 521–3.

Majerus, M., O'Donald, P. and Weir, J. (1982b). Evidence for preferential mating in *Adalia bipunctata*. *Heredity* **49**: 37–49.

Majumder, P. P. and Chakraborty, R. (1981). Mean and variance of the number of samples showing heterozygote excess or deficiency. *Heredity* **47**: 259–62.

Malthus, T. R. (1798). *An Essay on the Principles of Population*. J. Johnson, London.

Mandarino, L. and Cadien, J. D. (1974). Use of ranked migration estimates for detecting natural selection. *Amer. J. Hum. Genet.* **26**: 108–12.

Mani, G. S. (1980). A theoretical study of morph ratio clines with special reference to melanism in moths. *Proc. Roy. Soc. Lond.* **B210**: 299–316.

Manly, B. F. J. (1972). Estimating selective values from field data. *Biometrics* **28**: 1115–25.

Manly, B. F. J. (1973a). A note on the estimation of selective values from recaptures of marked animals when selection pressures remain constant over time. *Res. Pop. Ecol.* **14**: 151–8.

Manly, B. F. J. (1973b). A linear model for frequency-dependent selection by predators. *Res. Pop. Ecol.* **14**: 137–50.

Manly, B. F. J. (1974a). Estimating survival from a multi-sample single recapture census where recaptures are not made at release times. *Biom. J.* **16**: 185–90.

Manly, B. F. J. (1974b). A model for certain types of selection experiments. *Biometrics* **30**: 281–94.

Manly, B. F. J. (1975a). Estimating survival from a multi-sample single recapture census – the case of constant survival and recapture probabilities. *Biom. J.* **17**: 431–5.

Manly, B. F. J. (1975b). A second look at some data on a cline. *Heredity* **34**: 423–6.

Manly, B. F. J. (1975c). The measurement of the characteristics of natural selection. *Theor. Pop. Biol.* **7**: 288–305.

Manly, B. F. J. (1976). Some examples of double exponential fitness functions. *Heredity* **36**: 229–34.

Manly, B. F. J. (1977a). A note on the design of experiments to estimate survival and relative survival. *Biom. J.* **19**: 687–92.

Manly, B. F. J. (1977b). A model for dispersion experiments. *Oecologia* **31**: 119–30.

Manly, B. F. J. (1977c). The estimation of the fitness function from several samples taken from a population. *Biom. J.* **19**: 391–401.

Manly, B. F. J. (1977d). Examples of the use of GLIM. *New Zealand Statist.* **12**: 26–42.

Manly, B. F. J. (1977e). A new index for the intensity of natural selection. *Heredity* **38**: 321–8.

Manly, B. F. J. (1978a). A simulation study of three methods for estimating selective values and survival rates from recapture data. *Res. Pop. Ecol.* **20**: 15–22.

Manly, B. F. J. (1978b). Regression models for proportions with extraneous variance. *Biom. Praxim.* **18**: 1–18.

Manly, B. F. J. (1980). A note on a model for selection experiments. *Biometrics* **36**: 9–18.

Manly, B. F. J. (1981a). Estimation of absolute and relative survival rates from the recoveries of dead animals. *New Zealand J. Ecol.* **4**: 78–88.

Manly, B. F. J. (1981b). The estimation of a multivariate fitness function from several samples taken from a population. *Biom. J.* **23**: 267–81.

Manly, B. F. J. (1983). Analysis of polymorphic variation in different types of habitat. *Biometrics* **39**: 13–27.

Manly, B. F. J. (1985a). A test of Jackson's method for separating death and emigration with mark–recapture data. *Res. Pop. Ecol.* (to appear).

Manly, B. F. J. (1985b). Detecting and measuring stabilizing selection. *Evol. Theor.* (to appear).

Manly, B. F. J., Miller, P. and Cook, L. M. (1972). Analysis of a selective predation experiment. *Amer. Nat.* **106**: 719–36.

Mantel, N. (1967). The detection of disease clustering and a generalised regression approach. *Cancer Res.* **27**: 209–20.

Mantel, N. and Valand, R. S. (1970). A technique for nonparametric multivariate analysis. *Biometrics* **26**: 547–58.

Marcus, L. F. (1964). Measurement of natural selection in natural populations. *Nature* **202**: 1033–4.

Marcus, L. F. (1969). Measurement of selection using distance statistics in the prehistoric orang-utan *Pongo pygmaeus palaeosumatrensis*. *Evolution* **23**: 301–7.

Markow, T. A., Richmond, R. C., Mueller, L., Sheer, I., Roman, S., Laetz, C. and Lorenz, L. (1980). Testing for rare male mating advantage among various *Drosophila melanogaster* genotypes. *Genet. Res., Camb.* **35**: 59–64.

Maruyama, T. and Yamazaki, T. (1974). Analysis of heterozygosity in regard to the neutrality theory of protein polymorphism. *J. Mol. Evol.* **4**: 195–9.

Mascie-Taylor, C. G. N., McManus, I. C., Golding, J., Hicks, P., Butler, N. R., Hawkins, J. D., Valenzuela, C. Y., Hartung, J., Beardmore, J. A. and Karmi-Booshehri, F. (1984). Blood group and social class. *Nature* **309**: 395–9.

Mayr, E. (1954). Change of genetic environment and evolution. In *Evolution as a Process* (eds J. S. Huxley, A. C. Hardy and E. B. Ford), pp. 157–80. Allen and Unwin, London.

Mayr, E. (1982). Speciation and macroevolution. *Evolution* **36**: 1119–32.

Mendel, G. J. (1865). Versuche uber pflanzen-hybriden. Verh. Naturf. Ver. in Brunn. 4. Reprinted in an English translation in *Experiments in Plant Hybridisation* (ed. J. H. Bennett), 1965, Oliver and Boyd, Edinburgh.

Merrell, D. J. (1981). *Ecological Genetics*. Univ. of Minnesota Press, Minneapolis.

Mielke, P. W. (1978). Classification and appropriate inferences for Mantel and Valand's nonparametric multivariate analysis technique. *Biometrics* **34**: 277–82.

Mielke, P. W., Berry, K. J. and Johnson, E. S. (1976). Multi-response permutation procedures for a priori classifications. *Commun. Statis. Theor. Meth.* **A5**: 1409–24.

Moran, P. A. P. (1975). Wandering distributions and the electrophoretic profile. *Theor. Pop. Biol.* **8**: 318–30.

Mourant, A. E., Kopec, A. C. and Domaniewska-Sobczak, K. (1978). *Blood Groups and Diseases*. Oxford Univ. Press, Oxford.

Muggleton, J. (1978). Selection against the melanic morphs of *Adalia bipunctata* (two-spot ladybird): a review and some new data. *Heredity* **40**: 269–80.

Muggleton, J. (1979). Non-random mating in wild populations of polymorphic *Adalia bipunctata*. *Heredity* **42**: 57–65.

Muggleton, J. (1983). Relative fitness of malathion-resistant phenotypes of *Oryzaephilus surinamensis* L. (Coleoptera: silvanidae). *J. Appl. Ecol.* **20**: 245–54.

Muggleton, J., Lonsdale, D. and Benham, B. R. (1975). Melanism in *Adalia bipunctata* L. (Col., Coccinellidae) and its relationship to atmospheric pollution. *J. Appl. Ecol.* **12**: 465–71.

Mukai, T. and Yamazaki, T. (1980). Test for selection on polymorphic isozyme genes using the population cage method. *Genetics* **96**: 537–42.

Murray, J. and Clarke, B. (1978). Changes of gene frequency of *Cepaea nemoralis* over fifty years. *Malacologia* **17**: 317–30.

Nei, M. (1973). The theory and estimation of genetic distance. In *Genetic Structure of Populations* (ed. N. E. Morton), pp. 45–54. Univ. of Hawaii Press, Honolulu.

Nei, M. (1975). *Molecular Population Genetics and Evolution*. North Holland, Amsterdam.

Nei, M., Chakraborty, P. A. and Fuerst, P. A. (1976). Infinite allele model with varying mutation rate. *Proc. Nat. Acad. Sci., USA* **73**: 4164–8.

Nei, M. and Maruyama, T. (1975). Lewontin–Krakauer test for neutral genes. *Genetics* **80**: 395.

Nei, M. and Tajima, F. (1981). Genetic drift and estimation of effective population size. *Genetics* **98**: 625–40.

Nei, M. and Tateno, Y. (1975). Interlocus variation of genetic distance and the neutral mutation theory. *Proc. Nat. Acad. Sci., USA* **72**: 2758–60.

Nei, M. and Yokoyama, S. (1981). Estimation of fitness reduction due to a chronic disease in man. *Ann. Hum. Genet.* **45**: 261–5.

Nelder, J. A. (1974). Log-linear models for contingency tables – a generalisation of classical least squares. *Appl. Statist.* **23**: 323–9.

Nelder, J. A. (1975). *Glim Manual, Release 2*. Numerical Algorithms Group, 13 Banbury Rd., Oxford OX2 GNN, England.

Nelder, J. A. and Wedderburn, R. W. M. (1972). Generalised linear models. *J. Roy. Statist. Soc.* **A135**: 370–84.

North, P. M. and Cormack, R. M. (1981). On Seber's method for estimating age-specific bird survival rates from ringing recoveries. *Biometrics* **37**: 103–12.

North, P. M. and Morgan, B. J. T. (1979). Modelling heron survival using weather data. *Biometrics* **35**: 667–81.

O'Donald, P. (1968). Measuring the intensity of natural selection. *Nature* **220**: 197–8.

O'Donald, P. (1970). Change of fitness by selection for a quantitative character. *Theor. Pop. Biol.* **1**: 219–32.

O'Donald, P. (1971). Natural selection for quantitative characters. *Heredity* **27**: 137–53.

O'Donald, P. (1973). A further analysis of Bumpus' data: The intensity of natural selection. *Evolution* **27**: 398–404.

O'Donald, P. (1980). *Genetic Models of Sexual Selection*. Cambridge Univ. Press, Cambridge.

O'Donald, P. (1983). Do flies choose their mates? A comment. *Amer. Nat.* **122**: 413–6.

O'Donald, P. and Davis, J. W. F. (1975). Demography and selection in a population of Arctic Skuas. *Heredity* **35**: 75–83.

O'Donald, P. and Davis, J. W. F. (1976). A demographic analysis of the components of selection in a population of Arctic Skuas. *Heredity* **36**: 343–50.

O'Donald, P. and Muggleton, J. (1979). Melanic polymorphism in ladybirds maintained by selection. *Heredity* **43**: 143–8.

Ohta, T. (1973). Slightly deleterious mutant substitutions in evolution. *Nature* **246**: 96–8.

Ohta, T. (1974). Mutational pressure as the main cause of molecular evolution and polymorphism. *Nature* **252**: 351–4.

Ohta, T. (1976). Role of very slightly deleterious mutations in molecular evolution and polymorphism. *Theor. Pop. Biol.* **10**: 254–75.

Ohta, T. (1982). Linkage disequilibrium with the island model. *Genetics* **101**: 139–55.

Ohta, T. and Kimura, M. (1973). A model of mutation appropriate to estimate the number of electrophoretically detectable alleles in a finite population. *Genet. Res., Camb.* **22**: 201–4.

Orzack, S. H. (1981). The modern synthesis in partly Wright. *Paleobiology* **7**: 128–31.

Østergaard, H. and Christiansen, F. B. (1981). Selection component analysis of natural polymorphisms using population samples including mother–offspring combinations, II. *Theor. Pop. Biol.* **19**: 378–419.

Pamilo, P. and Varvio-Aho, S. L. (1980). Estimation of population size from allele frequency changes. *Genetics* **95**: 1055–7.

Parker, R. P. (1971). Size selective predation among juvenile salmonid fishes in a British Columbia inlet. *J. Fish. Res. Bd. Canada* **28**: 1503–10.

Partridge, G. G. (1979). Relative fitness of genotypes in a population of *Rattus norvegicus* polymorphic for warfarin resistance. *Heredity* **43**: 239–46.

Pearre, S. (1982). Estimating prey preference by predators: uses of various indices, and a proposal of another based on chi-square. *Can. J. Fish. Aquat. Sci.* **39**: 914–23.

Pearre, S. (1983). Reply to Vanderploeg and Scavia. *Can. J. Fish. Aquat. Sci.* **40**: 250–1.

Pearson, E. S. and Hartley, H. O. (1966). *Biometrika Tables for Statisticians*, Vol. 1. Cambridge Univ. Press, London.

Pearson, K. (1903). Mathematical contributions to the theory of evolution. XI. On the influence of natural selection on the variability and correlation of organs. *Phil. Trans. Roy. Soc. Lond.* **A200**: 1–66.

Petit, C. (1951). Le role de l'isolement sexuel dans l'evolution des populations de *Drosophila melanogaster*. *Bull. Biol. France Belgique* **85**: 392–418.

Petit, C. (1954). L'isolement sexuel chez *Drosophila melanogaster*. Etude du mutant white et de san allemorphe sauvage. *Bull. Biol. France Belgique* **88**: 435–43.

Petit, C. (1958). Le determinisme genetique et psychophysiologique de la sexuelle chez *Drosophila melanogaster*. *Bull. Biol. France Belgique* **92**: 248–329.

Phelps, R. J. and Clarke, G. P. Y. (1974). Seasonal elimination of some size classes in males of *Glossina morsitans morsitans* Westw. (Diptera, Glossinidae). *Bull. Ent. Res.* **64**: 313–24.

Phillips, P. R. and Mayo, O. (1981). Problems in statistical studies on protein polymorphisms in natural populations. Letter to the Editor, *Genetics* **97** (after p. 494).

Piazza, A. (1980). Evolution in human populations: data and models. In *Vito Voltera Symposium on Mathematical Models in Biology* (ed. C. Barigozi), pp. 98–132. Springer Verlag, Berlin.

Pollak, E. (1983). A new method for estimating the effective population size from allele frequency changes. *Genetics* **104**: 531–48.

Popham, E. J. (1941). The variation in the colour of certain species of *Arctocorisa* (Hemiptera, Corixidae) and its significance. *Proc. Zool. Soc. Lond.* **A111**: 135–72.

Popham, E. J. (1943). Further experimental studies on the selective action of predators. *Proc. Zool. Soc. Lond.* **A112**: 105–17.

Popham, E. J. (1944). A study of the changes in an aquatic insect population using minnows as the predators. *Proc. Zool. Soc. Lond.* **A114**: 74–81.

Popham, E. J. (1947). Experimental studies of the biological significance of non-cryptic pigmentation with special reference to insects. *Proc. Zool. Soc. Lond.* **A117**: 768–83.

Popham, E. J. (1966). An ecological study on the predatory action of the three spined stickleback (*Gasterosteus aculeatus* L.). *Arch. Hydrobiol.* **62**: 70–81.

Powell, J. R. and Dobzhansky, Th. (1976). How far do flies fly? *Amer. Sci.* **64**: 179–85.

Press, S. J. (1972). *Applied Multivariate Analysis*. Holt, Rinehart and Winston, New York.

Prout, T. (1965). The estimation of fitnesses from genotypic frequencies. *Evolution* **19**: 546–51.

Prout, T. (1969). The estimation of fitnesses from population data. *Genetics* **63**: 949–67.

Prout, T. (1971a). The relation between fitness components and population prediction in *Drosophila*. I. The estimation of fitness components. *Genetics* **68**: 127–49.

Prout, T. (1971b). The relation between fitness components and population prediction in *Drosophila*. II. Population prediction. *Genetics* **68**: 151–67.

Ramshaw, J. A. M., Coyne, J. A. and Lewontin, R. C. (1979). The sensitivity of gel electrophoresis as a detection of genetic variation. *Genetics* **93**: 1019–37.

Ramshaw, J. A. M. and Eanes, W. F. (1978). Study of the charge-state model for electrophoretic variation using isoelectric focusing of esterase-5 from *Drosophila pseudoobscura*. *Nature* **275**: 68–70.

Rao, C. R. (1952) *Advanced Statistical Methods in Biometric Research*. Wiley, New York.

Raup, D. M. and Sepkoski, J. J. (1984). Periodicity of extinctions in the geologic past. *Proc. Nat. Acad. Sci., USA* **81**: 801–5.

Reed, T. E. (1969a). Genetic experience with a general maximum likelihood estimation program. In *Computer Applications in Genetics* (ed. N. E. Morton), pp. 27–29.

Univ. of Hawaii Press, Honolulu.

Reed, T. E. (1969b). Caucasian genes in American Negroes. *Science* **165**: 762–8.

Reed, T. E. (1975). Selection and blood group polymorphisms. In *The Role of Natural Selection in Evolution* (ed. F. M. Salzano), pp. 231–45. Plenum Press, New York.

Reed, T. E. and Schull, W. J. (1968). A general maximum likelihood estimation program. *Amer. J. Hum. Genet.* **20**: 579–80.

Reyment, R. (1982a). Application of quantitative genetics to evolutionary series of microfossils. In *Nordic Symposium in Applied Statistics and Data Processing*, pp. 307–25. Forlag NEUCC, Copenhagen.

Reyment, R. (1982b). Phenotypic evolution in a cretaceous foraminifer. *Evolution* **36**: 1182–99.

Reyment, R. (1983). Phenotypic evolution in microfossils. *Evol. Biol.* **16**: 209–54.

Rhodes, F. H. T. (1983). Gradualism, punctuated equilibrium and the Origin of Species. *Nature* **305**: 269–72.

Ricker, W. E. (1969). Effects of size-selective mortality and sampling bias in estimates of growth, mortality, production, and yield. *J. Fish. Res. Bd. Canada* **26**: 479–541.

Ripley, B. D. (1981). *Spatial Statistics*. Wiley, New York.

Roberson, P. K. and Fisher, L. (1983). Lack of robustness in time-space clustering. *Comm. Statist. Simul. Comput.* **12**: 11–22.

Roberts, D. F. (1973). The origins of genetic variation in Britain. *Symp. Soc. Study Hum. Biol.* **12**: 1–16.

Robertson, A. (1955). Selection in animals: synthesis. *Cold Spring Harbour Symp. Quant. Biol.* **20**: 225–9.

Robertson, A. (1975a). Remarks on the Lewontin–Krakauer test. *Genetics* **80**: 396.

Robertson, A. (1975b). Gene frequency distributions as a test of selective neutrality. *Genetics* **81**: 775–85.

Robson, D. S. and Chapman, D. G. (1961). Catch curves and mortality rates. *Trans. Amer. Fish. Soc.* **90**: 181–9.

Rohlf, F. J., Gilmartin, A. J. and Hart, G. (1983). The Kluge–Kerfoot phenomenon – a statistical artefact. *Evolution* **37**: 180–202.

Rosenzweig, M. (1968). The strategy of body size in mammalian carnivores. *Amer. Midl. Nat.* **80**: 299–315.

Ross, G. J. S. (1980). *MLP, Maximum Likelihood Program*. Rothamstead Experimental Station, Harpenden, UK

Royaltey, H. H., Astrachan, E. and Sokal, R. R. (1975). Tests for patterns in geographic variation. *Geog. Anal.* **7**: 369–95.

Samollow, P. B. (1980). Selective mortality and reproduction in a natural population of *Bufo boreas*. *Evolution* **34**: 18–39.

Sansing, R. C. and Chinnici, J. P. (1976). Optimal and discriminating birth weights in human populations. *Ann. Hum. Genet.* **40**: 123–31.

Schaffer, H. E., Yardley, D. and Anderson, W. W. (1977). Drift or selection: a statistical test of gene frequency variation over generations. *Genetics* **87**: 371–9.

Schindel, D. E. and Gould, S. J. (1977). Biological interaction between fossil species: character displacement in Bermudian land snail. *Paleobiology* **3**: 259–69.

Schultz, B. (1983). On Levene's test and other statistics of variation. *Evol. Theory* **6**: 197–203.

Searcy, W. A. (1982). The evolutionary effects of mate selection. *Ann. Rev. Ecol. Syst.* **13**: 57–85.

Seber, G. A. F. (1962). The multi-sample single recapture census. *Biometrika* **49**: 339–50.

Seber, G. A. F. (1965). A note on the multiple-recapture census. *Biometrika* **52**: 249–59.

Seber, G. A. F. (1970). Estimating time-specific survival and reporting rates for adult birds from band returns. *Biometrika* **57**: 313–8.

Seber, G. A. F. (1971). Estimating age-specific survival rates from bird-band returns when the reporting rate is constant. *Biometrika* **58**: 491–7.

Seber, G. A. F. (1982). *Estimation of Animal Abundance and Related Parameters*, 2nd Edit. Griffin, London.

Sheppard, P. M. (1951a). Fluctuations in the selective value of certain phenotypes in the polymorphic land snail *Cepaea nemoralis* (L.). *Heredity* **5**: 125–34.

Sheppard, P. M. (1951b). A quantitative study of two populations of the moth *Panaxia dominula* (L.). *Heredity* **5**: 349–78.

Sheppard, P. M. and Cook, L. M. (1962). The manifold effects of the medionigra gene of the moth *Panaxia dominula* and the maintenance of a polymorphism. *Heredity* **17**: 415–26.

Sheppard, P. M., MacDonald, W. W., Tonn, R. J. and Grab, B. (1969). The dynamics of an adult population of *Aedes aegyptyi* in relation to dengue heamorrhagic fever in Bangkok. *J. Anim. Ecol.* **38**: 661–702.

Simpson, G. G. (1944). *Tempo and Mode in Evolution*. Columbia Univ. Press, New York.

Simpson, G. G. (1953). *The Major Features of Evolution*. Columbia Univ. Press, New York.

Singh, R. S., Lewontin, R. C. and Felton, A. A. (1976). Genetic heterogeneity within electrophoretic "alleles" of xanthine dehydrogenase in *Drosophila pseudoobscura*. *Genetics* **84**: 609–29.

Slatkin, M. (1981). Estimating levels of gene flow in natural populations. *Genetics* **99**: 323–35.

Slatkin, M. (1982). Testing neutrality in sub-divided populations. *Genetics* **100**: 533–45.

Smith, H. D., Jorgensen, C. D. and Tolley, H. D. (1972). Estimation of small mammal using recapture methods: partitioning of estimator variables. *Acta Theor.* **17**: 57–66.

Smouse, P. E. (1974). Likelihood analysis of recombinational disequilibrium in multiple-locus gametic frequencies. *Genetics* **76**: 557–65.

Smouse, P. E. and Neel, J. V. (1977). Multivariate analysis of gametic disequilibrium in the Yanomama. *Genetics* **85**: 733–52.

Smouse, P. E., Neel, J. V. and Liu, W. (1983). Multiple-locus departures from panmictic equilibrium within and between village gene pools of Amerindian tribes at different stages of agglomeration. *Genetics* **104**: 133–53.

Smouse, P. E. and Williams, R. C. (1982). Multivariate analysis of HLA-disease associations. *Biometrics* **38**: 757–68.

Sokal, R. R. (1978). Populations differentiation: something new or more of the same? In *Ecological Genetics: the Interface* (ed. P. F. Brussard), pp. 215–39. Springer-Verlag, New York.

Sokal, R. R. (1979). Testing statistical significance of geographical variation patterns. *Syst. Zool.* **28**: 227–32.

Sokal, R. R. and Oden, N. L. (1978a). Spatial autocorrelation in biology. 1. Methodology. *Biol. J. Linn. Soc.* **10**: 199–228.

Sokal, R. R. and Oden, N. L. (1978b). Spatial autocorrelation in biology. 2. Some biological implications and four applications of evolutionary and ecological interest. *Biol. J. Linn. Soc.* **10**: 229–49.

Sokal, R. R. and Wartenberg, D. E. (1983). A test of spatial autocorrelation analysis using an isolation-by-distance model. *Genetics* **105**: 219–37.

Southwood, T. R. E. (1978). *Ecological Methods*. Chapman and Hall, London.

Spiess, E. B. (1977). *Genes in Populations*. Wiley, New York.

Spiess, E. B. (1982). Do females choose their mates? *Amer. Nat.* **119**: 675–93.

Stalker, H. D. (1942). Sexual isolation studies in the species complex *virilis*. *Genetics* **27**: 238–57.

Stanley, S. M. (1979). *Macroevolution*. W. H. Freeman, San Francisco.

Stanley, S. M. (1981). *The New Evolutionary Timetable*. Basic Books, New York.

Stebbins, G. L. (1982). Perspectives in evolutionary theory. *Evolution* **36**: 1109–15.

Steel, R. G. D. and Torrie, J. H. (1980). *Principles and Procedures of Statistics*. McGraw-Hill, New York.

Sternburg, J. G., Waldbauer, G. P. and Jeffords, M. R. (1977). Batesian mimicry: selective advantage of colour pattern. *Science* **195**: 681–3.

Stewart, F. M. (1977). Computer algorithm for obtaining a random set of allele frequencies for a locus in an equilibrium population. *Genetics* **86**: 482–3.

Tajima, F. and Nei, M. (1984). Note on genetic drift and estimation of population size. *Genetics* **106**: 569–74.

Takahata, N. and Kimura, K. (1981). A model of evolutionary base substitution and its application with special reference to rapid change of pseudogenes. *Genetics* **98**: 641–57.

Templeton, A. R. (1974). Analysis of selection in populations observed over a sequence of consecutive generations. I. Some one locus models with a single, constant fitness component per genotype. *Theor. Appl. Genet.* **45**: 179–91.

Templeton, A. R. (1983a). Phylogenetic inference from restriction endonuclease cleavage map sites with particular reference to the evolution of humans and the apes. *Evolution* **37**: 221–44.

Templeton, A. R. (1983b). Convergent evolution and nonparametric inferences from restriction data and DNA sequences. In *Statistical Analysis of DNA Sequence Data* (ed. B. S. Weir), pp. 151–79. Marcel Dekker, New York.

Terrenato, L. (1983). Natural selection associated with birth weight. IV. U.S.A. data from 1950 to 1976. *Ann. Hum. Genet.* **47**: 67–71.

Terrenato, L., Gravina, M. F. and Ulizzi, L. (1981a). Natural selection associated with birth weight. I. Selection intensity and selective deaths from birth to one month of life. *Ann. Hum. Genet.* **45**: 55–63.

Terrenato, L., Gravina, M. F., San Martini, A. and Ulizzi, L. (1981b). Natural selection associated with birth weight. III. Changes over the last twenty years. *Ann. Hum. Genet.* **45**: 267–78.

Terrenato, L., Ulizzi, L. and San Martini, A. (1979). The effects of demographic transition on the opportunity for selection: changes during the last century in Italy. *Ann. Hum. Genet.* **42**: 391–9.

Thomson, E. Y., Bell, J. and Pearson, K. (1911). A third cooperative study of *Vespa vulgaris*. Comparison of queens of a single nest with queens of the general autumn population. *Biometrika* **8**: 1–12.

Thomson, G. (1977). The effect of a selected locus on linked neutral loci. *Genetics* **85**: 753–88.

Thomson, G. (1981). A review of theoretical aspects of HLA and disease associations. *Theor. Pop. Biol.* **20**: 168–208.

Thomson, G., Bodmer, W. F. and Bodmer, J. (1976). The HLA system as a model for studying the interaction between selection, migration and linkage. In *Population Genetics and Ecology* (eds S. Karlin and E. Nevo), pp. 465–99. Academic Press, New York.

Thorpe, J. P. (1982). The molecular clock hypothesis: biochemical evolution, genetic differentiation and systematics. *Ann. Rev. Ecol. Syst.* **13**: 139–68.

Tilling, S. M. (1983). An experimental investigation of the behaviour and mortality of artificial and natural morphs of *Cepaea nemoralis* (L.). *Biol. J. Linn. Soc.* **19**: 35–50.

Timofeeff-Ressovsky, N. W. (1940). Zur analyse des polymorphismus bei *Adalia bipunctata* L. *Biol. Zbl.* **60**: 130–7.

Trueman, A. E. (1916). Shell-banding as a means of protection. *Ann. Mag. Nat. Hist.* **18**: 341–2.

Turner, J. R. G. (1968). On supergenes. II. The estimation of gametic excess in natural populations. *Genetica* **39**: 82–93.

Turner, J. R. G. (1972). Selection and stability in the complex polymorphism of *Moraba scurra*. *Evolution* **26**: 334–43.

Ulizzi, L., Gravina, M. F. and Terrenato, L. (1981). Natural selection associated with birth weight. II. Stabilizing and directional components. *Ann. Hum. Genet.* **45**: 207–12.

Van Valen, L. (1963). Selection in natural populations: *Merychippus primus*, a fossil horse. *Nature* **197**: 1181–3.

Van Valen, L. (1964). Age in two fossil horse populations. *Acta Zool.* **45**: 93–106.

Van Valen, L. (1965a). Selection in natural populations. III. Measurement and estimation. *Evolution* **19**: 514–28.

Van Valen, L. (1965b). Selection in natural populations. IV. British housemice (*Mus musculus*). *Genetica* **36**: 119–34.

Van Valen, L. (1978). The statistics of variation. *Evol. Theory* **4**: 33–43. (Correction in *Evol. Theory* **4**: 202.)

Van Valen, L. and Mellin, G. W. (1967). Selection in natural populations. 7. New York babies (fetal life study). *Ann. Hum. Genet.* **31**: 109–27.

Van Valen, L. and Weiss, R. (1966). Selection in natural populations. V. Indian rats (*Rattus rattus*). *Genet. Res., Camb.* **8**: 261–7.

Vanderploeg, H. A. and Scavia, D. (1983). Misconceptions about estimating prey preference. *Can. J. Aquat. Sci.* **40**: 248–50.

Verspoor, E. (1983). Allozyme frequencies in western European populations of *Asellus aquaticus* (L.) Isopoda and their associations with water pollution. *Biol. J. Linn. Soc.* **19**: 275–93.

Vithayasai, C. (1973). Exact critical values of the Hardy–Weinberg test statistic for two alleles. *Comm. Statist.* **1**: 229–42.

Waddington, C. H. (1948). The genetic control of development. *Symp. Soc. Exp. Biol.* **2**: 145–54.

Waldbauer, G. P. and Sternburg, J. G. (1975). Saturniied moths as mimics: an alternative interpretation of attempts to demonstrate mimetic advantage in nature. *Evolution* **29**: 650–8.

Waldbauer, G. P. and Sternburg, J. G. (1983). A pitfall in using painted insects in studies of protective colouration. *Evolution* **37**: 1085–6.

Wall, S., Carter, M. A. and Clarke, B. (1980). Temporal changes in gene frequencies in *Cepaea hortensis*. *Biol. J. Linn. Soc.* **14**: 303–17.

Wallace, B. (1981). *Basic Population Genetics*. Columbia Univ. Press, New York.

Watterson, G. A. (1977). Heterosis or neutrality? *Genetics* **85**: 789–814.

Watterson, G. A. (1978a). An analysis of multi-allelic data. *Genetics* **88**: 171–9.

Watterson, G. A. (1978b). The homozygosity test of neutrality. *Genetics* **88**: 405–17.

Watterson, G. A. (1979). Estimating and testing selection: the two allele, genetic selection diffusion model. *Adv. Appl. Prob.* **11**: 14–30.

Watterson, G. A. (1982). Testing selection at a single locus. *Biometrics* **38**: 323–31.

Watterson, G. A. and Anderson, R. (1980). Detecting natural selection: the stepwise mutation model with heterosis. *Aust. J. Statist.* **22**: 125–42.

Wehrhahn, C. F. (1975). The evolution of selectively similar electrophoretically detectable alleles in finite natural populations. *Genetics* **80**: 375–94.

Weir, B. S. (1979). Inferences about linkage disequilibrium. *Biometrics* **35**: 235–54.

Weir, B. S., Allard, R. W. and Kahler, A. L. (1972). Analysis of complex allozyme polymorphisms in a barley population. *Genetics* **72**: 505–23.

Weir, B. S., Brown, A. H. D. and Marshall, D. R. (1976). Testing for selective neutrality of electrophoretically detectable protein polymorphisms. *Genetics* **84**: 639–59.

Weir, B. S. and Cockerham, C. C. (1978). Testing hypotheses about linkage disequilibrium with multiple alleles. *Genetics* **88**: 633–42.

Weir, B. S. and Cockerham, C. C. (1979). Estimation of linkage disequilibrium in randomly mating populations. *Heredity* **42**: 105–11.

Weldon, W. F. R. (1895). An attempt to measure the death rate due to the selective destruction of *Carcinus maenas* with respect to a particular dimension. *Proc. Roy. Soc.* **57**: 360–82.

Weldon, W. F. R. (1901). A first study of natural selection in *Clausilia laminatu* (montagu). *Biometrika* **1**: 109–124.

Weldon, W. F. R. (1903). Note on a race of *Clausilia itala* (Van Marteus). *Biometrika* **3**: 299–307.

White, E. G. (1971a). A versatile fortran program for the capture–recapture stochastic model of G. M. Jolly. *J. Fish. Res. Bd. Canada* **28**: 443–5.

White, E. G. (1971b). A computer program for capture–recapture studies: A fortran listing for the stochastic model of G. M. Jolly. *Tussock Grassland and Mountain Land Inst., Spec. Publ.* 8.

White, M. J. D. (1957). Cytogenetics of the grasshopper *Moraba scurra*. II. Heterotic systems and their interaction. *Aust. J. Zool.* **5**: 305–37.

White, M. J. D., Lewontin, R. C. and Andrew, L. E. (1963). Cytogenetics of the grasshopper *Moraba scurra*. VII. Geographical variation of adaptive properties of inversions. *Evolution* **17**: 147–62.

White, R. J. and White, R. M. (1981). Some numerical methods for the study of genetic changes. In *Genetic Consequence of Man Made Change* (eds J. A. Bishop and L. M. Cook), pp. 295–342. Academic Press, London.

Whittam, T. S. (1981). Is a negative regression of delta *p* on *p* evidence of a stable polymorphism? *Evolution* **35**: 595–6.

Whittam, T. S., Ochman, H. and Selander, R. K. (1983). Multilocus genetic structure in natural populations of *Escherichia coli. Proc. Nat. Acad. Sci., USA* **80**: 1751–5.

Williamson, M. H. (1960). On the polymorphism of the moth *Panaxia dominula* (L.). *Heredity* **15**: 139–51.

Wilson, J. (1970). Experimental design in fitness estimation. *Genetics* **66**: 555–67.

Wilson, S. (1980). Analysing gene-frequency data when the effective population size is finite. *Genetics* **95**: 489–502.

Wong, B. and Ward, F. J. (1972). Size selection of *Daphnia publicaria* by yellow perch (*Perca flavescens*) fry in West Blue Lake, Manitoba. *J. Fish. Res. Bd. Canada* **29**: 1761–4.

Woolf, B. (1955). On estimating the relationship between blood group and disease. *Ann. Hum. Genet.* **19**: 251–3.

Workman, P. L. (1968). Gene flow and the search for natural selection in man. *Hum. Biol.* **40**: 260–79.

Workman, P. L., Blumberg, B. S. and Cooper, A. J. (1963). Selection, gene migration and polymorphic stability in U.S. White and Negro populations. *Amer. J. Hum. Genet.* **15**: 429–37.

Wright, S. (1948). On the roles of directed and random changes in gene frequency in the genetics of populations. *Evolution* **2**: 279–94.

Wright, S. (1978). *Evolution and the Genetics of Populations*, Vol. 4. Univ. Chicago Press, Chicago.

Wright, S. (1982). The shifting balance theory and macroevolution. *Ann. Rev. Genet.* **16**: 1–19.

Wright, S. and Dobzhansky, Th. (1946). Genetics of natural population. XII. Experimental reproduction of some of the changes caused by natural selection in certain populations of *Drosophila pseudoobscura. Genetics* **31**: 123–56.

Yamazaki, T. and Maruyama, T. (1972). Evidence for the neutral hypothesis of protein polymorphism. *Science* **178**: 56–7.

Yamazaki, T. and Maruyama, T. (1973). Evidence that enzyme polymorphisms are selectively neutral. *Nature New Biol.* **245**: 140.

Yamazaki, T. and Maruyama, T. (1974). Evidence that enzyme polymorphisms are selectively neutral but blood group polymorphisms are not. *Science* **183**: 1091–2.

Yardley, D. G., Anderson, W. W. and Schaffer, H. E. (1977). Gene frequency changes at the amylase locus in experimental populations of *Drosophila pseudoobscura. Genetics* **87**: 357–69.

Yasuda, N. and Kimura, M. (1968). A gene counting method of maximum likelihood for estimating gene frequencies in ABO and ABO-like systems. *Ann. Hum. Genet.* **31**: 409–20.

Zaret, T. M. and Kerfoot, W. C. (1975). Fish predation on *Bosmina longirostris*: body size selection versus visibility selection. *Ecology* **56**: 232–7.

Zouros, E., Golding, G. B. and Mackay, T. F. C. (1977). The effect of combining alleles into electrophoretic classes on detecting linkage disequilibrium. *Genetics* **85**: 543–50.

Name index

Subject index

482   Subject index